SIKU: Knowing Our Ice

Igor Krupnik · Claudio Aporta · Shari Gearheard ·
Gita J. Laidler · Lene Kielsen Holm
Editors

SIKU: Knowing Our Ice

Documenting Inuit Sea Ice Knowledge and Use

Editors
Dr. Igor Krupnik
Smithsonian Institution
National Museum of Natural History, Dept. Anthropology
10th and Constitution Ave. NW.,
Washington DC 20013-7012
USA
krupniki@si.edu

Dr. Claudio Aporta
Carleton University
Dept. Sociology & Anthropology
1125 Colonel By Dr.
Ottawa ON K1S 5B6
B349 Loeb Bldg.
Canada
Claudio_aporta@carleton.ca

Dr. Shari Gearheard
University of Colorado, Boulder
National Snow & Ice Data Center
Clyde River NU
X0A 0E0
Canada
shari.gearheard@nsidc.org

Dr. Gita J. Laidler
Carleton University
Dept. Geography & Environmental Studies
1125 Colonel By Drive
Ottawa ON K1S 5B6
Canada
gita_laidler@carleton.ca

Lene Kielsen Holm
Inuit Circumpolar Council, Greenland
Dr. Ingridsvej 1, P. O. Box 204
Nuuk 3900
Greenland
lene@inuit.org

This book is published as part of the International Polar Year 2007–2008, which is sponsored by the International Council for Science (ICSU) and the World Meteorological Organization (WMO)

ISBN 978-90-481-8648-8 e-ISBN 978-90-481-8587-0
DOI 10.1007/978-90-481-8587-0
Springer Dordrecht Heidelberg London New York

Library of Congress Control Number: 2010920470

Front cover photo: Lucian Read, Qaanaaq, North Greenland, April 2008
Back cover photo: Gita J. Laidler, Cape Dorset, Nunavut, January 2005
Cover design: Anya Vinokour, 2009

Printed on acid-free paper

Springer is part of Springer Science+Business Media (www.springer.com)

Foreword

Helena Ödmark

Ice and traditional knowledge of ice continue to play a role in modern society in ways that may surprise many people who live in warmer regions. Ice is still a big factor of life in my native area of northern Sweden, even though the large ice sheet that covered Scandinavia and neighboring regions during the last ice age had melted away several thousand years ago. Long after the ice sheet has melted, the land here is still rising from the sea, almost 1 cm a year in some areas. Certain ports along the Swedish Baltic Sea coast have been moved several times because of that land rise and are now located many kilometers away from where they were originally built.

When I was growing up in a small town on that coast, the ice was used for several months every year for winter roads across lakes and to the offshore islands. People went fishing on the ice on weekends. We had to be very careful in the areas where big icebreakers kept water channels open for commercial shipping into our town's port and to the surrounding industrial harbors. We knew how to "read" the ice in the fall, in the winter, and in the spring to determine if it was safe to use. That was a part of my life and it is still a very important part of everyday knowledge to millions of people who live along the ocean, lakes, and rivers that are commonly covered with ice for some portion of the year. The climate warming may shorten or shift the ice season, but frozen water and ice-covered sea will continue to be our shared reality for generations to come.

I first heard about the "International Polar Year 2007–2008," or "IPY," at the Arctic Science Summit Week in Kiruna, Sweden, in March 2003. There, a presentation was made on the initial plans for a new IPY. One PowerPoint slide in particular caught my attention. It listed six key themes in a left-hand column and an additional explanatory wording in a right-hand column. The last entry in the left-hand column was "human dimension." That made sense to me. But I was confused and very surprised to see a big question mark, a "?", in the corresponding box in the right-hand column. The explanation was simple. The presenters did not know what to write in that "human dimension" box.

At that time, I was not aware of the almost total lack of communication within the polar research community between scientists working in natural and physical sciences and those working in the field of social and human sciences. But I knew that a priority for the Arctic Council was to "strengthen cooperation in Arctic research."

The Arctic Council had also started a very ambitious project to compile an *Arctic Human Development Report* (eventually published in 2004), which would rely heavily on research from the social and human sciences. Trying to be helpful, I showed the IPY presenters a folder that outlined the chapter headings for that report and suggested that they might provide inspiration for a text that could replace the "?" in the box on their slide.

At the next meeting of the Senior Arctic Officials in April 2003, I informed the chair and my colleagues on the Arctic Council that an "International Polar Year" was being planned. We all found the issue to be of great interest. It was difficult for us to envisage a successful "IPY 2007–2008" without the active participation of the Arctic Council. After all, the scientific community was planning an "International Polar Year," not an "International Polar Research Year."

At the Senior Arctic Officials meeting in October 2003, we decided to invite the IPY planners. Dr. Chris Rapley made an excellent presentation about the emerging IPY science program at our meeting in May 2004. The reaction in the room confirmed the strong interest from the Arctic Council member states, permanent participants, and observers in the IPY planning process. In particular, the group was pleased to hear that the once brief reference to "human dimension" would be translated into substantive input by social and human sciences, as well as the full attention to the needs and interests of Arctic residents. The planners were encouraged to involve indigenous and other local communities in IPY research activities, to appreciate the value of traditional knowledge, and to share the results of their work with Arctic residents. The meeting adopted a special statement to reiterate the Arctic Council's commitment to IPY. We also launched a public diplomacy effort to emphasize the importance of including a "human dimension" in IPY. The Declaration adopted at the Arctic Council Meeting of Foreign Ministers in Reykjavik in October 2004 welcomed "the continuing contribution of indigenous and traditional knowledge to research in the Arctic" and "recognized the IPY 2007–2008 as a unique opportunity to stimulate cooperation and coordination on Arctic research."

The Declaration adopted at the next Arctic Council Meeting of Foreign Ministers in Salekhard, Russia, in October 2006 specifically endorsed "(the) expansion of the IPY to include the human dimension." The Ministers also supported "the inclusion of (research) programs initiated by Arctic residents, the effective involvement of Arctic indigenous peoples in IPY activities and recognize(d) that their traditional and indigenous knowledge is an invaluable component of IPY research." They also emphasized "the importance of climate change in the context of the IPY, and to achieve a legacy of enhanced capacity of Arctic peoples to adapt to environmental, economic and social changes in their regions, and enabling Arctic peoples to participate in and benefit from scientific research."

Almost 3 years later, at the conclusion of the IPY observational period in spring 2009, the Declarations adopted at the Antarctic Treaty Consultative Meeting–Arctic Council Joint Meeting in Washington, DC, and at the Arctic Council Ministerial Meeting in Tromsö in April 2009 highlighted "that for the first time [the IPY] considered the human dimension and concerns of local and indigenous peoples and engaged Arctic residents." The Ministers specifically urged for "continued international coordination to maximize the legacy of IPY," including "benefits to local and indigenous peoples."

In short, for all these years, the Arctic Council has championed the very type of scientific work that the SIKU project has accomplished in four of the Arctic Council member states. I am honored to be associated with that work and I have found this book to be fascinating and extremely inspirational reading. It conveys valuable new knowledge that needs to be presented to a wide audience and also to be incorporated into the further work of the Arctic Council.

Our works goes on, and it will be reinforced by the outcomes and the legacy of the SIKU project. Thank you, the SIKU team, for your eye-opening book. This is exactly what we had in mind when we were voicing Arctic Council's support for IPY 2007–2008 and argued for the value of bringing the social and human studies fully into its science program.

Lasalie Joanasie (*left*) and Shari Gearheard, an Inuit hunter and a researcher, both residents of Clyde River, Nunavut, keep an eye on a passing polar bear while traveling the sea ice off the coast of Baffin Island near Clyde (Photo: Edward Wingate, 2009)

Preface

Joelie Sanguya and Shari Gearheard

There is a special time of the year, just before freeze up, when our community is buzzing with anticipation. We are waiting for ice. This is "the waiting season" and it goes back to traditional times. Families would gather at certain camps in the fall to await the arrival of sea ice. When the *siku* (sea ice) finally formed, it meant freedom. Families could once again travel the great networks of ice trails and hunt animals in their ice habitats. The ice reconnected us to people and places. Even our name for the month of November, *Tusaqtuut*, recognizes the end of the anticipation, for *Tusaqtuut* means "the news season", when we are able to travel and hear from other camps. It is still like this today as we enjoy visiting other communities *sikukkut* (via the sea ice) and taking those short cuts that return across rivers, bays, and fjords.

Not only people wait for ice. The animals and even the land itself seem wishful for the sea ice in fall time. Sled dogs are restless, yearning to pull a *qamutiik* again. Polar bears pace the coastlines and are the first to venture onto the new grey ice, eager to follow the growing edge out to sea. Seals, sea gulls, and foxes are also among the first to welcome the ice, still too thin for people. The expanding new ice becomes an extension of the land which seems to exhale with a great stretch at freeze up, happy to finally reach out again into the distance.

Not long after the bears take their first steps out to sea, with their great weight distributed evenly and safely on huge and ingenious paws, the hunters follow. Few people but the most experienced hunters travel the very new ice, their snowmachines creating a supple wave of plastic-like ice ahead of them as they carefully navigate the ocean's new elastic cover.

Soon grey turns to white and the *tuvaq* (first-year ice) is here. This is what we have been waiting for. The ice transforms not only the physical landscape around us but also the emotional landscape within us. We welcome the return of the sights, sounds, and even smells and tastes that come with the sea ice. We appreciate the icescapes and ice features, new and reoccurring ice places, that appear. We begin to come across our friends on the ice, again, recognizing them from afar by their winter clothing, their particular *qamutiik*, or their lead dog. We hear our sea ice language being spoken, our complex terminology for sea ice and its use. We hear the music of the sea ice, the *niiqquluktuq* (ice rubbing sound), the puff of air when a seal rises to breath in its *aglu*, and the "Uuuua, uuuua, hut-hut-hut!" of the *qimuksiqtiit* (dog teamers). We smell the fresh air way out on the ice and at the start of the season always enjoy catching a fresh *nattiq* (ringed seal), eating it on the ice, and drinking tea.

Along with the senses of sea ice come the stories, the journeys, and the memories. As we travel we tell about old adventures and make new ones. We remember friends and relatives in the old times and we take our children and grandchildren along so that they can hear the stories and have the journeys that they will someday tell their own children about. The sea ice has always been a place for the making and telling of good stories, from the fantastic travels of great shamans to the mythical beings that live on and under sea ice, like the *Qalupaliit* ("people snatchers"). All the ice stories have their lessons, and they are still important today.

When visitors come to experience or study sea ice, they can not help but see and understand the ice through its physical properties. To most visitors, it is the frozen ocean and their interest is in its features, characteristics, and processes. Perhaps they might broaden their understanding to related environmental features like winds, snow, atmosphere, land, and currents. Of course this is their starting point. To know and use sea ice and experience it the way we do, through its music, stories, journeys, emotions, and memories, takes years and lifetimes. As the sea ice changes across the Arctic, these are the things that people who live with ice face losing, not only the physical aspects and environmental functions of sea ice, but the intangible soul-filling stuff that we collect through our life with it.

The icescape is our stage for old and new performances, for stories and songs created and shared by people, animals, and the ice itself. It is our workspace, our picnic area, our classroom, our highway, and our home.

As we become more involved in the study of ice, our unique perspectives as Arctic residents are starting to be recognized. We can provide detailed knowledge of its physical properties, its dynamics and characteristics, its connections to other environmental features, but also to its important role in the lived (and living) history of Arctic peoples, and as a bringer of joy, wonder, and fulfillment to human life.

The SIKU book is an important initiative that begins to bring some of this understanding to light. Many of the projects described here worked in collaboration with Arctic residents and share local wisdom. The book will be a lasting record of what can be achieved when science and indigenous knowledge are brought together. It starts the journey toward linking science with soul. Like the sea ice, may it be a platform for learning and inspiration.

Contents

Contributors

Gary Aipellee works part-time as the constituency assistant to the Member of the Legislative Assembly of Nunavut for the Uqqummiut Riding (Clyde River and Qikiqtarjuaq), James Arreak. He also works part-time as the liaison officer for Biogenie S.R.D.C/Aarruja Development Corporation's DEW line clean-up of FOX-3 Dewar Lakes. In his spare time, Gary enjoys working with his traditional language, Inuktitut. He has been providing translation services, mainly targeting documents and texts, for the last 8 years. Born and raised in Clyde River, Aipellee still lives there with his partner Debbie and their three beautiful daughters.

Eli Angiyou lives in Akulivik, Nunavik. He is an avid hunter and fisherman and is employed by Kativik Regional Government as an environmental technician. Mr. Angiyou was the mayor of his community from 1998 to 2000. He has also been a member of the Kativik Environmental Advisory Committee since 2001 and is now the vice-chairman. Eli has been involved in the Kativik Regional Government climate change monitoring and research program since 2005.

Leonard Apangalook, Sr. *(Peetkaq)* was born in the Yupik community of Gambell (*Sivuqaq*) on St. Lawrence Island, Alaska. He graduated from Mt. Edgecombe high school in Sitka, Alaska, and continued with classes at the Sheldon Jackson College and military school, where he received training as weather serviceman. He worked for many years as a village agent and weather observer for local air companies; he also served for 14 years on the Alaska National Guard. He is an active hunter, whaler, and boat captain, and he served as the vice-chairman of the Alaskan Eskimo Whaling Commission. Upon his initiative, he kept records of weather and ice conditions in Gambell for many years, prior to becoming the first ice observer for the SIKU project in 2006.

Paul Apangalook (*Siluk*) was born in Gambell (*Sivuqaq*) on St. Lawrence Island, Alaska. Upon graduating from Mt. Edgecombe high school in Sitka, Alaska, he returned to his native community and worked in construction and various administrative positions. Over the years, he has held many public offices, including being the mayor of the Town of Gambell, the president of the Sivuqaq, Inc., and the president of the Tribal Government, as well as the member of the City Council, Tribal Government, and local Corporation Board. He also served for 24 years on

Alaska National Guard. He is an active subsistence hunter and whaler, and he is very passionate about the preservation of Yupik language and culture in his native community.

Claudio Aporta is an associate professor in the Sociology and Anthropology Department at Carleton University, Ottawa. He is a faculty member of the Geomatics and Cartographic Research Centre and the principal investigator of the Inuit Sea Ice Use and Occupancy Project (ISIUOP). Born in Argentina, Aporta received his BA in Communication from the Universidad Nacional de Cuyo and his PhD in Anthropology from the University of Alberta. Aporta has been working not only in Nunavut since 1998, with most of his research taking place in Igloolik, but also in other communities and regions of the Canadian Arctic. He has published extensively on issues of Inuit wayfinding, indigenous cartography, Inuit place names and routes, and Inuit knowledge of the sea ice. He also has an interest in anthropology of technology and has written on the use of GPS by Inuit hunters.

Michael Bravo is senior lecturer at the University of Cambridge and is head of the Circumpolar History and Public Policy Research Group at the Scott Polar Research Institute. Bravo is contributing through research collaborations and supervision to help develop the next generation of researchers "writing from the Arctic." He played a key role in writing the humanities theme for International Polar Year (2007–2009), the first such "big science" polar event to include explicitly the importance of citizenship as well as the participation of northern peoples and social scientists. Following the International Polar Year, Bravo was the rapporteur for the international report on the *Canadian Arctic Research Initiative* (2008) commissioned by the Council of Canadian Academies. The panel's recommendations were taken up by the Government of Canada in its recent decision to invest in the renewal of research infrastructure in the Arctic. Bravo is coeditor of *Narrating the Arctic* (2002) and the author of numerous scholarly and popular articles.

Lewis Brower was born in July 1964 and is a lifelong resident of Barrow Alaska. He is the full-time logistical manager for the Barrow Arctic Science Consortium, where he also advising many visiting scientists on a range of topics. He is a subsistence hunter and has many hobbies. He loves to go hunting, loves life with family, and likes the winter more than summer. Lewis uses the Arctic Ocean as a playground to survive the brutal winters by hunting and traveling to recover subsistence food catch.

Matthew L. Druckenmiller is a PhD candidate in geophysics at the University of Alaska Fairbanks (UAF) and a fellow in UAF's Resilience and Adaptation Program. With a background in engineering and glaciology, he is interested in researching and monitoring coastal sea ice as an environment that supports and provides for the activities of indigenous communities as well as for other groups. Most of his experiences on sea ice have been near and with the community members from Barrow and Wales, Alaska. Prior to his time at UAF, he worked to survey changes in mountain glacier volume in Alaska and as a science policy graduate fellow at the U.S. National Academies' Polar Research Board.

Hajo Eicken studies sea ice as a professor of geophysics at the Geophysical Institute, University of Alaska Fairbanks (UAF). Before joining UAF, he led the sea ice and remote-sensing group at the Alfred Wegener Institute for Polar and Marine Research in Bremen, Germany. Dr. Eicken's research examines the growth, evolution, and properties of sea ice in the polar regions. During the past 10 years he has learned much about uses of sea ice in a changing climate through work in coastal Alaska communities. He is heading several research programs aimed at enhancing ice-forecasting service and exchange between scientists, different sea ice users, and the public at large.

Pootoogoo Elee was born in Iqaluit, Nunavut, in 1959. He lived there until he was 6 years old and spent the next few years moving between Iqaluit and Cape Dorset. He completed high school in Iqaluit and then spent the next 12 years living and training in Yellowknife, Edmonton, and Ottawa, until he moved back to Cape Dorset, where he lives now. The first time he went on the ice was with his grandfather by dog team when he was 6 years old and he recognizes the ongoing importance of sea ice to his community. He has worked as an independent guide and interpreter with many tourist and research groups over the years and continues to do so between his local construction and legal services support work.

Ann Fienup-Riordan is a cultural anthropologist who has lived and worked in Alaska since 1973. Her books include *The Nelson Island Eskimo* (1983), *Eskimo Essays* (1990), *Boundaries and Passages* (1994), *The Living Tradition of Yup'ik Masks* (1996), and *Wise Words of the Yup'ik People: We Talk to You Because We Love You* (2005). In 2000, she received the Alaska Federation of Natives President's Award for her work with Alaska Native People and in 2001 the Governor's Distinguished Humanist Award. At present she works with the Calista Elders Council, helping to document traditional knowledge.

Chris Furgal is an associate professor in the Indigenous Environmental Studies program at Trent University in Peterborough, ON, Canada. He is also the co-director of the Nasivvik Centre for Inuit Health and Changing Environments at Laval University (Quebec City, Canada) and Trent University. He conducts research on the health impacts of environmental change in circumpolar communities. He has been a lead author or contributor to several international assessments and scientific articles on the topics of climate change and contaminants and their impacts on northern health, including the Arctic Climate Impact Assessment, IPCC Fourth Assessment Report, Canadian climate, impacts and vulnerability assessments among others.

Shari Gearheard is a geographer and research scientist with the National Snow and Ice Data Center (NSIDC) at the University of Colorado at Boulder. Since 1995, she has been working with Inuit communities in Nunavut on a variety of environmental research projects, in particular, on Inuit knowledge of climate and environmental change. She has also worked on collaborative projects at the territorial and international levels, including the Government of Nunavut's Inuit Qaujimajatuqangit of Climate Change Study (North and South Baffin) and the Arctic Climate Impact Assessment (ACIA). Her current work focuses on bringing

Inuit and scientific knowledge together to study sea ice, as well as weather patterns and weather forecasting. Gearheard is a passionate qimuksiqti (dog teamer) and lives with her husband and 20 Inuit sled dogs in Clyde River, Nunavut, where she bases her work.

John "Craig" George has worked as a wildlife biologist with the North Slope Department of Wildlife Management in Barrow, Alaska, for 25 years. Craig earned a BS in Wildlife Biology from the Utah State University in 1976 and just finished a PhD in bowhead whale energetics, age estimation, and morphology from the University of Alaska Fairbanks in 2009. In 1977, Craig moved to Barrow to work as a research technician at the Naval Arctic Research Laboratory. Beginning in 1982, Craig worked on and later coordinated the bowhead whale ice-based population assessment project on the sea ice near Point Barrow for nearly two decades. He also has conducted over 200 postmortem exams on bowheads harvested by Alaskan Eskimos (since 1980) and published a number of papers on this work. Craig has attended International Whaling Commission meetings since 1987 focusing mainly on management of aboriginal whaling. Craig is married to Cyd Hanns, a wildlife technician, and they enjoy community and outdoor activities with their two sons Luke and Sam.

Amos Hayes is the technical manager for the Geomatics and Cartographic Research Center (GCRC) at Carleton University. He has designed, assembled, troubleshot, and ran scalability and performance tests on a wide variety of large information infrastructure systems in public and private sector environments before joining the GCRC. His attention to real-world deployment issues as well as a keen interest in open standards and open source software is helping the GCRC build an innovative and robust framework for participatory atlas building based on research activities at the center.

Anne Henshaw is a program officer with the Oak Foundation U.S. in their Environmental Program in Portland, ME, and a research associate with the Peary-MacMillan Arctic Museum at Bowdoin College in Brunswick, ME. Before joining the Oak Foundation, she was a visiting professor in Anthropology at Bowdoin College from 1996 to 2007 and director of Bowdoin's Coastal Studies Center from 2000 to 2007. Anne has been conducting community-based anthropological research in Nunavut since 1990. She currently serves on the Advisory Committee for the Office of Polar Programs at the U.S. National Science Foundation.

Lene Kielsen Holm is the director for Research and Sustainable Development at the Inuit Circumpolar Council, Greenland. She is also the current chair of the Board for the Greenland Institute of Natural Resources. Holm received her masters degree in Social and Cultural History from Ilisimatusarfik (University of Greenland) in 2002. She has worked on several research projects within the Arctic Council's Working Group for Sustainable Development, particularly on gender and resource issues. Holm is originally from Qaqortoq, south Greenland. She is the mother of two and lives in Nuuk, Greenland's capital.

Henry P. Huntington is director of science, Arctic Program, with the Pew Environment Group. He lives in Eagle River, Alaska, with his wife and two sons. He received his PhD from the University of Cambridge. He has been involved in Arctic research for over two decades, focusing primarily on various aspects of human–environment interactions, including traditional knowledge, impacts of climate change, and conservation efforts. Recently he has also done research in Nepal. He has written many academic and popular articles, as well as two books.

Theo Ikummaq was born off Igloolik, in an igloo in 1955. He has lived in Igloolik off and on for most of his life, but full time in the community since around 1990. He grew up on the land in various camps around Foxe Basin, spending winter months at boarding school in Chesterfield Inlet and summers with his family in Hall Beach. After school he moved back with his family to their camp in Steensby Inlet for another 8 years followed by several more years of living in various parts of the Canadian North: in Fort Smith, for his grade 12 diploma and a Renewable Resources Technology Program, and a few years in Arctic Bay, Qikiqtarjuaq, and Iqaluit with work as a wildlife officer, and one summer in Ottawa with the Canadian Museum of Nature. He is now working as the conservation officer in Igloolik as well as continuing to consult with researchers, tourists, film crews, and mining companies in the area.

Eric Joamie is a 52-year-old Inuk living in Pangnirtung, Nunavut. He was born on the land and educated through Federal Day School and territorial education system, a system foreign to his own society. He has been involved with numerous community and research projects over the years including work with the Qikiqtani Inuit Association and an ongoing contract with the Hamlet of Pangnirtung. He now owns a communication and consulting business and has worked on this SIKU project since its inception.

Alana Johns is professor of linguistics at the Department of Linguistics, University of Toronto, Canada. She combines research into the theoretical syntax and morphology of complex words in Inuktitut and other languages with community-based language maintenance and online language support efforts. She has studied Inuktitut dialects in Nunatsiavut (Labrador), Qamani'tuaq (Baker Lake), and across Baffin Island (Nunavut). She is currently working with a team of linguists to produce a dictionary of Utkuhiksalingmiutitut, a dialect of Inuktitut once spoken in the Utkuhiksalik area of Nunavut, and now preserved by elders in Uqsuqtuuq (Gjoa Haven) and Qamani'tuaq (Baker Lake).

Marie-Luise Kapsch is currently completing her diploma studies in meteorology at the University of Karlsruhe in Germany. She was a visiting student at the University of Alaska Fairbanks in 2008/2009 and worked on a project linking indigenous ice observations, weather and sea ice data, and walrus harvest records. As part of this work she communicated with several Inupiat and Yupik sea ice experts, entering their observations into a database and had a chance to see the ice-covered ocean at Barrow during a winter field trip.

Igor Krupnik is cultural anthropologist and curator of Arctic ethnological collections at the National Museum of Natural History, Smithsonian Institution in Washington, DC. For over 30 years, he has studied cultural heritage, impacts of climate change, and ecological knowledge of the Arctic people, particularly in Alaska and in the Russian Arctic. He published several books and exhibit catalogs, and he was the curator of the exhibit, *Arctic: A Friend Acting Strangely* (2006), focused on the Arctic people's observations of climate change. He was among the first champions of bringing the social and indigenous topics to the program of International Polar Year (IPY) 2007–2008 and was the lead coordinator (with Claudio Aporta) of the SIKU project (IPY #166). He currently serves on the Joint Committee for IPY 2007–2008 and as co-editor of the IPY "Overview" Report *Understanding Earth's Polar Challenges: International Polar Year 2007–2008.*

Gita J. Laidler is assistant professor in the Department of Geography and Environmental Studies at Carleton University in Ottawa. She has been working in northern Canada since 1998 after first being introduced to northern research in Labrador and, later on Boothia Peninsula, Nunavut. Her current research focuses on cultural and environmental geography, particularly on the local importance of sea ice processes, use, and change based on Inuit expertise. She has been working collaboratively with the Nunavut communities of Pangnirtung, Cape Dorset, and Igloolik since 2003; this work has developed into ongoing, long-term partnerships to explore human–environment interactions and understandings at local and regional scales.

A. Chase Morrison is doing his anthropology MA in the Department of Sociology and Anthropology of Carleton University, Ottawa. He graduated from Wilfrid Laurier University in Waterloo, where he majored in anthropology and biology. As a research assistant with ISIUOP, Chase has been exploring the social dimensions of sea ice use. His research focuses on conceptualizations of marine mammal hunting by younger Inuit people in the Lancaster Sound region.

Ludger Müller-Wille (Dr. phil., Ethnology, Münster, Germany, 1971), professor in geography at McGill University in Montréal, Canada (1977–2008), has conducted research with the Sámi and Finns in Finland, Inuit, and Dene in Canada, on ethnic identity, reindeer herding and caribou hunting, indigenous place names, and the legacy of Franz Boas' pioneer work in the Canadian Arctic. He was the founding president of the International Arctic Social Sciences Association, IASSA (1990–1995) and director of the Arctic Center in Rovaniemi, Finland (1994–1996).

Helena Ödmark is senior Arctic official for Sweden in the Arctic Council. She graduated from the Stockholm School of Economics and served at numerous Swedish diplomatic missions in Africa, Central America, the Middle East, and Europe, most recently as ambassador of Sweden to Mozambique and Swaziland, and also at the Ministry for Foreign Affairs in Stockholm. In 2004–2008, she was the president of the Scientific Advisory Board for the Abisko scientific station of the Swedish Academies of Science. She is a member of the Swedish National Committee for International Polar Year (2007–2008) and Arctic Council's representative on the IPY Joint Committee. She also serves as Representative of Sweden at the Antarctic Treaty Consultative Meetings and Member of the Commission for the Conservation of Antarctic Marine Living Resources.

Kyle O'Keefe is an assistant professor of Geomatics Engineering at the University of Calgary, in Calgary, Alberta, Canada. He completed BSc in physics at the University of British Columbia in Vancouver in 1997 followed by BSc and PhD degrees in Geomatics Engineering at the University of Calgary in 2000 and 2004. He has worked in positioning and navigation research since 1996 and in satellite navigation since 1998. His major research interests are GNSS (Global Navigation Satellite System) simulation and assessment, space applications of GNSS, carrier phase positioning, and local and indoor positioning with ground-based ranging systems.

Peter L. Pulsifer is a research associate with the Geomatics and Cartographic Research Center, Department of Geography, Carleton University. Dr. Pulsifer's research focuses on the link between scientific knowledge and environmental policy and the related use of spatial data infrastructures and computer-based forms of geographic information representation. Much of his work focuses on the polar regions, and he has been active in contributing to the development of data management theory and practice in the Arctic and Antarctic regions. At present, he is leading the data and knowledge stewardship aspects of IPY Inuit Sea Ice Use and Occupancy Project (ISIUOP http://gcrc.carleton.ca/isiuop/) while actively developing a research program focused on geographic information management in polar environments.

Alice Aluskak Rearden from Napakiak, Alaska, is a graduate in anthropology and history from the University of Alaska Fairbanks. She is a fluent Yup'ik speaker committed to working with regional tradition bearers to gather, translate, and share their knowledge. Her first bilingual book is *Yupiit Qanruyutait: Yup'ik Words of Wisdom* (2005), and she has recently finished work on *Qanruyuteput Iinruugut/Our Teachings Are Medicine* (2009).

Martin Robards studies the challenges associated with balancing Indigenous Rights and Species Conservation in policies governing marine mammals. He is currently a National Research Council (NRC) post-doctoral research fellow hosted by the U.S. Marine Mammal Commission in Washington, DC. Before his post-doctoral Research, he studied Pacific Walrus co-management in Alaska where he had lived and studied for nearly 20 years. Prior research focused on ecosystem studies of Alaska's coastal environment with particular emphasis on forage fish and seabirds.

Joelie Sanguya is a hunter and qimuksiqti (dog teamer) from Clyde River, Nunavut. As a child he lived on the sea ice in an igloo for 2 or 3 months of the year just after Christmas, before moving back to the family camp in late March or April. He is a film producer and co-owner of Piksuk Media. His recent film projects include "The Mystery of Arqioq" and "Qimmiit: the Clash of Two Truths." He is a lead researcher in the international SIKU–Inuit–Hila (Sea ice–people–weather)

Project. Sanguya has been active in leading and facilitating many territorial initiatives including the Government of Nunavut's Inuit Qaujimajatuqangit of Climate Change Study (North and South Baffin) and the establishment of Niginganiq, the first Inuit-initiated whale sanctuary and Canada's first National Marine Wildlife Area. Sanguya lives in Clyde River with his wife Igah and together they have 5 children, 9 grandchildren, and 24 Inuit sled dogs.

Pierre Taverniers is meteorologist at Meteo France in Limoges, France, with training in anthropology and Inuit languages (Institut National des Langues et des Civilisations Orientales, INALCO, Paris). He started his research on sea ice and polar weather in 1983, when wintering at the French polar station Dumont d'Urville in Antarctica. In 1985 he visited Canadian Inuit community of Pangnirtung and in 1986, Greenlandic community of Ilulissat. In 1987–1988 he spent 11 months in Qeqertaq, small Inuit community in Northwest Greenland, studying weather, sea-ice, and the Qeqertamiut knowledge of ice and snow. He currently works on the issues of climate change in the Arctic and he regularly visits Inuit communities in Greenland and Canada.

D.R. Fraser Taylor is a distinguished research professor of International Affairs and Geography and Environmental Studies at Carleton University in Ottawa, Canada as well as director of the Geomatics and Cartographic Research Center. Dr. Taylor's main research interests lie in the application of geomatics to the understanding of socioeconomic issues. Dr. Taylor is active in a number of CODATA committees and working groups and is a Board member of the OGC (Open Geospatial Consortium) Interoperability Institute. He chairs the International Steering Committee for Global Mapping (ISCGM). He was secretary-treasurer of the Canadian Association of African Studies for 15 years, president of the Canadian Cartographic Association, and president of the International Cartographic Association from 1987 to 1995.

Nicole Tersis is a specialist in the Eskimo-Aleut languages at the Center for the Study of the Indigenous Languages of the Americas (CELIA), at the CNRS (National Center for Scientific Research) in Paris, France. Her fields of interest include syntax, morphology, and information structure; semantics, lexicology, and typology of languages. She has done field research in East Greenland since 1990 that resulted in her recent dictionary (lexicon) of the East Greenlandic/Tunumiisut dialect (2008) and also among the speakers of the Nilo-Saharien languages Zarma and Dendi in Africa. She is also the member of the French *Fédération Typologie et Universaux Linguistics* of the CNRS.

Martin Tremblay completed a PhD in physical geography at the Nordic Studies Center (Laval University) and completed a postdoctoral fellowship on climate change and traditional knowledge at the Nasivvik Center at Laval University in 2006. Between 2006 and 2009, he worked for Kativik Regional Government in Nunavik on climate change projects as well as other environmental issues (environmental impact assessment and rehabilitation of abandoned mine sites in Nunavik).

Martin now works for Indian and Northern Affairs Canada in the Climate Change Division as an environmental policy analyst.

Winton Weyapuk, Jr. was born in Wales, Alaska and has lived there almost all of his life except for several years while he was attending college. He has been a subsistence hunter since he was a young boy and he now serves as a whaling captain and the chair of the Wales Whaling Captains Association. He received a BA in Rural Development and a BA in Inupiaq Eskimo language from the University of Alaska Fairbanks in the 1970s. He currently serves on the Wales Native Corporation, Board of Directors, and on the Native Village of Wales, IRA Council. In winter 2009 he started the first Inupiaq adult language classes for his fellow villagers to support the preservation of the Inupiaq language in his community.

Josh Wisniewski is a PhD candidate in anthropology at the University of Alaska Fairbanks. His dissertational study of subsistence hunting and of the use and knowledge of sea ice was based on 3 years of participation in subsistence activities with family marine mammal hunting crews in Shishmaref, Alaska. Wisniewski's ongoing research examines natural resource uses by Alaska Native communities through time and the factors that shape ethno-ecological knowledge and subsistence cultural practices. Specifically, he focuses on the pre-contact, historic, and contemporary Inupiaq and Yupik relations with marine mammals in the Bering and Chukchi seas.

List of Figures

List of Tables

Chapter 1
SIKU: International Polar Year Project #166 (An Overview)

Igor Krupnik, Claudio Aporta, and Gita J. Laidler

Natural sciences should no longer dictate the Earth system research agenda; social sciences will be at least as important in its next phase

(Reid, Bréchignac, and Lee in Science 2009, 245[1])

Abstract The SIKU (Sea Ice Knowledge and Use) project emerged in response to the growing public and scholarly attention to the environmental knowledge of the Arctic residents, as well as to the rising concerns about the impact of climate change on Arctic environment and polar sea ice. The special momentum for the SIKU project was created by International Polar Year (IPY) 2007–2008 that launched a new era of international and interdisciplinary collabration and partnership with northern communities. This introductory chapter tells how the SIKU project has originated and developed in 2004–2005; it reviews its structure made of various regional and individual initiatives, and covers major activities undertaken by the team during 2006–2009. It summarizes the key scientific outcomes and public messages of the SIKU project, as well as its contribution to the overall science program of IPY 2007–2008. It ends up with the synopsis of the present volume with the acknowledgements to many institutions and individuals who were instrumental to the success of the SIKU project.

Keywords Sea ice · Indigenous knowledge · Inuit · International polar year

Modern science and, even more so polar research, are always an exercise in partnership. Conducting complex long-term studies in the distant polar regions is usually beyond the resources of individual projects, a single discipline, or even one nation. Partnership among scholars and nations has been a pillar of polar science since the first International Polar Year 1882–1883 and it has been epitomized in the past

I. Krupnik (✉)
Department of Anthropology, National Museum of Natural History, Smithsonian Institution, Washington, DC, 20013-7012, USA
e-mail: krupniki@si.edu

I. Krupnik et al. (eds.), *SIKU: Knowing Our Ice*,
DOI 10.1007/978-90-481-8587-0_1, © Springer Science+Business Media B.V. 2010

decade through the planning and implementation of the fourth International Polar Year (IPY) 2007–2008.

This book presents the story and the key outcomes of one of the IPY 2007–2008 collaborative projects called SIKU, "Sea Ice Knowledge and Use: Assessing Arctic Environmental and Social Change" (IPY #166). For over 3 years, 2006–2009, this venture inspired scholars and indigenous experts from 5 nations and more than 30 communities spread around the Arctic, from Greenland to Siberia. The SIKU project embodied the collaborative and cross-disciplinary spirit of IPY 2007–2008. It also broke new ground in polar residents' participation in research and in bridging local knowledge and observations of indigenous experts with the data, tools, and models used by academic scientists.

The Message of the Sea Ice

Sea ice, frozen saltwater that solidifies on the sea surface, is one of the key components of the global system. It is also the most powerful image in our common vision of the Earth's polar regions. It can be stiff and silent but also blasting and crushing with terrible noise, and it can advance and retreat as if a living being. Every year sea ice extends across 14–16 million km^2 in the Arctic (and 17–20 million km^2 in the Antarctic Southern Ocean)[2] and then it shrinks and withers by the end of the summer to a fraction of its winter might. As it forms, persists, advances, and melts, it changes the ocean circulation, regulates global climate, and affects the life in the polar regions. It is also the most powerful agent affecting the human occupancy of polar areas since time immemorial.

Everlasting as it might seem, the sea ice is a surprisingly sensitive indicator of the processes in the air above and in the ocean below and, thus, a rapid messenger of change. People who live with the sea ice always watch its fluctuations; they recount its movement in myths, stories, and historical records. In fact, the written record of the sea ice dynamics is remarkably extensive in many northern areas – more than 1,000 years around Iceland, more than 800 years in the Baltic Sea, more than 500 years in the Barents and White Seas off Northern Russia, and almost 250 years off West Greenland and Labrador. Arctic residents, polar explorers, and mariners have marveled about the forces that govern ice movements. Only in the last 20–30 years have people realized that they had become a powerful enough factor to directly affect the polar ice through what is now known as "global climate change."

Scientific evidence of human influences on the planetary climate system emerged in the international public sphere in 1979, at the First World Climate Conference in Geneva, Switzerland. That scientific gathering issued a declaration calling on the world's governments "to foresee and prevent potential man-made changes in climate that might be adverse to the well-being of humanity" (WCC 1979). Through this internationally broadcasted message, the science of global climate change that existed at least since the early 1900s (Fleming 1998) entered the public domain. Public awareness of climate change increased throughout the 1980s, as did government concerns. In 1988, the governing bodies of the United Nations

Environment Programme (UNEP) and the World Meteorological Organization (WMO) established the Intergovernmental Panel on Climate Change (IPCC). Through its four assessment reports (1990, 1995, 2001, and 2007), the IPCC confirmed that planetary climate changes are underway – and are influenced by human activities – and offered comprehensive evidence of the ongoing shifts in the global climate system (Solomon et al. 2007). In this modern science of climate change, the study of sea ice, and particularly of the Arctic ice, plays a crucial role.[3] The status of Arctic sea ice is now a matter of major concern to polar scientists and general public, especially after the three consecutive record summer reductions of the last 3 years, 2007, 2008, and 2009.

When planning for a new major program in polar research, the International Polar Year (IPY) 2007–2008 began in the early 2000s, one of the key priorities was the need to address the ongoing changes in polar sea ice (ICSU 2004). This has been further articulated in the IPY 2007–2008 science program that featured more than 30 projects studying sea ice, including ship expeditions, remote sensing, ice-related ecosystems, the role of ice in climate modeling, ice monitoring, and the role of ice in the life of polar residents.[4] It is to this latter aspect that this book is dedicated, as the polar ice formation, the health of northern wildlife, Arctic people's livelihoods and cultures, and the reliability of their age-old expertise may all be irreversibly altered by drastic reductions projected for polar ice by the end of this century.

As Sheila Watt-Cloutier, the former president of the Inuit Circumpolar Council (ICC) argues, "The culture, economy, and identity of the Inuit as an indigenous people depend upon the ice and snow. Nowhere on Earth has global warming had a more severe impact than the Arctic" (Watt-Cloutier 2005:1). Northern indigenous peoples are thus in a position to exert significant influence in future global debates, including those on climate change, as well as in science research, monitoring, and modeling. This breadth of sea ice expertise shared by the Inuit and other indigenous nations in the Arctic inspired this volume. As Theo Ikummaq, one of our contributors, reminds us, "ice has a life of its own." It has many stories and teachings to share; we attempt to convey its messages through the stories and lessons in the chapters to follow.

Indigenous Knowledge

For many Arctic indigenous peoples, particularly not only the Inuit but also the Chukchi, Koryak, Nenets, and Saami, despite pronounced changes in lifestyles and local cultures, the ice-covered sea continues to be an integral component of daily life.[5] It remains a productive and widely used habitat for 6–9 months every year. It is also the main platform for traveling and for observing weather, tidal and current cycles, marine mammals, and other biota; as well as for training in navigational and subsistence skills passed from elders to youth.[6]

Further, the status of local use and knowledge of sea ice is an indicator of social change, a function of new technologies, economic and dietary trends and of shifts in educational practices and cultural values. For many communities, the use and knowledge of the ice-covered sea remains the pillar of their identity and resilience, their most prized intellectual treasure, the best of their scholarship based upon generations of experience and achievements. For scientists it offers an invaluable vision on how changes in polar ecosystems can be thoroughly documented and internalized through "another form" of knowledge and observations.[7]

Inuit and other polar indigenous people have recently been observing changes in ice and weather patterns that may indicate long-term climatic trends and increasing climate variability.[8] The shrinking, thinning, and/or disappearance of Arctic sea ice could exacerbate long-term climate warming; it would also severely impact the social, economic, and cultural practices of many indigenous communities in the circumpolar Arctic. The effects are already beginning to be felt. Inuit and other polar indigenous people are increasingly voicing their concerns about the possible implications of global warming in high latitudes (see Chapter 19). These implications often include (but are not limited to) the alteration of travel routes and access to hunting grounds; seasonal subsistence cycle, marine mammal distribution; economic opportunities; safety assessments; increased risks to hunters and travelers; and cultural knowledge loss (Huntington and Fox 2005; Ford et al. 2009; Laidler et al. 2009; Chapters 2, 3, 4, 5, 6, 7, 9, 11, 12, 13, 19, this volume).

Because Arctic people's perceptions of sea ice and climate change develop from place-based knowledge and personal interaction with their environment, most studies focusing on local observations of change are necessarily community-based. Of course, the type, degree, and importance of change vary based on geography, culture, economy, and community dynamics (Duerden 2004). By 2003–2004, collaborative research in documenting local observations of climate change was developing rapidly and many of the contributors to this volume were already engaged in community-based studies relating to sea ice and indigenous knowledge.[9] This common drive to understand and compare variations among regions, ice types, cultures, and communities sparked the synergy resulting in the SIKU initiative. Research expertise and many personal ties on which the SIKU partnership would be built were mostly in place by 2005. However, it took the momentum and the spirit of IPY 2007–2008 to make it happen.

The Origins of the SIKU–ISIUOP Project (IPY #166): 2004–2005

The story of the origination of IPY 2007–2008 and its early planning in 2001–2004 has been reviewed elsewhere (Krupnik et al. 2005; Krupnik 2009). Much like its predecessors, the first IPY in 1882–1883, the second IPY in 1932–1933, and International Geophysical Year (IGY) 1957–1958, IPY 2007–2008 was proposed as a primarily geophysical initiative. True, early champions of IPY 2007–2008 expressed interest in adding socio-cultural ("human") research to its program; but little was accomplished until the International Arctic Social Science Association (IASSA) joined the IPY planning process in summer 2004. Following IASSA's recommendation, in September 2004 the IPY Planning Group approved the additional ("sixth") theme for IPY aimed "[...] to investigate the cultural, historical, and social processes that shape the sustainability of circumpolar human societies, and to identify their unique contributions to global cultural diversity and citizenship" (ICSU 2004:7). Shortly after, in November 2004, a call for pre-proposals ("Expressions of Intent", EoI) for future IPY projects was issued and the new sociocultural field had to be filled almost at a minute's notice. Here, previous experience and personal ties among many members of the soon-to-be SIKU team played a crucial role.

The first discussion on the prospective projects for IPY 2007–2008 took place at the symposium "Reversing Language and Knowledge Shift in the North?" in Québec City, Canada, in October 2004, where Igor Krupnik, Claudio Aporta, and Shari Gearheard presented papers on their knowledge documentation research (Dorais and Krupnik 2005). Out of that first exchange came three IPY pre-proposals submitted in December 2004–January 2005, "Inuit Sea Ice Use and Occupancy Project" (ISIUOP, Claudio Aporta, EoI #715), "Elders of the Northern Ice: Sea Ice, Knowledge and Change in the Arctic, 1957–2007" (Igor Krupnik, EoI #332), and "Arctic Peoples' Observation Center" (APOC, Shari Fox Gearheard, EoI #358).[10] Independently, Hajo Eicken produced his proposal for the "Pacific Biophysical Seasonal Ice Zone Observatory" (EoI #58).[11] Ideas, research plans, and a network of prospective collaborators from these four initiatives contributed to the later development of the SIKU project in 2005.

In March 2005, Krupnik, Aporta, Gearheard, and Eicken were notified by the IPY Joint Committee (JC) that their projects were endorsed and they "have the potential to make a significant contribution to IPY 2007–2008." All IPY lead investigators were encouraged to seek international collaboration and build alliances for "large coordinated initiatives" that would engage researchers from several nations working across polar regions, with diverse sources of funding.[12] That recommendation became our main strategy in planning for the SIKU initiative.

A new round of talks on how to combine our proposals was launched in April 2005. By that time, a new acronym SIKU ("Sea Ice Knowledge and Use") was proposed for this joint initiative reflecting its focus on both the "knowledge" ("Elders") and "use" (ISIUOP) of sea ice. That new project title, SIKU, was a blessing, as it is also the main Inuit (Eskimo) word for sea ice (*siku*) known in every Inuit (Eskimo) community from Greenland to Russia. More exchanges followed in spring and summer 2005 and more people were brought to discussion. On July 31, 2005, Igor Krupnik sent a new draft of the combined SIKU initiative to Claudio Aporta, Shari Gearheard, Gita Laidler and also to Michael Bravo, Karim-Aly Kassam, Lene Kielsen Holm, and Frank Sejersen. That letter initiated an eight-member "international steering" committee for the SIKU project and urged for the submission of a joint "full proposal" by September 30, 2005.

Two more months of intense online communication followed, until the proposal was finally submitted on September 29, 2005.[13] It formally combined three earlier "Expressions of Intent" (Nos. 332, 715, and 358) and listed 40 researchers from 7 nations: Canada, U.S., U.K., Russia, Greenland, Denmark, and France.[14] The proposal sought partnership with collaborators from 16 polar communities: Barrow, Gambell, Point Hope, Toksook Bay, and Wainwright in Alaska (U.S.); Clyde

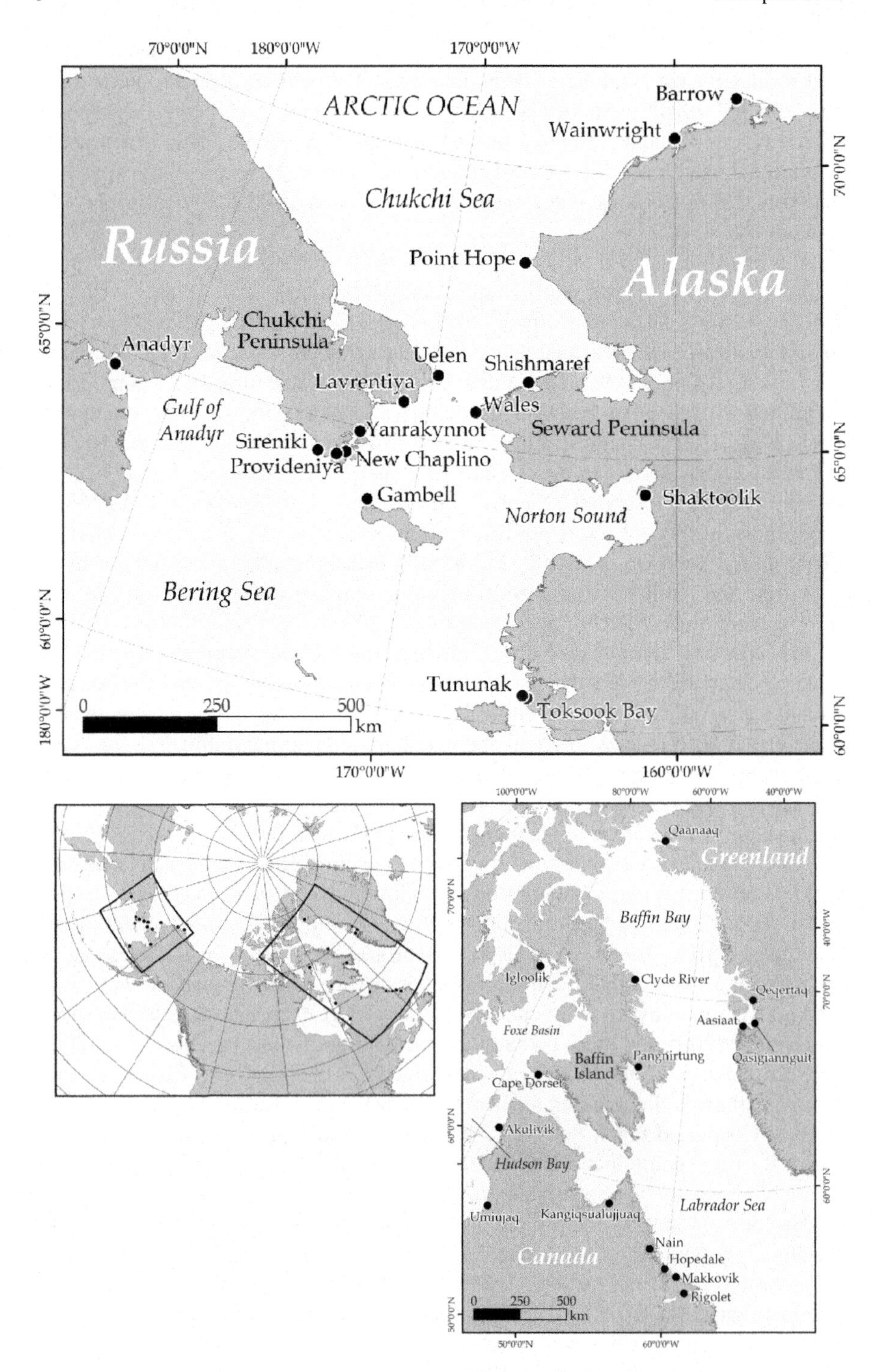

Fig. 1.1 Map of the SIKU study area. (**a**) Alaska-Chukotka; (**b**) Nunavut-Greenland

River, Grise Fjord, Igloolik, Pangnirtung, Ivujivik in Canada; Qaanaaq, Ilulissat, Upernavik, Tasiilaq in Greenland; and Inchoun and Novo-Chaplino in Russia. Not all of the researchers listed on the original proposal were able to obtain funding to take part in the SIKU studies, but several others joined the SIKU team in 2006–2008 (see later). Also, the list of collaborating indigenous communities in four Arctic nations was substantially expanded (Fig. 1.1).

On November 30, 2005, the IPY Joint Committee formally endorsed the SIKU Project as one of its international research initiatives (IPY #166) and notified the team that it should start seeking research funding.[15] By that time, several elements of the future collaborative network were already in place[16] and the new SIKU initiative was ready to be launched in full.

Overview of SIKU Activities: 2006–2009

The original SIKU science plan of 2005 argued for a coordinated international study of local knowledge and use of sea ice in several indigenous communities across the Arctic. It aspired to produce an impressive amount of new data, including logs of local observations of ice and weather conditions; maps of ice use and traveling; video and photo documentation; dictionaries of local sea ice terms and place names in indigenous languages; and other records, such as narratives, elders' memoirs, and oral histories. Topics of special interest listed in the SIKU science plan covered the status of today's ice expertise and daily use; contemporary ice hunting, traveling, and navigation methods; safety rules; patterns of knowledge transmission and training; interpretations of the ice characteristics and environmental variability; cultural responses to increase resilience; and community sustainability in the time of change.

These and other records were viewed as contributions to the first-ever international data set on indigenous knowledge and use of the Arctic sea ice during the IPY 2007–2008 era. The 2007–2008 SIKU data were thus foreseen as a crucial baseline to detect the evidence of change, past and present, environmental and social, for decades to come. It aspired to follow the model of the Inuit Land-Use and Occupancy Project (ILUOP) that produced a similar "snapshot" of the Inuit use of land and land resources across Arctic Canada in the 1970s (with portions of adjacent sea and ice-covered areas – see Freeman 1976). Much in the same vein, SIKU studies advocated the use of historical photography, early researchers' field notes, archival records, participatory subsistence and travel mapping, recording of indigenous ice terminologies and oral histories to document the scope of change in ice-associated knowledge, and subsistence triggered by social and environmental agents.

Due to its many goals, the SIKU project was conceived as a network ("family") of coordinated local initiatives led by teams of researchers, students, and indigenous contributors, such as elders, hunters, and youth. Research specialists included human geographers, anthropologists, ice scholars, marine biologists, indigenous knowledge experts, and personnel from local subsistence agencies. Each team had to seek funding through their respective national agencies and IPY programs, with an overall coordination of plans, activities, and data sharing among the constituent teams. In many ways, the structure of the SIKU project followed the overall organization of IPY 2007–2008, as it emerged by its official start in spring 2007.

First research for the SIKU project was conducted in late 2005 and early 2006, mostly as a continuation of the earlier activities focused on ice knowledge documentation.[17] The first new work explicitly listed as a part of the IPY "SIKU Initiative" was launched in spring 2006, when Leonard Apangalook, Sr., began his daily recording of ice and weather conditions in the Yupik community of Gambell, on St. Lawrence Island, Alaska (Chapter 4). In winter 2006–2007 local observations were also started in the community of Uelen, Russia, and in the Alaskan communities of Wales and Barrow, under the NSF-sponsored SIZONet project (which upon its submission to NSF in 2006 included links to SIKU).[18] Several other components of the SIKU network received funding and began operations in spring 2007. During the second SIKU and the first IPY winter 2007–2008, activities were conducted in more than 20 communities in Alaska, Canada, Greenland, and Russian Chukotka. Specific details on the implementation of individual projects are given in the related chapters in this volume.

A flexible and overlapping structure of the SIKU activities consisting of two dozen core and "associated" ventures evolved quickly and remained in place for the duration of the project, 2006–2010. Eventually, all SIKU-related operations split into four regional clusters called SIKU-Canada, SIKU-Alaska, SIKU-Chukotka, and SIKU-Greenland. Two major regional projects became the "pillars" of the SIKU program, the Canadian Inuit Sea Ice Use and Occupancy Project (ISIUOP, Claudio Aporta, PI) funded by the Government of Canada's International Polar Year (IPY) Program[19] and the Alaska-Chukotkan "SIKU: The Ice We Want Our Children to Know" project (Igor Krupnik and Lyudmila Bogoslovskaya, PIs) funded by the Shared Beringia Heritage Program of the US National Park Service, Alaska Office. Several other studies with their independent funding aligned with these "umbrella" projects under individual data- and resource-sharing agreements (Table 1.1).

Such a flexible multi-focal network for a large international project proved to be extremely vibrant; it was also highly adaptable to inevitable variability in funding, scheduling, and field operations among constituent projects. It also facilitated the initiation of new activities under the broad SIKU family and partnering with other efforts that opted to be associated with our project. As a result, we were able to launch certain new studies in winter 2008–2009, almost at the end of the last field year of our program (see Chapters 5, 17, and 18).

Over the course of our project, a number of team members met for the assessment of our operations at two ISIUOP project workshops held at Carleton University (Ottawa, Canada) in October 2007 and in November 2008 (Fig. 1.2) and at the all-day SIKU session at the 6th International Congress of Arctic Social Sciences (ICASS-6) in Nuuk, Greenland, in August 2008 (Figs. 1.3 and 1.4). Papers delivered at the Nuuk session, "SIKU (IPY #166): Polar Residents Document Arctic

Fig. 1.2 SIKU–ISIUOP workshop at Carleton University, November 2008

Fig. 1.3 SIKU session at the International Arctic Social Sciences Congress in Nuuk, August 2008

Fig. 1.4 SIKU session in Nuuk (Lene Kielsen Holm, Shari Gearheard, Gita Laidler, and Igor Krupnik)

Ice and Climate Change," eventually made the core of the present volume (see below).[20] During 2007–2009, team members gave over two dozen science and public presentations on the outcomes of their work at various venues,[21] including several public sessions in the participating indigenous communities (Figs. 1.5, 1.6, and 1.7) and at the big IPY "open science" conference "Polar Research –

Fig. 1.5 ISIUOP project team at a meeting in Iqaluit, February 2007

Fig. 1.6 SIKU observational seminar in Provideniya, Russia, March 2008

Fig. 1.7 SIKU project presentation in Uelen, Russia. Victoria Golbtseva (to the left) speaks to local hunters and elders, March 2008

Arctic and Antarctic Perspectives in the International Polar Year" in St. Petersburg, Russia.[22] The SIKU–ISIUOP activities have been featured in several online and public venues,[23] such as on the main IPY 2007–2008 web site, the special IPY "Sea Ice Day" (September 2007), during the IPY "Polar People Week" (September 2008),[24] and in a number of overview publications (Krupnik and Bogoslovskaya 2007; Bogoslovskaya et al. 2008). The SIKU web site designed by Gita Laidler in 2008, with the help of Kelly Karpala and Amos Hayes, remained an important platform for the dissemination of information on our project throughout 2009–2010.

The winter of 2008–2009, the last under IPY official observational period, featured scores of new activities, including observation, mapping, field trips, and interviews in many participating communities. As the third SIKU "ice-year" drew to a close in summer 2009, SIKU team members shifted their attention to the processing and analysis of the data collected and to writing papers and reports. Some of the latter became chapters in this volume (see below).

Outcomes and Messages of the SIKU Project

At the time of this writing (February 2010), the story of many efforts undertaken within the SIKU project and associated initiatives has been featured in about 50 scientific, popular, and online publications in at least four languages, English, Russian, Danish, and Inuktitut/Greenlandic (Appendix B). A special issue of *The Canadian Geographer/Le Géographe Canadien* dedicated to the SIKU-Canada initiative (ISIUOP) will appear in late 2010, with at least six papers describing project activities and outcomes. Another major publication, a bilingual English–Norwegian catalogue of historical photographs from Point Hope, Alaska, taken by Berit Arnestad Foote in 1959–1962 was produced in March 2009 (Arnestad Foote 2009). Several other publications are in press or in preparation, including at least three books featuring individual SIKU-related efforts scheduled for 2010, such as the *Wales Inupiaq Sea Ice Dictionary* (Chapter 14), *Ellavut/Our World and Weather* (Chapter 13), and the full story of the Sila–Inuit–Hila project (Chapter 11). The results of the Russian SIKU activities will be presented in a special Russian volume in 2011.

In addition, SIKU and associated ventures produced numerous educational and outreach materials for participating communities[25] and scores more will be delivered. The online *Cybercartographic Atlas of Inuit Sea Ice Knowledge and Use* featuring the data collected by the ISIUOP team (Chapter 10) is due to be released in early 2010. A sea ice and indigenous knowledge-focused exhibit, *Silavut: Inuit Voices in a Changing World* (curated by Shari Gearheard), was produced at the University of Colorado Museum of Natural History and was on display for a full year from April 15, 2008, to May 1, 2009 (Fig. 1.8).[26] Lastly, of no small importance are two Ph.D. dissertations at the University of Alaska Fairbanks by Josh Wisniewski (Chapter 12) and Matthew Druckenmiller (Chapter 9) and

Fig. 1.8 Inuit elder Ilkoo Angutikjuak and interpreter Geela Tigullaraq (*left*) from Clyde River meet with a visiting teacher and her students who came to see the *Silavut* exhibit in 2008 (Photo: Shari Gearheard)

three M.A. theses at Carleton University by Karen Kelley, Kelly Karpala, and Jennifer McKenzie, all of which rely heavily on data collected during the SIKU effort.

The last major assessment of SIKU activities will take place at the next IPY conference "Polar Science – Global Impact" in Oslo, Norway, in June 2010, as a part of its sessions on "Communities and Change"[27] and "New Approaches for Linking Science and Indigenous knowledge."[28] As the funding for many components of the SIKU network continues throughout 2010, the overall contribution of our project to IPY 2007–2008 may not be known in full for some time. Nonetheless, some of the key results of the SIKU initiative are already obvious.

First, it would have never happened in its current shape without enthusiastic contributions by many devoted local partners and without immense interest in collaborating communities. As polar research advanced in the last two decades and particularly during the preparation for IPY and other major interdisciplinary efforts, such as the *Arctic Climate Impact Assessment* (ACIA 2005), International Conference for Arctic Research Planning (ICARP-2), and International Study of Arctic Change (ISAC), it became more "human-oriented." It can be seen in the much stronger focus on human and societal issues in these and other major integrative studies, but also in the increased input by local knowledge experts to science planning, research, and writing. All our project teams included indigenous residents and/or locally based researchers, and they now make almost half of the contributors to this volume. We hope that our project's experience encourages more collaborative studies to follow.

Second, the organization of SIKU as a flexible decentralized network illustrates that building such research "alliances" may be a promising template for future international polar projects. We believe that our experience from the SIKU project concurs with the overall lesson of IPY 2007–2008. Carrying exciting pioneer research in the polar regions does not necessarily require huge oversight costs, as resources may be amplified via sharing and good coordination.

Third, an important lesson of the SIKU effort is that modern technologies create new openings for research and generate support for science collaboration in northern communities. Many of these distant small towns now have modern schools, computer facilities, and local residents interested in research tools and technology of the twenty–first century. Observations, diaries, maps, and photographs generated by our local partners, often on their home computers, were routinely transferred to us via the Internet, so that data and messages were often exchanged in the matter of minutes not months. New GPS technology, local ice-monitoring techniques, and cybercartographic tools were all embraced by communities seeking to document their sea ice expertise and to use it to increase safety and to educate their youth (see Chapters 3, 7, 8, 9, 10, and 15).

Fourth, as much as northern residents are anxious to participate in technologically advanced research, they also want research to support their languages, knowledge, and ways of living. Fancy computers, GPS, and other devices have not taken away people's concerns about the future of their language, subsistence, and cultural well-being. The lesson we learned from our many initiatives is clear: local communities want their languages used and accessible on the Internet, CD-ROMs, electronic maps, and computer screens (Chapters 3, 7, 8, 9, 10, 11, 14, and 15). Converting science records into electronic public domain(s), thus, should be recognized as capacity building for future research and language/heritage preservation rather than an element of "science outreach." That may require a re-assessment of our priorities and resource allocations for future projects.

Fifth, distant northern communities are clearly feeling the pressure of public anxiety about the impacts of planetary climate change (see Marino and Schweitzer 2009). During 3 years of the SIKU project we have talked to and encountered many of journalists and television and documentary film crews. All were coming to the North to broadcast a message of rapid change from small local towns to the global audiences. Of course, we welcome public concern; but broad media exposure often produces community fatigue and "stomping" of indigenous observations, an unwarranted factor to our collaborative assessment of climate shifts in the North.

SIKU Legacies

By the time this book is published, the SIKU-related activities will be in the final stage, that is, in data assessment and management, publication, and the production of materials for participating communities. These efforts will help extend our partnership far beyond the years of IPY 2007–2008. Nonetheless, we may already identify certain elements of the SIKU "legacy" and of its general input to IPY 2007–2008. A preliminary list (to be certainly expanded) looks as follows.

IPY made a difference. Whereas many of the projects that contributed to the international SIKU initiative almost certainly would have taken place independently of IPY, only something on the scale of IPY could bring together such a diverse team and so many places. The IPY momentum also brought with it a strong collaborative spirit, international attention, focus on outreach and communication with the public, and the resources that no single project could ever generate. This was a widely shared experience across the diverse SIKU team.

Another question is whether such an internationally coordinated network approach could be repeated or updated in the future. The crucial issue here is whether something comparable to SIKU can be produced in 10 or 20 years from now to capture rapid change in polar sea ice and local ice use that is underway? We believe that the outcomes of the SIKU project will have a longer life span, because they were a part of IPY 2007–2008, and that invoking the "IPY legacy" would be the best chance to revisit the SIKU records in the decades to come.

The power of "multiple perspectives." The SIKU project clearly demonstrated what one of the chapters calls "the power of multiple perspectives" (Chapter 11 by Huntington et al. this volume) in modern interdisciplinary science. These days, many large programs, like IPY 2007–2008, are designed to include social science; but too often its contribution is seen in only illuminating the "human dimension," namely the impact upon people of certain environmental conditions and processes studied by physical and natural scientists, or societal response to these biophysical processes. By exploring indigenous peoples' knowledge and use of sea ice, the SIKU project introduced a new field of interdisciplinary research, the study of social (socio-cultural) aspects of the natural world, or what we may call the *social life* of sea ice (Chapter 20 by Bravo, this volume). It incorporates local terminologies and classifications, place names, personal stories, teachings, safety rules, historic narratives, and explanations of the empirical and spiritual connections that people create with the natural world. By recognizing that sea ice has a "social life" and emphasizing the value of indigenous perspectives we made a novel contribution to IPY, to science, and to the public.

Breaking new ground. Three years of research under SIKU and related projects produced a tremendous amount of new information produced by coordinated research in four polar nations. Again, we see this as the fulfillment of the very special mission of IPY identified as "an intense international campaign of coordinated observations and analysis across the polar regions ... [to] expand our ability to detect ongoing changes ... and to extend this knowledge to the public" (ICSU 2004; Krupnik et al. 2005:89). Within this vast body of new knowledge created by the SIKU project, certain fields stand as truly pioneer contributions. The main product of SIKU-Canada's ISIUOP, a new cybercartographic "Atlas" developed for the documentation and communication of regional Inuit knowledge and use of sea ice (Chapter 10), is a remarkable achievement. The very task of bringing modern technologies to assist polar residents in using the ice may be one of the landmark legacies of the SIKU project (see Chapters 8, 9, and 10). Three-year ice observation records in 11 communities (Chapters 4, 5, and 11) and almost 30 dictionaries of local indigenous ice terminologies, from Chukotka to Greenland (Chapters 3, 4, 11, 14, 16, 17, and 18), are certain to advance our understanding and databases to an entirely new level.

"Participatory science." As much as scientists cherish their engagement in high-level and cutting-edge research, so do our partners in northern communities. We believe that one of the key legacies of the SIKU initiative and of IPY 2007–2008 is the input from communities, particularly in providing valuable expertise and commitment to further research partnership with scientists. As Paul Apangalook, one of our partners, stated in his letter, "It has been an honor to take part in this project and a fortunate privilege for the village to have a voice in the effort. It provides us with an unbiased language platform outside of politics and special interests." Almost every chapter in this volume combines the voices of academic researchers and community experts and that spirit of "participatory" science excited everyone involved.

Crossing boundaries. IPY 2007–2008 was devised as a cross-disciplinary venture aimed at addressing questions and issues beyond the scope of individual science disciplines (Allison et al. 2007). It also aspired to cross national boundaries and to make research and dissemination of data a matter of international collaboration rather than of state science programming. In fulfilling this IPY mission the SIKU network did its best to advance partnerships across national and disciplinary boundaries as well as the university/agency and scientist/local expert divides. We worked in mixed teams, tackled issues beyond our professional fields, and published our results in many languages, styles, and formats. We hope that the readers will enjoy this aspect of our work that we have tried to convey in many chapters of this book.

The Structure of the Volume

The idea to publish a summary book on SIKU activities emerged in 2005, at the very first stage of our planning. As contributors to several earlier overviews of indigenous observations of Arctic environmental change (i.e., Huntington and Fox 2005; Krupnik and Jolly 2002), we were keen to produce an update on the next generation of research with northern communities, as a part of IPY 2007–2008. This plan gained momentum at the full-day SIKU session at the 6th International Arctic Social Sciences Congress (ICASS-6) in Nuuk in August 2008 (see earlier). Following ICASS-6, a proposal was submitted for a major SIKU book to be a part of Springer's publications related to IPY 2007–2008. Ten (out of 14) papers presented at the Nuuk session were eventually revised and expanded to become chapters in the present volume, and several more chapters were added in 2009.

By expanding the number of chapters and authors, we introduced a broader spectrum of the SIKU and related activities; we also changed the volume's coverage from geographic (i.e., Alaska, Canada, Greenland, and Russia, as in Nuuk) to thematic. Thus, the four parts of the book tell their stories from different thematic angles of the original research plan. The two major tasks of the SIKU initiative were the recording of local observations of climate and ice change (Part I, Chapters 2, 3, 4, 5, and 6) and the documentation of the current use of ice across the Arctic, particularly with the use of modern technologies (Part II, Chapters 7, 8, 9, and 10). Our expanded focus now includes learning and transmission of traditional knowledge to the younger generations is presented in Part III (Chapters 11, 12, 13, and 14), and the cross-disciplinary assessment of indigenous knowledge and use of sea ice from the perspectives of many science disciplines, i.e., ice studies, linguistics, history, and political sciences (Part IV, Chapters 15, 16, 17, 18, and 19).

Three additional perspectives were added to the book to broaden its overall message. In this volume's Foreword, Helena Ödmark comments on how the need to bring social and the human studies to IPY 2007–2008 was viewed and encouraged by the Arctic Council. The value of the sea ice to the Arctic residents is addressed by Joelie Sanguya and Shari Gearheard in the Preface. Lastly, Michael Bravo analyzes the humanistic aspects of polar science, its history, and evolving ethics in the book's Epilogue. Each contribution has its own story; together they embody the multi-vocal and cross-boundary nature of our work. We believe that the very composition of the book sends a strong message of the inclusive and collaborative spirit of the SIKU project and of this International Polar Year in general.

Table 1.1 Organizational structure of the SIKU (SIKU–ISIUOP) initiative, 2006–2010

Project title	Umbrella project	Chapter	Collaborating communities	Contributors	Funder(s)
SIKU-*Canada*	ISIUOP	3, 7, 8, 10, App. A			Government of Canada IPY Program
Mapping Inuit sea ice knowledge and use	ISIUOP	3	Cape Dorset, Igloolik, Pangnirtung, Nunavut, Canada	Claudio Aporta, Gita Laidler, Mark Kapfer, Tom Hirose, Roger DeAbreu, Kelly Karpala, Karen Kelley, Pootoogoo Elee, Eric Joamie, Theo Ikummaq	Government of Canada IPY Program
Igliniit ("trails") project	ISIUOP	8	Clyde River	Shari Gearheard, Gary Aipellee, Kyle O'Keefe, Apiusie Apak, Jayko Enuaraq, David Iqaqrialu, Laimikie Palluq, Jacopie Panipak, Amosie Sivugat, Desmond Chiu, Brandon Culling, Sheldon Lam, Josiah Lau, Andrew Levson, Tina Mosstajiri, Jeremy Park, Trevor Phillips, Michael Brand, Ryan Enns, Edward Wingate, Peter Pulsifer, Christine Homuth	Government of Canada IPY Program
Nunavik communities and ice: environmental change and community safety	ISIUOP	App. A	Kangiqsualujjuaq, Kangiqsujuaq, Akulivik	Chris Furgal, Martin Tremblay, Tuumasi Annanack, Eli Angiyou, Michael Barrett	Government of Canada IPY Program

Table 1.1 (continued)

Project title	Umbrella project	Chapter	Collaborating communities	Contributors	Funder(s)
Multi-media educational materials and sea ice cybercartographic atlas development	ISIUOP	10	Cape Dorset, Igloolik, and Pangnirtung	Fraser Taylor, Peter Pulsifer, Amos Hayes, Gita Laidler, Claudio Aporta, Shari Gearheard, Chris Furgal, Karen Kelley, Kelly Karpala, Christine Homuth	Government of Canada IPY Program
Innuttut sea ice terminology in Labrador	SIKU	17	Rigolet, Nain, Makkovik	Paul Pigott and local elders	SIKU
SIKU-*Alaska, 2006–2010*					
Community sea ice and weather observations	SIKU	4, 6	Gambell, Shaktoolik, Wales, Barrow	Igor Krupnik, Leonard Apangalook, Sr., Paul Apangalook, Clara Mae Sagoonick, Winton Weyapuk, Jr., Joe Leavitt, Arnold Brower, Sr.	National Park Service, Alaska Office; NSF (for Wales and Barrow)
Alaska sea ice dictionaries	SIKU	1, 12, 14, 16	Wales, Shishmaref, Shaktoolik, Barrow Wainwright	Igor Krupnik, Winton Weyapuk, Jr., Herbert Anungazuk, Faye Ongtowasruk Ronald Brower, Sr., Clara Sookiayak, Josh Wisniewski, Hajo Eicken, Matthew Druckenmiller, Lawrence Kaplan	National Park Service, Alaska Office
Point Hope: Historical ice photography, 1959–1962	SIKU		Point Hope	Berit Arnestad Foote and Rex and Piquk Tuzroyluk	National Park Service, Alaska Office

Table 1.1 (continued)

Project title	Umbrella project	Chapter	Collaborating communities	Contributors	Funder(s)
Integrated seasonal ice zone observing network (SIZONet)*	SIZONet	5, 9, 15	Barrow, Wales, Gambell	Hajo Eicken, Matthew Druckenmiller, Winton Weyapuk, Jr., Joe Leavitt, Howard Brower, Sr., Mette Kaufman, Marie-Luise Kapsch	NSF, University of Alaska
Shorefast sea ice in Alaska: An ecosystem services approach to improve monitoring*	SIZONet	1, 9, 15	Wales, Barrow	Matthew Druckenmiller, Hajo Eicken, and local collaborators	NSF; OSRI; Druckenmiller's Ph.D. dissertation
Hunting as being and knowing in Northwest Alaska*		12	Shishmaref	Josh Wisniewski and local collaborators	NSF, University of Alaska Fairbanks, Wisniewski's Dissertational Project
Nelson Island natural and cultural history project*		13	Toksook Bay, Chefornak, Nightmute, Nunapitchuk, Newtok	Ann Fienup-Rirodan, Mark John, Alice Rearden, and local elders	NSF-BEST (Bering Sea Ecosystem Program)
SIKU-*Chukotka*					
Ice and weather observations by community monitors	SIKU		Uelen, Novo-Chaplino, Yanrakinnot, Sireniki	Roman Armaergen, Victoria Golbtseva, Alexander Borovik, Arthur Apalu, Oleg Raghtilku, Natalya Kalyuzhina, Anatoly Kosyak, Igor Krupnik, Lyudmila Bogoslovskaya	National Park Service, Beringia Natural and Ethnic Park, Institute of Cultural and Natural Heritage, Chukotka Branch of Northeastern Research Institute

Table 1.1 (continued)

Project title	Umbrella project	Chapter	Collaborating communities	Contributors	Funder(s)
Indigenous sea ice dictionaries	SIKU	14, 16	Uelen, Yanrakinnot, Sireniki, Lavrentiya	Igor Krupnik, Lyudmila Bogoslovskaya, Roman Armaergen, Victoria Golbtseva, Elizaveta Dobrieva, Boris Alpergen, Arthur Apalu, Leonid Kutylen, Natalya Kalyuzhina, Michael Fortescue, Aron Nutwayi, Natalya Rodionova	National Park Service, Beringia Natural and Ethnic Park, Institute of Cultural and Natural Heritage, Chukotka Branch of Northeastern Research Institute
Historical and contemporary instrumental records of ice and climate change in Chukotka	SIKU			Boris Vdovin, Igor Zagrebin, Vladimir Struzhikov	National Park Service, Institute of Cultural and Natural Heritage
SIKU-*Greenland*					
Sila-Inuk*	SIKU	6		Lene Kielsen Holm	
Weather variability and changing sea ice use in Qeqertaq	SIKU	2	Qeqertaq	Pierre Taverniers and local collaborators	

Table 1.1 (continued)

Project title	Umbrella project	Chapter	Collaborating communities	Contributors	Funder(s)
Comparative initiatives					
Siku–Inuit–Hila: The dynamics of human–sea ice relationships: Comparing changing environments in Alaska, Nunavut, and Greenland*		11	Barrow, Clyde River, Qaanaaq	Shari Gearheard, Henry Huntington, Lene Kielsen Holm, Yvon Csonka, Mamarut Kristiansen, Joe Leavitt, Nancy Leavitt, Andy Mahoney, Warren Matumeak, Qaerngaaq Nielsen, Toku Oshima, Ilkoo Angutikjuak, Joelie Sanguya, Igah Sanguya, Geela Tigullaraq	NSF Health Canada Inuit Circumpolar Council-Greenland (Support for participation and travel of Lene Kielsen Holm)
Indigenous sea ice dictionaries	SIKU	14, 16, 17, 18, App. A		Igor Krupnik, Louis-Jacques Dorais, Michael Fortescue, Alana Johns, Steven Jacobson, Lawrence Kaplan, Ludger Müller-Wille, Paul Pigott, Gita Laidler, Martin Tremblay, Claudio Aporta, Shari Gearheard, and local collaborators	Combined funding from several sources

Associated projects are marked with an asterisk (*).

Acknowledgments The SIKU volume and the entire SIKU project network would never have become a reality without the support and dedicated input by many organizations, research institutions, and individuals. Many of them are acknowledged in individual chapters in the volume; here we would like to name people and agencies that helped build and implement the SIKU initiative.

We are grateful to the IPY 2007–2008 planners – Chris Rapley, Robin Bell, Eduard Sarukhanian, Ian Allison, Robert Bindschadler, and others – who worked with us to open the field of socio-cultural ("human") research in IPY 2007–2008, of which our SIKU project (IPY #166) became a vital part. We thank the Joint Committee for IPY 2007–2008 and the Canadian, Greenlandic, and Danish National IPY Committees that endorsed the SIKU project and its contributing initiatives. Many colleagues participated in the earlier planning of the SIKU activities; the role of Ernest S. Burch, Jr., Louis-Jacques Dorais, Milton Freeman, Karim-Aly Kassam, Richard Nelson, Rick Riewe, Frank Sejersen, Natasha Thorpe, Peter Usher, George Wenzel, and Stephen B. Young should be acknowledged with most gratitude.

In Canada, we are grateful to the IPY Canada Federal Program Office for their continuous and dynamic support of ISIUOP, particularly to Kathleen Fischer (executive director), Robert Fortin (director), Sarah Kalhok (manager, IPY Science), Katherine Wilson (coordinator of Logistics and Licencing), and Scott Tomlinson (coordinator of Data Management). We would like to thank Jamal Shirley and Mary-Ellen Thomas of the Nunavut Research Institute for their guidance and support through the IPY process. We are also indebted to the administrative and technical support we received from Carleton University, particularly to Barbara George, Amos Hayes, Sandra Nelson, and Darlene Gilson.

In the U.S., we are grateful to the "Shared Beringia Heritage Program" of the U.S. National Park Service, Alaska Office (Peter Richter and Katerina Solovjeva-Wessels) that provided funding for the SIKU-Alaska and SIKU-Chukotka activities in 2007–2010. Additional support was given by the National Museum of Natural History, Smithsonian Institution; the National Science Foundation (to the "Seasonal Ice Zone Observational Network," SIKU–Inuit–Hila Project ("The Dynamics of Human–Sea Ice Relationships: Comparing Changing Environments in Alaska, Nunavut, and Greenland") and the "Nelson Island Natural and Cultural History" projects), the University of Alaska Fairbanks, University of Alaska Press, and other institutions. Special thanks to William Fitzhugh, G. Carleton Ray, Sergei Bogojavlenski, Gary Hufford, Vera Metcalf, Martha Shulski, Martin Robards, Matthew Sturm, and others who offered valuable advice and shared their knowledge with the SIKU-Alaska team.

In Russia, the SIKU-Chukotka activities were conducted in collaboration with the Russian Institute of Cultural and Natural Heritage in Moscow (director, Dr. Yuri Vedenin), the Chukotka Branch of the Northeastern Research Institute (SVKNII) in Anadyr (director, Oleg Tregubov), and the "Beringia" Ethno-Cultural Park in Provideniya (director, Natalya Kalyuzhina). The crucial role of many Russian SIKU contributors, Lyudmila Bogoslovskaya, co-chair of the Russian SIKU program, Boris Vdovin, Victoria Golbtseva, Igor Zagrebin, Nadezhda Vukvukai, Anatoly Kosiak, Roman Armaergen, Arthur Apalu, Alexander Borovik, and many others should be specially acknowledged. Their contribution to the joint SIKU effort will be presented in a separate Russian volume to be published later.

In Greenland the Inuit Circumpolar Council provided funding and support for the Sila-Inuk Project (Chapter 2) and the SIKU–Inuit–Hila Project (Chapter 11). KNAPK (Kalaallit Nunaani Aalisartut Piniartullu Kattuffiat) and Greenland hunters and fishermen's organization also provided support and partnership to the Sila-Inuk Project.

We are grateful to the organizers of the 6th International Congress of Arctic Social Sciences in Nuuk (August 2008), Birger Poppel (chair), Yvon Csonka (IASSA president), Inge Seiding, and Janus Chemnitz Kleist, who assisted us in making our session on the SIKU project an inspirational event. Grete Hovelsrud helped us connect to Springer's IPY publication program. Margaret Deignan at Springer kindly navigated us through Springer's application and submission process. Milton Freeman and Lenore Grenoble offered valuable comments to the original manuscript and Aqqaluk Lynge endorsed its publication on behalf of the ICC-Greenland. Cara Seitchek and Chase Morrison were instrumental in editing and processing chapters and illustrations. Lucian Read kindly offered his photograph from Qaanaaq, North Greenland for the book cover and Anya Vinokour produced cover design. Shari Gearheard and Hajo Eicken offered valuable comments to the earlier versions of this chapter. We thank you all.

Notes

1. Walter V. Reid is chair of the ICSU Earth System Visioning Task Group and directed the Millenium Ecosystem Assessment. Catherine Bréchignac is the current president of ICSU (International Council for Sciences) and Yan Tseh Lee is the president-elect of ICSU.
2. http://nsidc.org/sotc/sea_ice.html
3. There is a rapidly growing literature on the sea ice as an indicator of climate change. Critical sources may be accessed in the respective sections of major international reports (Walsh 2005:189–196; Lemke et al. 2007:350–356) and also at http://nsidc.org/sotc/references.html
4. See http://www.ipy.org/index.php?option=com_k2&id=943&view=item&Itemid=47, Six more sea ice-related projects (## 81, 88, 107, 141, 258 and 313) were launched in Antarctica.
5. Sea ice is also an important element of life, culture, and economies of the Icelanders, Swedish, and Finnish residents along the northern Baltic Sea and many old-settlers' communities in the Russian Arctic (the Pomors) and Newfoundland-Labrador, as well as of thousands of other people who reside in towns and cities along the seasonally frozen ocean shores.
6. See earlier and recent studies by Nelson (1969), Freeman (1984), Riewe (1991), Nakashima (1993), McDonald et al. (1997), Huntington (2000), Aporta (2002, 2004, 2009), Jolly et al. (2002), Norton (2002), Krupnik (2002), George et al. (2004), Oozeva et al. (2004), Gearheard et al. (2006), Laidler (2006), Ford et al. (2009), and Laidler et al. (2009).
7. See Aporta (2002), Bogoslovskaya et al. (2008), Eicken et al. (2009), Freeman (1984), Gearheard et al. (2006), Laidler (2006), Nakashima (1993), Nelson (1969), and Norton (2002).
8. See McDonald et al. (1997), Ford (2000), Bogoslovskaya et al. (2008), Riedlinger and Berkes (2001), Fox (2002), Herlander and Mustonen (2003), Jolly et al. (2002), Kavry and Boltunov (2006), Krupnik (2002), Laidler and Elee (2008), Laidler and Ikummaq (2008), Laidler et al. (2008), Nickels et al. (2002), Nichols et al. (2004), Ford (2005), Ford et al. (2007), and Nickels et al. (2006).
9. Gearheard, Huntington, Krupnik, and Furgal were contributors to "The Earth Is Faster Now" volume (Krupnik and Jolly 2002). Henshaw started her study on Sikusilarmiut place name and climate change in Cape Dorset in 2000 (Chapter 19). Krupnik and Huntington had a collaborative project on ice and weather knowledge on St. Lawrence Island, Alaska, in 2000–2002 (Krupnik 2002; Oozeeva et al. 2004), Eicken and Huntington were involved (since 1999) in a historical ice study in Barrow (Huntington et al. 2001; George et al. 2004). Huntington and Gearheard started a pilot project to compare ice knowledge in Barrow and Clyde River in 2003 (Gearheard et al. 2006; Chapter 11), and they were lead authors for the ACIA chapter on indigenous observations of Arctic climate change (Huntington and Fox 2005). Also, Aporta, Gearheard, and Laidler defended their Ph.D. dissertations based on the Inuit sea ice use and climate change in Nunavut (Aporta 2003; Fox 2004; Laidler 2007).
10. See the original pre-proposals at http://classic.ipy.org/development/eoi/index.html
11. See http://classic.ipy.org/development/eoi/details.php?id=58
12. Letter from Ian Allison and Michel Beland, IPY Joint Committee (JC) co-chairs, to Igor Krupnik, March 30, 2005.
13. See http://classic.ipy.org/development/eoi/proposal-details.php?id=166
14. Igor Krupnik (U.S., principal investigator), PI, Herbert Anungazuk (U.S.), Claudio Aporta (Canada), Sergei Bogojavlensky (U.S.), Lyudmila Bogoslovskaya (Russia), Michael Bravo (U.K.), Roger DeAbreu (Canada), Louis-Jacques Dorais (Canada), Ann Fienup-Riordan (U.S.), Shari Fox Gearheard (Canada), Milton Freeman (Canada), Christopher Furgal (Canada), Rolf Gilberg (Denmark), Nicole Gombay (Canada), Anne Henshaw (U.S.), Carol

Jolles (U.S.), Karim-Aly Kassam (Canada), Darren Keith (Canada), Lene Kielsen Holm (Greenland), Gita Laidler (Canada), John Macdonald (Canada), James Maslanik (U.S.), Heather Meyers (Canada), Anna Motschenbacher (U.S.), Richard Nelson (U.S.), David Norton (U.S.), Carl Christian Olsen (Greenland), James Overland (U.S.), Lynn Peplinski (Canada), G. Carleton Ray (U.S.), Martin Robards (U.S.), Chie Sakakibara (U.S.), Frank Sejersen (Denmark), Nicole Tersis (France), Natasha Thorpe (Canada), Martina Tyrell (U.K.), Nadezhda Vukvukai (Russia), George Wenzel (Canada), Kevin Wood (U.S.), and Steven Young (U.S.).

15. Letter from Ian Allison and Michel Beland, JC co-chairs, November 30, 2005.
16. The Nelson Island Natural and Cultural Knowledge Project was submitted to the NSF BEST (Bering Sea Ecosystem Study) Program in late 2005 (Chapter 13); the Coupled Humans and Sea Ice Systems project (NSF OPP #0308493, Chapter 11) was in its second year; and the Sila-Inuk study under the ICC-Greenland was in planning since 2004 (Chapter 6).
17. Laidler's PhD research in Igloolik, Pangnirtung, and Cape Dorset (Chapter 3); comparative study of indigenous ice knowledge in Barrow and Clyde River (to become the SIKU–Inuit–Hila project, Chapter 11); SILA-Inuk survey in Greenland (Chapter 6), to name but a few.
18. SIZONet builds on the concept of sea ice system services (Eicken et al. 2009). The coastal observatory sites at Barrow and Wales combine geophysical measurements with Iñupiat ice observations to develop an integrated observing site (Druckenmiller et al. 2009).
19. http://gcrc.carleton.ca/isiuop
20. The session was co-chaired by Igor Krupnik, Shari Gearheard, Lene Kielsen Holm, and Gita Laidler. The full list of presentations for the ICASS-6 session 1.06 can be accessed at http://www.arctichost.net/ICASS_VI/
21. See Appendix B at the end of this volume.
22. Claudio Aporta's presentation, "The Ice We Want Our Children to Know: SIKU (IPY #166) Overview with an Emphasis on Canada" was given in Section 5.4, The Role of Indigenous Knowledge in Modern Polar Science, chaired by Victoria Gofman and Shari Gearheard – see conference program at http://www.scar-iasc-ipy2008.org/scar-iasc-ipy2008/program.xls
23. See http://www.youtube.com/watch?v=kpG3yZ-Hz8o; http://www.ipy.org/index.php?option=com_k2&id=309&view=item&Itemid=12; http://www.ipy.org/index.php?option=com_k2&id=943&view=item&Itemid=47
24. See more on the IPY "International People Day" at http://www.ipy.org/index.php?option=com_k2&id=1733&view=item&Itemid=47 and http://ipycanada.ca/web/guest/Polar_Day
25. That is, Anijaarniq: Introducing Inuit Landskills and Wayfinding, educational CD-ROM on the Canadian Inuit ice navigational skills (produced by Claudio Aporta in collaboration with the Nunavut Research Institute, 2006); several ISIUOP flyers and posters in English and Inuktitut produced for the communities of Igloolik, Cape Dorset, Pangnirtung, and Clyde River (see http://gcrc.carleton.ca/isiuop); Nunavik Ice Terminology, poster produced jointly by the Kativik Regional Government, Nunavik, and Trent University (Tremblay et al. 2008).
26. See http://cumuseum.colorado.edu/Exhibits/Traveling/Silavut/index.html. The exhibit was produced as a part of the IPY education and outreach project #410.
27. Convenors are Grete Hovelsrud and Igor Krupnik.
28. Convenors are Martin Nweeia and Shari Gearheard.

References

Allison, I., Béland, M., Alverson, K., Bell, R., Carlson, D., Darnell, K., Ellis-Evans, C., Fahrbach, E., Fanta, E., Fujii, Y., Glasser, G., Goldfarb, L., Hovelsrud, G., Huber, J., Kotlyakov, V., Krupnik, I., Lopez-Martinez, J., Mohr, T., Qin, D., Rachold, V., Rapley, C., Rogne, O., Sarukhanian, E., Summerhayes, C., and Xiao, C. 2007. The scope of science for the International Polar Year 2007–2008. *World Meteorological Organization, Technical Documents* 1364. Geneva.

Aporta, C. 2002. Life on the ice: Understanding the codes of a changing environment. *Polar Record* 38: 341–354.

Aporta, C. 2003. Old Routes, New Trails: Contemporary Inuit Travel and Orienting in Igloolik, Nunavut. Unpublished Ph.D. thesis, Department of Anthropology, University of Alberta, Edmonton.

Aporta, C. 2004. Routes, trails and tracks: Trail breaking among the Inuit of Igloolik. *Etudes/Inuit/Studies* 28: 9–38.

Aporta, C. 2009. The trail as home: Inuit and their pan-Arctic network of routes. *Human Ecology* 37(2): 131–146.

Arnestad Foote, B. 2009. *Point Hope. Life on Frozen Water/Tikigaq: En fotografisk reise blant eskimoene i Point Hope 1959–1962*. Fairbanks: University of Alaska Press/Oslo: Forlaget Press AS.

Bogoslovskaya, L., Vdovin, B., and Golbtseva, V. 2008. Izmeneniia klimata v regione Beringova proiliva: Integratsiia nauchnykh i traditisionnykh znanii (Climate change in the bering Strait Region: Integration of scientific and indigenous knowledge) (SIKU, IPY #166). *Ekonomicheskoe planirovanie i upravlenie* 3–4 (8–9):58–68. Moscow.

Druckenmiller, M.L., Eicken, H., Johnson, M.A., Pringle, D.J., and Willliams, C.C. 2009. Towards an integrated coastal sea ice observatory: System components and a case study at Barrow, Alaska. *Cold Regions Science and Technology* 56: 61–72.

Dorais, L.-J. and Krupnik, I. (eds.) 2005. Preserving language and knowledge of the North/Préserver la langue et les savoirs du Nord. *Etudes/Inuit/Studies* 29(1–2), Special issue.

Duerden, F. 2004. Translating climate change impacts at the community level. *Arctic* 57: 203–212.

Eicken, H., Lovecraft, A.L., and Druckenmiller, M. 2009. Sea-ice system services: A framework to help identify and meet information needs relevant for Arctic observing networks. *Arctic* 62: 119–136.

Fleming, J.R. 1998. *Historical Perspectives on Climate Change*. New York and Oxford: Oxford University Press.

Ford, N. 2000. Communicating climate change from the perspective of local people: A case study from Arctic Canada. *The Journal of Development Communication* 11: 92–108.

Ford, J. 2005. Living with climate change in the Arctic. *World Watch* 18: 18–21.

Ford, J.D., Gough, W.A., Laidler, G.J., MacDonald, J., Irngaut, C., and Qrunnut, K. 2009. Sea ice, climate change, and community vulnerability in northern Foxe Basin, Canada. *Climate Research* 38: 137–154.

Ford, J., Pearce, T., Smit, B., Wandel, J., Allurut, M., Shappa, K., Ittusujurat, H., and Qrunnut, K. 2007. Reducing vulnerability to climate change in the Arctic: The case of Nunavut, Canada. *Arctic* 60: 150–166.

Fox, S. 2002. These are things that are really happening: Inuit perspectives on the evidence and impacts of climate change in Nunavut. In *The Earth Is Faster Now: Indigenous Observations of Arctic Environmental Change*. I. Krupnik and D. Jolly (eds.), Fairbanks: Arctic Research Consortium of the United States.

Fox, S. 2004. When the Weather is Uggianaqtuq: Linking Inuit and Scientific Observations of Recent Environmental Change in Nunavut, Canada. Ph.D. dissertation, Department of Geography, University of Colorado at Boulder, Boulder, CO.

Freeman, M.M.R. 1976. *Inuit Land Use and Occupancy Project*. Ottawa: Department of Indian and Northern Affairs.

Freeman, M.M.R. 1984. Contemporary Inuit exploitation of the sea ice environment. *Sikumiut: "The People Who Use the Sea Ice."* Montreal: Canadian Arctic Resources Committee, pp. 73–96.

Gearheard, S., Matumeak, W., Angutikjuaq, I., Maslanik, J., Huntington, H.P., Levitt, J., Kagak, D.M., Tigullaraq, G., and Barry, R.G. 2006. "It's not that simple": A collaborative comparison of sea ice environments, their uses, observed changes, and adaptations in Barrow, Alaska, USA, and Clyde River, Nunavut, Canada. *AMBIO* 35: 203–211.

George, J.C., Huntington, H.P., Brewster, K., Eicken, H., Norton, D.W., and Glenn, R. 2004. Observations on shorefast ice dynamics in Arctic Alaska and the responses of the Iñupiat hunting community. *Arctic* 57: 363–374.

Herlander, E. and Mustonen, T. (eds.) 2004. *Snowscapes, Dreamscapes. Snowchange Book on Community Voices of Change*. Study Materials 12. Tampere: Tampere Polytechnic Publications.

Huntington, H. and Fox, S. 2005. *The Changing Arctic: Indigenous Perspectives: Arctic Climate Impact Assessment*. Cambridge: Cambridge University Press, pp. 61–98.

Huntington, H.P. 2000. Using traditional ecological knowledge in science: Methods and applications. *Ecological Applications* 10(5): 1270–1274.

Huntington, H.P., Brower, H., Jr., and Norton, D.W. 2001. The barrow symposium on sea ice, 2000: Evaluation of one means of exchanging information between subsistence whalers and scientists. *Arctic* 54(2): 201–206.

ICSU. 2004. *A Framework for the International Polar 2007–2008*. Produced by the ICSU IPY 2007–2008 Planning Group. Paris: International Council for Science (ICSU).

Jolly, D., Berkes, F., Castleden, J., Nichols, T., and The Community of Sachs Harbour 2002. We can't predict the weather like we used to: Inuvialuit observations of climate change, Sachs Harbour, Western Canadian Arctic. In *The Earth Is Faster Now: Indigenous Observations of Arctic Environmental Change*. I. Krupnik and D. Jolly (eds.), Fairbanks: Arctic Research Consortium of the United States, pp. 92–125.

Kavry, V. and Boltunov, A. 2006. Observations of Climate Change Made by Indigenous Residents of Coastal Regions of Chukotka Autonomous Okrug. WWF Climate Change Report. Moscow, WWF-Russia http://www.wwf.ru/resources/publ/book/eng/196

Krupnik, I. 2002. Watching ice and weather our way: Some lessons from Yupik observations of sea ice and weather on St. Lawrence Island, Alaska. In *The Earth Is Faster Now: Indigenous Observations of Arctic Environmental Change*. I. Krupnik and D. Jolly (eds.), Fairbanks: Arctic Research Consortium of the United States, pp. 156–197.

Krupnik, I. 2009. IPY 2007–2008 and Social Sciences: A Challenge of Fifty Years. http://www.iassa.gl/icass6/publications/Igor%20PlenaryFin_March%202009_InclFig.pdf. Plenary paper presented at the 6th International Congress of Arctic Social Sciences, Nuuk Greenland, August 24, 2008 (accessed September 6, 2009).

Krupnik, I. and Bogoslovskaya, L. 2007. Izmeneniie klimata i narody Arktiki. Proekt SIKU v Beringii (Climate change and Arctic people: SIKU Project in Beringia). *Ekologicheskoe planirovanie i upravlenie* 4(5): 77–84.

Krupnik, I., Bravo, M., Csonka, Y., Hovelsrud-Broda, G., Müller-Wille, L., Poppel, B., Schweitzer, P., and Sörlin, S. 2005. Social sciences and humanities in international polar year 2007–2008: An integrating mission. *Arctic* 58(1): 89–95.

Krupnik, I. and Jolly, D. (eds.) 2002. *The Earth Is Faster Now: Indigenous Observations of Arctic Environmental Change*. Fairbanks: Arctic Research Consortium of the United States.

Laidler, G.J. 2006. Inuit and scientific perspectives on the relationship between sea ice and climate change: The ideal complement? *Climatic Change* 78: 407–444.

Laidler, G.J. 2007. Ice, Through Inuit Eyes: Characterizing the Importance of Sea Ice Processes, Use, and Change Around Three Nunavut Communities. Ph.D. Dissertation, University of Toronto.

Laidler, G.J., Dialla, A., and Joamie, E. 2008. Human geographies of sea ice: Freeze/thaw processes around Pangnirtung, Nunavut, Canada. *Polar Record* 44: 335–361.

Laidler, G.J. and Elee, P. 2008. Human geographies of sea ice: Freeze/thaw processes around cape Dorset, Nunavut, Canada. *Polar Record* 44: 51–76.

Laidler, G.J., Ford, J.D., Gough, W.A., Ikummaq, T., Gagnon, A.S., Kowal, S., Qrunnut, K., and Irngaut, C. 2009. Travelling and hunting in a changing Arctic: Assessing Inuit vulnerability to sea ice change in Igloolik, Nunavut. *Climatic Change* 94: 363–397.

Laidler, G.J. and Ikummaq, T. 2008. Human geographies of sea ice: Freeze/thaw processes around Igloolik, Nunavut, Canada. *Polar Record* 44: 127–153.

Lemke, P., Ren, J., Alley, R.B., Allison, I., Carrasco, J., Flato, G., Fujii, Y., Kaser, G., Mote, P., Thomas, R.H., and Zhang, T., 2007. Observations: Changes in snow, ice and frozen ground. In *Climate Change 2007: The Physical Science Basis*. Contribution of Working Group I to the Fourth Assessment Report of the Intergovernmental Panel on Climate Change. S. Solomon, D. Qin, M. Manning, Z. Chen, M. Marquis, K.B. Averyt, M. Tignor and H.L. Miller (eds.), Cambridge, UK and New York, Cambridge University Press.

Marino, E. and Schweitzer, P.P. 2009. Talking and not talking about climate change in Northwestern Alaska. In *Anthropology and Climate Change: From Encounters to Actions*. M. Nuttall and S. Crates (eds.), Walnut Creek: Left Coast Press, pp. 209–217.

McDonald, M., Arragutainaq, L., and Novalinga, Z. 1997. *Voices from the Bay – Traditional Ecological Knowledge of Inuit and Cree in the Hudson Bay Bioregion*. Ottawa and Sanikiluaq: Canadian Arctic Resources Committee and The Municipality of Sanikiluaq.

Nakashima, D.J. 1993. Astute observers on the sea ice edge: Inuit knowledge as a basis for Arctic co-management. In *Traditional Ecological Knowledge: Concepts and Cases*. J.T. Inglis (ed.), Ottawa: International Program on Traditional Ecological Knowledge and International Development Research Centre, pp. 99–110.

Nelson, R.K. 1969. *Hunters of the Northern Ice*. Chicago: The University of Chicago Press.

Nichols, T., Berkes, F., Jolly, D., Snow, N.B., and The Community of Sachs Harbour. 2004. Climate change and sea ice: Local observations from the Canadian Western Arctic. *Arctic* 57: 68–79.

Nickels, S., Furgal, C., Buell, M., and Moquin, H. 2006. *Unikkaaqatigiit – Putting the Human Face on Climate Change: Perspectives from Inuit in Canada*. Ottawa: Inuit Tapiriit Kanatami, Nasivvik Centre for Inuit Health and Changing Environments at Universite Laval, and the Ajunnginiq Centre at the National Aboriginal Health Organization.

Nickels, S., Furgal, C., Castleden, J., Moss-Davies, P., Buell, M., Armstrong, B., Dillon, D., and Fonger, R. 2002. Putting the human face on climate change through community workshops: Inuit knowledge, partnerships, and research. In *The Earth Is Faster Now: Indigenous Observations of Arctic Environmental Change*. I. Krupnik and D. Jolly (eds.), Fairbanks: Arctic Research Consortium of the United States, pp. 301–333.

Norton, D. 2002. Coastal sea ice watch: Private confessions of a convert to indigenous knowledge. In *The Earth Is Faster Now: Indigenous Observations of Arctic Environmental Change*. I. Krupnik and D. Jolly (eds.), Fairbanks: Arctic Research Consortium of the United States, pp. 127–155.

Oozeva, C., Noongwook, C., Noongwook, G., Alowa, C., and Krupnik, I. 2004. *Watching Ice and Weather Our Way*. Washington: Arctic Studies Center, Smithsonian Institution.

Riedlinger, D. and Berkes, F. 2001. Contributions of traditional knowledge to understanding climate change in the Canadian Arctic. *Polar Record* 37(203): 315–328.

Riewe, R. 1991. Inuit use of the sea ice. *Arctic and Alpine Research* 23(1): 3–10.

Solomon, S., Qin, D., Manning, M., Chen, Z., Marquis, M., Averyt, K.B., Tignor M., and Miller H.L. (eds.) 2007. *Climate Change 2007: The Physical Science Basis. Contribution of Working Group I to the Fourth Assessment Report of the Intergovernmental Panel on Climate Change*. Cambridge and New York: Cambridge University Press. 966pp.

Tremblay, M., Furgal, C., Tookalook, P., Annanack, T., Qiisiq, M., Angiyou, E., and Barrett, M. 2008. La terminologie inuite de la glace: Une valeur intrinsèque pour un accès sécuritaire aux ressources et au territoire au Nunavik. 3e Symposium scientifique Ouranos. Montréal, QC. Poster.

Walsh, J.E. 2005. Cryosphere and hydrology. In *Arctic Climate Impact Assessment*. C. Simon (ed.), Cambridge: Cambridge University Press, pp. 184–242.

Watt-Cloutier, S. 2005. Connectivity: The Arctic–The planet. *Silarjualiriniq/Inuit in Global Issues* 20: 1–4.

World Climate Conference. 1979. Conférence mondiale sur le climat. Conférence d'experts sur le climat and l'homme. WMO proceedings of the World Climate Conference: a conference of experts on climate and mankind. Geneva, 12–23 February 1979. WMO No. 537.

Part I
Recording the Knowledge: Inuit Observations of Ice, Climate and Change

Chapter 2
Weather Variability and Changing Sea Ice Use in Qeqertaq, West Greenland, 1987–2008

Pierre Taverniers

Abstract This chapter reviews changes in local weather, sea ice, and ice use in the small Greenlandic hunting community of Qeqertaq (population 147) located in the northeast section of Disko Bay, Northwest Greenland. In the 1980s, the island was surrounded by shorefast ice during 6–8 months of the year. Traveling on ice by dogsleds used to be the only way to go hunting, fishing, and to connect with other nearby communities. The Qeqertamiut are highly dependent on sea ice to maintain their traditional subsistence knowledge and economy based on extensive use of the ice-dominated marine environment. In 1987–1988, the author stayed in Qeqertaq conducting meteorological observations and studying the impact of weather variability on sea ice extent, thickness, and quality. Since 1987, the average annual temperature in the Qeqertaq area has increased by more than 3°C, resulting in major impacts to the local sea ice regime. In 2008, the author revisited the community to document how the Qeqertamiut continue to use the sea ice today, under a much warmer climate and higher weather variability.

Keywords Sea ice and weather change · Ice use · Greenland · Qeqertaq

Qeqertaq ("the island" in Kalaallisut, the language of Western Greenlandic Inuit) is a rural Greenlandic community located at 70°N, on a small island off the coast of West Greenland, in the northeast section of Disko Bay, 90 km north of the town of Ilulissat, the area hub (Fig. 2.1). The island that bears the same name, *Qeqertaq*, is 6 km long from north to south and only few hundred meters wide in its narrowest section. The settlement (Fig. 2.2) is located at the southernmost part of the island. Because of two rocky hills, the island is not easy to cross. The island is located in a bay about 1 km off the main coast of Greenland, near the mountainous Nuussuaq peninsula, close to the mouth of the Torsukatak Fjord. At the end of the 40 km long

P. Taverniers (✉)
Meteo France, Aeroport de Limoges, Limoges 87100, France
e-mail: pierre.taverniers@meteo.fr

I. Krupnik et al. (eds.), *SIKU: Knowing Our Ice*,
DOI 10.1007/978-90-481-8587-0_2,

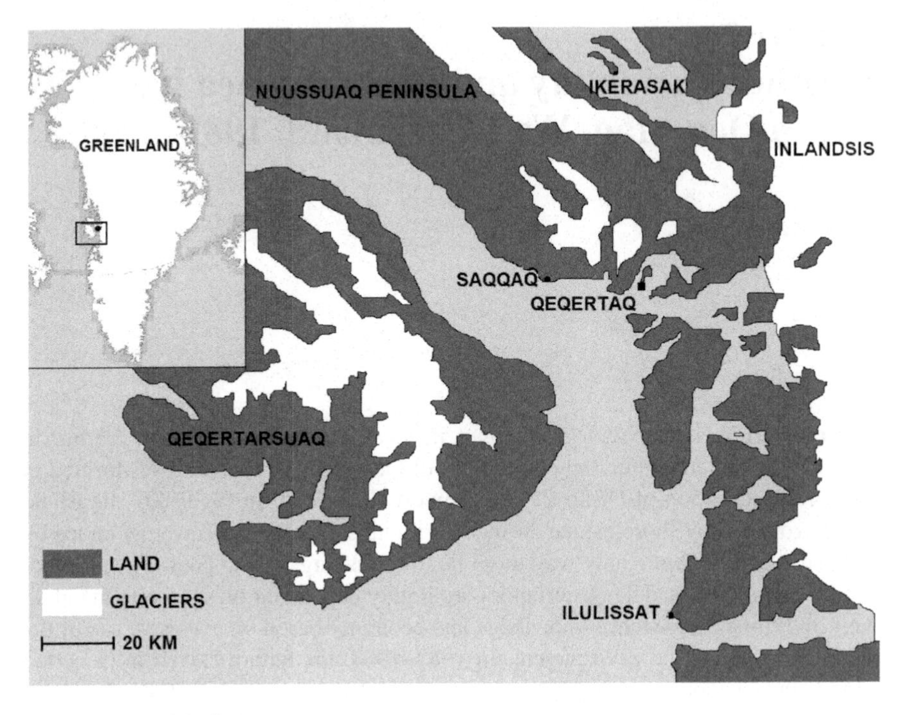

Fig. 2.1 Map of the Qeqertaq area

fjord, a glacier calves large icebergs and smaller pieces of ice that drift down to the main bay pushed by winds and currents.

Qeqertamiut, "people of Qeqertaq" (population 100 in 1988, 147 in 2008) are indigenous Greenlandic (Kalaallit) marine hunters and fishers. The hunting is primarily for subsistence, though some sealskins are also sold for commercial processing. Seal hunting is the most important activity throughout the year, but most notably during the ice season that lasted from December to May in the 1980s and even from November to June, during "good" winters. Seal meat – mostly of ringed seal, secondly of harp seal, sometimes of bearded seal – is very important to the Qeqertamiut diet. Sealskins are used to make winter boots and winter clothing. Qeqertamiut also hunt beluga (white whale); but this hunt is not successful each year because belugas' usual migration routes are far from Qeqertaq (25–50 km). Qeqertamiut can catch many belugas only when they are trapped by sea ice (*sassat*), an event that was documented 25 times in Disko Bay from 1899 to 1990 (Dahl 2000). Hunting of terrestrial animals is a minor activity. On the small island, the Qeqertamiut can find some ptarmigans and hare, only. On the Nuussuaq peninsula, on the Greenland mainland, they may find caribou, but in 1987–1988 the annual quota was limited to one caribou per hunter for the year. Many migratory birds are also available in the area from late spring to the fall for subsistence hunting.

The Qeqertamiut also fish Greenland cod, capelin, and wolf fish, primarily for domestic consumption and to feed their dogs. Fishing for Greenlandic halibut is a

Fig. 2.2 View of Qeqertaq (Photo: Pierre Taverniers; May 2008)

major commercial activity, with fishermen using long lines with several hundred hooks. For hunting marine mammals and fishing during the open-water season, the Qeqertamiut use small open boats with outboard motors. The boats made of fiber glass are imported from Denmark via the commercial supply system; they have long replaced indigenous skin-covered boats, qajat (*kayaks)* and umiat (*umiaks*), that Greenlanders produced themselves (the last kayak was seen in early 1980s). During the open-water hunting and fishing season, the Qeqertamiut have to deal with the calving icebergs that can collapse at any moment and damage small fishing boats.

Qeqertamiut Use of Sea Ice in 1987–1988

From July 1987 to June 1988, the author, then a student in Inuit language and culture at INALCO (Institut National des Langues et Civilisations Orientales, Paris), was stationed in Qeqertaq to study the Qeqertamiut use of ice and snow-covered environment and to conduct daily meteorological observations (Taverniers 2005). The study was supported by the French National Meteorological Office. Thanks to the installation of standard meteorological instruments, the author was able to measure air, water, and soil temperature, wind speed and direction, and the status and variation of the ice cover (fast ice) on a regular basis. That 11-month record, supported by the long-term meteorological data from the nearby full weather station at

Ilulissat, offered insight into the weather and ice conditions in West Greenland in the 1980s, prior to the onset of the current climate warming (see below).

During winter 1987–1988, the sea surface temperature reached its freezing point (–1.8°C) on October 1. A few days later the weather became anticyclonic; there was no wind and the sea was calm. With clear sky, the first sea ice formed on the bay around Qeqertaq on October 6. That first thin ice was broken by waves as soon as the wind started to blow. Between October 6 and December 22, the sea froze nine times, during anticyclonic periods, and broke up eight times because of strong winds generated by repeated low-pressure systems. As soon as the sea ice was safe to walk on, even for a short period, the Qeqertamiut went onto it. The first to go on the young ice on October 21, were the children, who had no flat area to play on the island. Also, those Qeqertamiut, who had no private boats ventured on ice as soon as the condition permitted. During the open-water season, boat owners commonly shared their boats with relatives and friends. With the sea ice established around the island, any Qeqertamioq ("resident of Qeqertaq") could go hunting and fishing on one's own. Using a long-handled wooden ice-chipping tool (called *tooq*), the Qeqertamiut made holes in young ice, usually not far from the island, where the water is not very deep. Through those ice holes, they fished for the Greenland cod by jigging, mostly to feed their dogs.

In mid-November, more solid sea ice formed and thickened to 16 cm. Usually, there is only shore-fast ice off the island during the main ice season, since no drift or pack ice reaches the Qeqertaq area deep in the Disko Bay (in summer, there are also floating icebergs calving from the glacier in the fjord). The shore-fast ice is normally flat, with the ridged area formed along the shore only, where the tides moved the sea ice against the ice foot (a narrow fringe of ice attached to the coast). In Qeqertaq the tidal range is about 2.40 m. In winter sea ice moves about 1 m so the ridged area is not as important as in other Arctic areas. Shortly after local shore-fast ice was established, some Qeqertamiut started to set nets under the ice to catch seals (Fig. 2.3). They usually chose an area further from the village, because seals are afraid of noise from dogs and human activities. The best places for seal nets were at the edge of small icebergs or between a small iceberg and the ice foot where seals come to breathe through cracks. Hunters usually made three holes in the ice with their *tooq*. Then the *tooq* was pushed under the ice from one hole to another. A string was tied to the *tooq* to draw the net under the ice. Hunters visited their nets regularly to check for seals. Unfortunately, in late November 1987, a strong gale broke the thin ice once again and the use of ice was delayed for 2 more weeks.

On November 25, 1987, the sun disappeared for 2 months. Finally, solid shore-fast ice was formed on December 22, to stay for 5 months, until May 20, 1988. It thickened rapidly. Eventually, solid shorefast ice covered the area around the island and the surrounding sections of Disko Bay. The new ice was very slippery and the Qeqertamiut had to wait for a major snowfall before they could use their dogsleds to travel on ice. Qeqertamiut are deeply attached to their dogsleds. In 1987–1988 the dogsled was the cheapest way to travel because dogs were fed with local resources; in fact, using dogsleds on ice was the Qeqertamiut's only way to travel in winter.[1] Dogsled travel was considered very safe, since trained dogs were able to detect dangerous cracks covered by snow. Dogsled is also a cultural heritage and today's

Fig. 2.3 Seal hunting with sea ice nets in Qeqertaq, 1987 (Photo: Pierre Taverniers)

Qeqertamiut are very proud about their knowledge on how to breed Greenlandic sled dogs and to use dogsleds, just like their ancestors did for many generations.

As soon as the young sea ice was thick enough to walk on, the Qeqertamiut installed long lines under the ice to fish for Greenlandic halibut (Fig. 2.4). The best fishing grounds were inside the Torsukatak Fjord, because the halibut prefers deep-water habitats (up to few hundred meters). During the winter-ice season, fishing was easier, because there were no drifting icebergs to damage fishing lines, like it often happened in summertime. Qeqertamiut fishermen made holes in the ice and set lines (up to 500–800 m long) with several hundred hooks that were drawn to the bottom of the fjord by a metal glider. To haul the lines, the fishermen used hand-made rollers set on the sea ice above the hole. Fishermen also used cracks in the sea ice to set fishing nets to fish for Greenland cod. Those nets were hauled by dogsled.

When the sun came back in February, the Qeqertamiut started to hunt seals that basked in the sun on the shore-fast ice, called *uuttut*. To approach a seal close enough for a good shot, the hunters used a white screen set on a small sled (60 cm long), which is a well-known and widely common Inuit hunting technique. The *uuttut* hunting area was usually far from the village and required dogsleds to reach the hunting grounds. Many *uuttut* were caught at the upper reaches of the Torsukatak Fjord, about 40 km from the village. During the winter-ice season, the Qeqertamiut hunters caught more seals than during the open-water season, from June to September. In 1987–1988, for traveling, hunting, and fishing on ice, the Qeqertamiut used seal skin clothing and boots that were made locally by women in the community.

Fig. 2.4 Qeqertamiut fishermen and their fishing hole (*alluaq*) to fish for Greenlandic halibut. The ice thickness is over 40 cm (Photo: Pierre Taverniers; March 1988)

Another common form of using sea ice in 1987 was for drinking water (Fig. 2.5). To get fresh water during the summer season, the Qeqertamiut used to collect small floating iceberg blocks drifted to the shore by winds and currents. When there were no floating icebergs, the Qeqertamiut had to travel on a boat across the bay to a river located far away on the mainland shore. During the winter ice season, pieces of freshwater (calved) icebergs were commonly frozen into the shore-fast ice, often close to the village. Thus, people could easily obtain good drinking water by chipping ice blocks from the nearby icebergs.

With the sea ice firmly established for 5 months of the year, the Qeqertamiut did not think of themselves as "islanders." Even those who had no boat or dogs could reach the neighboring coast. Using dogsleds over sea ice, the Qeqertamiut traveled regularly to other nearby settlements (Fig. 2.6), such as Saqqaq, the nearest neighboring community 25 km west of Qeqertaq or the town of Ilulissat located 90 km south of Qeqertaq. The Nuussuaq peninsula, which is mountainous and difficult to cross during summertime, became easier to reach and to cross when covered by snow.

In March 1988, when the ice was thick and the days were longer, almost half the community of Ikerasak (217 inhabitants), located 55 km north of Qeqertaq in the Uummannaq area, crossed the Nuussuaq Peninsula by dogsleds and reached Qeqertaq to visit relatives and to play community soccer matches on sea ice. Those winter travels by dogsleds supported close relationship among Greenlandic communities. As there was no flat area on the island, the soccer team from Qeqertaq trained and played the game only on sea ice (as is also done in some other Greenlandic

Fig. 2.5 Collecting iceberg blocks for drinking water (Photo: Pierre Taverniers; February 1988)

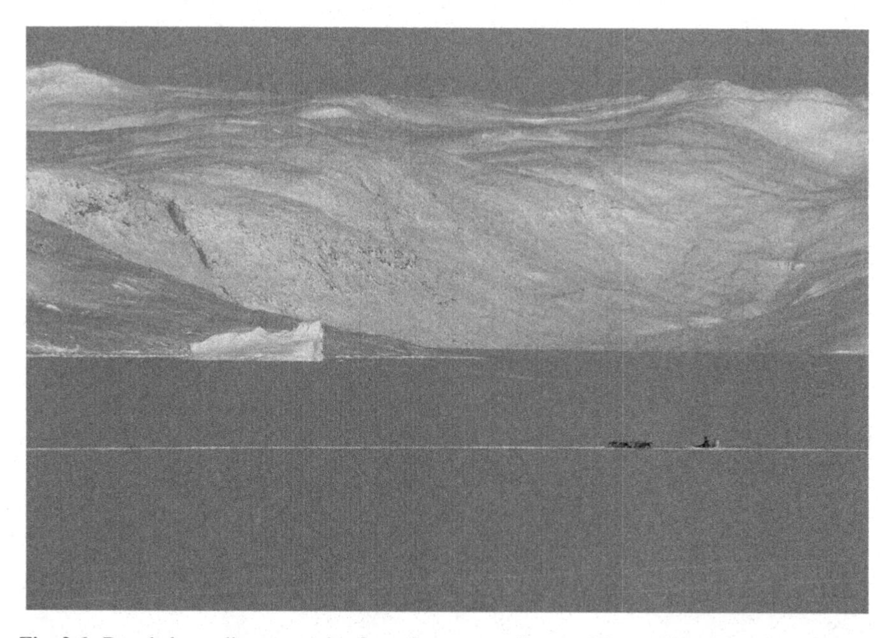

Fig. 2.6 Dogsled traveling on sea ice from Qeqertaq to Saqqaq (Photo: Pierre Taverniers; March 1988)

communities). Sea ice was also a place for sharing knowledge and training of young hunters and fishermen. When the weather was good, young boys 5- or 6 years old went with relatives on ice to learn hunting and fishing techniques. In March 1988, a 10-year-old boy set a sealing net on his own and caught his first seal through ice holes.

Sharing ice knowledge means also using specific terms. Between them, the Qeqertamiut speak exclusively Kalaallisut (western Greenlandic language) and they know and use more than 100 terms associated with the sea ice, according to the author's count (see Chapter 18 by Tersis and Taverniers this volume). In 1987, many Qeqertamiut also spoke some Danish, but only one person in the village knew some English. As such, traditional terminology for sea ice was still strong and in regular use.

In April and May 1988, dogsled races for women and children took place over the frozen bay around Qeqertaq (Fig. 2.7). All of the Qeqertamiut went on the sea ice to participate in that major annual event. As dogsled races on ice were the only event gathering the entire community, they were socially important.

In the last days of April 1988, a large recurring polynya opened 4 km southeast of Qeqertaq. Thousands of migratory birds (black guillemot, common eider, glaucous gull, black-legged kittiwake) stopped to feed in the polynya. Qeqertamiut reached the polynya by dogsled and then used small boats to hunt birds.

In May 1988, the sea ice broke up in Disko Bay and the floe edge (demarcation between the open sea and shore-fast ice) was established 10 km off the village. Qeqertamiut then regularly traveled to the ice edge by dogsled carrying small boats

Fig. 2.7 Women dogsled race in Qeqertaq (Photo: Pierre Taverniers; May 1988)

to hunt seals. The Qeqertamiut used the sea ice until May 20, 1988. During the last weeks of the ice season, the ice was covered with water puddles before it broke up and drifted away with currents and winds. The last Qeqertamiut to use sea ice were children playing on drifting ice floes near the shore. A dangerous game of jumping from floe to floe was allowed by adults, because it was considered valuable training to develop skills needed for ice hunting. Altogether, during the winter 1987–1988, the sea ice thickened up to 40 cm (sea ice was twice as thick during the previous winter of 1986–1987, according to the Qeqertamiut). The maximal thickness was reached on March 25 and was maintained until April 10. In 1987–1988, the Qeqertamiut used the ice during almost 7 months, from November till late May; and they traveled, fished, and hunted on approximately 600 km^2.

Recent Changes in Weather and Ice Conditions, 1987–2007

Between 1987 and 2007, the climate warmed substantially in the Disko Bay area. According to data from the Danish Meteorological Institute, the average annual temperature at Ilulissat, the closest weather station to Qeqertaq, has increased by 3.5°C in the last 20 years (1987–2007, see Fig. 2.8) and the average winter temperature increased by 7°C; those trends are also representative for the conditions in Qeqertaq.[2] Relatively cold winters predominated from 1991 to 1995, with the coldest winter of 1994–1995 (when the average temperature was –20.9°C), followed by the rapid warming after 1996, with record warm winters throughout 2002–2007. This warming trend is confirmed by the recent analysis of meteorological data

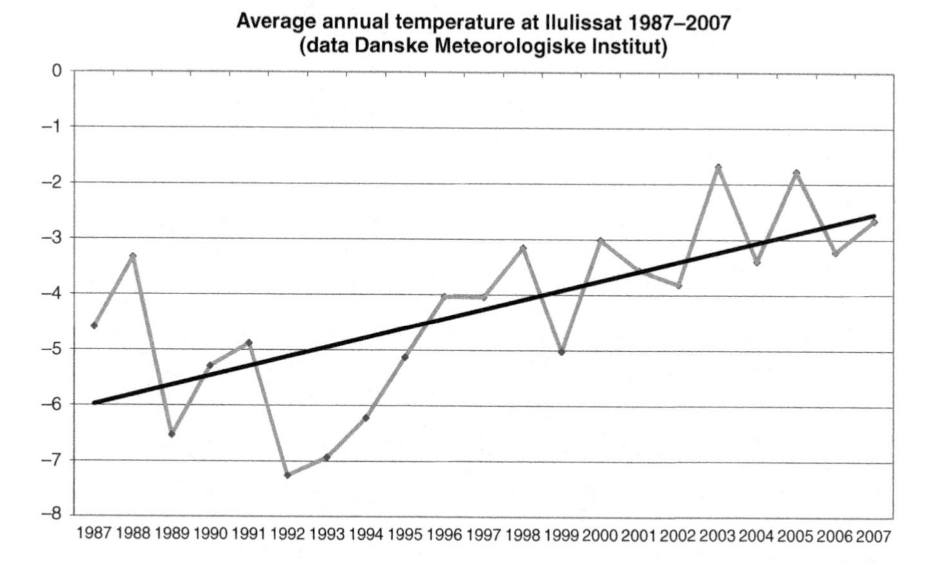

Fig. 2.8 Temperature change in Ilulissat, 1987–2007 (Danish Meteorological Institute)

from the Arctic Station, located 120 km far from Qeqertaq, at the southernmost part of Qeqertarsuaq (Disko Island), where the mean annual air temperatures have increased by 0.4°C per year between 1991 and 2004 (Hansen et al. 2006). During the same period, observations at the Arctic Station indicated that the sea ice cover has decreased by 50% in the vicinity of Disko Island. Ice charts published by the Danish Meteorological Institute also show that the sea ice cover in the Disko Bay has decreased significantly.

In May 2008, the author had a chance to revisit Qeqertaq and to document changes in sea ice conditions and ice use via personal observation and interviews with local hunters. Within the 20-year period since 1987, many hunters remember several cold winters in a row from 1991 to 1995 and, particularly, the winter of 1990–1991, because of its especially good ice cover and a rare whale hunt. Sea ice formed rapidly that year in the entire Disko Bay area and many white whales (belugas) migrating south were trapped in a polynya between Ilulissat and Qeqertarsuaq, approximately 100 km from Qeqertaq. As soon as they have heard about the whale ice entrapment (*sassat*), several Qeqertamiut hunters crossed the Disko Bay by dogsled to participate in the hunt and to get meat and *mattak* (whale skin). For Qeqertamiut, beluga *mattak* is one of their most favored meals. When *sassat* occurs in the Disko Bay area, Qeqertamiut are able to travel by sled over

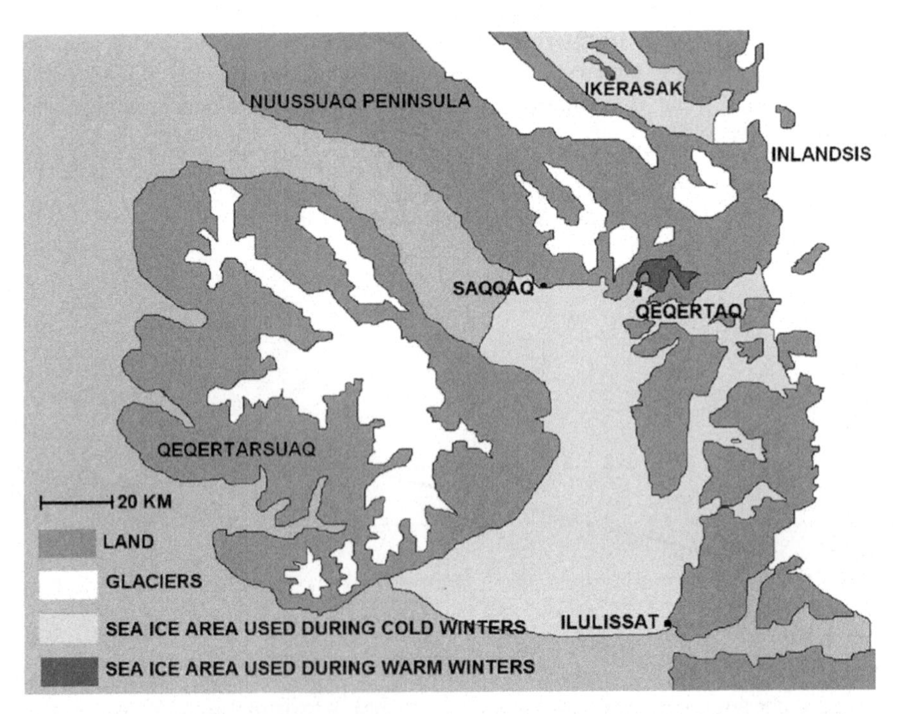

Fig. 2.9 Changes in the ice-covered area used by Qeqertamiut, 1987–2007. (See also Color Plate 1 on page 471)

sea ice, so that their total exploited ice-covered area that year could reach 3,500 square km. On the coldest winters, like that of 1994–1995, the sea ice thickened up to 1 m in the bays and fjords.

Since 1995, the Qeqertamiut have observed winters becoming warmer and much more windy. The bay started to freeze later, in late November and even December, so that the sea ice cover is being formed much later. It is also thinner and can be easily broken by strong winds. Southwest winds (generally linked with low-pressure systems) became stronger and often broke whatever little fast ice was built off the island. Winter weather also became much more unpredictable. Thus, decrease in sea ice was triggered not only by rising winter temperature but also by stronger winds and more clouds, as a result of many more low-pressure systems going through the area in wintertime. From 2003 to 2007 there was no shore-fast ice built around the settlement in winter, just a thin ice cap (10 cm) in the northern portion of the bay. As a result, the formerly used sea ice area of 600–3,500 km^2 has been reduced in recent years to less than 50–70 km^2 (Fig. 2.9). Many more accidents have occurred involving hunters and fishermen falling through the thin ice. Some hunters indicate that new currents undermine the sea ice stability and make it unsafe to travel.

Qeqertamiut Use of Sea Ice in 2008

Since 1995, new ice conditions triggered significant economic, social, and cultural changes in Qeqertamiut's life style. During warm winters, the ice is more difficult to access in front of the village; so, the Qeqertamiut now have to cross the island to the northern side through hills and rocks. Dogsleds are now used for 1 or 2 months only. Since they have to feed their dogs year-round and can use them for a much shorter time, some hunters stopped using dogsleds altogether or have built new types of lighter sleds to travel over thin sea ice. In 1988 there were about 200 adult dogs per 100 inhabitants in Qeqertaq. In May 2008 there were 240 adult dogs (24–30 sleds) per 147 inhabitants. The rate of dogs per inhabitant in Qeqertaq has decreased from 2.0 to 1.6. As the use of dogsleds decreases, the use of motorboats increased substantially and the number of boats in use grew from 12 in 1987 to 47 in 2008. Those who had no boat 20 years ago were forced to invest in imported and expensive equipment, because of a much longer period of open-water hunting and fishing during the year. At the same time, due to change in glacier activity, winds and currents create more and more icebergs and smaller floating pieces of ice of land origin that are now being drifted toward Qeqertaq area. Hulls made of fiber glass and propellers are more often damaged by ice of land origin. Also, local boat owners have to cope with the increasing repair costs and gasoline and motor oil prices.

In 2008, most of the Qeqertamiut succeeded to cope with the raising hunting costs thanks to the increased income provided by Greenlandic halibut fishing. A small fish factory (*Qeqertaq Fish*) has been opened in Qeqertaq, so that the processed halibut (frozen and packaged) can be exported. That project initiated by the Qeqertamiut with partly local investments gives more income to the fishermen and creates new job opportunities for the Qeqertamiut (such as plant manager, laborers, mechanic, electrician).

Because of warmer winters and less stable sea ice, fewer seal nets are now set under sea ice and less *uuttut* are hunted on ice in late winter and early spring. As a result, fewer seals are caught in the community in wintertime. As less seal meat is now available in winter, there is an increasing dependence upon imported and expensive food. Seal skin clothes once used to hunt, fish, and travel on sea ice are less used these days. In 2008, hardly any new sealskin clothing was made in Qeqertaq. Sealskins are now frozen and sent to a big factory in south Greenland to be processed for souvenir and commercial production. No *sassat* (whale entrapping events) were observed in recent years, which transforms in less whale meat and *mattak* available to the community. With the lack of sea ice there is no more winter polynya southeast of Qeqertaq and hunters catch less migratory birds because they are rare in the area or avoid it altogether.

There are also substantial changes in the community social life as well. The local schoolteacher complains that during warm winters he cannot go with the students on sea ice to educate them on ice hunting, fishing, and traveling techniques and safety. There are no more dogsled races in the community over ice-frozen bay. The town soccer team cannot train on sea ice and cannot play at home any more; so, the regular soccer ice tournaments, visitations, and major community events conducted on sea ice ceased to exist.

Fig. 2.10 The author with Aka and Zacharias Tobiassen, Qeqertaq, May 2008

Traditional Qeqertamiut sea ice knowledge is now more difficult to share with younger hunters, because the ice is less used and certain knowledge, for example, regarding safe areas for ice crossing and hunting around the island, is not relevant to today's ice conditions. Nonetheless, with the assistance of Aka, Thora, and Zacharias Tobiassen (Fig. 2.10), the author was able to collect the Qeqertamiut vocabulary associated with the sea ice of more than 120 terms (see Chapter 18 by Tersis and Taverniers in this volume). This Qeqertamiut ice vocabulary makes a substantial addition to an earlier list of local sea ice terms that the author recorded during his work in 1987–1988, with the assistance of local hunters Jorgen and Halfdan Jensen.

Conclusion

In the past, Greenlandic communities in the Disko Bay and elsewhere in Greenland had learned how to cope with warmer periods and decrease or even lack of sea ice. There is a long history of climate and ice change, through which indigenous Greenlanders had to live and to adapt throughout their history (Vibe 1967). In fact, many of today's Qeqertamiut still believe that the current string of warm winters is an episodic phenomenon and that the winter sea ice will recover in the coming years. That is why they still keep dogs and do whatever little dogsled driving and hunting on ice they can afford under today's ice condition. But the sea ice decrease observed in the Disko Bay area over the past decade seems to be linked with the general winter sea ice decrease in the Arctic observed since 1978 (Richter-Menge et al. 2008). According to the last IPCC report (Solomon et al. 2007), the Arctic is going to warm significantly during this century and winter sea ice will continue to decrease. As the southern extent of seasonal winter ice is now located not far from the Disko Bay and is shifting northward because of warm winters, sea ice may become a short seasonal event or completely disappear around Qeqertaq in the next decades, with more dramatic cultural changes for the Qeqertamiut to face.

Acknowledgments Thanks to the Qeqertamiut community and especially to Hans Barlaj, Kristian Grønvold, Johannes Jerimiassen, Augo Lange, Jakob Jonatansen, Aka, Jakob, Thora and Zacharias Tobiassen, and Lars Wille for their help in my research and for kindly sharing their knowledge about the changing ice and weather conditions over the past 20 years.

Notes

1. In 1987 there was only one snow-scooter in Qeqertaq. In this part of Greenland north of Sisimiut snow-scooters were not allowed to be used for hunting. In 2008 there were five snow-scooters in town.
2. As records from 1987–1988 indicate, the difference between the average temperature in Ilulissat and Qeqertaq was only 0.46°C; so the Ilulissat data are representative for Qeqertaq.

References

Dahl, J. 2000. *Saqqaq: An Inuit Hunting Community in the Modern World*. Toronto: University of Toronto Press.

Hansen, B.U., Elberling, B., Humlum, O., and Nielsen, N. 2006. Meteorological trends (1991–2004) at Arctic Station Central West Greenland (69°15'N) in a 130 years perspective. *Danish Journal of Geography* 106(1): 45–55.

Richter-Menge, J., Comiso, J., Meier, W., Nghiem, S., and Perovich, D. 2008. Sea ice cover. *Arctic Report Card 2008*.

Solomon, S., Qin, D., Manning, M., Chen, Z., Marquis, M., Averyt, K.B. , Tignor, M., and Miller, H.L. 2007. Climate Change 2007: *The Scientific Basis*. Contribution of Working Group I to the Fourth Assessment Report of the Intergovernmental Panel on Climate Change. New York and Cambridge: Cambridge University Press.

Taverniers, P. 2005. Les quatre saisons de Qeqertaq. Climat, banquise et vie quotidienne à Qeqertaq, village inuit. *Met Mar – Revue de Météorologie Maritime* 206: 20–24.

Vibe, C. 1967. Arctic animals in relation to climatic fluctuations. *Meddelelser om Grønland, udg. af Kommissionen for videnskabelige underøgleser i Grønland* 170 (5).

Chapter 3
Mapping Inuit Sea Ice Knowledge, Use, and Change in Nunavut, Canada (Cape Dorset, Igloolik, Pangnirtung)

Gita J. Laidler, Pootoogoo Elee, Theo Ikummaq, Eric Joamie, and Claudio Aporta

Contributing Authors: Karen Kelley and Kelly Karpala.

Abstract This chapter reviews the efforts under SIKU-ISIUOP to expand upon previous research that characterized the importance of sea ice processes, use, and change around the Baffin Island communities of Cape Dorset, Igloolik, and Pangnirtung, Nunavut. In these three communities, local ice conditions are intertwined with daily activities and provide a means of traveling and hunting, as well as sustaining marine wildlife and aspects of Inuit culture. In order for people to effectively travel and hunt on the sea ice, they have to become knowledgeable about the complexity and dynamism of the oceanic environment. Through these understandings and long-term experience and observation, local experts (such as Inuit elders and active hunters) are acutely aware of the local and regional manifestations of climate change, as indicated by long-term changes and increased unpredictability of sea ice. Specifically, Inuit have observed changes in floe edge position, weather, the timing of freeze-up and breakup, ice thickness, and the presence of multi-year ice. This chapter reviews specific indicators used to evaluate sea ice changes, offers a regional comparison of sea ice changes in the three communities, and provides an overview of some of the local implications of sea ice changes.

Keywords Sea ice · Inuit knowledge · Climate change · Nunavut · Baffin Island

Introduction

Sea ice, climate change, and northern communities are intimately linked when considering arctic marine environments. Sea ice is an integral part of Inuit life, while also being the pre-eminent focus of scientific research (Norton 2002; George et al.

G.J. Laidler (✉)
Department of Geography & Environmental Studies, Carleton University, Ottawa, ON K1S 5B6, Canada
e-mail: gita_laidler@carleton.ca

I. Krupnik et al. (eds.), *SIKU: Knowing Our Ice*,
DOI 10.1007/978-90-481-8587-0_3,

2004; Nichols et al. 2004; Nickels et al. 2006; Gearheard et al. 2006; Laidler 2006a). Scientific and Inuit interests have become increasingly intertwined with an explosion of sea ice and climate change observations, in-depth studies, and political debate in the last two decades. The far-reaching implications of changing sea ice thickness, distribution, and extent render this dynamic environment a primary research target in efforts to model or project future scenarios of global climatic conditions, as influenced by feedbacks at high latitudes (Ledley 1988; Ingram et al. 1989; Bintanja and Oerlemans 1995; Curry et al. 1995; Lohmann and Gerdes 1998; Lemke et al. 2000; Holland and Bitz 2003). While this type of research has raised the global profile of circumpolar regions, it has also sparked investigations into the human dimensions of climate change (Ford 2000; Cruikshank 2001; Fenge 2001; Riedlinger and Berkes 2001; Berkes 2002; Berkes and Jolly 2002; Fox 2002; Huntington 2002; Duerden 2004; Ford 2005). Comparatively little is known about the implications of changing sea ice extent, distribution, and thickness for daily life in arctic communities. As scientific assessments of change move from documentation to exploration of adaptive and mitigative strategies (from environmental, economic, and cultural perspectives), scientific and Inuit expertise are increasingly being considered alongside one another to improve our understanding of the relationships between sea ice, climate change, and northern communities (Ford 2000; Ford and Smit 2004; Symon et al. 2005; Furgal et al. 2006; Ford et al. 2007; Paci et al. 2008; Eicken et al. 2009; Ford et al. 2009; Laidler et al. 2009). This integration is necessary to adequately incorporate the multiple stressors of northern life into assessments of vulnerability, or resilience, to climate change (McCarthy et al. 2005). Indeed, Inuit want to be involved in this process in order to share their observations, have their voices heard, and be taken seriously (Ashford and Castleden 2001; Kusugak 2002; ITK 2005; Nickels et al. 2006; NTI 2005; Laidler 2006b). It is within this context, that the Inuit Sea Ice Use and Occupancy Project (ISIUOP) was conceived (see Chapter 1, Introduction). In particular, this chapter focuses on the ISIUOP sub-project "Mapping Inuit Sea Ice Knowledge and Use," which was designed by the co-authors to expand upon previous research that characterized the importance of sea ice processes, use, and change around the communities of Cape Dorset, Igloolik, and Pangnirtung, Nunavut (Figure 1.1) (Laidler and Elee 2006; Laidler 2007; Laidler and Elee 2008; Laidler and Ikummaq 2008; Laidler et al. 2008, 2009).

Indigenous experts, such as Inuit elders and active hunters, have developed detailed and sophisticated local- and regional- scale knowledge about the complexity and dynamism of the marine environment through long-term experience. This knowledge has been developed and transmitted from generation to generation through personal experience, and by oral means, to the point that Inuit knowledge of sea ice could in fact occupy several written volumes if documented comprehensively. Inuit experts are acutely aware of the local and regional manifestations of climate change, as indicated by long-term changes and increased unpredictability of sea ice conditions. Specifically, changes are observed using key indicators such as the floe edge position, weather, the timing of freeze-up and breakup, ice thickness, and the presence of multi-year ice (MYI). In order to address some of the gaps in local-scale scientific knowledge, this chapter reviews specific indicators used to evaluate sea ice changes, offers a regional comparison of sea ice changes in the three communities, and provides an overview of some of the local implications of sea ice changes. By sharing these local perspectives, we hope to contribute to more comprehensive assessments of community vulnerability/resilience to change. Thus our aim is to document Inuit perspectives to help support local adaptive strategies for ongoing ice use and to provide additional considerations for the safe use of ice for future generations.

Methods

Results presented in this chapter reflect a compilation of more than 6 years of collaborative community-based research undertaken in Cape Dorset, Igloolik, and Pangnirtung, Nunavut. We sought to work with community members, and to learn about Inuit expertise, based on the premise that those who live in – and use – the sea ice environment on a daily basis are most knowledgeable of their local and regional contexts. The co-authors began working together in 2003 during the early stages of Laidler's doctoral research (2002–2006). This project then evolved to become the foundation of one of the four ISIUOP sub-projects (2006–2010, see Chapter 1, Introduction). Throughout this process we have attempted to learn from, adhere to, and contribute to the principles of collaborative northern research (see ACUNS 2003; ITK and NRI 2007; Laidler 2007; Gearheard and Shirley 2007; Pearce et al. 2009). Our approach and methods have been previously outlined elsewhere (Laidler and Elee 2008; Laidler and Ikummaq 2008; Laidler et al. 2008, 2009) and in great detail in Laidler (2007). Therefore, in this chapter we provide a summary of techniques used, as they relate specifically to results presented below.

Field Work

We use the term "field work" for lack of a better term, but it is more appropriate to interpret field work trips as the time of intensive research collaboration in each of the communities, when university and community-based researchers[1] are able to work most closely together. However, work was ongoing in various forms within each community, as undertaken by the community researchers and by university researchers at their home institutions between trips, with frequent contact maintained throughout. So, a total of approximately one year has been spent together in the three communities to facilitate this research, but more than six years of cumulative effort has been invested in the project to yield current results.[2] The amount of time spent and the number of research visits were critical to collaboratively developing this project and maintaining community interest and commitment.

Semi-directed Interviews

Interviews were used to gain in-depth understanding of individual experiences, observations, and expertise in relation to sea ice around their communities. No fixed questionnaire was administered and no time limit was placed on discussions (average interview duration was 2 h). An interview guide was used to help cover important themes of interest and those specifically related to this chapter include (i) the importance and uses of sea ice; (ii) rare or notable sea ice events; (iii) timelines of observed changes; (iv) indicators (and influences) of sea ice changes; and (v) locations, geographic representations of sea ice features, uses, and changes.

Inuit elders and hunters deemed to be ice experts in each community were invited to participate based on recommendations provided by community organizations, community researchers, elders or hunters previously interviewed, and other community members. Interviews were documented with audio, video, and digital photo recordings – wherever consent was provided. Honoraria[3] and small gifts were provided to interviewees as a token of appreciation for the time taken out of their daily schedule. In total, 88 people contributed to this project through their participation in interviews (35 in Cape Dorset, 26 in Igloolik, and 27 in Pangnirtung) based upon local recommendations of approximately 40 experts to consult in each community. Some of the individuals were interviewed several times, so overall 120 interviews were conducted. However, only those individuals who specifically discussed observations or implications of change are referenced in this chapter (Appendix 1).

Participatory Mapping

One characteristic of ISIUOP is the use of maps in the documentation process. We found maps to be excellent tools in the context of Inuit environmental knowledge research, particularly because they help trigger memories and facilitate conversations, as well as being effective in spatially and visually synthesizing diverse observations. Several Canadian National Topographic Service (NTS) map sheets were incorporated in interviews to (i) facilitate knowledge sharing; (ii) enhance explanations of ice conditions or uses; (iii) enable visual identification of key ice conditions or uses; and (iv) promote discussion and spark memories. For each community a combination of three to five 1:250 000 scale NTS map sheets were employed to represent the surrounding coastline and ocean areas. A clear plastic overlay (similar to Mylar) was placed on top of each map and registered for future digitizing. Common features drawn on the maps (and later digitized) include travel routes, floe edge positions, tidal cracks, polynyas, areas that melt early in the spring, wildlife harvesting areas, and traditional or current camps. Such features were indicated whenever the participant felt it helped enhance discussion, explain a condition/process, or in response to a question posed during the interview.

Sea Ice Trips

It was imperative that the community researchers had a chance to share their knowledge and experience in context, on the ice, so a number of sea ice trips were organized to facilitate in situ knowledge exchanges through sea ice travel, observation/evaluation of conditions, and participation in related activities.[4] We were fortunate to undertake 20 trips to experience and learn about ice terminology, navigation, conditions, and change firsthand. Trip destinations in each community included the floe edge, nearby polynyas, fishing lakes, hunting cabins, and unique ice features, and ranged from day trips to week-long trips. All these experiences were documented with photo and video cameras and a GPS was used to track our travel routes.

Focus Groups

We undertook 13 focus group sessions in order to (i) link Inuktitut terminology for ice conditions to pictures taken on ice trips; (ii) develop and verify terminology links within a sequential order of seasonal ice formation and decay; (iii) verify and/or clarify maps of ice feature positions and labels; (iv) discuss indicators of sea ice/weather changes (the primary group session results reflected in this chapter); and (v) define plans and areas of interest for the expansion of the Polar View Floe Edge Service. These sessions were jointly facilitated by a community and a university researcher, whereby an average of three to six elders or hunters participated in each one. These sessions ranged between half and full days and comprised guided but relatively unstructured discussions. Photographs were taken to document the general setup and discussions were recorded similar to the semi-directed interviews.

Data Analysis

The participatory nature of diverse components of this research approach is well suited to undertake a multi-faceted, qualitative analysis of the knowledge-sharing results and processes. The major components of the analyses include transcript (content analysis), map (visual spatial analysis), and focus group interpretation (see Laidler 2007 for full description).

Results

Local Importance and Uses of Sea Ice

Sea ice is very important to each of the partner communities involved in this project. It is a key means of traveling to access animals, engaging in subsistence or commercial hunting/harvesting, and enjoying leisure time. With Igloolik and Cape Dorset located on islands, sea ice is one of the most important means of connecting with other communities, camps, or land-based resources.

> Sea ice is very useful, in that we are on an island. [When it freezes], now we're not on an island anymore, we're now connected to everywhere (Qamaniq 2004; Igloolik).

In Cape Dorset, there is significant emphasis on the year-round presence of open water, as a defining characteristic of unique local ice conditions.

> [The] way this island is built, around Hudson [*Strait*] . . . that's why the ice never goes any further than where it goes. That's why it's called Sikusilaaq [where there is no ice], but if you go further southeast of here, like 100 miles from here, it could take you a whole day's trip to get to the floe edge from the [the coastline]. But over here it's not like that, it's only like a 10 minute [snowmobile] drive from this town (Peter 2004; Cape Dorset).

The strong Hudson Strait currents prevent solid ice formation from extending far offshore, creating a dynamic sea ice/open water environment around Cape Dorset. Regardless of the significance of open water, sea ice remains an important part of life for this island community.[5] For Pangnirtung, sea ice is of equal importance, enabling people to avoid the surrounding mountainous terrain during travel outside the community, while also being described as a deeply entrenched part of Inuit culture.[6]

> It's very very important to Inuit, because it's our *qaujiti*, which means we were born to it and we've always lived in it... If the sea ice doesn't form anymore, although we still get snow, our life would drastically change (Maniapik 2004b; Pangnirtung).

Once the sea ice forms solidly there is little distinction between land and ocean as the ice essentially becomes an extension of the land, an important "highway" and preferred shortcut to many popular destinations. Then people are free to travel wherever they wish, as long as they are aware of the dangerous areas and local ice dynamics. Sea ice acts as an essential travel and hunting platform, as well as valued ecosystem component supporting a diversity of arctic marine mammals (e.g., seals (ringed, bearded, harbor, hooded, etc.), beluga whales, narwhals, walrus, as well as polar bears) and a variety of fish and migratory marine birds. Previously, marine mammals were a means of survival (i.e., clothing, fat/oil for light and heat, food for people and dogs, and equipment). Although much has changed since Inuit moved to permanent settlements, the sea still provides important food sources and/or income potential (from skins or related clothing/craft products) in most Nunavut communities.

> [The ice is] part of the hunter's life and it can have an effect on his livelihood. [As] a hunter, [I] have to hunt out there, and [I] depend on the animals to bring in food as subsistence and also skins, depending on what kind of animal it is, to be able to make money off that. So it has an impact on how much a hunter, not only [myself] but other hunters, as to how much money [we] can bring into the family (Qappik 2004a; Pangnirtung).

These aspects are inseparable components of Inuit sea ice use because travel on the sea ice is mainly undertaken to access hunting, fishing, or harvesting grounds (inland or marine).

Local Observations of Change

Inuit are accustomed to yearly variations in ice conditions and timing, and yet, increasingly consistent shifts beyond expected variability are being noted – affecting both sea ice use and safety. In all three communities people have observed, and are experiencing, considerable change in local climatic and sea ice conditions. Furthermore, while many of the indicators used to evaluate change locally may be similar between communities, they are described and interpreted differently, highlighting geographic variations in physical conditions as well as sea ice uses and regional cultural differences.

Timelines

The changes noted by local elders and hunters as "unique" or "unexpected" were described as being most prominent starting around the year 2000. The time frames varied by the person's age and experience, but generally focused on the past 5–10 years, compared to observations and experiences accrued over their lifetime. Unusual or extraordinary conditions tended to be compared to what were previously considered "normal" (i.e., within expected ranges of seasonal variation). Due to the age and experience of most of the people interviewed, the periods identified as "normal" tended to be the 1960s or earlier,[7] although the 1980s were noted in Pangnirtung as having anomalously high ice extents. Thus, in the following sections references to "recent" indicate conditions/processes observed since the year 2000, in comparison with "past" or "previous" years indicating the 1960s, which was taken as a general baseline for individual evaluations. Particularly distinct years are then identified specifically, such as the spring and fall of 2004 in Igloolik and Pangnirtung (unusual breakup and freeze-up processes) and the winters of 2006–2007 (unusually low ice extents) and 2007–2008 (higher ice extents than the past several years) in all three communities.

Indicators of Change

In all three communities, evaluations of change frequently centered around a similar set of indicators that include (i) the position of the floe edge; (ii) weather predictability; (ii) freeze-up timing; (iv) breakup timing; (v) ice thickness; and (vi) presence/absence of multi-year ice (MYI) (Fig. 3.1). Keep in mind throughout this

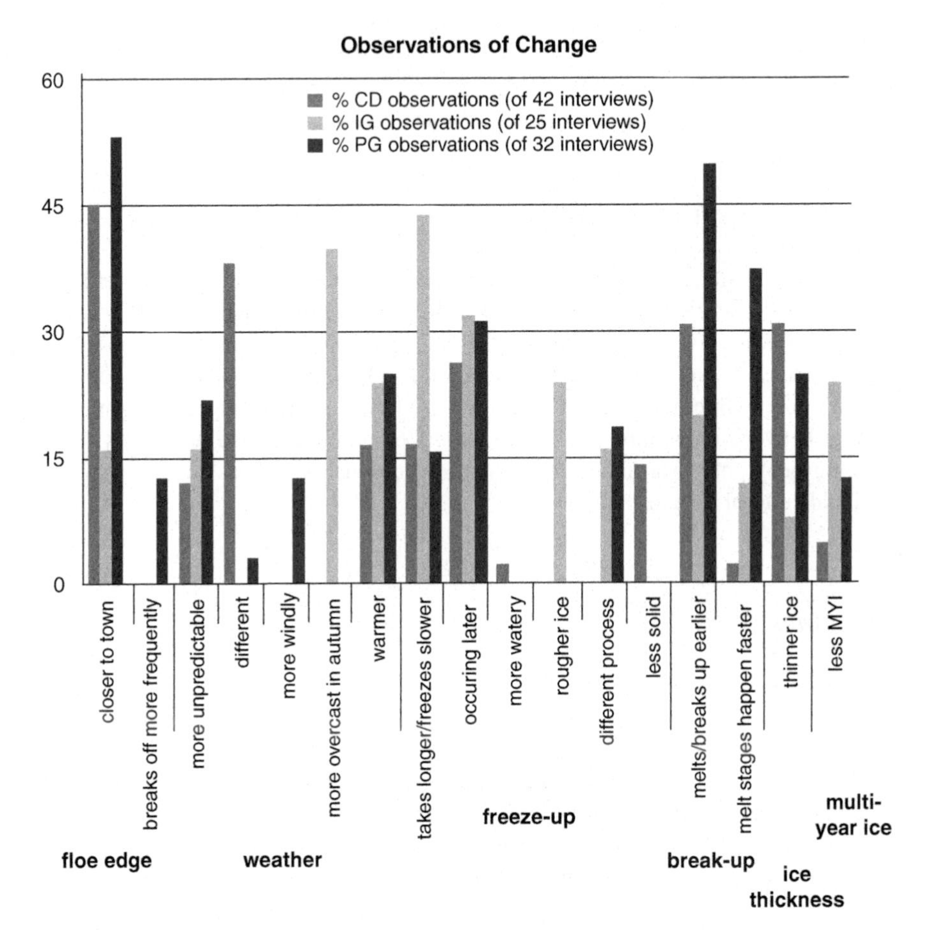

Fig. 3.1 Summary of observed sea ice changes as highlighted by contributors in Cape Dorset (CD), Igloolik (IG), and Pangnirtung (PG)

section that Igloolik experiences the greatest extent of ice formation in Fury and Hecla Strait (Laidler and Ikummaq 2008), followed by Pangnirtung whose ice conditions are dominated by the geography of local fiords and strong current and wind influences in Cumberland Sound (Laidler et al. 2008). Cape Dorset has the least ice extent, as influenced by its more southerly location and strong currents of Hudson Strait (Laidler and Elee 2008). Thus the temporal and spatial aspects of change affect communities differently (see Discussion), although many of the indicators used are similar.

Floe edge. The floe edge (edge of the land-fast ice where it intersects with open water) is a prominent indicator of sea ice change because its position is easily compared with land features. Furthermore, the floe edge is frequently used to determine spatial frames of reference (see Chapter 7 by Aporta, this volume), so its influence

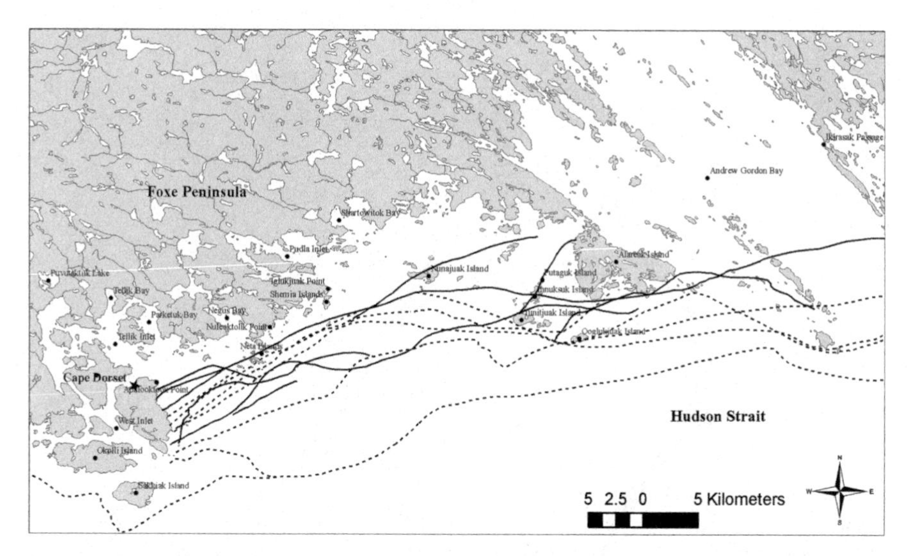

Fig. 3.2 Map of floe edge changes near Cape Dorset. *Solid lines* indicate recent floe edges (i.e. since 2000); *dashed lines* indicate past floe edges (i.e., approximately in the 1960s); *star indicates* the location of the Hamlet of Cape Dorset.
Sources: Alasuaq (2004), Kelly (2004), Petaulassie (2004b), Pootoogook (2004), and Saila (2004)

on sea ice travel means that it is regularly observed and evaluated for safety. Change in the floe edge position is typically gauged by its proximity to the community and it is the most frequently referenced indicator of change in Cape Dorset[8] and Pangnirtung[9] (Fig. 3.1). Generally speaking, a closer floe edge indicates warmer weather (less sea ice), although other factors such as winds and currents should also be taken into account.

> [T]he ice from years ago would form all the way down past the point... where the cracks would start from. But since the ice hasn't formed down there anymore, the floe edge will be around *Aupaluqtuq* area, and then just before that is where cracks would be forming, depending on the month (Saila 2004; Cape Dorset) (Fig. 3.2).

The alterations in floe edge delineation are most dramatic in Cumberland Sound (Fig. 3.3). One of the furthest floe edges occurred in the mid-1980s, around 1984 (potentially associated with weaker currents that year).[10] However, it is rare these days for Cumberland Sound to become *nunniq* (a term used to refer to the extent of freezing in Cumberland Sound – typically when the floe edge is more than half way out to the mouth of the Sound), which has altered the "definition" of *nunniq*.[11]

> When it's a *nunniq* period it would be normally here [pointing at the map], like when it's called *nunniq*, [this area] would be frozen over, the floe edge would be [far], and that would be called *nunniq*.... Normally [now] it's even closer, [so] hunters describe it as the floe edge being far away when it's like this close... So if a hunter says the floe edge is far, it's only over there (Qappik 2004b; Pangnirtung).

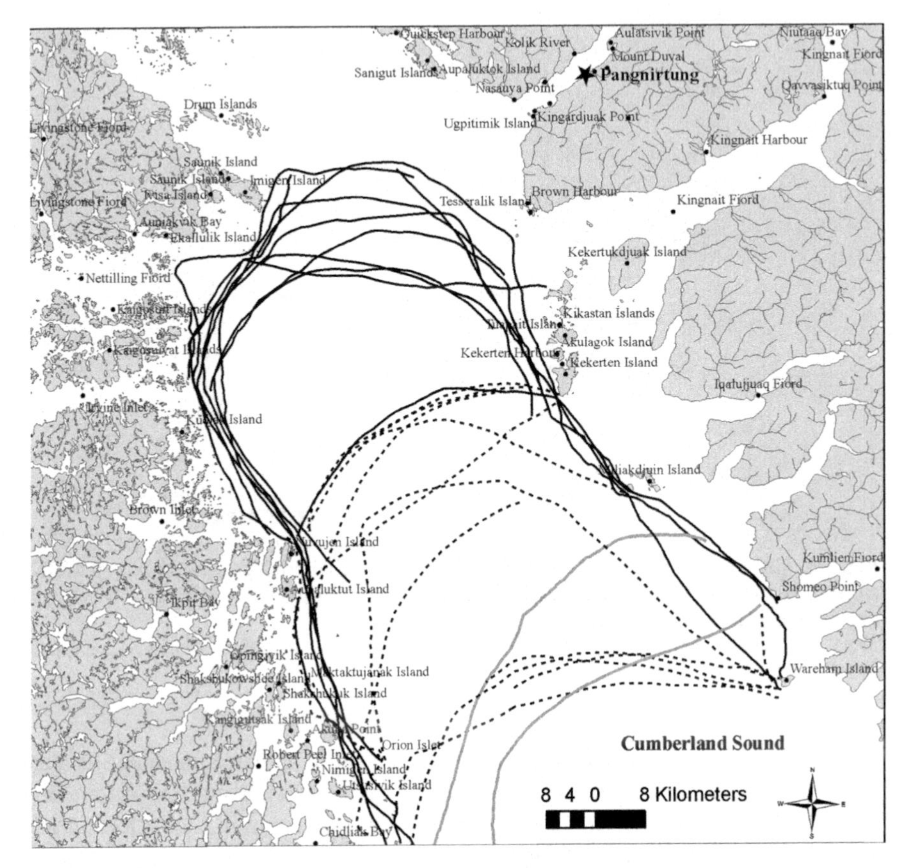

Fig. 3.3 Map of floe edge changes near Pangnirtung. *Solid lines* indicate recent floe edges (i.e., since 2000); *dashed lines* indicate past floe edges (i.e., approximately in the 1960s); *gray lines* indicate anomalous ice extents in the mid-1980s; *star* indicates location of the Hamlet of Pangnirtung. *Sources*: Anonymous (2004), Evic (2004), Ishulutak (2004a), Kisa (2004), Mike (2004), Noah (2004), Nowyook (2004), Qappik (2004b), Soudluapik (2004), and Nowdlak (2005)

In Igloolik, the floe edge is described mainly in the context of yearly variations, which are highly dependent on the grounding and piling of MYI on three key reefs in Fury and Hecla Strait, between Melville Peninsula (mainland Canada) and Baffin Island (Fig. 3.4).[12] There was less emphasis placed on floe edge change in Igloolik (Fig. 3.1), but still a general consensus on closer floe edge proximity to town. The MYI does not seem to be piling up on the reefs as much as in the past, and without this *kikiak* (anchor) the floe edge breaks off more easily.

> When [the floe edge] is smooth all the way, that could be one of the factors in that the ice breaks easily, maybe the current is stronger in recent times. But it just breaks off. And then, *Ivunirarjuq*, this reef is a *kikiak*, in that it nails the ice, *kikiak* meaning "a nail." It stops this ice, so once it freezes it stays there, because this [reef] prevents it from breaking off. But in the last few years [I] notice that even though this [reef piling] is there, it breaks off here. It's not doing what it used to anymore (Kunuk 2004; Igloolik).

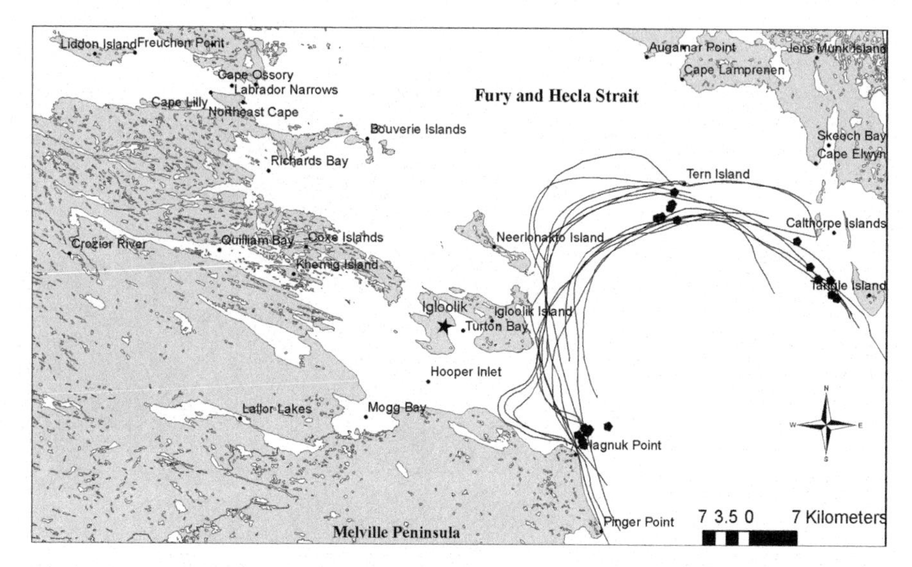

Fig. 3.4 Map of floe edge changes near Igloolik. *Black lines* indicate floe edge variations in relation to the locations of three nearby reefs (*black hexagons* showing reefs as drawn by various interviewees).
Sources: Ammaq (2004), Angutikjuaq (2004), Ikummaq (2004), Ipkanak (2004), Irngaut (2004), Kunuk (2004), Qamaniq (2004), and Qattalik (2004)

Because recent local floe edges have not established "properly," certain areas are more prone to breaking off and increased break-off events were noted in all three communities.[13]

Despite trends toward a reduced sea ice extent that translate into closer floe edge proximity, stark yearly variations highlight the intense fluctuations in ocean and atmospheric processes at play. The winter of 2006–2007 was noted as having low ice extent in Pangnirtung, with highly unstable conditions, such as the ice breaking off even in the dead of winter.[14] In winter 2007–2008 experts from Cape Dorset and Pangnirtung[15] noted an increase in ice extent as cold conditions that year led to more ice growth and extensive distribution than had been observed in several years prior (Fig. 3.5). Nevertheless, with no icebergs in Cumberland Sound the floe edge near Pangnirtung was still not stable and it broke up quickly in the spring.[16]

Weather. Emphasis was commonly placed on the variability of weather, and its influence on sea ice conditions, rather than on consistent uni-directional climate change. No two winters are the same, just as the ice conditions are different from year to year. However, within the inter-annual variability there is still a sense of expected conditions, processes, and timing. There are many nuanced weather indicators described as being linked to changes in climate and ice, so only a summary of those that were beyond anticipated seasonal or annual variability is highlighted here (Table 3.1).

Fig. 3.5 Hunters evaluate the spring 2009 floe edge in Cumberland Sound. Like 2007/2008, the 2008–2009 ice season was also one of the furthest floe edge formations in a number of years as indicated by the hour-and-a-half drive to reach this floe edge in Cumberland Sound. However, the area at the mouth of Pangnirtung Fiord was still melting and opening early, making it difficult/dangerous to access the still ice-covered areas of the Sound (Photo: G.J. Laidler; May, 2009)

> As far back as [I] can remember winds usually came from the north. But today even in fall, summer, spring, winds come from all directions, not from one direction anymore. If it was 20 years ago [I] would predict what kind of weather it would be in the next couple days. But if [you] were to ask [me] what weather we're going to have like next couple days, and if [I] tell [you] from today's weather conditions, [I] would probably be lying because weather changes in a matter of more like minutes than days now (Tapaungai 2004; Cape Dorset).

The most common observation of recent weather change was a warming of winter temperatures (Fig. 3.1).[17] This is partially indicated by (i) the decrease in ice crystal formation on people's faces and parka hoods; (ii) fewer days of extreme cold as indicated by the lack of ice fog presence and diesel fuel no longer becoming gelatinous in the winter; (iii) more overcast conditions; (iv) the decreased need to wear caribou skin clothing; (v) the use of canvas tents in January and February (where an igloo was required for adequate shelter in the past); and (vi) people's breath while exhaling no longer crackles in mid-winter. Another commonly shared observation was the increased unpredictability of weather, and frequency of weather shifts, linked to changes in prevailing winds (Fig. 3.1, Table 3.1).

Table 3.1 Summary of weather indicators and related sea ice changes observed in Cape Dorset, Igloolik, and Pangnirtung, Nunavut

Weather indicator	Related changes observed in Cape Dorset	Related changes observed in Igloolik	Related changes observed in Pangnirtung
Winds	• *Prevailing winds shifted* (more SE winds, less NW winds) – more southerly winds push pack ice into the floe edge, leads to ice instability • No consistent prevailing wind direction, *frequent directional shifts* – leads to ice instability and break-off events • *Colder winds* – can facilitate ice formation • *Warmer winds* – can facilitate ice deterioration • *Windier fall* conditions – affect ice formation	• *Severe winds* come up suddenly and die down suddenly • *Weaker winds* blow for longer periods of time • *Prevailing winds shifted* (more SE winds, less NW winds) – more southerly winds push pack ice into the floe edge, leads to ice instability, snowdrift formations no longer reliable for directional navigation • *Windier conditions* lead to rougher ice conditions – harder to use for travel • *More W, SW wind conditions* experienced – contributes to ice deterioration with storm events and warmer air • *Winter winds* have been less severe – less blowing snow • *Summer winds* seem to have increased in prevalence	• *Prevailing winds shifted* (more SE winds, less NW winds) – more southerly winds push pack ice into the floe edge, leads to ice instability, warmer weather • More prevalent *windy* conditions (especially in winter) • *Less consistent freeze-up* due to wind shifts – softer ice consistency • More *sudden and frequent wind shifts* mean more sudden and frequent weather shifts
Clouds	• Cloud formations formerly used to predict weather conditions are *no longer reliable*	• Cloud formations formerly used to predict weather conditions are *no longer reliable* • More *overcast conditions* in autumn	• Cloud formations formerly used to predict weather conditions are *no longer reliable* – timing is off, increased inconsistency, increased challenges
Storms	• Fewer *blizzards*	• *Storms* come up quickly, but do not last as long as expected	
Precipitation	• *Less snow* accumulation	• *Increased autumn snowfall* and accumulation	• *Increased autumn snowfall* and accumulation

Cape Dorset references: Joanasie (2004), Mikigak (2004a), Nuna (2004), Ottokie (2004), Petaulassie (2004a, 2004b), Peter (2004), Saila (2004), Solomonie (2004), Tapaungai (2004), Ezekiel (2005), CD WKSP (2007), Etungat (2008), Mikigak (2008), Oshutsiaq (2008a), Pootoogook (2008), Ragee (2008), Saila (2008), Shaa (2008), Takiasuk (2008), and Tukiki (2008).

Igloolik references: Ammaq (2004), Arnatsiaq (2004), Ikummaq (2004), Ipkanak (2004), Kunuk (2004), Paniaq (2004), Taqqaugak (2004), Uttak (2004), Ivalu (2005), Palluq (2005), Qaunaq (2005), IG WKSP (2007), Ammaq (2008), Paniaq (2008), and Qaumaq (2008).

Pangnirtung references: Keyuajuk (2004), Ishulutak (2004c), Nashalik (2004), Nuvaqiq (2004), Qappik (2004a, 2004b), Soudluapik (2004), Nuvaqiq (2005), PG WKSP (2007), and Yank (2008).

Unique to Cape Dorset[18] was a general description of the weather being "different" than expected, while in Igloolik[19] more overcast conditions were noted in the fall and in Pangnirtung[20] more frequent windy episodes (i.e., occurrences of moderate to high wind speeds) have been experienced. These shifting weather conditions were noted for their influence on longer transitional freeze-up and breakup periods and on the stability of winter ice conditions.

> Our weather has changed a great deal from the past weather we used to have. Because of the lack of cold weather in our community the sea ice has not been able to get thicker for a number of years now. In the past, in the 1960s the sea ice would be usable for travel just over night. What we call the newly formed ice near the floe edge ... hunters would be able to start travelling on it [as it was freezing], but today it takes days and days for it to form properly to travel on (Mike in PG WKSP 2007; Pangnirtung).

Despite some comments on general warming of the weather, there were also several postulations that it is the ocean – and not the air – that may in fact be warming.[21] Such observations were noted based on (i) warmer sensation to touch; (ii) the cold water layer being deeper in the ocean resulting in warmer surface layers, as evaluated through seal breathing holes; (iii) the slower speed of re-freezing after a surface layer of ice is broken in the winter; and (v) some localized water temperature measurements in Fury and Hecla Strait. This increased water temperature

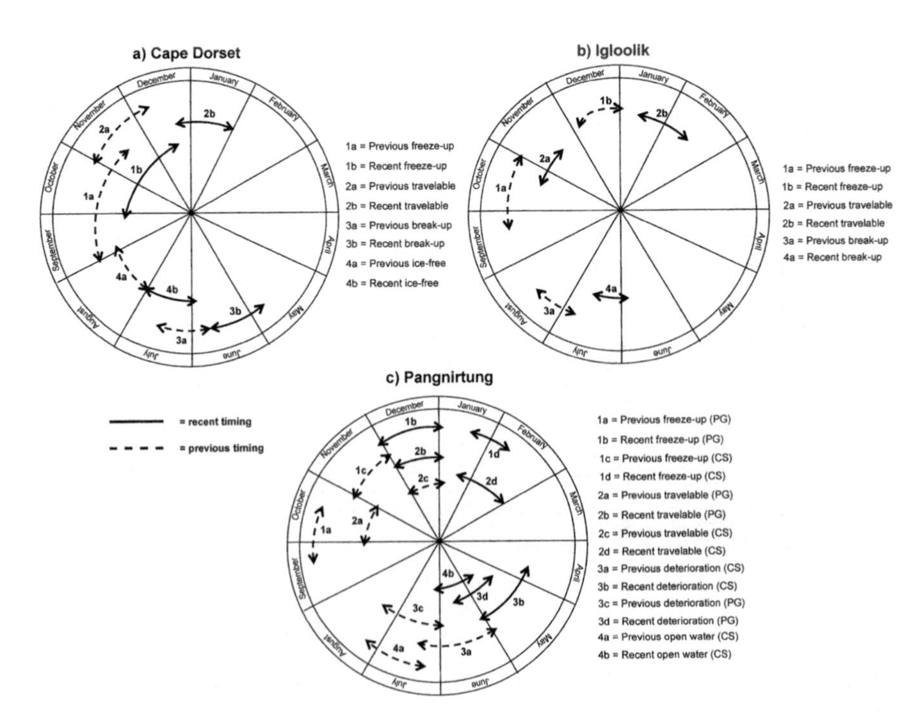

Fig. 3.6 Summaries of temporal change in freeze-up and breakup timing around (**a**) Cape Dorset (CD); (**b**) Igloolik (IG); and (**c**) Pangnirtung (PG).

may be a key contributor to closer floe edge proximity, alterations in freeze-up and breakup timing, and thinning sea ice.

Freeze-up. Experts in all communities noted that freeze-up is occurring later each year, and that the freezing process is taking longer (Fig. 3.1).[22]

> What [I] have noticed is more in the last 5–8 years, when it should be freezing up [I] have noticed that it becomes overcast, snow starts falling for a long period of time. Whenever it's overcast the temperature rises a bit, freeze-up doesn't occur as quickly, or it doesn't even occur at all at some times when you have clouds and the wind working together . . . It might pile up at places and then it disperses . . . as opposed to other years where freeze-up was quite constant, it was the same year after year pretty much. But in the last few years . . . it clears up later, so freeze-up occurs a little later (Arnatsiaq 2004; Igloolik).

The timing described for recent and past freeze-up was not as different between communities as initially anticipated. However, not as many people in Igloolik stressed specific changes in the timing of freeze-up, compared to those interviewed in Cape Dorset and Pangnirtung. Figure 3.6 provides a visual depiction

Fig. 3.6 (continued) *CD Freeze-Up:* The beginning of the freezing processes occurs approximately 1 month later than in the past (i.e., October–December instead of September–November). Similarly, the ice is not formed solidly enough for travel until later in December in recent years, while previously people were traveling on the ice as early as late October.
CD Breakup: The offset in breakup timing is approximately 1 month, as well as for the ice-free season. It used to be possible to travel on the sea ice in June and into July, whereby boat travel (especially in relation to accessing a soap stone mine eastward along the coast) was not possible until August. However, in recent years the ice begins to breakup in May, is rarely travelable in June, and the ocean is ice-free in July. This allows earlier boat access to the soap stone mine.
CD References: Joanasie (2004), Nunguisuituq (2004), Kelly (2004), Mikigak (2004a), Ottokie (2004), Parr (2004), Peter (2004), Saila (2004), Suvega (2004), and Manning (2005).
IG Freeze-up: Early freezing processes used to begin in late September or early October, but especially in recent times has been forming only around November. An important gauge of freeze-up timing is when it is possible to cross Ikiq (Fury and Hecla Strait) to reach Baffin Island. This sea ice crossing used to be undertaken before Christmas but is no longer possible until late January or early February.
IG Breakup: Breakup was discussed as occurring at a different time every year, although it was considered "normal" to break off around August whereas in the spring of 2004 the ice broke up in early July.
IG References: Angutikjuaq (2004), Aqiaruq (2004), Ikummaq (2004a), Ipkanak (2004), and Qattalik (2004).
PG Freeze-up: The beginning of the freezing processes occurs approximately 2 months later than in the past (i.e., December instead of September/October). Similarly, the ice is not formed solidly enough for travel across Cumberland Sound until January or February, whereas people from the other side of Cumberland Sound used to cross to Pangnirtung to celebrate Christmas.
PG Breakup: The offset in dangerous travel, and breakup, timing is approximately 2 months. For example, crossing Cumberland Sound was still possible in May/June previously, whereas now travel becomes dangerous around April. Furthermore, fiords were still travelable in June and are now only safe until around May.
PG References: Alivaktuk (2004), Evic (2004), Ishulutak (2004a, 2004b), Keyuajuk (2004), Maniapik (2004b), Mike (2004), Noah (2004), Nuvaqiq (2004), Qappik (2004a), Soudluapik (2004), Vevee (2004), Evic (2005), and Nuvaqiq (2005)

summarizing the temporal change in freeze-up timing, typically gauged according to when (i) the very early signs of freezing are visible and (ii) the ice becomes travelable (i.e., thick enough for travel by dog team or snowmobile). The shifts identified in Cape Dorset (Fig. 3.6a) and Igloolik (Fig. 3.6b) were quite similar. However, in Igloolik travel referred to much larger expanses of sea ice used to cross Fury and Hecla Strait, while in Cape Dorset travel meant crossing the relatively narrow Tellik Inlet, in both cases seeking to reach Baffin Island. In Pangnirtung, the timing of freeze-up was described distinctly for the fiord and within Cumberland Sound (Fig. 3.6c). The most drastic shifts in freeze-up have been noted in this area, with almost a 2-month shift in the timing of ice formation. These later freeze-up conditions and different ice consistency (Fig. 3.1) were frequently linked to the detrimental influences of shifting and unpredictable wind and weather (Table 3.1). Specific indicators used to identify changes in the freezing process are highlighted in Table 3.2. The shifting timing of ice formation has implications not only for winter ice conditions but for spring as well.

Breakup. All three communities have noted a shift toward earlier spring ice breakup, and an increased speed with which ice deteriorates in spring (Fig. 3.1).[23]

> [I]n June [we] used to go by dog team. But now [we] don't even snowmobile in this month of June, it doesn't look like it's going to happen again (Ottokie 2004; Cape Dorset).

Table 3.2 Summary of local indicators used to evaluate altered freezing processes around Cape Dorset, Igloolik, and Pangnirtung, Nunavut

Cape Dorset	Igloolik	Pangnirtung
• No *qanguqtuq* • No *qaikut* • *Ilu* not happening anymore • Ice no longer beginning to form from the bottom of the low tide zone • *Naggutiit* refreeze roughly	• Ice is more commonly freezing upwind (*aggurtipalliajuq*), creating ice conditions more like *qinu* than SIKU • *Qaingu* forming later • New *aukkarniit* (polynyas) forming	• *Sikuvaalluuti* does not form properly or stay until the next year • *Uiguaq* breaks off much easier and more frequently • Areas with strong currents are not freezing over where they used to, new *saqvait* (polynyas) forming • Some points are thin/dangerous where they used to be solid • Softer consistency of sea ice (more *sikurinittuq*)

See Laidler and Elee (2008), Laidler and Ikummaq (2008), and Laidler et al. (2008) for meanings of the Inuktitut terms used for ice forms listed in the table.

Cape Dorset References: Alasuaq (2004), Parr (2004), and Nunguisuituq (2004).
Igloolik references: Ammaq (2004), Kunuk (2004), Qattalik (2004), Palluq (2005), and Kunnuk (2008).
Pangnirtung References: Keyuajuk (2004), Kisa (2004), Nashalik (2004), Noah (2004), Nuvaqiq (2004), Qappik (2004a, 2004b), Nuvaqiq (2005), and Vevee (2004).

Breakup timing is often referred to as when the sea ice is no longer travelable, distinguished from when the ocean is ice-free (and boating is possible). Again, around Pangnirtung descriptions relate specifically to when the sea ice starts to become dangerous for travel and when the ice in Cumberland Sound and the surrounding fiords begins breaking up. Figure 3.6 summarizes the temporal change in breakup timing, alongside the freeze-up timing discussed earlier. Changes related to early breakup timing are most prominent in Cape Dorset and Pangnirtung (Figs. 3.1 and 3.6a, c), while in Pangnirtung and Igloolik there is more emphasis on faster melt stages (notably that some stages are being skipped) (Figs. 3.1 and 3.7).[24]

> This year [2004] for some odd reason, [the cracks never formed to create open leads]. And then you find that all of a sudden the ice wears out faster ... [My] theory is that because the cracks are not there, the drainage doesn't take place, so once the snow melted the water stayed on [the ice] ... and the ice just wore out, it just melted. It didn't break up and go like in normal years, it just melted. And that could be a factor because of that film of water that's on the ice. Normally the stage would be first there's water, and then it drains along the tidal cracks or breathing holes, and then [the water on the ice] gets deep. We didn't get to the deep stage this year... because of the absence of these cracks (Qattalik 2004; Igloolik).

Fig. 3.7 Spring ice conditions in Fury and Hecla Strait. With earlier or more unpredictable spring breakup, there is a greater chance of people being stranded on moving ice or on land away from the community (Photo: G.J. Laidler; June, 2005)

In all three communities there are commonly known areas that tend to melt earlier than others (usually associated with strong currents or nearby polynyas); these areas have now been observed to begin melting and opening up to a month earlier than expected, based on 1960s conditions.[25] Furthermore, the increased speed of ice deterioration is suggested to link to the generally thinner and weaker (softer) ice resulting from the shorter ice season[26] and other weather-related factors (Table 3.1).

Ice Thickness. Thinning ice conditions were most prominently noted in Cape Dorset[27] (Fig. 3.8), closely followed by Pangnirtung.[28] This trend was mentioned to a lesser degree in Igloolik[29]; instead more focus was placed on risks generally associated with thin ice. Indicators used to evaluate changes in ice thickness are summarized in Table 3.3. Thickness evaluations vary according to a person's height and visual depth perception, as well as the initial ice conditions that they take as a baseline. For example, areas with strong currents would be thinner to begin with, but may be experiencing a greater amount of thickness change. Nevertheless, there is a strong relationship between thinning ice and earlier spring breakup, as well as the opening of new polynyas, as the ice is worn away more easily by the currents from underneath. Thinner ice conditions could also be created by increased precipitation (more snowfall could mean more insulation on the ice surface, and thus preventing

Fig. 3.8 Hunters at the edge of a polynya near Cape Dorset. With thinning ice conditions, these areas tend to be more dangerous and melt earlier in the spring. However, hunters are used to these dynamic conditions so they always bring boat along for seal retrieval, as well as being a safety precaution (Photo: G.J. Laidler; January, 2005)

Table 3.3 Summary of local indicators used to evaluate ice thickness around Cape Dorset, Igloolik, and Pangnirtung, Nunavut

Cape Dorset	Igloolik	Pangnirtung
• Seal breathing holes are no longer as deep or tunnel-like • Ice at open cracks is no longer as deep • Ice is thinner in comparison to personal height – usually gauged when drilling a fishing hole or setting seal nets • Travel routes are not as sturdy or solid as previous years (e.g., more open water on the winter soap stone mine route, which can also link Cape Dorset to the community of Kimmirut) • MYI are smaller and not present as frequently • Formation of new *saqvait* (polynyas) or presence of more open water	• Seal breathing holes used to evaluate thickness changes, generally no longer as deep or tunnel-like • Ice at open cracks used to evaluate thickness changes	• Seal breathing holes are no longer as deep or tunnel-like, seals do not have to go straight down to get into the hole • Melt holes are no longer as deep or tunnel-like • Ice is thinner in comparison to personal height when chipping away a hole • a fishing spear (more than 10 ft long) is no longer necessary to retrieve seals from their breathing holes (now only about 1.5 ft thick) • *Tuvaq* (landfast ice) is thinner

Cape Dorset references: Alasuaq (2004), Nuna (2004), Petaulassie (2004b), Peter (2004), Pootoogook (2004), Ezekiel (2005), and Manning (2005).
Igloolik references: Palluq (2005).
Pangnirtung references: Keyuajuk (2004), Ishulutak (2004a, 2004b), and Qappik (2004b).

thickening). Qualitative evaluations of ice thickness (i.e., to determine safety for use) thus require taking into account fall freeze-up conditions, geographic location (i.e., topographic, bathymetric, and current/tide effects), surface conditions, as well as seasonal weather.

Multi-year ice. The presence of multi-year ice (MYI) has also changed, with each community focusing on different aspects. In Cape Dorset,[30] less MYI has been seen drifting nearby, and people note that the pans seem smaller and they do not last through the summer. So, there have been fewer instances of boat travel being hampered by multi-year ice pushed into the floe edge; however, it still does occur with the right combination of wind and ice conditions.

> [Q]avvaq [MYI], they're big, like Hudson Strait polar ice, they're really clear white blue colours ... when people see that say 'oh oh, we're going to get trapped'. Because when they start to see that *qapvaq* [it] means a big area of ice like that can move into your area

> and block off all the shoreline. That has happened here, we got stranded out right here one time, with *qapvaq*. And then it stayed there over almost two weeks. (Manning, 2005, Cape Dorset)

Unique to Igloolik assessments of sea ice change is the importance of MYI and moving ice conditions. Most notably, there are two commonly expected types of MYI in *Ikiq* (Fury and Hecla Strait): "dirty" (i.e., yellowish-brownish colored ice coming from the sandy areas to the south and east, from Steensby Inlet and Cape Dorset) and "clean" (i.e., white MYI coming from the north through Labrador Narrows). Now both of these ice types are only concentrated in certain areas and they are generally further from town or virtually absent, and thus have less influence on summer temperatures and fall freeze-up.[31] For example, the yellowish-brownish "dirty" MYI, from the south, has been late in moving toward town in the spring.[32] Furthermore, the "clean" MYI, from the north, has been more common in recent years and is potentially linked to the shifts in prevailing wind directions.[33] In general, the moving ice does not appear as solid, there are more cracks present, and the ice pans are smaller/thinner.[34] Combined, these have important consequences for the formation of the floe edge.[35]

> But for the sea ice, again this year is unique in that we don't have multi-year ice. And with the lack of multi-year ice it took forever to freeze over to the point, so that February 14 was the only time you could cross Fury and Hecla to Baffin Island . . . normally it freezes earlier than that (Ulayuruluk 2008; Igloolik).

In Pangnirtung, there have been fewer glacier-calved icebergs noted in the fiord. Furthermore, less MYI coming from the north has been collecting in Cumberland Sound and it melts much faster than previously when it does enter the Sound.[36]

> And the *qavvaq* (MYI), that [ice that] comes from up there [North], when they come into Cumberland Sound they melt a lot faster than they used to. . . Like for example if the ice came into Pang Fiord, none of them would come back out, they would all melt in there. It's a one-way trip when ice comes in . . . you won't see it again, it's gone, it's going to melt in there (Nuvaqiq 2005; Pangnirtung).

Discussion

Implications of Change

After comparing observations of change, the following question remains: What do these changing ice conditions mean for each community in the context of their local ice processes and uses? Many of the changes described exacerbate the risks inherent in sea ice travel and hunting and render safety indicators less reliable than they were in previous years. Sea ice changes are also affecting subsistence and commercial harvesting success and the health and well-being of the animals themselves.[37] Figure 3.9 presents an overview – admittedly highly generalized – of environmental (and related sea ice) changes and their implications for people, animals, and associated adaptive strategies that were commonly highlighted in all three communities.

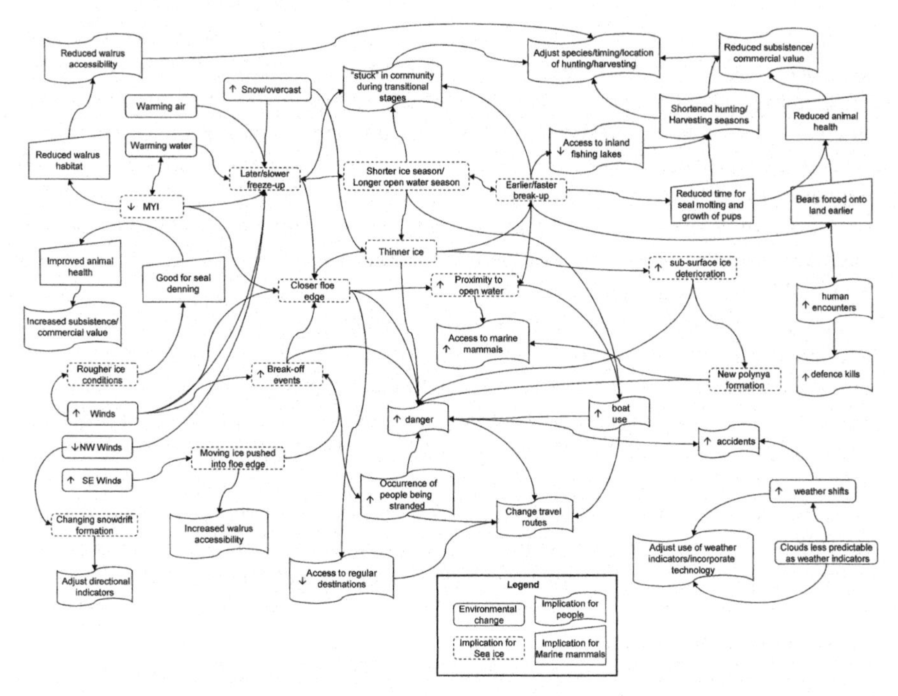

Fig. 3.9 Conceptual model of common observations of environmental changes in Cape Dorset, Igloolik, and Pangnirtung, and their implications for sea ice, people, and animals

Floe edge. In all three communities, the increased proximity of the floe edge to the settlement leads to an increased potential of break-off events and people getting stranded on moving ice. Usually, smooth ice extending to the floe edge would indicate a higher likelihood of the ice breaking off, but in recent years it seems to break off more frequently and unpredictably, regardless of ice surface topography. A closer floe edge also means that key travel routes are being compromised, especially in Cape Dorset and Pangnirtung, so people are increasingly forced onto land or have to take long detours due to poor ice conditions. In Igloolik there was less emphasis on change and more on variability of the floe edge in relation to other ice stability indicators (i.e., the amount of ice piled up on reefs). Generally, a closer floe edge means increased access to marine wildlife (associated with open water), but this is accompanied by enhanced danger of sea ice travel to hunting destinations. Even terminology references or uses are being changed (see the changed definition of *nunniq* in Pangnirtung), which has important implications for language tied to environmental conditions.

Perhaps most influential in an economic sense is the effect of shorter ice seasons, thinner ice cover, and a closer floe edge on the commercial turbot fishery in Cumberland Sound. Several Pangnirtung fishermen make a living off-seasonal long-line fishing and solid, extensive ice cover is essential to reach the most desirable and productive (i.e., deepest, farthest offshore) fishing spots in the Sound.

In the past few years (with the exception of 2008–2009), with closer floe edge proximity and higher frequency of storms and high winds, the fishing season has been drastically shortened. As such, there is greater risk involved in this activity due to an enhanced likelihood of the ice breaking up. Fishermen need to be more cautious because, in such an event, they can lose all their equipment at once. This had happened to some of the people interviewed and a number of others whom we met on the ice or around town. In many cases the equipment was simply irretrievable and irreplaceable, causing several people to abandon fishing completely due to the financial burden of trying to start anew.

Weather. Warmer winters are generally shortening the ice season, which means that hunters do not have as much time to use the ice. Potential benefits of this warming are that hunters do not have to wear the same heavy clothing and that tents suffice as shelter (i.e., igloo building may no longer be a necessity). This concerns some people as they fear that the ice may no longer form at some point in the future, while others feel that the air temperatures are not warming (rather that the water temperatures have increased or that more windy conditions are preventing adequate ice formation). Therefore, while there is consensus on changing ice conditions, there is no clear agreement on the relationship between temperature and the abbreviated ice season. Increased snowfall was noted in the fall as contributing to more dangerous ice conditions, but this can also be considered a positive implication for ringed seals that depend on snow cover (among other things) to build and maintain safe dens on the sea ice. However, a decrease in snowfall could be detrimental for seals as it could hinder their creation of adequate shelter for themselves or their pups in the winter months.

The increased unpredictability of weather shifts and traditional weather prediction techniques are rendering sea ice travel more dangerous. The diminished prevalence of northwest winds has affected not only these shifting weather patterns but also the ability to navigate accurately using snowdrifts. Shifting winds create differently shaped and oriented snowdrifts, thus hunters must recognize these changes to continue using snowdrifts as a navigational aid. Moreover, increased prevalence of southeast winds means, in some cases, more open water or moving ice is being blown toward the communities, making sea ice travel more difficult. At the same time it renders marine mammals more accessible where open water is maintained, but more frequent wind shifts increase the likelihood of ice breaking off from the floe edge. This (lack of) pattern in wind direction and strength has (i) restricted walrus hunting opportunities in Igloolik; (ii) influenced the amount of moving ice that concentrates around the communities; (iii) rendered ice conditions less stable and more prone to breaking up; (iv) sped up spring melt; and (v) undermined traditional weather prediction skills (e.g., by gauging wind and cloud conditions). Community members are thus increasingly turning to weather forecasts, satellite imagery, and GPS technology to help evaluate ice conditions prior to travel or to maneuver ice conditions in poor visibility. However, even weather forecasts received over the local radio are not deemed consistently reliable. They often contradict the current observable conditions and, combined with less predictable traditional indicators, are leading to more community members being stranded on the ice or land away from the community, getting lost, and/or having more accidents. Since favorable weather systems (i.e., clear and calm) are not remaining as long as they had previously, hunters have to be prepared for anything when they embark on longer sea ice trips.

Freeze-up timing. Each community noted later and slower freeze-ups, which lead to thinner and non-uniform winter ice conditions. Inconsistent freezing progression was described in comparison to previous years, as the ice tends to form, break up/get blown out, and then start forming again. This cycle goes on several times before the ice actually solidifies, leading to rougher ice and a softer consistency when it does form. This makes it more dangerous to begin traveling on the ice in the fall and less predictable as changed weather could drastically affect freeze-up progression. In Cape Dorset and Igloolik sea ice travel has been delayed nearly a month in recent years and in Pangnirtung nearly 2 months. Hunters and other community members are essentially stuck in town as they wait for ice to form solidly, when new ice prevents boating but is not yet thick enough for travel. People express their unhappiness and impatience with this delay, as they are eager to be on new ice hunting. In addition, altered travel routes tend to be longer and rougher as zigzag trails follow thicker ice areas or shoreline contours, affecting access to both sea ice and inland hunting destinations.

The slower freeze-up progression has also extended the boating season, while substantially delaying the start of the turbot fishing season (in Pangnirtung), causing seals to go longer without their prime habitat (land-fast ice), and polar bears to go longer without sea ice to hunt seals. Consequently, hunters around Igloolik have noted less fat on the seals they are catching in the fall (and a "healthy seal is a fat seal") and polar bears becoming increasingly lean, lacking the healthy fat they need to survive the winter and to support their young. This has also been noted around Pangnirtung, where substantial changes in ice conditions are translating into negative effects for the local seal population (e.g., poorer health and fur conditions, lower survival rate of seal pups, and generally decreasing numbers in Cumberland Sound).

Breakup timing. A shift toward earlier spring ice breakup was observed in each community. Some melt stages are being skipped and the ice is also deteriorating more easily from underneath. Consequently, melt stages are occurring more rapidly as the ice begins breaking up soon after the water on the ice drains away. This has implications for travel risks, prediction of ice breakup, and use of sea ice terminology. Specific areas open up 2–3 weeks earlier than others, so non-uniform melt conditions translate into dangerous travel, altered routes, and increased potential to be stranded away from town if snow/ice melt is sudden. Earlier breakup has enabled earlier boat travel and a longer boating season, which increases access to soap stone mines (Cape Dorset), walrus hunting areas (Cape Dorset), and visiting other communities (all), while markedly shortening the turbot fishing season (Pangnirtung).

Hunters in all three communities, but particularly in Pangnirtung, emphasized that with earlier breakup ringed seals do not get enough time to bask on the ice in the spring. This hampers seal hunting (by people and polar bears) and prevents the seals from molting adequately. Young seals then have a harder time ridding themselves of their white fur (to allow their dark, spotted fur to come in). Because molting cannot occur on land, more brown seals are being caught in the spring. These are considered bad skins, meaning that hunters are affected economically because they cannot sell the furs. Furthermore, important denning areas are breaking up early, which forces the pups into the water too early (i.e., they have not accumulated enough fat). This may increase the mortality rate of young seals, in turn affecting both the seal population and hunting success.

Ice thickness. Observations of thinning ice were mentioned in all three communities. This greatly enhances the risks associated with sea ice travel. Thinner ice is less resistant to the influences of currents, lunar cycles, winds, and snowfall. In other words, thinner ice deteriorates more easily from currents underneath (rendering it deceptively solid) or can be broken up more easily from waves or winds. Thinner ice can also initiate the enlargement or new occurrence of polynyas, increased break-off events at the floe edge, and decreased access to popular hunting and fishing destinations. In all three communities, thinning sea ice was also linked to later freeze-up (i.e., not as much time for the ice to thicken) and earlier breakup (i.e., the ice wears away faster) and thus an overall abbreviated ice season.

In general, increased presence of open water could be considered a positive influence of sea ice change as it allows marine mammals easier access to air and food and decreases travel time for hunters to reach areas where animals congregate. But, it also makes for potentially dangerous travel, restricted hunting options, and diminished habitat for some key marine species (e.g., ringed seals and the polar bears who prey on them). Community members also enjoy longer summer boating seasons. However, boat travel is more sensitive to wind, current, and weather conditions than sea ice travel due to the dangers of rough open seas or navigating through moving ice. In addition, the financial implications of boat ownership and maintenance are higher than for snowmobiles, so fewer people may be able to engage in boat hunting activities.

Multi-year ice. Less MYI around each community was linked to potentially warmer water or air temperatures, especially in the summer. This may, in turn, contribute to the delayed freezing in the fall, thinner ice conditions overall, and a closer/less stable winter floe edge. Less MYI present in the summer also renders boating conditions more difficult as waves propagate higher and further when there are no large ice pans to shelter boats from the wind or to dampen the effects of waves. Around Cape Dorset, less MYI was explained as potentially being part of an ice movement cycle where the ice pans might congregate in one area one year, and in another area another year, depending on wind and current circulation. Decreased presence of MYI around Igloolik has meant that walrus are further from the community. As a mainstay for Igloolik, this makes it difficult for local hunters to continue practicing their moving ice hunting techniques and to successfully reach and catch walrus.

Considerations for Assessments of Change

An assessment of the influence of sea ice change on northern communities must consider the nature and degree of the environmental changes and local implications, simultaneously with social and lifestyles changes that people are experiencing. Individuals may perceive or react to – or indeed understand or observe – changes differently depending on their economic status, hunting methods, interest in news media, family background, to name a few. These affect both peoples' exposure to, and perspectives on, environmental changes and how they are assessed.

Exposure to Change

Considerations for the perception/influence of sea ice change relate to the degree to which community members participate in hunting or other traditional/cultural activities. Compared to the more traditional and mobile lifestyles prior to permanent community settlement, people are no longer traveling as much on the sea ice. Country foods are an important part of the local diet but are no longer essential for survival. Because dog teams are rarely used for travel in these communities, the requirements for hunting have greatly diminished along with the necessity of feeding dog teams. Therefore, people are not hunting (or not hunting certain animals) as frequently as they would have in the past. In addition, some people are too busy with employment to be able to use the ice on a regular basis, while others are aging and not as able to travel (and the younger generations in their family are not as interested). Therefore, ice conditions may seem different or the weather may feel warmer, simply because community members are not spending as much time outside. The sea ice used to be the "classroom" where knowledge and skills related to traveling and hunting were traditionally shared. Now, without using the ice regularly, some community members may be less aware of the environment and less adept at sea ice travel. As such, accidents may happen more frequently not only because of sea ice change, but because (i) less people are checking the ice properly; (ii) snowmobile travel lacks the helpful security instincts of dog teams; (iii) travel is faster by snowmobile so people take less time to observe the weather and their surroundings or are relying more on weather forecasts and GPS navigational technology; (iv) people are not waiting long enough before traveling on the sea ice in the fall; and (v) younger people are not as familiar with sea ice routes, traditional indicators of safety, or effective techniques for survival in harsh or unexpected conditions. Nevertheless, during the spring season sea ice travel remains popular. It is an important time when extended families travel together to go hunting or fishing. In contrast, winter ice travel is mainly undertaken by hunters for subsistence or commercial harvesting.

Long-Term Cycles

A few elders explained that they were warned by their parents and grandparents that the weather and ice would be very different when they became adults and had their own grandchildren. In Pangnirtung, Anglican ministers were said to have foretold that Cumberland Sound would not freeze over at some point in the future. Having now reached that age, the seemingly unbelievable predictions of warmer winters and shorter ice seasons are actually coming true. Because they expected this type of change, some elders are not surprised or concerned, they accept that whatever is destined to happen will happen. Moreover, some people feel that the current changes are part of a long-term (approximately 50-year) cycle and that a cooler phase will return in the future. Therefore, long-term change seems to have become part of the ever-variable arctic sea ice system, whereby Inuit continue to adapt as best they can.

These are only a few of the potential influences of social, technological, or lifestyle changes on Inuit expertise or local perception of change. Ice conditions and processes are changing physically, but the degree to which they are noticed and are affecting people is also a function of who is using the sea ice, how the ice is being used, what time of year the ice is being used, and where. Many of the observations of change described in interviews depend on the initial ice conditions and the areas used, around each community. For example, a similar reduction in floe edge extent could minimally influence sea ice travel around Igloolik, but could prevent community members in Cape Dorset from getting off the island. Another example is that while new polynyas along the shores of Cumberland Sound prevent access to key hunting grounds outside Pangnirtung, new polynyas in Hudson Strait make it easier to access wildlife near Cape Dorset despite declining ice extents. Furthermore, changes occurring in areas that are not frequently used may be dramatic (i.e., too dangerous to venture near or too far to regularly access), but these do not affect people as much as areas that are frequently used and are experiencing moderate change (see Ford et al. 2007, 2008; Laidler et al. 2009 for other examples related to differential exposure). Therefore, due to the relative nature of changes (as linked to local and regional geographies, local and regional sea ice use, and qualitative assessments of change), the utility of some standardized measures to quantitatively assess physical change is also recognized.

Conclusions

Sea ice travel is hazardous in any season, yet experienced hunters are skilled in managing the associated risks. Adapting to yearly variations in sea ice cycles and conditions is incorporated in Inuit sea ice knowledge and their respective use of the sea ice environment (Nelson 1969; Freeman 1984; Riewe 1991; Aporta 2002). Plans to hunt, travel, or camp at particular times/locations are continually altered according to conditions at the time, but these are not consciously delineated adaptations to change. They are a reflection of Inuit flexibility and skill in dealing with the variable and extreme nature of the arctic ecological system (Jolly et al. 2002; Nichols et al. 2004; Ford 2005; Ford et al. 2006). Nevertheless, such skills are being undermined by environmental uncertainty, alongside changing social and economic realities. In the past, before Inuit moved to settlements, changing ice conditions would not necessarily have restricted hunting opportunities. Families would have simply moved to other camps alongside areas of open water or solid ice conditions. Now, with a heavy reliance on fuel and equipment to undertake hunting and harvesting activities, as well as temporal restrictions for those who are part of the wage economy, the costs (economic or personal) to adapt can become prohibitive to some. Local adaptive strategies are thus a combination of traditional skills, income availability, dietary preferences, incorporation of technological advancements, community capacity, and so on. In a sense, life within northern communities could be considered an overarching limitation on adaptive capacity (i.e., in terms of limiting opportunities to acquire country food, maintain a healthy lifestyle, and engage in cultural activities). However, it can also be considered as an important means of minimizing vulnerability (i.e., community stores ensure access to food – even if it is not preferred – search and rescue operations can be launched from the community with ground, air, or water support, and community organizations and households provide a communication hub to initiate support where needed). There are many complex factors at play, so the interconnections between social, cultural, economic, political, and environmental changes need to be further investigated.

Characterizing a population's vulnerability to climatic, or related, change has become an important element of the UNFCCC, IPCC, ACIA, IPY, and other large international efforts to understand or characterize the influence of climate change on human systems (Smit et al. 2000; Ford and Smit 2004; McCarthy et al. 2005). The concept of resilience is also used in a complementary manner to characterize areas in which a community has a high adaptive capacity, specifically in the consideration of social–ecological systems (Berkes and Jolly 2002; Davidson-Hunt and Berkes 2003; Turner et al. 2003; Adger et al. 2005; Berkes 2007). Therefore, assessing community vulnerability/resilience to sea ice change becomes implicated in politics, research, economics, and environmental change at every level from local to global.

At the local scale we have much to learn from the long-term experiences and accumulated knowledge that Inuit elders and hunters have gained regarding sea ice and its complex links with wind, weather, and tidal conditions. Thus, we have sought to systematically document and communicate Inuit expertise of sea ice processes, use, and change in order to complement other studies of a physical, cultural, health, or socio-economic nature. It is true that there are many more pressing issues facing northern communities than climate change (Duerden 2004). However, the strong local interest in ISIUOP is a clear indication of how the sea ice has direct implications for multiple aspects of Inuit community health, safety, and well-being. Therefore, we hope that this initial local characterization of sea ice change, and related implications, provides a valuable foundation to help move toward effectively evaluating the vulnerability or resilience of Cape Dorset, Igloolik, and Pangnirtung to such changes and to ultimately support locally viable adaptive strategies.

Acknowledgments We are grateful for the support and collaboration of community members and local organizations in Cape Dorset, Igloolik, and Pangnirtung that made this research possible. We would also like to thank our primary funder, the Government of Canada International Polar Year Programme, for supporting this project over the past 3 years, which enabled the fruitful collaborations presented here and in other chapters. Additional financial support (for earlier and current work contributing to project results) was received from the Social Sciences and Humanities Research Council, Environment Canada, Indian and Northern Affairs Canada, Natural Resources Canada, the Society of Women Geographers, the Ocean Management Research Network, and the

Association of Canadian Universities for Northern Studies/Canada Polar Commission. Finally, we would like to thank the book and section reviewers for their constructive comments, which led to substantial improvements in this chapter.

Notes

1. Community researchers involved in ISIUOP field work include (in alphabetical order) Pootoogoo Elee (Independent Interpreter, Cape Dorset) Theo Ikummaq (Wildlife Officer, Igloolik) and Eric Joamie (Joamie Communications and Consulting, Pangnirtung). Additional interpreters involved in ISIUOP field work include Eena Alivaktuk (Independent Interpreter, Pangnirtung) and Andrew Dialla (Qaqqaq Translation and Tours, Pangnirtung). University researchers involved in ISIUOP field work include Kelly Karpala (MA student, Carleton University, Anthropology), Karen Kelley (MA student, Carleton University, Geography), and Gita Laidler (Carleton University, Geography). Private sector researchers involved in ISIUOP field work include Tom Hirose (Noetix Research Inc., Ottawa) and Mark Kapfer (Noetix Research Inc., Ottawa).
2. Nearly 17 weeks have been spent working together in Pangnirtung on eight trips (ranging from 1 to 3 weeks each) from September 2003 until June 2008. Nearly 19 weeks have been spent in Cape Dorset on eight trips (ranging from 1 to 5 weeks each) from October 2003 until March 2009. Nearly 16 weeks have been spent in Igloolik on seven trips (ranging from 1 to 6 weeks each) from February 2004 until September 2009.
3. Honoraria for interview or workshop contributions are common practice in Nunavut and were provided in accordance with Nunavut Research Institute and International Polar Year recommended standards.
4. Various combinations of researchers involved in field work (see Note 1) participated in each trip, typically including one community researcher, one university researcher, and two or more local elders and hunters.
5. Suvega (2004) and Etidlouie (2005) (see Appendix 1 for full references of all interviews cited).
6. Kisa (2004), Maniapik (2004b), and Noah (2004).
7. This depiction of past "normal" ice conditions coincides with the time period of tremendous social change as Inuit families in the eastern Canadian Arctic were encouraged/coherced/forced to permanently settle into communities. The majority of elders and hunters who contributed to this project had grown up on the land or maintained strong connections to ice use through hunting and harvesting despite residential schooling. Therefore, they are using their extensive – and ongoing – experiences and observations to explain the changes occurring in recent years. However, younger or less experienced hunters do not have the same comparative experience and technological changes may have also influenced the ways in which they observe and use the sea ice environment, so differential exposure to changes may impact results if a similar project were to be undertaken with a younger population or with future generations of hunters (see Exposure to Change).
8. Mikigak (2004b), Petaulassie (2004a), Peter (2004), Saila (2004), Shaa (2008), and Takiasuk (2008).
9. Ishulutak (2004b, 2004c), Maniapik (2004b), Nuvaqiq (2004), Mike (2005), Papatsie (2005), Kanayuk (2008), and Qappik (2008b).
10. Ishulutak (2004b), Mike (2005), and Qappik (2008b).
11. Anonymous (2004), Kisa (2004), Noah (2004), Qappik (2004b), and Kakkee (2008).
12. Irngaut (2004), Kunuk (2004), Ivalu (2005), Paniaq (2008), and Ulayuruluk (2008).
13. Kelly (2004), Kisa (2004), Kunuk (2004), and Ivalu (2005).
14. PG WKSP (2007).
15. Anonymous (2008b), Kanayuk (2008), Qappik (2008a, 2008b), and Ragee (2008).
16. Anonymous (2008a) and Qappik (2008b).

17. Ipkanak (2004), Kelly (2004), Nuna (2004), Ottokie (2004), Paniaq (2004), Petaulassie (2004b), Qattalik (2004), Taqqaugak (2004), Uttak (2004), Saila (2008), and Shaa (2008).
18. Qattalik (2004), Ezekiel (2005), and Takiasuk (2008).
19. Aqiaruq (2004), Arnatsiaq (2004), Kunuk (2004), Qattalik (2004), Qrunnut (2004), Taqqaugak (2004), Ulayuruluk (2004), Palluq (2005), Paniaq (2008), and Qaumaq (2008).
20. Nashalik (2004), Soudluapik (2004), Nuvaqiq (2005), and PG WKSP (2007).
21. Kelly (2004), Nashalik (2004), Nuvaqiq (2004), Nunguisuituq (2004), Parr (2004), Petaulassie (2004b), Qappik (2004b): Ezekiel (2005), PG WKSP (2007), Anonymous (2008a), and Kakkee (2008).
22. Angutikjuaq (2004), Aqiaruq (2004), Evic (2004), Ikummaq (2004), Ipkanak (2004), Joanasie (2004), Kelly (2004), Kisa (2004), Maniapik (2004b), Mike (2004), Mikigak (2004a), Nunguisuituq (2004), Nuvaqiq (2004), Parr (2004), Peter (2004), Qappik (2004a), Qattalik (2004), Qrunnut (2004), Saila (2004), Soudluapik (2004), Suvega (2004), Taqqaugak (2004), Vevee (2004), Manning (2005), Palluq (2005), CD WKSP (2007), Alasuaq (2008), Anonymous (2008a), Kanayuk (2008), Kunnuk (2008), Paniaq (2008), Qaumaq (2008), Ragee (2008), Shaa (2008), and Takiasuk (2008).
23. Alivaktuk (2004), Aqiaruq (2004), Evic (2004), Irngaut (2004), Ishulutak (2004a, 2004b), Keyuajuk (2004), Kisa (2004), Noah (2004), Nuvaqiq (2004), Ottokie (2004), Petaulassie (2004a), Qappik (2004a, 2004b), Qattalik (2004), Vevee (2004), Evic (2005), Ezekiel (2005), Nuvaqiq (2005), Anonymous (2008a, 2008b), Ashoona (2008), Ezekiel (2008), Ikummaq (2008), Kanayuk (2008), Kakkee (2008), Mikigak (2008), Pootoogook (2008), Ragee (2008), Takiasuk (2008), Ulayuruluk (2008), and Yank (2008).
24. Qappik (2004b), Qattalik (2004), Palluq (2005), IG WKSP (2007), PG WKSP (2007), Anonymous (2008b), Kakkee (2008), and Yank (2008).
25. CD WKSP (2007), IG WKSP (2007), Anonymous (2008a), and Oshutsiaq (2008a).
26. Maniapik (2004a, 2004b), Qappik (2004a), Qattalik (2004), and Palluq (2005).
27. Alasuaq (2004), Nuna (2004), Petaulassie (2004b), Peter (2004), Pootoogook (2004), Ezekiel (2005), Manning (2005), Manumee (2008), Mikigak (2008), Oshutsiaq (2008a, 2008b), Pootoogook (2008), Saila (2008), and Tukiki (2008).
28. Ishulutak (2004a, 2004b), Keyuajuk (2004), Mike (2004), Nuvaqiq (2004), Qappik (2004b), Evic (2005), PG WKSP (2007), and Kakkee (2008).
29. Qrunnut (2004), Palluq (2005), IG WKSP (2007), Ammaq (2008), Amarualik (2008), Qaumaq (2008), and Ulayuruluk (2008).
30. Nuna (2004) and PG WKSP (2007).
31. Ikummaq (2004), Uttak (2004), and Ivalu (2005).
32. Ulayuruluk (2004).
33. Qaunaq (2005).
34. Ivalu (2005) and Qaunaq (2005).
35. IG WKSP (2007).
36. Mike (2004, 2005) and Nuvaqiq (2005), PG WKSP (2007).
37. While a combination of environmental and human-induced factors are contributing to changes in harvest timing, success, as well as wildlife health and behavior, only those aspects directly related to a changing sea ice environment are discussed here.

Appendix 1: Interview References (Contributors to "Mapping Inuit Sea Ice Knowledge and Use," ISIUOP Sub-project #1)

Alasuaq, A. 2004. Interview in Cape Dorset. Interviewed by Gita Laidler. Translated by Pootoogoo Elee. April 27.

Alasuaq, A. 2008. Interview in Cape Dorset. Interviewed by Karen Kelley. Translated by Pootoogoo Elee. May 28.

Alivaktuk, J. 2004. Interview in Pangnirtung. Interviewed by Gita Laidler. December 9.
Ammaq, S. 2004. Interview in Igloolik. Interviewed by Gita Laidler. Translated by Theo Ikummaq. November 9.
Ammaq, M. 2008. Interview in Igloolik. Interviewed by Kelly Karpala. June 2.
Amarualik, J. 2008. Interview in Igloolik. Interviewed by Kelly Karpala Translated by Theo Ikummaq. June 3.
Angutikjuaq, D. 2004. Interview in Igloolik. Interviewed by Gita Laidler. Translated by Theo Ikummaq. November 2.
Anonymous. 2004. Interview in Pangnirtung. Interviewed by Gita Laidler. Translated by Andrew Dialla. December 4.
Anonymous. 2008a. Interview in Pangnirtung. Interviewed by Gita Laidler. June 16.
Anonymous. 2008b. Interview in Pangnirtung. Interviewed by Gita Laidler. June 12.
Aqiaruq, Z. 2004. Interview in Igloolik. Interviewed by Gita Laidler. Translated by Theo Ikummaq. November 5.
Arnatsiaq, M. 2004. Interview in Igloolik. Interviewed by Gita Laidler. Translated by Theo Ikummaq. November 4.
Ashoona, S. 2008. Interview in Cape Dorset. Interviewed by Karen Kelley. Translated by Pootoogoo Elee. May 25.
CD WKSP. 2007. Workshop in Cape Dorset. Participants: Atsiaq Alasuaq, Lucassie Aningmiuq, Etulu Etidlouie, Matthew Saveakjuk Jaw, Serge Lampron. Facilitated by Gita Laidler and Mark Kapfer. Translated by Pootoogoo Elee. November 1.
Etidlouie, E. 2005. Interview in Cape Dorset. Interviewed by Gita Laidler. Translated by Pootoogoo Elee. January 15.
Etungat, P. 2008. Interview in Cape Dorset. Interviewed by Karen Kelly. Translated by Pootoogoo Elee. June 1.
Evic, M. 2004. Interview in Pangnirtung. Interviewed by Gita Laidler. Translated by Andrew Dialla. December 7.
Evic, L. 2005. Interview in Pangnirtung. Interviewed by Gita Laidler. Translated by Andrew Dialla. February 10.
Ezekiel, A. 2005. Interview in Cape Dorset. Interviewed by Gita Laidler. Translated by Pootoogoo Elee. January 18.
Ezekiel, A. 2008. Interview in Cape Dorset. Interviewed by Karen Kelley. Translated by Pootoogoo Elee. May 22.
IG WKSP. 2007. Workshop in Igloolik. Participants: Levi Qaunaq, David Irngaut, Sidonie Ungalaq, Mike Immaroituk. Facilitated by Gita Laidler and Tom Hirose. Translated by Theo Ikummaq. November 15.
Ikummaq, T. 2004. Interview in Igloolik. Interviewed by Gita Laidler. November 1.
Ikummaq, T. 2008. Interview in Igloolik. Interviewed by Kelly Karpala. May 27.
Ipkanak, E. 2004. Interview in Igloolik. Interviewed by Gita Laidler. Translated by Theo Ikummaq. November 9.
Irngaut, D. 2004. Interview in Igloolik. Interviewed by Gita Laidler. Translated by Theo Ikummaq. November 12.
Ishulutak, L. 2004a. Interview in Pangnirtung. Interviewed by Gita Laidler. Translated by Eric Joamie. May 6.
Ishulutak, L. 2004b. Interview in Pangnirtung. Interviewed by Gita Laidler. Translated by Eric Joamie. May 7.

Ishulutak, J. 2004c. Interview in Pangnirtung. Interviewed by Gita Laidler. Translated by Andrew Dialla. December 8.
Ivalu, A. 2005. Interview in Igloolik. Interviewed by Gita Laidler. Translated by Theo Ikummaq. June 10.
Joanasie, M. 2004. Interview in Cape Dorset. Interviewed by Gita Laidler. November 20.
Kakkee, L.-M. 2008. Interview in Pangnirtung. Interviewed by Gita Laidler. Translated by Eena Alivaktuk. June 6.
Kanayuk, P. 2008. Interview in Pangnirtung. Interviewed by Gita Laidler. June 12.
Kelly, S. 2004. Interview in Cape Dorset. Interviewed by Gita Laidler. November 29.
Keyuajuk, M. 2004. Interview in Pangnirtung. Interviewed by Gita Laidler. Translated by Andrew Dialla. December 16.
Kisa, M. 2004. Interview in Pangnirtung. Interviewed by Gita Laidler. Translated by Andrew Dialla. December 6.
Kunuk, E. 2004. Interview in Igloolik. Interviewed by Gita Laidler. Translated by Theo Ikummaq. November 10.
Kunnuk, J. 2008. Interview in Igloolik. Interviewed by Kelly Karpala. May 30.
Maniapik, J. 2004a. Interview in Pangnirtung. Interviewed by Gita Laidler. Translated by Eric Joamie. May 13.
Maniapik, M. 2004b. Interview in Pangnirtung. Interviewed by Gita Laidler. Translated by Andrew Dialla. December 14.
Manning, J. 2005. Interview in Cape Dorset. Interviewed by Gita Laidler. January 19.
Manumee, T. 2008. Interview in Cape Dorset. Interviewed by Karen Kelley. Translated by Pootoogoo Elee. May 31.
Mike, J. 2004. Interview in Pangnirtung. Interviewed by Gita Laidler. Translated by Eric Joamie. May 11.
Mike, J. 2005. Interview in Pangnirtung. Interviewed by Gita Laidler. Translated by Andrew Dialla. February 7.
Mikigak, O. 2004a. Interview in Cape Dorset. Interviewed by Gita Laidler. Translated by Pootoogoo Elee. November 26.
Mikigak, O. 2004b. Interview in Cape Dorset. Interviewed by Gita Laidler. Translated by Pootoogoo Elee. November 27.
Mikigak, O. 2008. Interview in Cape Dorset. Interviewed by Karen Kelley. Translated by Pootoogoo Elee. May 24.
Nashalik, E. 2004. Interview in Pangnirtung. Interviewed by Gita Laidler. Translated by Andrew Dialla. May 5.
Noah, M. 2004. Interview in Pangnirtung. Interviewed by Gita Laidler. Translated by Andrew Dialla. December 15.
Nowdlak, J. 2005. Interview in Pangnirtung. Interviewed by Gita Laidler. February 17.
Nowyook, L. 2004. Interview in Pangnirtung. Interviewed by Gita Laidler. Translated by Andrew Dialla. December 13.
Nuna, A. 2004. Interview in Cape Dorset. Interviewed by Gita Laidler. April 15.
Nunguisuituq, I. 2004. Interview in Cape Dorset. Interviewed by Gita Laidler. Translated by Pootoogoo Elee. April 19.
Nuvaqiq, M. 2004. Interview in Pangnirtung. Interviewed by Gita Laidler. Translated by Eric Joamie. May 6.

Nuvaqiq, M. 2005. Interview in Pangnirtung. Interviewed by Gita Laidler. Translated by Andrew Dialla. February 8.
Oshutsiaq, O. 2008a. Interview in Cape Dorset. Interviewed by Karen Kelley. Translated by Pootoogoo Elee. May 31.
Oshutsiaq, S. 2008b. Interview in Cape Dorset. Interviewed by Karen Kelley. Translated by Pootoogoo Elee. May 31.
Ottokie, O. 2004. Interview in Cape Dorset. Interviewed by Gita Laidler. Translated by Pootoogoo Elee. April 27.
Palluq, J. 2005. Interview in Igloolik. Interviewed by Gita Laidler. Translated by Theo Ikummaq. June 13.
Paniaq, H. 2004. Interview in Igloolik. Interviewed by Gita Laidler. Translated by Theo Ikummaq. November 2.
Paniaq, H. 2008. Interview in Igloolik. Interviewed by Kelly Karpala. Translated by Theo Ikummaq. June 9.
Papatsie, J. 2005. Interview in Pangnirtung. Interviewed by Gita Laidler. February 9.
Parr, A. 2004. Interview in Cape Dorset. Interviewed by Gita Laidler. Translated by Pootoogoo Elee. November 30.
Petaulassie, E. 2004a. Interview in Cape Dorset. Interviewed by Gita Laidler. Translated by Pootoogoo Elee. November 25.
Petaulassie, Q. 2004b. Interview in Cape Dorset. Interviewed by Gita Laidler. Translated by Pootoogoo Elee. November 29.
Peter, N. 2004. Interview in Cape Dorset. Interviewed by Gita Laidler. Translated by Pootoogoo Elee. November 22.
PG WKSP. 2007. Workshop in Pangnirtung. Participants: Jooeelee Papatsie, Jamesie Mike, Peterosie Qappik, Lena Angnako, Sakiasie Sowdlooapik, Sheena Machmer. Facilitated by Gita Laidler and Mark Kapfer. Translated by Eric Joamie. November 8.
Pootoogook, P. 2004. Interview in Cape Dorset. Interviewed by Gita Laidler. Translated by Pootoogoo Elee. April 23.
Pootoogook, K. 2008. Interview in Cape Dorset. Interviewed by Karen Kelley. Translated by Pootoogoo Elee. June 9.
Qamaniq, N. 2004. Interview in Igloolik. Interviewed by Gita Laidler. Translated by Theo Ikummaq. November 4.
Qappik, J. 2004a. Interview in Pangnirtung. Interviewed by Gita Laidler. Translated by Eric Joamie. May 17.
Qappik, P. 2004b. Interview in Pangnirtung. Interviewed by Gita Laidler. Translated by Andrew Dialla. December 8.
Qappik, M. 2008a. Interview in Pangnirtung. Interviewed by Gita Laidler. Translated by Eena Alivaktuk. June 20.
Qappik, Z. 2008b. Interview in Pangnirtung. Interviewed by Gita Laidler. Translated by Eena Alivaktuk. June 4.
Qattalik, D. 2004. Interview in Igloolik. Interviewed by Gita Laidler. Translated by Theo Ikummaq. November 11.
Qaumaq, L. 2008. Interview in Igloolik. Interviewed by Kelly Karpala. Translated by Theo Ikummaq. June 5.

Qaunaq, L. 2005. Interview in Igloolik. Interviewed by Gita Laidler. Translated by Theo Ikummaq. June 9.

Qrunnut, A. 2004. Interview in Igloolik. Interviewed by Gita Laidler. Translated by Theo Ikummaq. November 1.

Ragee, K. 2008 Interview in Cape Dorset. Interviewed by Karen Kelley. Translated by Pootoogoo Elee. May 30.

Saila, M. 2004. Interview in Cape Dorset. Interviewed by Gita Laidler. Translated by Pootoogoo Elee. November 26.

Saila, P. 2008. Interview in Cape Dorset. Interviewed by Karen Kelley. Translated by Pootoogoo Elee. May 26.

Shaa, A. 2008. Interview in Cape Dorset. Interviewed by Karen Kelley. Translated by Pootoogoo Elee. May 23.

Solomonie, K. 2004 Interview in Cape Dorset. Interviewed by Gita Laidler. Translated by Pootoogoo Elee. April 28.

Soudluapik, J. 2004. Interview in Pangnirtung. Interviewed by Gita Laidler. December 10.

Suvega, S. 2004. Interview in Cape Dorset. Interviewed by Gita Laidler. April 28.

Takiasuk, I. 2008. Interview in Cape Dorset. Interviewed by Karen Kelley. Translated by Pootoogoo Elee. May 30.

Tapaungai, Q. 2004. Interview in Cape Dorset. Interviewed by Gita Laidler. Translated by Pootoogoo Elee. November 29.

Taqqaugak, A. 2004. Interview in Igloolik. Interviewed by Gita Laidler. Translated by Theo Ikummaq. November 5.

Tukiki, Q. 2008. Interview in Cape Dorset. Interviewed by Karen Kelley. Translated by Pootoogoo Elee. May 31.

Ulayuruluk, A. 2004. Interview in Igloolik. Interviewed by Gita Laidler. Translated by Theo Ikummaq. November 2.

Ulayuruluk, A. 2008. Interview in Igloolik. Interviewed by Kelly Karpala. Translated by Theo Ikummaq. May 27.

Uttak, L. 2004. Interview in Igloolik. Interviewed by Gita Laidler. Translated by Theo Ikummaq. November 1.

Vevee, P. 2004. Interview in Pangnirtung. Interviewed by Gita Laidler. Translated by Andrew Dialla. December 13.

Yank, M. 2008. Interview in Pangnirtung. Interviewed by Gita Laidler. June 12.

References

ACUNS. 2003. *Ethical Principles for the Conduct of Research in the North*. Ottawa: Association of Canadian Universities for Northern Studies (ACUNS).

Adger, N.W., Hughes, T.P., Folke, C., Carpenter, S., and Rockstrom, J. 2005. Social-ecological resilience to coastal disasters. *Science* 309: 1036–1039.

Aporta, C. 2002. Life on the ice: Understanding the codes of a changing environment. *Polar Record* 38: 341–354.

Ashford, G. and Castleden, J. 2001. *Inuit Observations on Climate Change: Final Report*. Winnipeg: International Institute for Sustainable Development.

Berkes, F. 2002. Epilogue: Making sense of arctic environmental change? *The Earth Is Faster Now: Indigenous Observations of Arctic Environmental Change*. I. Krupnik and D. Jolly (eds.), Fairbanks: Arctic Research Consortium of the United States in cooperation with the Arctic Studies Center, Smithsonian Institution.

Berkes, F. 2007. Understanding uncertainty and reducing vulnerability: Lessons from resilience thinking. *Natural Hazards* 41: 283–295.

Berkes, F. and Jolly, D. (2002). Adapting to climate change: Social-ecological resilience in a Canadian Western Arctic Community. *Conservation Ecology* 5: 18p [online].

Bintanja, R. and Oerlemans, J. 1995. The influence of the albedo-temperature feed-back on climate sensitivity. *Annals of Glaciology* 21: 353–360.

Cruikshank, J. 2001. Glaciers and climate change: Perspectives from oral tradition. *Arctic* 54: 372–393.

Curry, J.A., Schramm, J.L., and Ebert, E.E. 1995. Sea ice-albedo climate feedback mechanism. *Journal of Climate* 8: 240–247.

Davidson-Hunt, I. and Berkes, F. (2003). Learning as you journey: Anishinaabe perception of social-ecological environments and adaptive learning. *Conservation Ecology* 8: 21p [online].

Duerden, F. 2004. Translating climate change impacts at the community level. *Arctic* 57: 203–212.

Eicken, H., Lovecraft, A.L., and Druckenmiller, M.L. 2009. sea ice system services: A framework to help identify and meet information needs relevant for Arctic observing networks. *Arctic* 62: 119–136.

Fenge, T. 2001. The Inuit and climate change. *Isuma* 2: 79–85, Winter.

Ford, N. 2000. Communicating climate change from the perspective of local people: A case study from Arctic Canada. *The Journal of Development Communication* 11: 92–108.

Ford, J. 2005. Living with climate change in the Arctic. *World Watch* 18: 18–21.

Ford, J.D. and Smit, B. 2004. A framework for assessing the vulnerability of communities in the Canadian Arctic to risks associated with climate change. *Arctic* 57: 389–400.

Ford, J.D., Smit, B., Wandel, J., and MacDonald, J. 2006. Vulnerability to climate change in Igloolik, Nunavut: What we can learn from the past and present. *Polar Record* 42: 127–138.

Ford, J., Pearce, T., Smit, B., Wandel, J., Allurut, M., Shappa, K., Ittusujurat, H., and Qrunnut, K. 2007. Reducing vulnerability to climate change in the Arctic: The case of Nunavut, Canada. *Arctic* 60: 150–166.

Ford, J.D., Smit, B., Wandel, J., Allurut, M., Shappa, K., Ittusarjuat, H., and Qrunnut, K. 2008. Climate change in the Arctic: Current and future vulnerability in two Inuit communities in Canada. *The Geographical Journal* 174: 45–62.

Ford, J.D., Gough, W.A., Laidler, G.J., MacDonald, J., Irngaut, C., and Qrunnut, K. 2009. Sea ice, climate change, and community vulnerability in northern Foxe Basin, Canada. *Climate Research* 38: 137–154.

Fox, S. 2002. These are things that are really happening: Inuit perspectives on the evidence and impacts of climate change in Nunavut. In *The Earth Is Faster Now: Indigenous Observations of Arctic Environmental Change*. I. Krupnik and D. Jolly (eds.), Fairbanks: Arctic Research Consortium of the United States in cooperation with the Arctic Studies Center, Smithsonian Institution.

Freeman, M.M.R. 1984. *Contemporary Inuit Exploitation of the sea ice Environment. Sikumiut: "The People Who Use the Sea Ice"*. Montreal: Canadian Arctic Resources Committee, pp. 73–96.

Furgal, C., Fletcher, C., and Dickson, C. 2006. *Ways of Knowing and Understanding: Towards the Convergence of Traditional and Scientific Knowledge of Climate Change in the Canadian North*. Toronto: Environment Canada.

Gearheard, S., Matumeak, W., Angutikjuaq, I., Maslanik, J., Huntington, H.P., Levitt, J., Kagak, D.M., Tigullaraq, G., and Barry, R.G. 2006. "It's not that simple": A collaborative comparison of sea ice environments, their uses, observed changes, and adaptations in barrow, Alaska, USA, and Clyde River, Nunavut, Canada. *AMBIO* 35: 203–211.

Gearheard, S. and Shirley, J. 2007. Challenges in community-research relationships: Learning from natural science in Nunavut. *Arctic* 60: 62–74.

George, J.C., Huntington, H.P., Brewster, K., Eicken, H., Norton, D., and Glenn, R. 2004. Observations on shorefast ice dynamics in Arctic Alaska and the responses of the Inupiat hunting community. *Arctic* 57: 363–374.

Holland, M.M. and Bitz, C.M. 2003. Polar amplification of climate change in coupled models. *Climate Dynamics* 21: 221–232.

Huntington, H.P. 2002. Preface: Human understanding and understanding humans in the Arctic System. In *The Earth Is Faster Now: Indigenous Observations of Arctic Environmental Change*. I. Krupnik and D. Jolly (eds.), Fairbanks: Arctic Research Consortium of the United States in cooperation with the Arctic Studies Center, Smithsonian Institution.

Ingram, W.J., Wilson, C.A., and Mitchell, J.F.B. 1989. Modeling climate change: An assessment of sea ice and surface albedo feedbacks. *Journal of Geophysical Research* 94: 9609–9622.

ITK. 2005. Climate change. *Inuit Tapiriit Kanatami (ITK) Environment Bulletin* 2: 1–18.

ITK and NRI. 2007. *Negotiating Research Relationships with Inuit Communities: A Guide for Researchers*. S. Nickels, J. Shirley, and G. Laidler (eds.), Ottawa and Iqaluit: Inuit Tapiriit Kanatami (ITK) and Nunavut Research Institute (NRI).

Jolly, D., Berkes, F., Castleden, J., Nichols, T., and The Community of Sachs Harbour. 2002. We can't predict the weather like we used to: Inuvialuit observations of climate change, Sachs Harbour, Western Canadian Arctic. In *The Earth Is Faster Now: Indigenous Observations of Arctic Environmental Change*. I. Krupnik and D. Jolly (eds.), Fairbanks: Arctic Research Consortium of the United States in cooperation with the Arctic Studies Center, Smithsonian Institution.

Kusugak, J.A. 2002. Foreword: Where a storm is a symphony and land and ice are one. In *The Earth Is Faster Now: Indigenous Observations of Arctic Environmental Change*. I. Krupnik and D. Jolly (eds.), Fairbanks: Arctic Research Consortium of the United States in cooperation with the Arctic Studies Center, Smithsonian Institution.

Laidler, G.J. 2006a. Inuit and scientific perspectives on the relationship between sea ice and climate change: The ideal complement? *Climatic Change* 78: 407–444.

Laidler, G.J. 2006b. Some Inuit perspectives on working with scientists. *Meridian* 1: 3–10, Spring/Summer.

Laidler, G.J. and Elee, P. 2006. Sea ice processes and change: Exposure and risk in Cape Dorset, Nunavut. In *Climate Change: Linking Traditional and Scientific Knowledge*. J. Oakes and R. Riewe (eds.), Winnipeg: University of Manitoba Aboriginal Issues Press and ArcticNet Theme, p. 3.

Laidler, G.J. (2007) *Ice, Through Inuit Eyes: Characterizing the Importance of Sea Ice Processes, Use, and Change Around Three Nunavut Communities.* Dissertation, University of Toronto.

Laidler, G.J. and Elee, P. 2008. Human geographies of sea ice: Freeze/thaw processes around Cape Dorset, Nunavut, Canada. *Polar Record* 44: 51–76.

Laidler, G.J. and Ikummaq, T. 2008. Human geographies of sea ice: Freeze/thaw processes around Igloolik, Nunavut, Canada. *Polar Record* 44: 127–153.

Laidler, G.J., Dialla, A., and Joamie, E. 2008. Human geographies of sea ice: Freeze/thaw processes around Pangnirtung, Nunavut, Canada. *Polar Record* 44: 335–361.

Laidler, G.J., Ford, J.D., Gough, W.A., Ikummaq, T., Gagnon, A.S., Kowal, S., Qrunnut, K., and Irngaut, C. 2009. Travelling and hunting in a changing Arctic: Assessing Inuit vulnerability to sea ice change in Igloolik, Nunavut. *Climatic Change* 94: 363–397.

Ledley, T.S. 1988. A coupled energy balance climate-sea ice model: Impact of sea ice and leads on climate. *Journal of Geophysical Research* 93: 15919–15932.

Lemke, F., Harder, M., and Hilmer, M. 2000. The response of Arctic Sea ice to global change. *Climatic Change* 46: 277–287.

Lohmann, G. and Gerdes, R. 1998. Sea ice effects on the sensitivity of the thermohaline circulation. *Journal of Climate* 11: 2789–2803.

McCarthy, J.J., Martello, M.L., Corell, R., Selin, N.E., Fox, S., Hovelsrud-Broda, G., Mathiesen, S.D., Polsky, C., Selin, H., and Tyler, N.J.C. 2005. *Chapter 17 – Climate Change in the Context of Multiple Stressors and Resilience. Arctic Climate Impact Assessment*. New York: Cambridge University Press.

Nelson, R.K. 1969. *Hunters of the Northern Ice*. Chicago: The University of Chicago Press.

Nichols, T., Berkes, F., Jolly, D., Snow, N.B., and The Community of Sachs Harbour 2004. Climate change and sea ice: Local observations from the Canadian Western Arctic. *Arctic* 57: 68–79.

Nickels, S., Furgal, C., Buell, M., and Moquin, H. 2006. *Unikkaaqatigiit – Putting the Human Face on Climate Change: Perspectives from Inuit in Canada*. Ottawa: Inuit Tapiriit Kanatami, Nasivvik Centre for Inuit Health and Changing Environments at Universite Laval, and the Ajunnginiq Centre at the National Aboriginal Health Organization.

Norton, D. 2002. Coastal sea ice watch: Private confessions of a convert to indigenous knowledge. In *The Earth Is Faster Now: Indigenous Observations of Arctic Environmental Change*. I. Krupnik and D. Jolly (eds.), Fairbanks: Arctic Research Consortium of the United States in cooperation with the Arctic Studies Center, Smithsonian Institution.

NTI. 2005. *What If the Winter Doesn't Come? Inuit Perspectives on Climate Change Adaptation Challenges in Nunavut (Summary Workshop Report)*. Iqaluit: Nunavut Tunngavik Incorporated (NTI).

Paci, C., Hodgkins, A., Katz, S., Braden, J., Bravo, M., Gal, R.A., Jardine, C., Nuttall, M., Erasmus, J., and Daniel, S. 2008. Northern science and research: Postsecondary perspectives in the northwest territories. *Journal of Northern Studies* 1: 23–52.

Pearce, T.D., Ford, J.D., Laidler, G.J., Smit, B., Duerden, F., Allarut, M., Andrachuk, M., Baryluk, S., Dialla, A., Elee, P., Goose, A., Ikummaq, T., Joamie, E., Kataoyak, F., Loring, E., Meakin, S., Nickels, S., Shappa, K., Shirley, J., and Wandel, J. 2009. Community collaboration and climate change research in the Canadian Arctic. *Polar Research* 28: 10–27.

Riedlinger, D. and Berkes, F. 2001. Contributions of traditional knowledge to understanding climate change in the Canadian Arctic. *Polar Record* 37(203): 315–328.

Riewe, R. 1991. Inuit use of the sea ice. *Arctic and Alpine Research* 23(1): 3–10.

Smit, B., Burton, I., Klein, R.J.T., and Wandel, J. 2000. An anatomy of adaptation to climate change and variability. *Climatic Change* 45: 223–251.

Symon, C., Arris, L., and Heal, B. 2005. *Arctic Climate Impact Assessment*. Cambridge: Cambridge University Press.

Turner, N.J., Davidson-Hunt, I., and O'Flaherty, M. 2003. Living on the edge: Ecological and cultural edges as sources of diversity for social-ecological resilience. *Human Ecology* 31: 439–461.

Chapter 4
"It's Cold, but Not Cold Enough": Observing Ice and Climate Change in Gambell, Alaska, in IPY 2007–2008 and Beyond

Igor Krupnik, Leonard Apangalook Sr., and Paul Apangalook

Abstract The chapter discusses the main outcomes of 3 years (2006–2007, 2007–2008, 2008–2009) of systematic observation of ice and weather conditions in the community of Gambell (*Sivuqaq*) on St. Lawrence Island, Alaska. The 3-year recording of ice and weather in Gambell by local monitors was a part of a larger observation effort under the SIKU project. Observers from eight communities in Alaska and Russian Chukotka took daily notes of ice and weather around their home areas for several consecutive winters. Data from Gambell are the longest and the most comprehensive within this larger SIKU data set. Observations by local monitors reveal a very complex signal of change that often differs by season or location, even among the nearby communities. The 3-year record of ice and weather observations offers new insight to Arctic climate and ice scientists. It will also help Arctic residents document their cultural tradition, ice use, and knowledge in the time of rapid environmental and social change.

Keywords Sea ice · Local observations · Gambell · Alaska · Indigenous knowledge

Arctic climate and sea ice are changing rapidly, as attested by numerous science experts and local residents. Nonetheless, the signal of change differs in its scope and intensity across the polar regions, particularly when assessed by various sources of data, types of observation, and proxy records. Whereas both scientists and indigenous experts are in agreement about the overall direction of change (ACIA 2005), many perspectives are to be taken into account to get a full picture of the ongoing processes.

Climate scientists realized that the polar regions were heading toward a major climate warming phase since the mid–late 1990s, but the first models were contradictory and many researchers remained unconvinced (cf. Serreze 2008/2009:1–2).

I. Krupnik (✉)
Department of Anthropology, National Museum of Natural History, Smithsonian Institution, Washington, DC 20013-7012, USA
e-mail: krupniki@si.edu

I. Krupnik et al. (eds.), *SIKU: Knowing Our Ice*,

DOI 10.1007/978-90-481-8587-0_4,

It took almost a decade for the science community to reach across-the-board consensus regarding the scope of change and about the speed of Arctic warming (ACIA 2005; Grebmeier et al. 2006, 2009; Meier et al. 2007; Pachauri and Reisinger 2007; Richter-Menge 2006, 2008; Stroeve et al. 2007; Serreze et al. 2007; Walsh 2009). Arctic residents have been similarly pointing to the many signals of environmental change in their areas since the early 1990s (cf. McDonald et al. 1997). By the late 1990s, several studies in the documentation of indigenous knowledge were in place; they unequivocally confirmed that polar residents were observing a strong signal of warming across the Arctic region (Fox 2002; Helander and Mustonen 2004; Huntington et al. 2001; Jolly et al. 2002; Krupnik 2000; Krupnik and Jolly 2002; Thorpe et al. 2003). The first systematic recording of the daily ice and weather patterns by indigenous village monitors was started in winter 2000–2001 on St. Lawrence Island, Alaska (Oozeva et al. 2004; Krupnik 2002).[1] This and other pioneer efforts offer the first glance into the rigorous ice and weather watch that goes on in many Arctic indigenous communities on a daily basis.

The Arctic Climate Impact Assessment report created a new momentum in understanding the value of indigenous knowledge (Huntington and Fox 2005). It was soon expanded during the preparation for IPY 2007–2008, particularly when the documentation of indigenous observations of Arctic climate change was added to its science program (Allison et al. 2007). The task to launch systematic ice and weather observations in local communities became one of the prime goals of the SIKU project, particularly, of its Alaskan and Russian components (see Chapter 1, Introduction).

SIKU Ice Observation Program, 2006–2009

The first local observations under the SIKU science plan were started in spring 2006 in Gambell, Alaska, by Leonard Apangalook, Sr., with a small pilot grant from the Smithsonian Institution's National Museum of Natural History. In November–December 2006, local observations of ice and weather conditions began in Wales (by Winton Weayapuk, Jr., 56) and Barrow (by Joe Leavitt, 52 and the late Arnold Brower, Sr., 84) in Alaska under the National Science Foundation-funded *SIZONet* project and in Uelen, Chukotka, Russia (by Roman Armaergen, 69 – see Krupnik and Bogoslovskaya 2008). In spring 2007, the daily ice and weather recording began in Shaktoolik, Alaska (by Clara Mae Sagoonick, 55); in the fall 2007, the Russian "Beringia" Nature and Ethnographic Park offered three of its indigenous rangers as local monitors in the communities of New Chaplino (Alexander Borovik, 52), Yanrakinnot (Arthur Apalu, 42), and Sireniki (Oleg Raghtilkon, 33), plus the ongoing set of ice observations in Provideniya Bay and Emma Harbor by Igor Zagrebin (2000–2009). Altogether, monitors from nine communities in Alaska and Chukotka were engaged in observation over three winters in 2006–2009 (Fig. 4.1 and Table 4.1).[2] The SIZONet team took responsibility for the maintenance of the common database at the Geophysical Institute, University of Alaska Fairbanks.

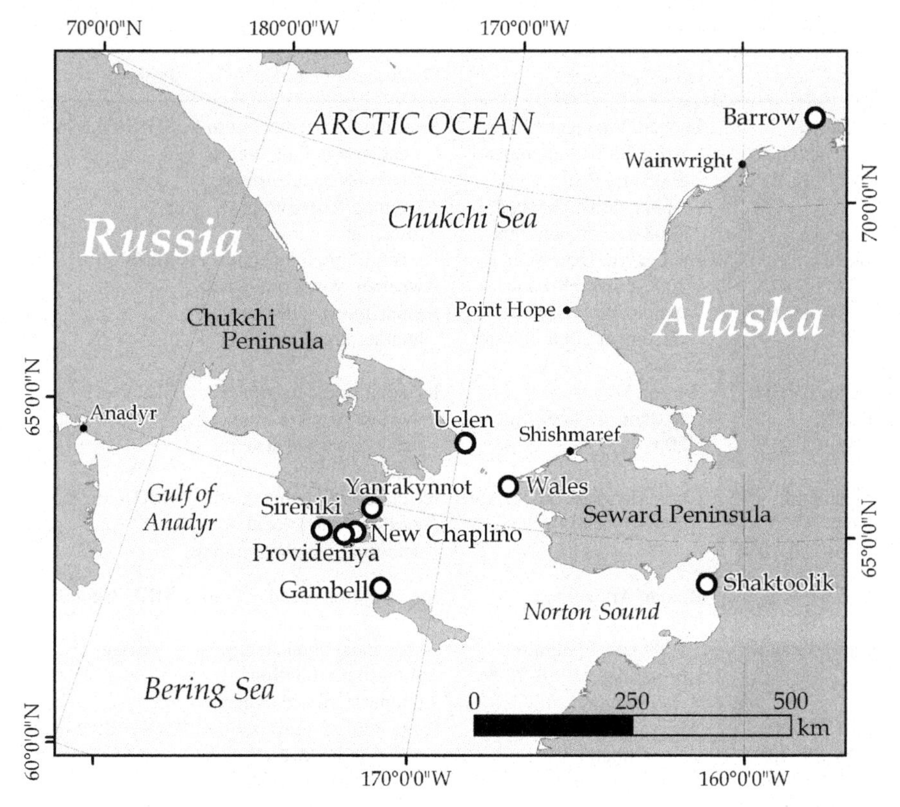

Fig. 4.1 Communities in the northern Bering Sea–southern Chukchi sea region with local ice and weather observation in 2006–2009 (produced by Matt Druckenmiller)

When starting local ice observations for the SIKU project, we deliberately avoided setting a standard template for each village monitor, except for the request to report daily temperature, wind direction, wind speed, and the ice condition at each location at least once a day. Instead, the observers were encouraged to add local details they believed were important, such as data on subsistence activities; marine mammals, birds, and terrestrial species; community events; personal travel across the observation area; and other comments. All village monitors were also asked to include local terms, place names, and key descriptions in their respective native languages, whenever possible. Chukotka observations were recorded in Russian[3]; all Alaskan village records were written in English.

Altogether, local SIKU observations in 2006–2009 produced a data set of more than 150 monthly logs from eight communities totaling several hundred pages (Table 4.1). It constitutes a unique database of its kind on local ice and weather conditions not only on the ground but also on subsistence activities, communal life, personal traveling, and the status of environmental knowledge in several communities during the IPY era. Some outcomes of this effort have been presented elsewhere[4]; this chapter introduces the longest and the most comprehensive of the SIKU observational records from Gambell, Alaska, 2006–2009.

Table 4.1 Community ice and weather observations under the SIKU project, 2006–2009

Site	Observer, year	Phenomena recorded	Program
Gambell 63° 46′N, 171° 42′W	Leonard Apangalook, Sr., April 2006–September 2008 and Paul Apangalook, October 2008–December 2009	Ice condition, temperature, weather wind, biological phenomena, subsistence hunting, local events, traveling	SIKU-Alaska
Barrow 71° 17′ 44N, 156° 45′ 59W	Joe Leavitt, December 2006–June 2009 and Arnold Brower, Sr., December 2006–Spring 2008	Ice condition, temperature, weather, wind, biological phenomena, subsistence hunting, local traveling	SIZONet
Wales 65° 36′ 44N, 168° 5′ 21W	Winton Weyapuk, Jr., November 2006–June 2009	Ice condition, temperature, weather, wind, biological phenomena, subsistence hunting, ice photography	SIZONet
Shaktoolik 64° 21′ 20N, 161° 11′ 29W	Clara Mae Sagoonick, April 2007–June 2009	Ice condition, temperature, weather, wind, local activities, ice photography	SIKU-Alaska
Uelen 66° 09′ 34N, 169° 48′ 33W	Roman Armaergen, November 2006–June 2009 and Vladimir Struzhikov, 1988–2009	Ice condition, wind direction, weather biological phenomena, subsistence hunting, temperature, ice dynamics	SIKU-Chukotka Local weather station
New Chaplino 64° 29′ 30N, 172° 54′W	Alexander Borovik, November 2007–June 2009	Ice condition, temperature, weather, wind, biological phenomena, subsistence hunting, traveling, ice photography	SIKU-Chukotka *Beringia* Park
Yanrakinnot 64° 54′N, 172° 32′W	Arthur Apalu, November 2006–June 2009	Ice condition, temperature, weather, wind, biological phenomena, subsistence hunting, traveling, ice photography	SIKU-Chukotka *Beringia* Park
Sireniki 64° 30′ 10N, 174° 08′W	Oleg Raghtilkun, Spring 2007, November 2008–June 2009	Ice condition, temperature, wind, marine mammals	SIKU-Chukotka *Beringia* Park
Provideniya 64° 25′ 24N, 173° 13′ 33W	Igor Zagrebin, 2000–2009	Ice development and dynamics, ice photography	SIKU-Chukotka Personal observations

Materials and Methods

The town of Gambell (population 649, U.S. 2000 census, currently about 700) is located at the northwestern tip of St. Lawrence Island, in the northern Bering Sea. Barely 58 km (36 miles) of water separates Gambell from nearby Chukchi Peninsula, Siberia, and the only other community on St. Lawrence Island, Savoonga (*Sivungaq*, population 710) is separated by almost the same distance. Most of the

Fig. 4.2 The Yupik village of Gambell on St. Lawrence Island, Alaska, one of the key research sites for the SIKU project. (Photo: Igor Krupnik, February 2008)

local residents are Yupik Eskimo; they call their village *Sivuqaq* and themselves *Sivuqaghmiit* when speaking in their native Yupik language.

Gambell/Sivuqaq has been permanently populated for the past 2,000 years and its rich cultural history is revealed in numerous ancient sites.[5] Today, it is a rural Alaskan community (Fig. 4.2) with mixed economy based upon subsistence marine hunting and fishing, supplemented by wage works, pensions, small development grants, ivory carving, and sale of beached and archaeological ivories to visitors and collectors. Almost all adult men in Gambell hunt in their spare time or year-round; during the spring hunting season in April–May as many as 200 men and teenage boys may be boating in the drifting ice. The islanders' knowledge of sea ice was extensively documented in recent years; Yupik hunters and elders have long voiced concerns about the shifts in local environment they observe and the growing impact of climate change upon their economy and way of living (Krupnik 2000, 2002, 2009; Krupnik and Ray 2007; Metcalf and Krupnik 2003; Metcalf and Robards 2008; Noongwook 2000; Noongwook et al. 2007; Oozeva et al. 2004; Robards 2008). This chapter introduces the first assessment of what has been recorded over more than 3 years of the SIKU observation program in Gambell.

One of the co-authors (LA, Fig. 4.3) started recording ice and weather condition and local subsistence activities in Gambell on April 1, 2006. He took daily records for 25 months, from April 2006 until June 2007 and again from September

Fig. 4.3 Leonard Apangalook, Sr., SIKU participant and local observer in the village of Gambell, St. Lawrence Island (courtesy of Leonard Apangalook)

2007 until June 2008. His monthly logs first came as handwritten entries of 30–40 words per day that were later replaced by computer-typed files. It included readings of the daily temperature, wind direction, and wind speed (taken in the morning from the outdoor thermometer) and supplemented by comments on local ice condition, hunting, and village activities. In addition, Leonard commonly checked the beach in Gambell twice a day and he traveled extensively around the area for hunting, fishing, and beach combing. Apangalook's data were supplemented with extensive interviews with Krupnik taped in Gambell in September 2007 and February 2008.

Since late September 2008 and until June 2009, another co-author (PA) (Fig. 4.4) continued the observation in Gambell.[6] Altogether, two observers contributed 33 monthly logs or more than 120 pages of written records. Electronic files of monthly observational logs were uploaded to the SIKU-SIZONet observational database at the Geophysical Institute, University of Alaska Fairbanks, and already used extensively (see Chapter 5 by Kapsch et al. this volume). Gambell 2006–2009 observational records feature tremendous amount of information beyond the sea ice and weather data, including references to marine mammal, bird, fish, and plant species; local subsistence activities, community events, and observers' personal comments. Small gaps in daily observations were covered using meteorological records from the Gambell airport automatic weather station.[7] Throughout March–June 2008 and, again, in September 2008–June 2009, the logs were checked against satellite imagery from the northern Bering Sea area (provided by Gary Hufford, Alaska Weather Service – see *Acknowledgements*). We also used historical weather records from Gambell that cover two periods, 1948–1953 and 1987–2008,

Fig. 4.4 Paul Apangalook, SIKU monitor in Gambell, 2008–2010 (courtesy of Paul Apangalook)

on file at the Geophysical Institute, UAF in Fairbanks. Using these data we selected two comparative weather sets, 1949–1952 and 1989–1992, so that 2006–2009 observations can be put into a broader historical perspective (see below).

Another set of historical weather data comes from daily journals of the first teachers and missionaries at Gambell, William Fuhrman Doty, 1898–1899 (Doty 1900), and P.H.J. Lerrigo, 1899–1900 and 1900–1901 (Lerrigo 1901, 1902). Unlike later meteorological records, they include numerous references to ice condition and subsistence hunting in 1898–1901. We used the sets of weather records during 1898–1901, 1949–1952, 1988–1992, and 2006–2009 to calculate the average monthly temperature and some other parameters for four short periods extending over 110 years. These data are organized and discussed below by three seasons, *Fall* (September–December), *Winter* (January–March), and *Spring* (April–June), or from the first signs of fall freeze-up and until the spring retreat of ice. In this first assessment of Gambell ice and weather conditions in 2006–2009, the seasonal records are analyzed along four major themes or topics: (1) *indicators* that people use to assess the seasonal dynamics; (2) *shifts*, that is, the vector of change; (3) *mismatches*, unusual or conflicting signals of change; and (4) *scope of change*, the assessment of today's condition against earlier historical data.

Fall Freeze-Up and Early Ice Formation

St. Lawrence Islanders have highly nuanced views on how the new sea ice is formed around their island each year. Their native Yupik language has more than 20 terms for many local types of young ice and freezing conditions (Oozeva et al. 2004); people also use various natural phenomena as indicators in the freeze-up process (see below). Also, hunters watch carefully for new ice and for the signs of its safety, so that they can assess its condition with great precision.

The process of the fall freeze-up and new ice buildup – or, rather, the way it *used* to happen in the "old days" – can be described as follows. Winds and currents would drive first small chunks of floating ice called *kulusiit* (in Yupik or "icebergs" in English) from the north; this would be an indicator of the approaching main polar pack ice. It usually happened in October or even in late September (Oozeva et al. 2004:133–134). Many would melt or would be washed ashore; but some would freeze into the locally formed slush (*qenu*), grease (*saagsiqu*), or frazil ice (*pugsaqa*). At this time, local slush ice (*qerngayaak*) starts forming in bays, lagoons, and creek estuaries at low temperature. Eventually, the slush ice that is washed onshore and the frozen beach spray on rocks and sand, *qayemgu* will solidify and build up to a bank (*sigugneq*). If the freezing occurs in quiet weather, *qenu* (slush ice) expands rapidly offshore and various local forms of thin new ice, *allungelquq*, *qateghrapak*, *sallgaaq*, *sallek* and, later, *salegpak*, may form quickly.[7]

Around the time as the first local ice was established (usually in early to mid-November) (Fig. 4.5), the prevailing winds would shift direction, from primarily

Fig. 4.5 "Locally built" fall ice off St. Lawrence Island. (Photo: Hiroko Ikuta, 2007)

southerly to northerly, followed by a drop in daily temperatures. Under the prevailing north–northeasterly winds, local ice and chunks of drifting ice from the north would combine to form a solid mass of ice by November, enough so that it was safe to go out on young ice for seal hunting (Noongwook 2000:22). Depending upon year-to-year variability, but usually by late November, the thick Arctic ice pack called "true ice" (*sikupik*) would arrive from the north, often merging with or smashing the young locally formed ice. Crashing, breaking, and refreezing would continue through December, until a more solid winter ice is formed to last until spring (Fig. 4.6; Oozeva et al. 2004:136–137).

Indicators. Besides major visible stages in the freeze-up and new ice formation, Gambell hunters use numerous other indicators to track the advancement of the fall and winter conditions. Leonard Apangalook produced a list of some 30 indicators ("benchmarks") that he commonly used to describe the fall/early winter condition, such as freezing of the puddles and lakes, buildup of the ice on rocks and in the tidal zone, first major snow on the ground, shifts in local subsistence activities, the arrival of particular species of marine mammals and birds (Table 4.2). Most of those benchmarks were traced for three consecutive fall seasons: 2006, 2007, and 2008. Of course, many indicators used by Gambell hunters are "proxies" and are not related to the building of the ice itself. Nonetheless, they create a set of well-defined markers to trace and to forecast the approaching winter season.

Fig. 4.6 Ice pressure ridges are built along Gambell beaches, as thin young ice solidifies offshore. (Photo: Igor Krupnik, February 2008)

Table 4.2 Fall and early winter indicators ("benchmarks") used by Gambell hunters

Indicator/"benchmark"	2006	2007	2008	Comments
1. People remove and pack summer fishing nets	September 2	September 4	September 5	This is a common sign of the end of the summer season
2. Night freezing of the ponds and puddles	September 18	September 30		
3. First snow on the Gambell mountain	September 25	October 11		
4. Foxes getting white	September 15–20	September 23		
5. Weathering of plants for winter	~September 10	September 9		
6. Certain species of birds arrive from the north (i.e., golden plover)	October 1	September 25		
7. Foam/dew in the morning		September 3, regular since September 28		This happens when the temperature falls below 45°F in the morning. You can see the lake water starting to foam along the edges
8. First freezing of the lakes	October 8	October 14	October 11	
9. Ice spray and cover on the rocks	November 8	October 24		
10. Gravel starts freezing in the tidal zone		October 15		
11. First snow on the ground; first major snowfall of the season	September 22; October 6	October 5; October 13	October 2; October 3	That first snow, even a major one usually melts quickly. In the old days, we used to have snow on the ground by October 5
12. First freezing of the land surface	October 5	October 13	October 6	Like the snow, it freezes and melts a few times before it gets frozen for the winter
13. Snow geese arrive on their main southward migration	~September 15 Mass flight on September 20–30	Not yet on October 2	September 30	
14. Fall seal hunting from shore blinds starts	~September 15	September 5; seals still sinking by September 25	September 3	This happens when the seals are fat enough to float when killed in the fall; usually last until November

Table 4.2 (continued)

Indicator/"benchmark"	2006	2007	2008	Comments
15. First sightings of walrus	September 27 (dead walrus ashore) November 11 first kill	September 29	October 22	We often see some walruses coming first, ahead of the main run. They are usually bulls, very lean; may be lame or sick. They come before the main healthy walruses get here several weeks later
16. Small icebergs floating in from the north	–	–	–	Stopped coming as of several years ago
17. First sightings of whales offshore	October 15 (gray), December 12 (bowhead)	November 20 (humpback)	Late October (bowhead)	
18. Sea lions are migrating back from Siberia	October 10, November 21 major haul out	September 27	Early November	New event
19. People collect washed out sea plants on the beach	September 22	September 9	September 15	By December, most of the sea plants lose some of the edible parts from winter setting in. We will not pick these until next fall
20. First major accumulation of snow to use snowmobile	November 30	November 21 (on the lake); December 10	October 20	People try to use four wheelers as long as possible, until they start bogging down in the gullies. When we shift to snowmobiles this is a good sign of winter snow conditions. In the old days, dog sleds were used much earlier, even with a bit of powdery snow
21. Flocks of cormorants migrating south	November 7	October 30	Signals serious freeze-up is coming	
22. First slush ice, *qenu* formation	November 20 on the southern side, December 4	November 9 in shallow bays, November 20 offshore	November 21 – west side	It normally starts on shallow water; easy to see when walking the beach

Table 4.2 (continued)

Indicator/"benchmark"	2006	2007	2008	Comments
23. Rocks start freezing over	November 8	November 11		People check it carefully, because it gets slippery. May be dangerous if you have to retreat a seal
24. Seals and birds seen on the surface of ice	December 14	December 22	December 20 (too thin for walrus to haul out)	This is a good indicator that the ice starts to solidify
25. People start walking on new ice	–	December 22(?)	–	This new ice is very dangerous; we are told to be careful when walking on it
26. Slush ice expands and solidifies quickly	December 11	December 17		
27. Major walrus migration hits the island	December 22–25; December 27 major hunting	December 17; December 26: hundreds of walrus hauled on new ice	November 25, first sighting; December 6 arrived, December 18 all people hunting	Normally we would have pack ice and hunt walrus on Thanksgiving Day, but not anymore, as the run is getting later every year
28. Bowhead whales seen/hunted	–	December 23	Late October; killed November 30, December 1	
29. Chukotka pack ice arrives	–	Yet to arrive on December 19	December 6?	
30. "Winter" ice conditions established	December 28	December 30	December 25	

> Some of these benchmarks that I am aware of as an example are: *Uksungnaqaa*, meaning (literally) "it's attempting to become winter." As Ralph Apatiki related, this time is distinctly marked by high winds of longer duration and dense blizzard periods like squalls, not comprised of snowflakes but that of fine ice crystals probably formed in vast open seas in below 0 °F temperatures. [. . .]. Older men expected the arrival of the main ice pack during this time. Winter in our custom officially arrives with the ice pack. [. . .] That westerly wind on December 24 (2008) [was] clearly the benchmark, which marked the beginning of winter despite the arrival of the ice before that time. (Paul Apangalook, January 2009).

Shifts. Since the 1980s, hunters started to observe changes to the previously common pattern that they have viewed as "normal." First, *kulusiit*, drifting chunks of the polar pack ice were late to arrive, often by a full month, and by the late 1990s, they have stopped coming altogether. The new ice is now being formed primarily out of local frozen slush or frazil ice, via its thickening, consolidation, repeated breakups, and refreezing. That usually happens in mid- or even late December, when slush ice turns into thick new ice, often about 10 in. thick, and may even pile up into onshore pressure ridges, *vuusleq* (sing.), on the wind-exposed beaches. At that point, the main pack ice ceased to arrive until January or even February; and in the last few years it did not arrive at all. Because of this new dynamics, the onset of winter ice conditions on St. Lawrence Island is now being delayed by 6–8 weeks, until late December or even early January (December 28, 30, and 25 in 2006, 2007, and 2008, respectively).[8] Even when the pack ice finally arrives from the north and gets frozen off Gambell, it is not the solid thick polar ice, *sikupik*, of the "old days" but rather a mass of broken new ice that was formed further north, in the bays and tidal zone between the Chukotkan and Alaskan coasts.

Observations in 2006–2009 help document this new pattern of the ice formation (Table 4.2). During these 3 years, no drifting icebergs (*kulusiit*) were seen arriving from the north. Although in 2007 the formation of slush ice was reported more than 2 weeks earlier than in 2006 and 2008, the slush ice was quickly broken up by a warming spell, so that on November 22, 2007, Leonard Apangalook reported "Thanksgiving day with no ice in the ocean; normally [we] would have ice and hunt walrus on Thanksgiving Day but not anymore." In the three SIKU observation years, the bulk of migrating walrus did not arrive to St. Lawrence Island until mid-December, so that the main walrus hunting in Gambell started on December 18, only.

The buildup of the new ice cover continued via expansion, rafting, and hardening of the local ice. On December 16, 2006, Apangalook reported that "when locally formed ice gets thick and encompasses the entire Bering Sea around our island, our elders used to say that we have a winter that is locally formed." On December 29, 2006, he stated "No polar pack ice yet, but locally formed ice is getting thicker and heavier. Some of the ice is getting to be almost three feet thick." The solidification of this locally formed, primarily coastal ice now marks the onset of winter condition and its movement dominates the local icescape for the duration of the winter.

Mismatches. Recent climate change has clearly impacted the detailed set of hunters' indicators inherited from the "old days." Some like the first major snow on the ground or the first freezing of the land surface or puddles are confusing, as today's weather instability commonly leads to several cycles of snowing/freezing and melting during the fall. Other former benchmarks, like the arrival of floating "icebergs"[9] (*kulusiit*), of the pieces of broken ice floes (*uughhutraghet*), or the advance of the main pack ice, *sikupik* (that ceased to arrive), have lost their observational value. Instead, new indicators emerged, like the mass arrival of the Steller sea lions prior to the main walrus migration that was unknown of in the "old days" (Table 4.2).

Hunters are also puzzled by the mix-up ("mismatch") of many events or stages in the fall–winter transition that used to be sequenced differently. A good example is the fall appearance of bowhead whales, often in advance of the migrating walrus. In the "old days," bowhead whales were rarely spotted in the fall and, and unlike in Siberia, were never hunted on St. Lawrence Island in November or December. In the past decade, the fall occurrence of, and hunting for, bowhead whales prior to the walrus migration in mid-December became quite common. In 2008, two whales were killed off Gambell and butchered on November 30 and December 1, a full week before the migrating walrus arrived from the north. Also, many earlier calendar dates (like "we used to have walrus here by Thanksgiving") are not working anymore.

Scope of change. Earlier sets of historical temperature records (1898–1901, 1948–1951, and 1987–1992) create a confusing picture, as inter-annual oscillations in the temperature regime are higher than any statistical difference among the four sets of historical data (Table 4.3). What is clear is that there is not much change

Table 4.3 Average fall–early winter air temperatures (F°) in Gambell, Alaska: 1898–2008

Year	September	October	November	December	First major cold spell (5 days below 0°F)	Start of active walrus hunting
1898						
1899	42.4	28.6	21.2	17.4	January 6	
1900	39.5	32.0	27.1	9.9	December 15	December 15
1948	38.4	29.1	19.1	11.9	December 27	...
1949	40.6	34.9	24.5	15.0	December 2[a]	...
1950/1951	40.4	35.0	29.5	20.0	January 3	...
1987	38.5	32.5	18.4[b]	5.7[c]	November 27	...
1988	39.6	29.6	22.7[d]	–	–	...
1991	39.2	29.5	20.0	7.9	December 20	...
1992	36.7	30.5	20.8	6.5	December 11	...
2006	41.1	35.2	30.0	12.4	December 27	December 22
2007/2008	46.4	32.7	27.6	18.9	January 8	December 27
2008/2009	39.5	29.9	21.8	17.7	January 3	December 18

[a]After a short cold spell between December 2 and December 8, the temperature did not drop again below 0 until March 18.
[b]26 days only.
[c]29 days only.
[d]First 13 days of the month only.

observed either in the average monthly air temperature for the fall season or in the timing of the early fall phenomena leading to the eventual freeze-up, like the freezing of the land, first major snow, the formation of slush ice. The difference is remarkable, though, in the timing of the events preceding the main ice formation and, particularly, with regard to the arrival (or lack) of the polar pack ice. In 2006 and 2007, the temperature dropped below freezing on the first week of December, 2–4 weeks later than a century ago. Even though conditions in Gambell were significantly colder in 2008, it was hardly comparable to the marked onset of winter, as described on December 5, 1894:

> A thick *poorga*, or snowstorm has whirled down upon us the day before from the north; a foot of snow had fallen, and great drifts nearly buried the village and blocked up the windows of the schoolhouse. Ice floes, packing against the coast, pressed great masses ashore in hummocks twenty or thirty feet in height. The entire sea, across to Siberia, was covered with ice (Gambell 1910:8).

The only explanation is that the ice formation is delayed by the warming of the areas north of Gambell and by the significant delay in the Arctic pack ice formation in the Chukchi Sea. Therefore, the fall temperature in Gambell may be not that different from the decades past but the ice formation follows a very different pattern. Citing elders' words, they now have ice and winters "that are locally born."

Winter Ice and Weather Regime

As the ice solidifies in early winter, the advantage of Gambell as the prime subsistence-hunting site becomes quite obvious. Thanks to the village location at the protruding northwestern tip of St. Lawrence Island, Gambell hunters have access to three types of ice-covered marine habitats, each with its specific type of ice formations. The western beach (*Ughqaa*, literally, "south side," "lee side") is a dynamic habitat dominated by the drifting ice that moves back and of winds and currents (Fig. 4.7). During the wintertime, hunters constantly watch for walrus and seals on the drifting floes, ready to launch their boats, when an opening between the shore and pack ice allows them to get closer to the game. Alternatively, hunters can go to the northern beach (*Aywaa*) that has much heavier pack ice and shore-fast ice, with higher and more solid pressure ridges (Fig. 4.8). The solid ice off the northern beach allows access to the game for hunters moving on foot; it used to be the most common way of winter hunting in Gambell up to the recent decades (Hughes 1960; Oozeva et al. 2004). Lastly, a short 5 km (3 miles) snowmobile journey to the east offers access to the solid shore-fast ice that normally forms along the northern shore of St. Lawrence Island. From the camp called *Qayilleq* (*Qayillghet*) hunters venture onto the ice seeking for seals and walrus in the open offshore leads.

The area off *Qayilleq* is also known to be a wintering habitat for a local population of bull walrus called *ayughayaak*. In the "old days" people used to hunt the "winter" walrus off the northern side and at *Qayilleq* by walking on foot to the dense pack ice and by scouting for walrus breathing holes where the animals may

Fig. 4.7 Drifting pack ice with patches of open water off the western beach in Gambell. (Photo: Igor Krupnik, February 2008)

be harpooned and killed with lances (Hughes 1960:103–104). This does not happen anymore, as both the winter ice and the walrus distribution around the island have changed significantly over the past few years (see below).

Indicators. Gambell hunters believe that the "genuine winter" in their area starts when the solid thick ice (*sikupik*) develops off the northern shore. That now commonly happens in late December or early January, at least 2–3 weeks later than in the decades past (Table 4.2; Oozeva et al. 2004:185). The first indicator of the transition to the spring conditions is the formation of the leads off the western and northern shore. That normally takes place in early to mid-March. Other than that, there are no particular stages or phases within the winter season identified by hunters. It normally progresses as a sequence of cold days dominated by northerly winds and interrupted by periods of often-violent snowstorms and/or by brief warm spells; the latter are usually brought by strong southerly winds. Strong wind-driven storms pushed fields of drifting pack ice back and forth and close or open the ice, particularly off the western beach. It may even destroy the beach pressure ridges or break off the strap of shore-fast ice off the northern shore. Then the wind subsides, the northerly current prevails again, and the temperature goes down. The ice solidifies quickly and the cycle repeats itself until the next storm. It is the frequency of the storms (see below) and the quality of the ice that have changed in recent years and transformed the dynamics of the winter season, as hunters, old and young, see it.

Fig. 4.8 Two Gambell hunters are looking for seals from ice pressure ridges. The ridges look high and solid, but they can be quickly destroyed under the impact of strong wind or storms. (Photo: Igor Krupnik, February 2008)

Shifts. The first reported major change is the later formation of solid ice, which is now delayed until very late December or even early January. As a result, the period of active walrus hunting in the dense floating ice at the end of the fall walrus migration often extends into the first week of January. That happened in all 3 years of observation in January 2007, 2008, and 2009. As the solid dense pack ice finally closes in (on January 3, 2007; January 17, 2008; and January 11, 2009, respectively), the winter finally prevails:

> Not much open water despite the wind direction. Rough broken ice with lots of pressure ridges (is) built up all around the west and north beach. Ralph Apatiki referred to this as the pack with lots of seals (January 17, 2009, Paul Apangalook).

Nonetheless, when the ice finally thickens and pressure ridges and shore-fast ice develop along the beach, the hunters report that it has thickness nowhere close to what was common in the past. That thin ice (about 30–50 cm) can be easily broken by storm or by strong wind and even be moved around by waves or currents. At least four episodes of complete ice disintegration off the western beach in Gambell were reported from December 2006 until March 2007. On January 10, 2007, LA wrote in his log: "It is unusual to have swells and to lose our ice in January compared to normal years in the past. Locally formed ice that covered our area to 9/10 easily disappears with rising temperatures and storm generated swells." And then on January 31, 2007, he noted the following:

> What a twist we have in our weather situation at the end of January! Wind driven waves cleared away pressure ridges on west side with open water west of the Island. Part of the shore-fast ice broke away on the north side beach also from the swells. [. . .] Much of the snow melted, especially along the top of the beach where we now have bare gravel. Undeniably, the climate change has accelerated over the past five years where severity of winds and erratic temperatures occur more frequently every year. Predictability of our game animals of the sea is inconsistent and erratic compared [to] how it used to be back in the normal seasons. We the hunters, along with the marine mammals we hunt, are truly at the mercy of our rapidly changing environment.

Today, the thin shore-fast ice and even substantial pressure ridges may be blown away any day in winter, just by heavy northeasterly winds (as happened on February 16, 2008, or January 21, 2009). Then, the wide-open water off the western beach can be used to launch the boats in the pursuits of walrus and seals that stay close to the pack ice. Boat hunting for walrus and seals was reported on 9 days in January 2007 and on 7 days in February 2008 (Table 4.5). It is a major shift from the years past, when boat hunting occurred hardly a few times during the entire winter and only after particularly strong storms. Instead, hunters used to drag their boats over shore-fast ice toward open leads, often for miles from shore (Hughes 1960:105; Oozeva et al. 2004:137–138, 142–143, 163–166). That type of hunting was still very common in the 1960s:

> They usually launch the boats from the west side. The ice was plentiful back then when we pulled our boats over the ice. You could go anywhere once you could go out offshore. We had good ice, good solid ice back then too. Some of it was much thicker; but it's the smooth young ice where you take your boats. That used to be fun when we were young. Just run with your boat over the ice. We used to go over the horizon from where you can't see the houses (in Gambell). [. . .] So, I figure we were more than eight miles away from here, so that we could not see the houses of the village anymore (LA, interview, February 15, 2008).

Hunting with skin boats dragged over the smooth ice (called *maklukestaq*, smooth ice with little bumps) cannot be practiced in recent years due to the lack of that type of ice. Instead, for a good portion of the season, Gambell hunters now face brittle young ice. That young thin ice (*ugmetaghaq* and *ugmelghu*) that LA also calls "flimsy ice" is too dangerous to walk on and too thin for walrus and big seals to haul out. So, both the hunters and the animals avoid it altogether. Today it is the most common icescape in Gambell in the middle of the winter (Fig. 4.7).

Mismatches. Winter conditions in Gambell, with the prevailing northerly winds, always used to be interrupted by a few days of violent storms and warm spells brought by southerly winds. This happened twice in the winters of 1899–1900 and 1900–1901 and three times in the winter of 1898–1899 (Oozeva et al. 2004:185). According to Conrad Oozeva, experienced Gambell elder, winter warm spells have been typical in his early days:

> We commonly have three waves of warm weather and thawing during the wintertime. After these warmings, we love to go hunting in boats on water opening, before it covers again

with the new ice. The only difference I see is that these warm waves were not long enough, just a few days only. We now have longer warming waves, often for several days (Oozeva et al. 2004:186).

These days, violent winter storms may come any time and often quite unpredictably, and the prevailing winter pattern of north–northeasterly winds can be broken any moment:

Our winds are so erratic now that severe winds may just come and crush the ice... Seems like the season is somewhat confused. It becomes unpredictable. Wind can blow from the south at up to 50–60 miles an hour one day and next morning it is from the west or from other direction. It just shifts around (Leonard Apangalook, February 15, 2009).

Elders on St. Lawrence Island unanimously point toward another big change in the winter condition. In the "old days," despite episodic snowstorms and warm spells, there were always extended periods of quiet, cold weather, with no winds. These long cold stretches were good for ice thickening, as well as for ice hunting and traveling; they also allowed hunters to predict the weather and ice for days in advance. These long calm periods do not happen anymore. In 2007, 2008, and 2009, there were few periods of relatively quiet weather (with winds up to 10 miles per hour) in either January or February and hardly any longer than a few days in a row. The weather was somewhat calmer in the month of March (Table 4.4).

Table 4.4 Average daily air temperature (F°) during winter season (January–March), Gambell, Alaska: 1899–2009 (... – no records)

Year	January	February	March	Longest cold spell (days below –10°F)	Number of major winter storms	Longest period of calm weather (days)[a]	Number of reported boat hunts (days)
1899	...	...	15.1	7(01.15–01.25)	...+...+3	...	0+0+4
1900	4.3	15.2	7.2	3(03.21–03.23)	...+2+2	3+9+4	0+1+0
1901	–5.2	–0.5	2.1	16(01.19–02.4)	2+3+2	...	0+0+1
1949	9.1	5.4	13.6	4(02.06–02.09)	3+2+3	2+2+3	...
1950	22.7	11.0	5.5	0	0+1+2	1+6+3	...
1951	–0.3	0.8	3.5	7(01.07–01.13)	1+2+1	1+1+1	...
1952	8.0	–0.8	4.6	5(02.11–02.15)	3+2+1	1+1+1	...
1988	12.6	3.8	3.7	5(02.17–02.21)	...	1+1+3	...
1991[b]	8.9	4.7	10.9	2(03.15–03.16)	...	2+1+2	...
1992[c]	5.4	8.8	–0.8	0	2+2+2	1+1+7	...
2007	5.0	15.1	5.9	0	5+7+1	3+3+6	9+3+6
2008	4.7	–4.0	–0.3	5(02.13–02.17)	3+3+0	2+2+7	3+7+6
2009	3.6	0	5.3	5 (02.8–02.12)	1+3+3	2+4+3	3+...+...

[a]Winds below 10 mph.
[b]For winter 1991, weather records are available for 24 days in January, 21 days in February, and 23 days in March.
[c]For winter 1992, weather records are available for 25 days in January and 23 days in February.

Besides thinner ice and more unpredictable weather, people see other confusing signals of change. On January 7, 2007, flying cormorants (evidently, the Pelagic cormorant, *Phalacrocorax pelagicus*) were observed in mid-winter, the first time in elders' memory. In February 2008, hunters watched the ribbon seals (*Phoca fasciata*) on ice off *Qayilleq*; that was also unheard of, as ribbon seals normally arrive to Gambell in late May, with the end of the spring retreat of the pack ice (see below). Evidently, certain shifts in the environment forced these and other species to show off St. Lawrence Island in the "wrong time" of the year.

By far the largest shift is a new possibility of encountering bowhead whales in mid-winter. Bowhead whales have a predictable seasonal migration pattern, with the main northward migration past Gambell in April and early May and the return southward migration in November–December. For the first time a bowhead whale was killed off Gambell in February 1992 (Noongwook et al. 2007:51); reportedly, some whales have been seen off Gambell in the wintertime in the following years. In February 2007, bowhead whales were observed in large numbers and for several days; on February 8, a whale was struck and lost in the ice. Hunters believe that in the absence of solid pack ice, many bowhead whales are now wintering off St. Lawrence Island and are not migrating south anymore. No whale sighting was reported in winter 2008 and 2009 that had substantially heavier ice than in 2007.

Scope of change. The 2006–2009 observation logs substantiate earlier statements by St. Lawrence Island hunters about the profound recent change in winter ice and weather regime (cf. Krupnik 2000; Noongwook 2000; Noongwook et al. 2007; Oozeva et al. 2004). Back in the "old days" (that now commonly means the 1940s and 1950s), as the pack ice arrived around the last week of November, it was made of solid multi-year ice. A long stretch of solid shore-fast ice was also built off the northern side at Gambell and at *Qayielleq*, and hunters would venture on ice on foot almost everyday in search for walrus and seals. That vision is confirmed by the elders' stories recorded in the 1970s and early 1980s (Sivuqam Naghnaghnegha 1985; Oozeva et al. 2004) and by the early description of winter hunting in Gambell (Hughes 1960:128).

In a broader historical perspective, however, the signal of today's winter change is mixed, particularly when compared to similar 3- to 4-year episodes (1899–1901, 1949–1952, and 1988–1992), for which we have observation records (Table 4.4). The average monthly winter temperatures were not substantially warmer in 2007–2009 and the variability of weather condition was almost the same as in 1949–1952, the earliest period for which we have solid weather data. Apparently, today's memories of the "good old years" refer to the 1960s and even the early 1970s, when conditions were indeed much colder on St. Lawrence Island and elsewhere in the Arctic. But there was hardly any difference in the winter temperatures in 1900, 1949, and 2007, and winters of 2008 and 2009 were even colder. Evidently, the signal of change is related to something other than slightly warmer average winter temperature.

Gambell hunters have their own explanation of the current change. They say that "it is not cold enough" in the wintertime these days to build solid local ice in the absence of the multi-year or thick pack ice from the north. They argue repeatedly about the role of long cold spells, of several very cold days *in a row*, particularly during the month of February in building thick shore-fast and pack ice. That ice is favored by the hunters and by their winter game, walrus and seals. Instead they face thin young ice ("flimsy ice"), strong erratic winds, and plenty of open water in the middle of the winter. Unfortunately, none of the three winters in 2007–2009 featured such long cold spells. With *not enough cold* during the wintertime (January–February), the Bering Sea system then progresses quickly into the spring breakup season and rapid disintegration of its seasonal ice cover.

Spring Breakup and Ice Retreat

As the days become longer and, usually, warmer in March, Gambell hunters start seeing the seasonal change. The first sign of spring is the formation of large leads between the shore-fast and pack ice on the west side of Gambell. These leads will soon be used by passing bowhead whales and, later, by walrus on their northward migration. In anticipation of the animals heading north, the hunters will get their spring equipment ready, including big whaling skin boats that have to be moved closer to the shore. The spring whaling season in April and May is an exciting time for the village and the climax of the annual subsistence cycle (Hughes 1960; Jolles 2003). The preparation for spring whaling normally starts in late March. Perhaps, the clearest indicators of the upcoming spring hunting are when boat-launching ramps are cut through the shore ice ridges and the big whaling boats are mounted on racks closer to the shore (Table 4.5).

Throughout the month of April, the unstable ice and weather make whale hunting unpredictable and often risky, as drifting ice may close in at any moment and strong wind creates high waves on open water. Thus, during four seasons of spring observation in 2006–2009, hardly a handful of days were favorable for whaling in April; in spring 2006 no whale was taken, due to bad weather and tight pack ice.[10]

The spring whaling usually lasts until the first week of May. Around May 5, hunters anticipate the first wave of walrus migrating north with the retreating pack ice. Around May 5–10, there is also a major change in the prevailing wind direction from primarily northerly to southerly followed by a 5–8°F spike in air temperature (Table 4.5). Commonly, it indicates the beginning of the walrus season, as hunters switch to the pursuit of walrus and bearded seals. Both animals are hunted from boats on the drifting ice floes, often in dense moving pack ice.

Gambell hunters report that the peak of the walrus migration usually occurs around May 15 (Table 4.5). As the ice retreats northward, hunters have to make ever-longer trips to get to the edge of the pack ice, where the game concentrates. By early June, they often have to travel up to 50–70 miles (80–110 km) north of Gambell to reach the ice. At this time, the shore-fast ice erodes rapidly and by early–mid-June the western beach in Gambell is cleared of ice. Shortly afterward, the northern shore-fast ice disintegrates. As the sea clears of ice, people start setting

Table 4.5 Spring–early summer observational indicators used by Gambell hunters

Indicator	2006	2007	2008	2009	Comments
1. Leads open off western beach	–	Early March	March 3	–	Common sign of the upcoming "spring" ice condition
2. Ice crabbing starts off the northern side	April 2	March 9	February 16	April 9	
3. Boat ramps cut in shore ridges for boat launching	–	March 19	March 31	–	Hunters start preparation for spring whaling season
4. Boats placed on racks close to the western shore	April 2	–	–	–	Hunters ready for spring whaling
5. First spring whaling (sailing) of the season	April 22	April 2	April 3	April 9	Bowhead whale migration northward has started; whales reached St. Lawrence Island
6. First spring whale killed	–	April 3	May 1	April 23	
7. *Kelliighineq* – large body of open water forms off the western shore	April 20	April 26	May 12	April 16	Huge polynya forms between Chukotka and St. Lawrence Island, as persistent northerly winds push floating ice to the south of St. Lawrence Island
8. First sighting of snow bunting	April 26	May 11	April 24	–	Beginning of bird spring migration
9. Shift in wind direction from northerly to southerly	May 10	May 4	May 15	May 14	With southerly winds, wind and current start pushing drifting pack ice northward toward Bering Strait
10. First report of female walrus with calves	May 3	May 4	May 15	April 24	Pregnant and calving female walrus with newborns reach St. Lawrence Island on the northward spring migration
11. Start of active walrus hunting	May 15	May 5	May 16	May 18	The bulk of the moving walrus hits St. Lawrence Island on its seasonal northward migration
12. Sighting of gray whales	May 10	May 11	May 5	May 29	Gray whales usually arrive *after* bowhead whales and, thus, indicate the end of the bowhead whale migration

Table 4.5 (continued)

Indicator	2006	2007	2008	2009	Comments
13. Peak of the walrus hunting activities	May 15–24	May 19–30	May 16–19	May 18–31	Major group of migrating walrus arrives with the pack ice moving northward from the central Bering Sea
14. Sighting of ribbon seals	June 13	–	–	May 29	"Tail end" of the main pack ice retreat; pack ice from the Gulf of Anadyr and western Bering Sea
15. Snow melting rapidly around the village	June 5	–	June 11 (melted)	–	
16. Last long boat trips NNE–NW to the edge of the pack ice	June 12	June 3	June 2	June 9	Indicates the final retreat of the pack ice past St. Lawrence Island
17. Cranes passing by on their northward flight	–	–	–	Late May (?)	Corresponds with the end of the main ice pack retreat past Gambel. In 2009, it happened slightly later than "normal"
18. Shore clears of ice; no ice in sight on the ocean	June 26	June 9	June 8	June 11	End of the sea ice "season"
19. Ice break-up on the Troutman Lake	June 26	June 17	–	–	Freshwater lakes and lagoons clear of ice for the summer
20. Fishing nets are put off western beach	June 15	June 18	–	–	Beginning of the summer season; no expectation of returning ice
21. People start picking greens	June 20	June 13	–	–	Ground has softened and warmed for the first green plants of the season
22. Last "Siberian" ice brought in by NW winds	–	–	–	–	Broken shore-fast ice from the Gulf of Anadyr and Kresta Bay pushed by strong NW wind in late June–early July

No March logs for 2006 and 2009; in 2008 and 2009 observation ceased after June 11.

fishing nets and prepare for bird hunting and collecting greens, all typical summer activities. With this, the spring season is over and the ice is usually gone until next fall.

The 3-month transition (mid-March to mid-June) from winter to spring to summer normally progresses through several ebbs and flows in ice concentration triggered by combination of winds, currents, and ice features (see Chapter 5 by Kapsch et al. this volume). Some of these spring ice/weather patterns are occasional and hardly predictable; others occur regularly and have established names. The most common feature in late March and April are short-term bouts of strong southerly winds that blow in the pack ice from the southern side of the island. This "southern" ice has its specific name, *ivghaghutkak* ("one brought around the SW Cape from the south side"[11]). That southern ice was blown in at least twice in spring 2006, once in 2007, three times in 2008, and four times in 2009. Commonly, hunters expect walrus and bearded seals to be abundant with the advent of southern ice, particularly in the later part of April and May.

A more predictable phenomenon is a large body of open water (polynya) that forms every year off the western side of St. Lawrence Island in late April and early May. It has its special Yupik name, *kelliighineq* ("large polynya," from *kellii*, meaning "open sea," "ocean"). The formation of *kelliighineq* is explained as follows:

> [...] A large body of open water beyond our shore packed ice was sighted on April 20th. [...] This large body of open water, "*kelliighineq*," is the result of moderating temperatures and prevailing northeasterly winds in late April and early May. The open leads south of the Siberian Peninsula cease to freeze over, thus expanding to our Island on the west and northwest side. The northerly currents that we refer to as *maaqneq* are much stronger in late March through June. Therefore the large open water that expands to our Island from the leeward side of the Siberian peninsula reaches us 90% of the time every spring. [...] When the large body of water extends northwest and north of Gambell, it also creates high winds that generate large swells. We stop boat hunting trips for a week or more during this winter/spring transitional occurrence. [...] Our elders would be excited about this condition as it creates good hunting following the higher winds, as it gets calmer. The whales not only migrate back up north through this open ocean but also stop and congregate to feed and mate off our shores. Because of higher winds, the walrus, bearded seal, and other animals haul out on the down wind ice pack and carried to our area when the winds calm down for the spring. This is *kelliighineq* and what it means to our people and our way of life (Leonard Apangalook, observation log, May 2006).

As the northern wind subsides and the polynya (*kellighineq*) is covered again with the floating pack from the south, the hunting resumes. Shortly after, usually in the last 2 weeks of May, Gambell hunters observe the main mass of the Bering Sea pack ice retreating north with the southerly winds and currents. That ice carries the bulk of the migrating walrus from the western and central Bering Sea. The retreat of the Bering Sea pack ice loaded with game usually around May 15–30 is the major event of the spring season; it has been recorded in every spring log in 2006–2009. By early June, the ocean clears up and no ice is visible around Gambell. However, if strong westerly wind blows for a few days it usually brings another type of pack ice that comes from the central part of the Gulf of Anadyr. Its main indicator is the sighting of ribbon seals that spend the wintertime in the dense pack ice of the northwestern Bering Sea. Hunters watch anxiously for ribbon seals, as an indicator of the last major pack ice of the season. That "western" pack ice was reported in the logs on June 13, 2006 (with ribbon seals); on May 30, 2007; on May 23, 2008 (without ribbon seals); and on May 29, 2009 (with ribbon seals).[12]

In certain years, Gambell may see one last type of drifting pack ice very late in the season, after the "southern," the "central Bering Sea," and the "Gulf of Anadyr" ice pass by and the sea is clear of ice for a few weeks. That last wave is made of remnants of the broken Siberian shore-fast ice from the mouth of the Anadyr River and from Kresta Bay, the northernmost portion of the Gulf of Anadyr. That section of the northern Bering Sea is always the last to clear of ice, usually in late June or even in the first weeks of July:

> If the southwesterly winds prolong we may not see ice from Anadyr Bay. As seen on satellite photo of ice conditions in our area, there is a little a bit of shore-fast ice in Anadyr Bay. If we can get northerly winds for an extended period, there is a chance we might see ice again in July. This has happened in the first week of July prior to 1990 (Leonard Apangalook, June 30, 2006).

That late "Siberian" shore-fast ice can be easily recognized, as it is dirty and even carries lumber and debris from Siberia.[13] It is also normally devoid of preferred game animals, walrus or bearded seals. No "Siberian" ice was recorded off Gambell in June 2006 and 2007 (in 2008 and 2009, the observation ceased after June 11).[14] Therefore, what scientists call the "spring retreat" of the sea ice is, in fact, the series of several individual ice "waves" recognized by local hunters.

Shifts. Hunters in many Alaskan communities report substantial changes in spring ice and weather conditions over the past several years. Most commonly they refer to an earlier and shorter breakup season, more rapid northern retreat of the seasonal pack ice, so that spring walrus migration now takes place 3–4 weeks earlier than in the previous decades (Metcalf and Krupnik 2003; Chapter 14 by Krupnik and Weyapuk this volume). Surprisingly, Gambell observations do not support such trend:

> Our temperatures [in the month of April] remained below what's considered 'normal,' as based on weather data collection available within the last twenty years, but more akin to that of the [19]60s, in my opinion. [. . .] This is clearly obvious in the abundance of ice. There seemed to be more game in the area as a result. All game taken appears to be healthy, which is attributable to the availability of the food supply. The frequent weather fronts in the area remains the norm making it difficult to read the weather trends based on our knowledge. As in the past trends, we continue to see fewer ideal hunting conditions. Most of the boat hunting [trips] were of short excursions and usually to go after game spotted from onshore (Paul Apangalook, May 2, 2009).

In four seasons in 2006–2009, the signs of warming in the springtime in Gambell were not that obvious and, certainly, not as dramatic, as during the fall and winter seasons. No major disruption of local hunting has been recorded, due to changed ice and weather conditions. Overall, hunting was quite good and the animals, particularly walrus, were abundant.[15] In fact, the trend of many local indicators is rather mixed and is clearly obscured by strong inter-annual variability (Table 4.5).

> This is the first time in 8 years when we had ice late this season. My yearly records show from 2000 where the pack ice left our vicinity in the last week of May. Although we had ice longer than the previous years [up to June 26], the animals were not present in the last few days [that] we had ice. Because the ice was able to stay in our area, everyone's meat racks are full this year. We attribute presence of ice longer this year to the high northeasterly winds we had most of the winter. Global warming weighed heavily in our minds when the pack ice left us early in the past eight years, but this season's condition kind of threw a curve in our apprehension of global warming. Will be interesting to see next season (Leonard Apangalook, June 2006).

Next year, on June 2, 2007 he reported "... We still have ice and surprisingly walrus and seals are still around. It would seem like back in the old days when our seasons were normal." Surprisingly, that very same June 2007 featured one of the most rapid sea ice retreats in the Bering Sea during the past 30 years that also shortened the walrus and bearded seal hunting window for other coastal communities (Hajo Eicken, pers. communication, August 13, 2009).

Scope of change. Available historical weather records confirm Gambell hunters' claim that, unlike the fall and winter season, no similar major shift in ice and weather has happened in the spring. True, the average monthly air temperatures for April, May, and June in four selected periods, 1899–1901, 1949–1952, 1988–1992, and 2006–2009, show no significant trend, as also do the dates of major weather shifts for the season (Table 4.6). Many reported dates look remarkably similar to the present-day condition (cf. Oozeva et al. 2004:187–191) and the sequence of events during spring 1899, as recorded by schoolteacher William Doty (1900:250–254), is almost as if taken from today's log.[16] Whatever trend can be retrieved from comparing old ice and hunting records may often be explained by the change in hunting techniques rather than in weather. In reviewing Gambell historical records from 1899 to 1901, Conrad Oozeva (born 1925) commented the following:

> Them people getting their first whales in 1899 is exactly like we do it today. We may even have some whales killed in the last days of March. But we are now getting whales much earlier than in my earlier years. The reason is that nowadays people want to hunt more walruses and *maklaks* (bearded seals) after the whales get there. So, they push it slightly earlier these days. Besides, in the old years, they used to hunt with big whaleboats. They needed more open water for that and that is why they probably started later. We also use lighter boats now, aluminum boats, and can use motors to pull the killed whale, while they had to paddle in their heavy boats. So, we can start whaling earlier even the same weather and ice they had. The year 1901, with their three whales killed in the end of May seem also quite normal (Oozeva et al. 2004:189).

Hunters' comments about minimal or no change in the timing of spring ice retreat and walrus migration off Gambell are confirmed by two other independent assessments (Robards 2008, Unpublished; Chapter 5 by Kapsch et al. this volume). In this regard, the situation in Gambell differs substantially from conditions in other Bering Strait communities, such as Savoonga, Diomede, and Wales (see Chapter 14 by Krupnik and Weyapuk this volume), where the peak of the spring walrus catch is now taking place almost 15 days earlier. It illustrates that the strength of the current warming trend clearly differs by place and season even within the same region.

Table 4.6 Average monthly air temperature (F°) during spring season (April–June) Gambell, Alaska: 1899–2009

Year	April	May	June	Shift in wind direction	Major spike in daily temperature[a]	Last cold spell[b]	Start of active walrus hunting
1899	14.8	31.7	37.3	May 8	May 8	05.29–06.02	May 8
1900[c]	6.7	30.3	44.2+	May 20?	May 20?	05.25–05.27	...
1901	24.1	29.9	37.1[d]	May 19	May 18–19	06.13–06.15	April 22; May 19
1949	13.8	26.1	34.6	...	May 12	06.11–06.14	...
1950	17.2	26.7	36.1	...	May 14–15	Up to 06.02	...
1951	16.8	29.7	37.3	...	May 2	Up to 05.27	...
1952	15.2	28.3	35.7	...	May 7	Up to 06.07	...
1988	19.7	29.5	38.7	...	May 9	Up to 06.03	...
1991	19.4	31.6	35.1	...	May 4	05.26–06.05	...
1992	23.5	28.6	34.4	...	May 7	06.2–06.07	...
2006	7.5	26.2	35.0	May 11	May 11	06.02–06.04	May 18
2007	17.5	30.5	40.7	...	May 14–18	Up to 05.24	May 5
2008	16.3	27.4	37.3	May 16	May 16	06.05–06.07	May 16
2009	18.2	31.8	37.6	May 7	...	05.22–05.24	May 18

[a]At least 5–7°F, usually an indicator to a major wind shift from northerly to southerly.
[b]Last cold spell: 3 days in a row with the average daily temperature below freezing.
[c]Records in May 1900 are for 20 days only; temperature readings in June 1900 for 18 days only, many seem unreliable.
[d]June 1–June 16, only.

Discussion

The data discussed in this chapter represents the first assessment of what has been reported during the 33-month observation in 2006–2009 and more analysis is to follow. Nonetheless, the message from this first evaluation of the Gambell SIKU observation records is "not that simple" (cf. Gearheard et al. 2006). On the one hand, it substantiates hunters' and scientists' claim for a strong warming signal in the northern Bering Sea that has been known at least since the late 1990s (Grebmeier et al. 2006, 2009; Kavry and Boltunov 2006; Krupnik and Ray 2007; Metcalf and Krupnik 2003; Noongwook 2000; Oozeva et al. 2004). On the other hand, the signal is nuanced, if not mixed. In Gambell the warming is the strongest in the fall and weaker in spring, whereas in other communities, like Diomede and Wales, barely 260 km (150 miles) to the north, it is more pronounced in the spring season (Robards 2008, Unpublished; Chapter 14 by Krupnik and Weyapuk this volume).

Hunters' logs, personal comments, and interviews recorded in 2006–2009 also indicate that the scope of the observed change, though significant, has not disrupted annual subsistence cycle in Gambell or elsewhere in the region. The productivity of subsistence hunting remains high and the most important game animals, bowhead whales, walrus, and seals, are plenty. So far Gambell bowhead and walrus catch records do not indicate any noticeable decline since 2000 (Garlich-Miller et al. 2006; Noongwook et al. 2007; Chapter 5 by Kapsch et al. this volume). Evidently, hunters are extremely adaptable and have been able to accommodate their subsistence strategies to the changes in ice and weather conditions they observe, at least for the time being.

SIKU observational logs also reveal a sophisticated set of local indicators ("benchmarks") that hunters use to track seasons transition, ice safety, and shifts in subsistence calendar. Here, hunters' approach clearly differs from that of scientists. In their assessment of change, scientists preferably cite what may be called *summer* indicators, like summer retreat of the Arctic pack ice ("annual ice minimum"); permafrost and glaciers' melt; increased storm activity and beach erosion; spread of forest fires; northward advance of southern marine and terrestrial species. These indicators naturally underscore the impact of the *warming* in the Arctic system.

Polar residents are focused more on a much longer period from October–November until May–June that is dominated by the seasonal ice cover, upon which

Fig. 4.9 West side of Gambell on February 5, 2010. There is no winter pack ice visible to the horizon, just *salleghhpuk*, new ice about 3–4 inches thick. The small ice floe near the shore is smooth, about a foot thick. One of the village boats returns from hunting along the northern side breaking *sallek* (thin new ice) as it approaches the shore. (Photo: Paul Apangalook)

their livelihood and subsistence economy depend. As such, their vision is that there is *not enough cold* in the system, particularly, during the peak of the winter season (January–February, see Fig. 4.9). In Gambell, it transforms into thinner and weaker, thus, more dangerous ice; more warm spells during the wintertime; and increased instability of winter weather condition. Hunters lament the *cold* of the former years, when the pack and shore-fast ice were solid, game arrived earlier, multi-year ice was common, and cooler weather could be predicted based upon age-old experience and teaching.

Gambell and other SIKU observational records demonstrate the intimacy of hunters' knowledge of the Arctic ice. Hunters have noticed the replacement of the northern pack ice (*sikupik*) by the "locally formed" types of ice in their area way before 2000 (Krupnik 2000; Noongwook 2000). They discerned the disappearance of the multi-year ice at least since the 1990s. Gambell hunters' vision of several (four or even five) types of sea ice of different origins passing through their area during the spring breakup season is still unsurpassed in the scientific ice models and their ability to track (and to name) the many types of ice they watch during the season is unmatched by the scientific ice nomenclatures.

Conclusions

The main lesson of 3 years of SIKU observations in Gambell and other northern towns is that by applying their traditional observational methods, local hunters have independently registered the signal of Arctic warming. They have traced it via several indicators; come to their own interpretation of its cause ("it's not cold enough"); and developed their set of "benchmarks" to compare seasonal, annual, and even decadal transitions. Of course, there are limits to what hunters can see, since they do not have access to global ice and climate databases, super-computers, and satellite imagery that would allow them to reach beyond their scale of knowledge. But this is where partnership with scientists can and should make a difference.

The success of Arctic residents' participation in IPY 2007–2008 via SIKU and similar joint projects demonstrates their eagerness to contribute time and knowledge to document the ongoing change in their home regions. This is a message that scientists and weather and ice-forecasting services cannot leave unanswered. There are plans to spearhead the momentum created by IPY 2007–2008 to build a new sustainable Arctic observing network that would serve science, government agencies, industries, and local communities alike (Eicken et al. 2009; SAON 2008). Indigenous residents should be an integral part of such observing network, both as its contributors and consumers. Their knowledge and observations are to be used and actively sought to calibrate the scientists' planetary and regional scenarios, ice and climate forecasts, and models. Arctic people's integrative vision is invaluable to our common understanding of the new polar system to be influenced by the climate warming in the decades to come.

Acknowledgments This chapter is a product of a 4-year partnership and a much longer collaboration between indigenous experts and scientists on St. Lawrence Island and Bering Strait region. It is also our tribute to the Yupik wisdom and teaching that help keep the Yupik knowledge of ice, animals, and hunting vigorous and strong. Elders Conrad Oozeva (*Akulki*), Willis Walunga (*Kepelgu*), Ralph Apatiki, Sr. (*Anaggun*), Chester Noongwook (*Tapghaghmii*) advised us at various stages of our project and hunters of today, George Noongwook (*Mangtaaquli*), Merle Apassingok (*Wawiita*), Branson Tungiyan (*Unguqti*), Aaron Iworrigan (*Yupestaaq*), and Edwin Campbell (*Iiyiitaq*) shared their knowledge in interviews from 2004 to 2008. Gary Hufford (Alaska Weather Service, Anchorage) and Martha Shulski (Geophysical Institute, University of Alaska Fairbanks) supplied historical weather records and ice satellite images. Our colleagues, G. Carleton Ray, Gary Hufford, Hajo Eicken, and James Overland offered valuable insight; Ray, Eicken, Claudio Aporta, Shari Gearheard, and Cara Seitchek commented on the first draft of this chapter. Matt Druckenmiller produced the map. We thank you all.

Notes

1. The small pilot project in 2000–2001 emerged as an outcome of a joint workshop on sea ice change (Huntington 2000), at which several indigenous participants expressed their concerns about the strong signal of warming in their home areas (Noongwook 2000; Pungowiyi 2000).
2. Eventually, all Russian observers and also Gambell and Shaktoolik were supported via the U.S. National Park Service SIKU grant; observations in Wales and Barrow were covered through the NSF-funded *SIZONet* project (OPP-0632398, PI Hajo Eicken, University of Alaska Fairbanks).
3. Observations in the Russian community of Uelen were originally reported in the local Chukchi language and later translated into Russian by SIKU team member Victoria Golbtseva.
4. Krupnik and Bogoslovskaya (2008); Bogoslovskaya et al. (2008); Chapter 15 by Eicken this volume; Chapter 5 by Kapsch et al. this volume; Chapter 14 by Krupnik and Weyapuk this volume; Druckenmiller et al. (2009). Russian SIKU data will be analyzed in full in a special publication.
5. See more on the history of Gambell in Hughes (1960); Jolles (2002); Krupnik (2004).
6. Observations in Gambell (by Paul Apangalook) were resumed in October 2009 and continued throughout winter-spring 2010, thus adding one more year to the Gambell observation dataset.
7. Records were provided by Martha Shulski, Geophysical Institute, UAF, and Gary Hufford, Alaska Weather Service.
8. For explanation of the Yupik ice terminology used here, see Oozeva et al. (2004:29–53).
9. Similar pattern of delayed freeze-up has been consistently reported by indigenous observers across the Arctic area – see Gearheard et al. (2006); Laidler and Elee (2006); Laidler and Ikummaq (2008); McDonald et al. (1997); Metcalf and Krupnik (2003).
10. "Iceberg" is the English word that the hunters use to describe what they call *kulusik/kulusiq* in Yupik (Oozeva et al. 2004:36). Strictly speaking, it is not the proper glaciological term, since in the science classification "iceberg" refers to the piece of floating ("calved") glacier ice.
11. Normally, Gambell hunters succeed in killing up to three bowhead whales in spring, with two to three more whales taken in the other island community, Savoonga (Noongwook et al. 2007).
12. From *ivgagh*, "to come or go around the corner" (Jacobson 2008:73).
13. Evidently, the same "western" pack ice was brought by westerly winds to Gambell on June 2, 1899 (Doty 1900:50), after several days of light or no ice in the ocean.
14. In a similar way, hunters in Uelen, Siberia, recognize the "Alaskan" ice that carries objects from Alaska, including broken crates, bags, clothing, and even shoes from the Alaskan side (Roman Armaergen – recorded by Victoria Golbtseva).

15. Evidently, this late "Siberian" ice blocked the beach in Gambell on June 19, 1900 (Lerrigo 1901:131) and was encountered by the *USCG* Bear on its way through the heavy masses of drifting ice floes from Gambell to Provideniya Bay on June 30, 1921 (Burnham 1929:3).
16. As Paul Apangalook noticed upon completing his June 2009 log, "... Game was plentiful as virtually all meat racks are full in the village. The condition of game harvested appears normal and healthy largely due to our selective hunting."
17. The first whaling in 1899 started on April 6; a whale was secured on May 10. By that time, the prevailing wind changed from N–NE to S–SW (on May 8) and the temperature jumped by almost 10°F. Last major walrus hunt occurred on June 4; the last major northward flow of drifting ice was between June 2 and June 6. Last ice jam on the west beach was on June 7 and people started storing walrus meat in underground cellars on June 16. Evidently, this was the end of the spring hunting season. These dates are very close to how it happened in 2007, 2008, and 2009 (see Table 4.5), though not in 2006, when the ice stayed until June 26.

References

ACIA. 2005. *Arctic Climate Impact Assessment (ACIA)*. Cambridge: Cambridge University Press.

Allison, I., Béland, M., Alverson, K., Bell, R., Carlson, D., Darnell, K., Ellis-Evans, C., Fahrbach, E., Fanta, E., Fujii, Y., Glasser, G., Goldfarb, L., Hovelsrud, G., Huber, J., Kotlyakov, V., Krupnik, I., Lopez-Martinez, J., Mohr, T., Qin, D., Rachold, V., Rapley, C., Rogne, O., Sarukhanian, E., Summerhayes, C., and Xiao., C. 2007. The Scope of Science for the International Polar Year 2007–2008. *World Meteorological Organization, Technical Documents* 1364. Geneva.

Bogoslovskaya, L.S., Vdovin, B.I., and Golbtseva, V.V. 2008. Izmeneniia klimata v regione Beringova proliva: Integratsiia nauchnykh i tradititsionnykh znanii (Climate change in the Bering Strait region: Integration of scientific and indigenous knowledge. SIKU, IPY #166). *Ekologicheskoe planirovanie i upravlenie* 3–4(8–9): 36–48, Moscow.

Burnham, J.B. 1929. *The Rim of Mystery. A Hunter's Wanderings in Unknown Siberian Asia*. New York and London: G.P. Putnam's Sons.

Doty, W.F. 1900. Log Book, St. Lawrence Island. *Ninth Annual Report on Introduction of Domestic Reindeer into Alaska 1899*. Washington: Government Printing Office, pp. 224–256.

Druckenmiller, M.L., Eicken, H., Johnson, M.A., Pringle, D.J., and Williams, C.C. 2009. Towards an integrated coastal sea ice observatory: System components and a case study at Barrow, Alaska. *Cold Regions Science and Technology* 56: 61–72.

Eicken, H., Lovecraft, A.L., and Druckenmiller, M.D. 2009. Sea ice system services: A framework to help identify and meet information needs relevant for Arctic observing networks. *Arctic* 62(2): 119–136.

Fox (Gearheard), S. 2002. These are things that are really happening: Inuit perspectives on the evidence and impacts of climate change in Nunavut. In *The Earth Is Faster Now: Indigenous Observations of Arctic Environmental Change*. I. Krupnik and D. Jolly (eds.), Fairbanks: ARCUS, pp.12–53.

Gambell, V.C. 1910. *The Schoolhouse Farthest West. St. Lawrence Island, Alaska*. New York: Woman's Board of Home Missions of the Presbyterian Church.

Garlich-Miller, J.L., Quakenbush, L.T., and Bromaghin, J.F. 2006. Trends in age structure and productivity of pacific walruses harvested in the Bering Strait region of Alaska, 1952–2002. *Marine Mammal Science* 22(4): 880–896.

Gearheard, S., Matumeak, W., Angutikjuaq, I., Maslanik, J., Huntington, H.P., Leavitt, J., Matumeak Kagak, D., Tigullaraq, G., and Barry, R.G. 2006. "It's not that simple": A collaborative comparison of sea ice environments, their uses, observed changes, and adaptations in Barrow, Alaska, USA, and Clyde River, Nunavut, Canada. *AMBIO* 35(4): 203–211.

Grebmeier, J., Overland, J., Moore, S., Farley, E., Carmack, E., Cooper, L., and Frey, K. 2009. Impact of Warming Temperatures on Marine Life and Fisheries in the Bering Sea.

http://www.akmarine.org/our-work/address-climate-change/fisheries-and-warming-oceans (accessed September 25, 2009).

Grebmeier, J.M., Overland, J.E., Moore, S.E., Farley, E.V., Carmack, E.C., Cooper, L.W., Frey, K.E., Helle, J.H., McLaughlin, F.A., and McNutt, S.L. 2006. A major ecosystem shift in the northern Bering Sea. *Science* 311: 1461–1464.

Helander, E. and T. Mustonen (eds.) 2004. *Snowscapes, Dreamscapes. Snowchange Book on Community Voices of Change. Study Materials.* Tampere: Tampere Polytechnic Publications. p. 12.

Hughes, C.C. 1960. *An Eskimo Village in the Modern World.* Ithaca: Cornell University Press.

Huntington, H.P. (ed.) 2000. *Impact of Changes in Sea Ice and Other Environmental Parameters in the Arctic.* Report of the Marine Mammal Commission Workshop. Girdwood, Alaska, 15–17 February 2000. Bethesda: Marine Mammal Commission.

Huntington, H.P., Brower, H., Jr., and Norton, D.W. 2001. The Barrow symposium on sea ice, 2000: Evaluation of one means of exchanging information between subsistence whalers and scientists. *Arctic* 54(2): 201–204.

Huntington, H.P., and Fox, S. Lead Authors 2005. The changing Arctic: Indigenous perspectives. In *Arctic Climate Impact Assessment.* Cambridge: Cambridge University Press.

Jacobson, S.A. (ed.) 2008. *St. Lawrence Island/Siberian Yupik Eskimo Dictionary*, vols. 1–2. Compiled by Linda Womkom Badten (Aghnaghaghpik), Vera Oovi Kaneshipr (Uqiitlek), Marie Oovi (Uvegtu), and Christopher Konooka (Petuwaq). Fairbanks: Alaska Native Language Center, University of Alaska Fairbanks.

Jolles, C.Z. 2002. *Faith, Food and Family in a Yupik Whaling Community.* Seattle and London: University of Washington Press.

Jolles, C.Z. 2003. When whaling folks celebrate: A comparison of tradition and experience in two Bering Sea whaling communities. In: *Indigenous Ways to the Present. Native Whaling in the Western Arctic.* A.P. McCartney (ed.), *Studies in Whaling* 6; *Occasional Publications* 54, Edmonton: *Canadian Circumpolar Institute*, pp. 209–254.

Jolly, D., Berkes, F., Castleden, J., Nichols, T., and The Community of Sachs Harbour 2002. We can't predict the weather like we used to. In *The Earth Is Faster Now: Indigenous Observations of Arctic Environmental Change.* I. Krupnik and D. Jolly (eds.), Fairbanks: ARCUS, pp. 92–125.

Kavry, V. and Boltunov, A. 2006. Observations of Climate Change Made by Indigenous Inhabitants of the Coastal Regions of the Chukotka Autonomous Okrug. Moscow, WWF-Russia, http://www.wwf.ru/resources/publ/book/eng/196/ (accessed October 4, 2009).

Krupnik, I. 2000. Native perspectives on climate and sea ice changes. In *Impact of Changes in Sea Ice and Other Environmental Parameters in the Arctic.* H.P. Huntington (ed.), Bethesda: Marine Mammal Commission, pp. 25–39.

Krupnik, I. 2002. Watching ice and weather our way: Some lessons from Yupik observations of sea ice and weather on St. Lawrence Island, Alaska. In *The Earth Is Faster Now: Indigenous Observations of Arctic Environmental Change.* I. Krupnik and D. Jolly (eds.), Fairbanks: ARCUS, pp. 156–197.

Krupnik, I. 2004. "The whole story of our land": Ethnographic landscapes in Gambell, St. Lawrence Island, Alaska. In *Northern Ethnographic Landscapes. Perspectives from Circumpolar Nations.* I. Krupnik, R. Mason, and T. Horton (eds.), Washington: Arctic Studies Center, Smithsonian Institution, pp. 203–227.

Krupnik, I. 2009. "The way we see it coming": Building the legacy of indigenous observations in IPY 2007–2008. In *Smithsonian at the Poles: Contributions to International Polar Year Science.* I. Krupnik, M. Lang, and S. Miller (eds.), Washington: Smithsonian Institution Scholarly Press, pp. 129–142.

Krupnik, I. and Bogoslovskaya, L.S. 2008. International Polar Year 2007–2008. Project SIKU in Alaska and Chukotka. In *Beringia: A Bridge of Friendship.* Tomsk: TSPU Press, pp. 196–204.

Krupnik, I. and Ray, G.C. 2007. Pacific walruses, indigenous hunters, and climate change: Bridging scientific and indigenous knowledge. *Deep-Sea Research II* 54: 2946–2957.

Krupnik I. and Jolly, D. (eds.) 2002. *The Earth Is Faster Now: Indigenous Observations of Arctic Environmental Change*. Fairbanks: ARCUS.

Laidler, G.J. and Elee, P. 2006. Sea ice processes and change: Exposure and risk in Cape Dorset, Nunavut. In *Climate Change: Linking Traditional and Scientific Knowledge*. Oakes, J. and R. Riewe (eds.), Winnipeg and Québec City: University of Manitoba Aboriginal Issues Press and ArcticNet, pp. 155–175.

Laidler, G.J. and Ikummaq, T. 2008. Human geographies of sea ice: Freeze/thaw processes around Igloolik, Nunavut, Canada. *Polar Record* 44(229): 127–153.

Lerrigo, P.H.J. 1901. Abstract of Journal, Gambell, St. Lawrence Island, Kept by P.H.J. Lerrigo, M.D. *Tenth Annual Report on Introduction of Domestic Reindeer into Alaska 1899*. Washington: Government Printing Office, pp. 114–132.

Lerrigo, P.H.J. 1902. Abstract of Daily Journal on St. Lawrence Island Kept by P.H.J. Lerrigo, M.D.Log Book, St. Lawrence Island. *Eleventh Annual Report on Introduction of Domestic Reindeer into Alaska 1899*. Washington: Government Printing Office, pp. 97–123.

McDonald, M., Arragutainaq, L., and Novalinga, Z. Comps. 1997. *Voices from the Bay. Traditional Ecological Knowledge of Inuit and Cree in the Hudson Bay Bioregion*. Ottawa: Canadian Arctic Resources Committee.

Meier, W.N., Stroeve, J., and Fetterer, F. 2007. Whither Arctic sea ice? A clear signal of decline regionally, seasonally, and extending beyond the satellite record. *Annals of Glaciology* 46: 428–434, doi:10.3189/172756407782871170.

Metcalf, V., and I. Krupnik (eds.) 2003. Pacific walrus. Conserving our culture through traditional management. Report produced by Eskimo Walrus Commission, Kawerak, Inc. under the grant from the U.S. Fish and Wildlife Service, Section 119, Cooperative Agreement # 701813J506

Metcalf, V. and Robards, M.D. 2008. Sustaining a healthy human-walrus relationship in a dynamic environment: Challenges for comanagement. *Ecological Applications* 18(2): 148–156.

Noongwook, G. 2000. Native observations of local climate changes around St. Lawrence Island. In *Impact of Changes in Sea Ice and Other Environmental Parameters in the Arctic*. H.P. Huntington (ed.), Bethesda: Marine Mammal Commission, pp. 21–24.

Noongwook, G., The Native Village of Savoonga, The Native Village of Gambell, Huntington, H.P., and George, J.C. 2007. Traditional knowledge of the Bowhead whale (*Balaena* mysticetus) around St. Lawrence Island, Alaska. *Arctic* 60(1): 47–54.

Oozeva, C., Noongwook, C., Noongwook, G., Aloowa, C., and Krupnik., I. 2004. *Watching Ice and Weather Our Way. Sikumengllu Eslamengllu Esghapaleghput*. Washington: Arctic Studies Center.

Pachauri, R. and Reisinger, A. (eds.) 2007. *IPCC, 2007. Climate Change 2007: Synthesis Report*. Contribution of Working Groups I, II and III to the Fourth Assessment Report of the Intergovernmental Panel on Climate Change. IPCC, Geneva.

Pungowiyi, C. 2000. Native observations of change in the marine environment of the Bering Strait Region. In *Impact of Changes in Sea Ice and Other Environmental Parameters in the Arctic*. H.P. Huntington (ed.), Bethesda: Marine Mammal Commission, pp. 18–21.

Richter-Menge, J., Overland, J., Proshutinsky, A., Romanovsky, V., Bengtsson, L., Brigham, L., Dyugerov, M., Gascard, J.C., Gerland, S., Graversen, R., Haas, C., Karcher, M., Kuhry, P., Maslanik, J., Melling, H., Maslowski, W., Morrison, J., Perovich, D., Przybylak, R., Rachold, V., Rigor, I., Shiklomanov, A., Stroeve, J., Walker, D., and Walsh, J. (2006). *State of the Arctic Report*. NOAA OAR Special Report. Seattle, NOAA/OAR/PMEL http://www.arctic.noaa.gov/soa2006/

Richter-Menge, J., Overland, J., Svoboda, M., Box, J., Loonen, M.J.J.E., Proshutinsky, A., Romanovsky, V., Russell, D., Sawatzky, C.D., Simpkins, M., Armstrong, R., Ashik, I., Bai, L.-S., Bromwich, D., Cappelen, J., Carmack, E., Comiso, J., Ebbinge, B., Frolov, I., Gascard, J.C., Itoh, M., Jia, G.J., Krishfield, R., McLaughlin, F., Meier, W., Mikkelsen, N., Morison, J., Mote, T., Nghiem, S., Perovich, D., Polyakov, I., Reist, J.D., Rudels, B., Schauer, U., Shiklomanov, A., Shimada, K., Sokolov, V., Steele, M., Timmermans, M.-L.,

Toole, J., Veenhuis, B., Walker, D., Walsh, J., Wang, M., Weidick, A., and Zöckler, C. (2008). *Arctic Report Card 2008*. http://www.arctic.noaa.gov/reportcard.

Robards, M.D. 2008. Perspectives on the Dynamic Human-Walrus Relationship. Unpublished Ph.D. Dissertation, University of Alaska Fairbanks.

SAON. 2008. Observing the Arctic. *Report of the Sustaining Arctic Observing Networks (SAON) Initiating Group*. www.arcticobserving.org (accessed August 5, 2009)

Serreze, M.C. 2008/2009. Arctic climate change: Where reality exceeds expectations. *Witness the Arctic* 13(1): 1–4.

Serreze, M.C., Holland, M.M., and Stroeve, J. 2007. Perspectives on the Arctic's shrinking sea ice cover. *Science* 315(5818): 1533–1536, Doi:10.1126/science.1139426.

Sivuqam 1985. *Sivuqam Nangaghnegha*. Lore of St. Lawrence Island. Echoes of Our Eskimo Elders, vol. 1. G.A. Apassingok, W. Walunga, and E. Tennant (eds.), Unalakleet: Bering Strait School District.

Stroeve, J., Holland, M.M., Meier, W., Scambos, T., and Serreze, M. 2007. Arctic sea ice decline: Faster than forecast. *Geophysical Research Letters* 34: L09501, doi: 10.1029/2007GL029703.

Thorpe, N., Hakongak, N., Eyegetok, S., and Elders, K. 2003. *Thunder on the Thundra. Inuit Qaujimajatukangit on the Bathurst Caribou*. Vancouver: Tuktu and Nogak Project.

Walsh, J.E. 2009. A comparison of Arctic and Antarctic climate change, present and future. *Antarctic Science* 21: 179–188.

Chapter 5
Sea Ice Distribution and Ice Use by Indigenous Walrus Hunters on St. Lawrence Island, Alaska

Marie-Luise Kapsch, Hajo Eicken, and Martin Robards

Abstract The hunting success of St. Lawrence Island walrus hunters from Savoonga (*Sivungaq*) and Gambell (*Sivuqaq*) is studied in relation to weather and sea ice conditions for the period 1979–2008. Satellite remote-sensing data, including ice concentration fields from passive-microwave radiometer data, have been examined over the entire time series in conjunction with walrus harvest data from two community-level monitoring programs. Important information to aid with interpretation of these data sets was provided by the hunters themselves, in particular through a log of ice conditions and ice use by L. Apangalook, Sr., of Gambell. From these data, we determined which ice conditions (concentrations >0 and <30%) and which wind speeds (1–5 m s^{-1} at Savoonga and 5–9 m s^{-1} at Gambell), temperatures (–5 to +5°C), and visibility (>6 km) provide the most favorable conditions for the walrus hunt. The research demonstrated that at the local level, though not necessarily at the region-wide scale, the sea ice concentration anomaly is a very good predictor of the number of favorable hunting days. With the exception of 2007 (and to a lesser extent, 2008), negative anomalies (less ice or earlier onset of ice retreat) coincided with more favorable (Savoonga) or near-average (Gambell) hunting conditions, controlled mostly by access to ice-associated walrus. Ice access and temporal variability differ significantly between Savoonga and Gambell; in contrast with northern Alaska communities, St. Lawrence hunters were able to maintain typical levels of harvest success during the recent record – low ice years of 2007 and 2008. We discuss the potential value of data such as assembled here in assessing vulnerability and adaptation of Arctic communities depending on marine-mammal harvests to climate variability and change.

Keywords Sea ice · Subsistence hunt · Ice conditions · Pacific walrus · Climate change

---

H. Eicken (✉)
Geophysical Institute, University of Alaska Fairbanks, Fairbanks, AK 99775-7320, USA
e-mail: hajo.eicken@gi.alaska.edu

I. Krupnik et al. (eds.), *SIKU: Knowing Our Ice*,
DOI 10.1007/978-90-481-8587-0_5, © Springer Science+Business Media B.V. 2010

Introduction

The Alaska Native population in coastal Alaska has relied on the marine environment as a source of food and sustenance for centuries to millennia. In the Bering Sea, and to a lesser extent off the North Slope of Alaska, Inupiaq and Yupik Eskimo have specialized in hunting walrus from small boats among the ice. The success of the hunt is directly linked to sea ice conditions, weather, the life cycle and population dynamics of the walrus, and the social and technological settings of the hunt (Krupnik and Ray 2007; Metcalf and Robards 2008).

The distribution of sea ice in the northern Bering Sea is influenced by several factors. It is linked to atmospheric circulation patterns, such as the Pacific Decadal Oscillation (PDO) or the Arctic Oscillation (AO; Stabeno and Hunt 2002). These two patterns are associated with variations in sea surface temperature and atmospheric pressure and thus influence ice conditions. Currently, atmospheric variability and larger scale change have resulted in a decline in fall and summer ice extent and have impacted sea ice properties, with more open water present, an earlier breakup and later arrival of the fall pack ice in the Bering Strait (Huntington 2000; ACIA 2005; Grebmeier et al. 2006; Stabeno et al. 2007). Such sea ice changes can affect walrus indirectly through changes in the distribution and abundance of their prey (Benson and Trites 2002). Direct impacts are mostly in connection with walrus' use of sea ice as a platform for resting, giving birth, and nursing. Walrus seek out ice floes that are thick enough to support their weight, surrounded by natural openings (leads) or thin ice that allow for access to the water column (Fay 1982). They typically gather on unconsolidated pack ice in late winter and spring, within 100 km of the leading edge of the ice pack (Burns 1970; Gilbert 1999). Hence, a change in the distribution and concentration of sea ice can affect timing and pathways of Pacific walrus and in turn impact walrus hunters' success and safety.

The interactions between walrus and sea ice are well understood by biologists (Fay 1982; Ray and Hufford 1989) and native hunters (Krupnik and Ray 2007). What is less clear at this time is how recent changes in surface climatology and sea ice at the local (Noongwook 2000), regional (Stabeno et al. 2007), and pan-Arctic scale (Stroeve et al. 2008) affect the walrus subsistence hunt. Clearly, hunters are more aware of such changes than anybody, as expressed by Leonard Apangalook from Gambell who comments that the "walrus season is very short now" (Oozeva et al. 2004). At the same time, large-scale changes as expressed in multiple decades of harvest observation records have been examined (Metcalf and Robards 2008). What is less understood are the mechanisms by which ice conditions and weather impact the hunt and in particular hunter's access to walrus. Furthermore, in order to effectively manage walrus harvest in a changing Arctic and provide information to coastal communities as they adapt to change, the local conditions governing access and hunting success need to be linked to variables such as ice concentration or weather patterns that can be tracked and possibly predicted on longer timescales.

This chapter is seen as a small contribution to help close this gap. It examines sea ice and weather conditions and their potential impact on the walrus hunt for the two neighboring villages of Savoonga (*Sivungaq*) and Gambell (*Sivuqaq*,

St. Lawrence Island; Fig. 5.1), two of the three primary walrus harvesting communities in Alaska over the past 60 years (Robards 2008). The study draws on a combination of ice observations by Siberian Yupik sea ice experts (gathered as part of a Sea Ice Knowledge and Use (SIKU) International Polar Year (IPY) project; Chapter 14 by Krupnik et al. this volume), weather records, and sea ice remote-sensing data obtained from different satellite sensors to provide both regional and local-scale perspectives on ice conditions. Information about the success and progression of the walrus hunt has been obtained from harvest records and local observations. The study focuses on three recent years (2006–2008), remarkable for their low summer minimum ice extent in the Pacific sector of the Arctic, and discusses the interplay between ice conditions, weather, and success of the walrus hunt in the context of long-term variability and change going back to 1979, the start of the systematic satellite record. The study is limited to the discussion of physical factors that control access to walrus and hence cannot provide insight into the role of variations in walrus distribution or population size nor the impact of sociological or technological change on the walrus hunt. Nevertheless, the study provides insight

Fig. 5.1 Map of the study region and the two villages on St. Lawrence Island, Alaska. The *black squares* identify the 75 × 75 km satellite subregions for Gambell and Savoonga and the *shaded areas* the ice edge for March, May, and July 2007

into what is regarded by many as the most important environmental constraints on the walrus hunt and their expression at the local level in relationship to large-scale, hemispheric climate change (Huntington et al. 2005); it is thus an important step in helping to downscale projections of future regional Arctic change to the community level at which the impact of such changes is manifested. At the same time, the detailed knowledge of local ice and walrus experts, also as expressed in hunting success, may help inform and guide future studies of the Arctic sea ice environment and walrus habitat.

Background – The Walrus Hunt on St. Lawrence Island

Walrus migration pathways are closely linked to the seasonal advance and retreat of the ice in the Bering and Chukchi Seas (Fay 1982; Krupnik and Ray 2007). Females and calves in particular remain within the margins of the ice pack as the ice edge moves north. Ice floes serve as platforms that allow walrus to feed over broad expanses of the shelf without having to retreat to land where access to food resources may be more limited and vulnerability to predation, disturbance, and trampling more exacerbated (Fay 1982). Bering Sea communities such as Gambell and Savoonga, that are located in the migration pathway of the walrus, thus have access to animals in close range to the village. Over the course of centuries, hunters from these communities have been able to develop and hone their hunting skills based on an intimate knowledge of the animal behavior and ice conditions.

Walrus hunting on St. Lawrence Island depends on several factors, including weather and ice conditions, the social and technological setting, and walrus ecology (Robards 2008). In Gambell and Savoonga the spring hunting season generally starts between the middle and end of April (Robards 2008, Benter and Robards 2009; Leonard Apangalook, personal communication 2009), when walrus begin to migrate northward with the thinning and retreat of the seasonal ice. Thus, hunters are strongly impacted by the rate of ice melt and retreat, which affects their ability to access and retrieve walrus. While walrus may migrate past the village it may be impossible for hunters under some conditions to pursue them because access to open leads is blocked by heavy near-shore ice conditions (Robards 2008). A further social constraint that may influence at least some hunters is the tradition or prioritization of postponing the focus on hunting walrus until the end of the subsistence whaling season in early May or until a whale has been harvested. While not all hunters may defer their hunt (Benter, personal communication 2009), the traditions of the whale hunt are one social factor that plays into the pursuit of walrus in early spring in some communities.

Irrespective of ice conditions, local weather conditions usually determine whether boats are able to safely hunt on a particular day. The hunters "rely on the mercy of the wind" (Leonard Apangalook, Gambell, personal communication 2009) and are also dependent on the visibility. If pack ice gets pushed against shore areas used to launch boats, there may be no other way to launch a boat without transporting boats to more distant locations across land. And if the visibility is poor, dangerous ice conditions are not readily apparent or predictable. Favorable weather is thus directly tied to safety on and in the ice and provides for better decision making by hunters. In recent years, hunters have also commented on how climate change has made it more difficult to apply traditional knowledge and anticipate changes in the weather or ice conditions (Krupnik and Jolly 2002). In prior years, "climate was more consistent, better predictable and people were able to plan their trips according to seasonal change. Now there is lots of wind, poor ice conditions and the polar pack has not been seen for the last 10–20 years. [. . .] Now there is just local grown ice around, which is less stable" (Leonard Apangalook, personal communication 2009).

In addition to the changes in weather and sea ice, the social and technological settings of the walrus hunt on the island have changed in many ways over past decades. Increases in the village populations, more numerous and faster boats (introduced in the 1970s), and better weapons (originally introduced in the 1800s; Robards 2008) have allowed Gambell and Savoonga to steadily increase their capacity to hunt walrus (Robards and Joly 2008). While many dramatic changes occurred prior to the start of the systematic remote-sensing record in 1979, social changes and technological advances continue and cannot be neglected in explaining the relationship between physical conditions and subsistence hunting.

Data and Methods

Sea Ice Data

To assess the potential for hunters to access ice-associated walrus and to link walrus harvest data to sea ice conditions, fields of daily sea ice concentration and total sea ice covered area have been obtained at the regional scale (75 × 75 km boxes and the whole Bering Sea area, Fig. 5.1), from satellite remote-sensing data for the time period 1979 to 2008. Specifically, data from the Scanning Multichannel Microwave Radiometer (SMMR) aboard the Nimbus-7 satellite from October 1978 to August 1987 and the Special Sensor Microwave/Imager (SSM/I) aboard the Defense Meteorological Satellite Program (DMSP) from July 1987 to present were used. Sea ice concentration at each grid cell is obtained from the corresponding brightness temperature and generated with the NASA Team algorithm (Eppler et al. 1992). For our calculations, final data released by the National Snow and Ice Center (NSICD) were used, including final NSIDC quality control (http://www.nsidc.org/data/nsidc-0051.html). The sea ice concentrations, averaged over a 75 × 75 km grid (Fig. 5.1), have an accuracy of ±15% during summer, when on ice melt ponds are present, and ±5% during winter (http://nsidc.org/data/docs/daac/nsicd0051_gsfc_seaice.gd.html). Multiple measurements are averaged to daily means and missing data are linearly interpolated.

The significance of sea ice concentration trends calculated in the results section has been tested with a single-sided t-test. This test is applicable to independent variables which holds for sea ice variables in the Bering Sea/St. Lawrence Island region, since the ice cover melts completely each summer with very little if any persistence of local anomalies into the spring of next year.

Weather Data

Hourly weather observations from the National Climatic Data Center (NCDC) have been examined (http://www.ncdc.noaa.gov/oa/land.html). The meteorological variables wind direction, wind speed, visibility, temperature, and cloud cover were chosen to assess whether hunting conditions were favorable. Many indigenous hunters report these variables as being of greatest importance when evaluating weather conditions for a hunt (Oozeva et al. 2004). The hourly measurements have been averaged to daily means (converted to and reported here in local time) in order to be able to compare them with daily sea ice concentration data.

Walrus Harvest Data

Daily walrus harvest data for the villages of Gambell and Savoonga have been analyzed between 1992 and 2008 for the spring hunting season, as well as for the years 1980–1984. Walrus harvest records for 1992–2008 were assembled by the U.S. Fish and Wildlife Service and the Alaska Eskimo Walrus Commission (AEWC). In the Walrus Harvest Monitoring Program (WHMP), during most of the spring the gender and age class for every walrus that had been retrieved were recorded by hunt monitors as hunters returned to the beach at their home village. The observation period for the years 1992–2004 and 1980–1984 amounted to 6–8 weeks each year between mid-April and mid-June and historically covered 90% of the total annual harvest numbers for Gambell and Savoonga (Benter, personal communication 2009). Only harvest data for which the date of harvest is known have been analyzed here. The same applies to the WHMP data between 2005 and 2008, when the observation period was reduced to 2 weeks, mostly in May. The monitors met most of the boats returning from walrus hunts, and hence the number of unrecorded walrus is thought to be relatively small (Garlich-Miller and Burn 1999). Since this study is not concerned with the age and gender distribution of walrus but focuses on the total number taken each day, all age and gender classes have been aggregated. The unpublished data from 1980 to 1984 were recorded during a similar harvest-monitoring program (Lourie 1982) and summarized by Robards (2008). To alleviate bias from the shortened season of direct observations used by the Walrus Harvest Monitoring Program since 2005, data from the Marking, Tagging, and Reporting Program (MTRP) for the years 2006–2008 have also been evaluated. The MTRP is a federally mandated program requiring all village hunters to report harvested walrus to the U.S. Fish and Wildlife Service within 30 days after the take, while the location of the hunt is unimportant. The numbers of walrus taken are aggregated into weekly values.

To examine the connection between these different data sets and observations by local hunters, favorable days for the walrus hunt were those days in the WHMP data on which at least one walrus had been taken. In contrast, days on which no walrus were taken were considered as unfavorable for the hunt, due to either limited access on account of ice conditions or hazardous weather or ice conditions. While we ignore any other factors that may have played into a decision for hunters not to go out, we only consider days between the onset of the spring hunting season (first walrus taken) and its end (last walrus taken, based on the harvest monitoring data). The time period between these two dates we consider as the length of the spring hunting season for a particular year, with some walrus taken outside of this time period in fall and winter. Since weather conditions may vary during a single day but we could only consider mean values for weather variables such as wind speed, this may have introduced small errors into assessments of suitable or unsuitable hunting conditions. At the same time, since hunters themselves rely on weather forecasts, erroneous forecasts may prevent hunting activity on suitable hunting days, potentially introducing small errors as well (Benter, personal communication 2009).

Local Observations of Ice Conditions and Hunting Activity

In order to help address the challenge of downscaling from satellite observations to local ice conditions and to aid in separating the different social, technological, and environmental factors impacting the walrus hunt, we rely on regular observations of ice conditions and sea ice use made by Leonard Apangalook from Gambell. Mr. Apangalook, a respected hunter with great knowledge of the local environment, has been keeping a daily log of weather and ice conditions, animal sightings, and activities associated with ice use (as part of the SIKU project under the leadership of Igor Krupnik; Chapter 14 by Krupnik et al. this volume); his records for the ice seasons in 2006–2008 have provided important insight into factors determining hunting success.

Results

Regional Ice Conditions, the Sea Ice Cycle, and Hunting Activities

Sea ice conditions around St. Lawrence Island are dominated by the seasonal cycle of temperature and associated weather and ice patterns. In recent years, the freeze-up of the ocean in front of Gambell (see Fig. 5.1) typically started in mid-December (Fig. 5.2). During the winter, ice concentration increases with an average (January

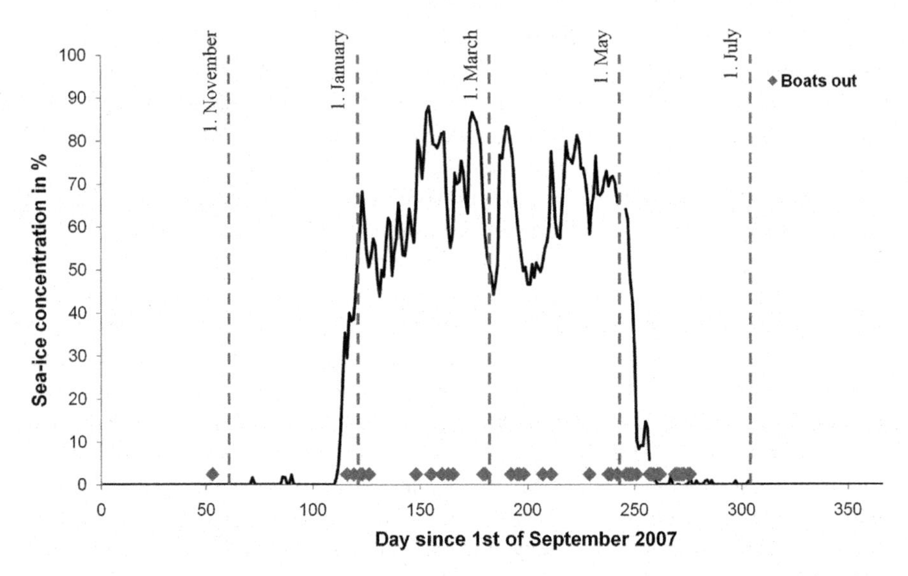

Fig. 5.2 Annual sea ice cycle in terms of ice concentrations derived from remote-sensing data (SSM/I) for Gambell, Bering Sea, from September 1, 2007 to August 31, 2008. The *diamonds* denote days on which one or several boats from Gambell were out hunting walrus and seals (based on observations by Leonard Apangalook)

through April) of 66% for the year 2007/2008. Ice concentrations remain high until the beginning of May, when they plummet with the start of breakup and melt, with ice completely absent starting in June. The start of freeze-up and breakup and the average winter sea ice concentration vary slightly between single years with about ±13 days for freeze-up and ±11 days for breakup in front of Gambell. A similar annual cycle is observed off Savoonga, though with higher wintertime ice concentrations (25% higher in 2007/2008) and an offset of the start of breakup of approximately 14 days due to a later ice melt.

Examining the dates on which hunters from Gambell were pursuing walrus or bearded seals by boat (L. Apangalook, unpublished observations 2008) indicates a close link between hunting activities and the ice cycle (Fig. 5.2). Thus, hunting by boat is mostly restricted to those times when sea ice – providing access to walrus that are migrating northward with the retreating ice – was present within a reasonable distance from the community while at the same time open leads allowed boats to pursue walrus. In 2007, Gambell hunters traveled on average between 13 and 146 km per harvested walrus, starting at the end of April with small distances, continuing through May with increasing travel distances between 32 and 80 km per captured walrus and finishing the season with larger distances in early June (Benter and Robards 2009). The most active time is during break-up, between the beginning of April and mid-May 2008 (Fig. 5.2). A similar conclusion is drawn from an analysis of the MTRP harvest records, which indicates that in the years 2006–2008 most walrus were taken between the end of April and early June (Fig. 5.3). In considering

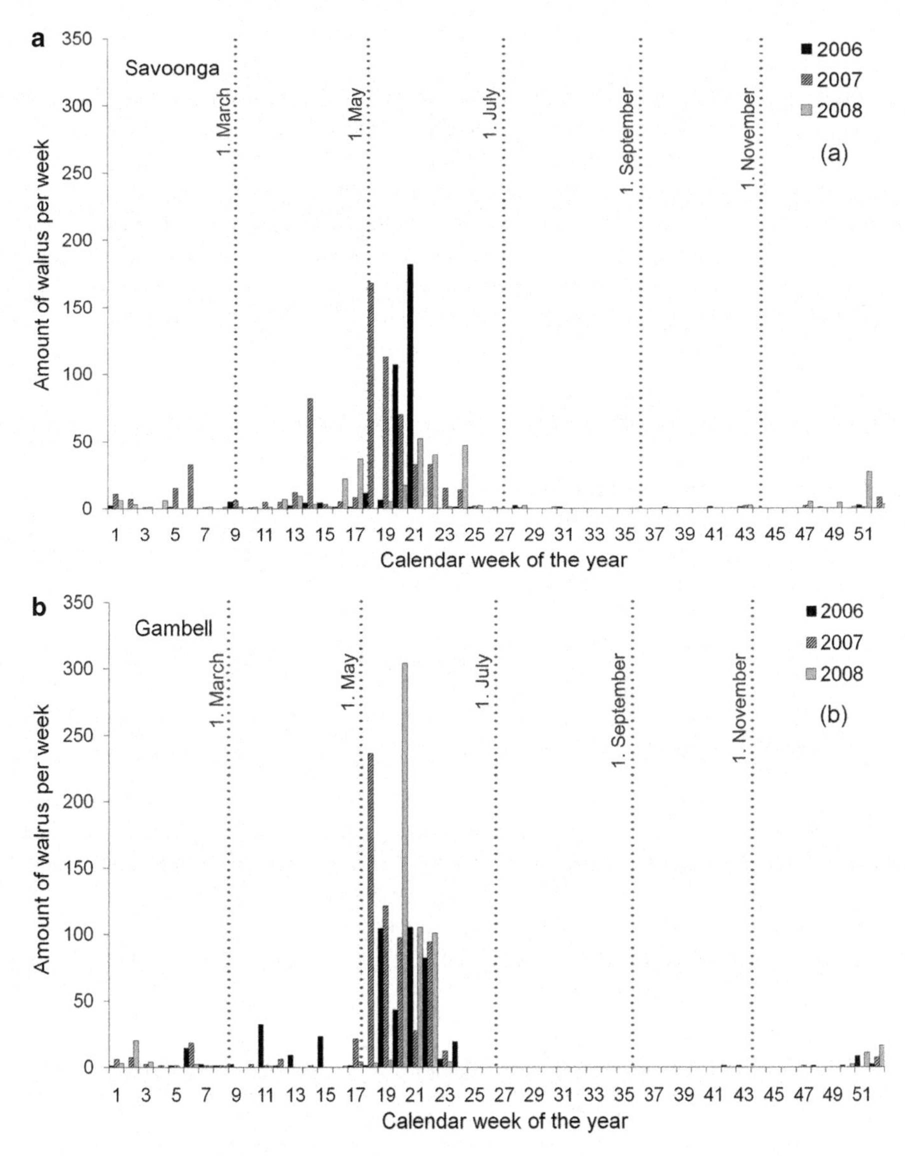

Fig. 5.3 Weekly takes of walrus harvested in (**a**) Savoonga and (**b**) Gambell based on MTRP data collected by the US Fish and Wildlife Service

MTRP data, it needs to be kept in mind that these data do not provide any insight into hunting effort, which is significantly larger based on observations by Leonard Apangalook (Fig. 5.2) or of an increased capacity to hunt (Robards 2008).

In addition to access afforded hunters by openings in the near-shore ice in early spring (Fig. 5.3a, 14th calendar week), in recent years they have also pursued walrus in the winter (November to January) in both Gambell and Savoonga, as evident from observations by Leonard Apangalook (Fig. 5.2) and MTRP data (Fig. 5.3). With wintertime ice concentrations mostly below 70–80% in Gambell (Fig. 5.2), the potential is high for a network of connected leads to provide access to walrus that overwinter close to the island – provided that near-shore ice conditions are favorable (see later). In Savoonga, the situation is somewhat different as ice concentrations are approximately 25% higher during winter, reducing the probability of finding leads and gaining access to walrus. However, Savoonga hunters use snowmobiles to hunt from landfast ice or haul boats over the island to gain access to open water in the polynya on the lee side of the island in the south.

Environmental Variables Controlling the Walrus Hunt

Weather Conditions and the Walrus Hunt

> North wind at 25 mph [11 m s^{-1}], 5 F [–15°C]. With large ice [floes] open leads extend to size of floes and good for boat hunting, but winds are high. Even walrus and seals don't like to haul on ice when it is windy (19th of March 2007).
>
> North wind at 21 mph [9 m s^{-1}], 4 F [–16°C]. Clear. Unable to get out by boat due to stronger winds and pack ice constantly moving down wind; with stronger wind the north current does not move ice north (20th of March 2007).
>
> Northeast wind at 20 mph [9 m s^{-1}]. 28 degrees [–2°C]. Heavy overcast with snow and fog. Everyone concerned about lack of favorable weather conditions and no hunting (10th of May 2008, 08:00 AKST).

These observations from the daily log of Leonard Apangalook from Gambell illustrate some of the weather conditions that impact the walrus hunt. Even if sea ice conditions would allow hunters to go out, weather may prevent that from occurring. To explore the extent to which weather impacts the hunt, National Weather Service observations from Gambell and Savoonga were compared to the walrus harvest records (WHMP data set, Tables 5.1 to 5.8). Based on statements by local

Table 5.1 Weather statistics for the spring season at Savoonga, 1993–2008

	Statistic	Wspd	Vis	SIC
Harvest days	Mean	4.0	12.4	25.0
	Number of days	210	195	215
	SD	2.1	3.6	26.8
	Max	20.2	16.1	97.6
	Min	0.4	0.8	0.0
Non-harvest days	Mean	5.2	11.0	37.5
	Number of days	329	195	334
	SD	3.1	4.1	34.7
	Max	20.2	16.1	97.6
	Min	0.0	0.1	0.0

Table 5.2 Weather statistics for the spring season at Gambell, 1992–2008

	Statistic	Wspd	Vis	SIC
Harvest days	Mean	5.0	13.0	17.2
	Number of days	239	219	238
	SD	2.2	3.3	19.6
	Max	11.7	16.1	92.3
	Min	0.0	0.8	0.0
Non-harvest days	Mean	7.9	11.6	22.3
	Number of days	343	219	342
	SD	3.2	3.9	22.8
	Max	16.5	16.1	92.3
	Min	0.0	1.3	0.0

Table 5.3 Walrus harvested in relation to wind direction in spring for Savoonga, 1993–2008

Wdir	Number of walrus	Fraction of harvest (%)	Wdir frequency (%)
NNE	52	1	3
NE	173	3	6
ENE	1503	28	21
ESE	615	12	14
SE	284	5	7
SSE	438	8	6
SSW	247	5	8
SW	461	9	8
WSW	803	15	14
WNW	477	9	5
NW	132	3	3
NNW	107	2	5

Table 5.4 Walrus harvested in relation to wind direction in spring for Gambell, 1992–2008

Wdir	Number of walrus	Fraction of harvest (%)	Wdir frequency (%)
NNE	2342	29	24
NE	802	10	15
ENE	956	12	9
ESE	446	6	6
SE	420	5	6
SSE	322	4	6
SSW	876	11	13
SW	661	8	8
WSW	556	7	6
WNW	125	21	1
NW	129	21	1
NNW	336	4	4

Table 5.5 Walrus harvested in relation to wind speed in spring for Savoonga, 1993–2008

Wspd (ms^{-1})	Number of walrus	Fraction of harvest (%)	Wspd frequency (%)
>1	395	8	5
1–5	3668	69	69
5–9	1142	22	25
>9	87	2	1

Table 5.6 Walrus harvested in relation to wind speed in spring for Gambell, 1992–2008

Wspd (ms^{-1})	Number of walrus	Fraction of harvest (%)	Wspd frequency (%)
>1	257	3	3
1–5	4660	58	47
5–9	3044	38	48
>9	33	0	3

Table 5.7 Walrus harvested in relation to visibility in spring for Savoonga, 1993–2008

Visibility (km)	Number of walrus	Fraction of harvest (%)	Visibility frequency (%)
0–3	76	2	2
3–6	72	1	3
6–9	96	19	14
9–12	1052	21	22
>12	2835	57	60

Table 5.8 Walrus harvested in relation to visibility in spring for Gambell, 1992–2008

Visibility (km)	Number of walrus	Fraction of harvest (%)	Visibility frequency (%)
0–3	29	0	1
3–6	352	5	4
6–9	500	7	8
9–12	917	12	18
>12	5758	76	69

experts wind direction, wind speed, visibility, cloud cover, and temperature were considered in the analysis. Ocean currents have not been considered because to our knowledge no systematically collected time series of currents (or other oceanographic data) is available for this region and this time period.

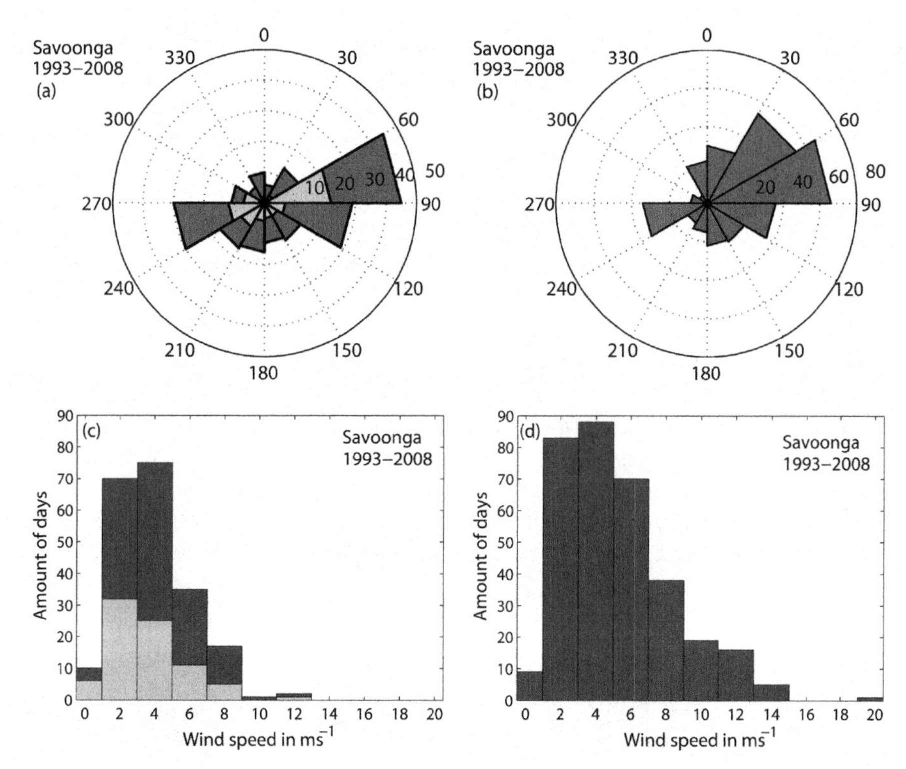

Fig. 5.4 Histogram of (**a**) wind direction and (**c**) wind speed for days on which walrus were taken, compared to (**b**) wind direction and (**d**) wind speed histograms for days (during the spring season) on which no walrus have been taken by Savoonga. The wind directions are summarized in 30° sections (0–30° = NNE, 30–60° = NE, 60–90° = ENE, etc.)

Wind Velocity

> Tunglu – When the wind blows and blows, it forms ice ridges in the northern side (in Gambell), and it also starts pushing ice south. Then the (wretched) ice covers the whole ocean with nowhere to go on boat (Oozeva et al. 2004, 49).

Wind direction and wind speed are very important factors determining whether crews go out to hunt. When the wind speed is low (less than about 2 m s^{-1}; George et al. 2003) open leads tend to freeze over in cold weather and are inaccessible; at very high wind speeds ice movement and deformation pose dangers, as indicated by Oozeva's description of the Yupik term *tunglu*. Since wind speed and direction also determine whether boats can be launched at village sites (ostensibly a north and west facing beach in Gambell and a north facing beach in Savoonga), weather station data have been analyzed for Savoonga and Gambell (Figs. 5.4 and 5.5).

Wind conditions on spring walrus harvest days between 1992 and 2008 differ between the villages, although the range of wind speeds during which walrus were taken is comparable for both (1–9 m s^{-1}; Tables 5.5 and 5.6; Figs. 5.4 and 5.5). In Savoonga, 69% of the walrus were taken when wind speed was between

1 and 5 m s^{-1}, corresponding to 69% of all harvesting days. In Gambell, more than half of the walrus (58%) were taken during the same wind speed interval, but these wind speeds only prevailed during 47% of all harvest days. On most hunting days, wind speed was between 5 and 9 m s^{-1} (48%). This is also evident when comparing mean wind speed on harvest dates. In Savoonga the mean overall wind speed was 4.0 ± 2.1 m s^{-1}, 1 m s^{-1} lower than in Gambell (5.0 ± 2.2 m s^{-1}; Tables 5.1 and 5.2). In Savoonga the mean wind speed for days when no walrus were taken was 5.2 ± 3.1 m s^{-1}, significantly higher (paired t-test with the null hypothesis that the means are equal can be rejected at the 5% level, $p = 0.05$) than during harvest days. The same holds for Gambell but the wind speeds on days without walrus takes are higher than in Savoonga (7.9 ± 3.2 m s^{-1}; Table 5.2; Fig. 5.5).

The wind directions favored by hunters from the two villages vary much more than the favored wind speeds, and both are correlated to some extent. In Savoonga, most walrus were harvested with winds from ENE, ESE, and WSW directions (Table 5.3). The prevailing wind direction in spring (month of March, April, May, and June) is from ENE, corresponding to 21% of harvest days (Table 5.5). In Gambell the preferred wind direction for the hunt is from NE (0–90°; 48% of all

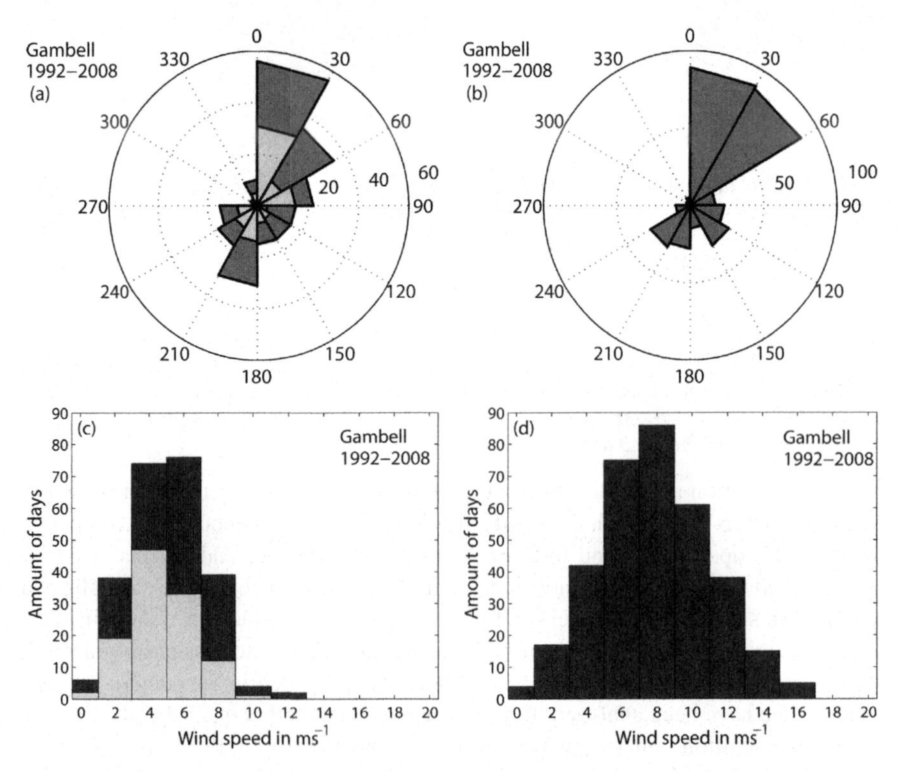

Fig. 5.5 Histogram of (**a**) wind direction and (**c**) wind speed for days on which walrus were taken, compared to (**b**) wind direction and (**d**) wind speed visibility histograms for days (during the spring season) on which no walrus have been taken by Gambell

hunting days) and from SSW. During NW winds (270–360°) hunters rarely went out (6%; Table 5.4).

Visibility

St. Lawrence Island hunters articulate that good visibility is key for being safe on the ice, to detect changes in ice conditions and to be able to spot walrus. Hence, hunters do not go out when they are unable to judge ice conditions or detect walrus due to poor visibility. This is reflected in the disproportionate number of harvests (and hunting days) for visibilities higher than 6 km (Tables 5.7 and 5.8). Harvest days with visibility below 6 km account for 5% or less of all successful harvest days in Gambell and Savoonga (Tables 5.7 and 5.8). In contrast, 60 and 69% of all days with walrus takes had visibility of more than 12 km in Savoonga and Gambell, respectively.

Cloud Cover and Temperature

Cloud cover and temperature play secondary roles for the walrus harvest during spring, other than through their indirect linkage to ice conditions and seasonal evolution of ice. At Savoonga, 57% of all walrus were harvested when the sky was overcast with 7–8 octas cloud cover, accounting for 69% of all harvesting days. At Gambell, the corresponding numbers account for 70% of the walrus harvested and 70% of all harvest days. This finding is explained by consistently cloudy conditions at both Savoonga and Gambell during the spring hunting season, with an average cloud cover of $7/8 \pm 2/8$.

The temperature ranged between –5 and +5°C for over 90% of the harvest days, and more than 95% of all walrus were harvested under these temperature conditions in both communities. On average the temperature was 1 K higher in Gambell and 2 K higher in Savoonga on days that walrus were taken compared to those days that no hunt took place.

Ice Conditions and the Walrus Hunt

The WHMP walrus harvest data indicate that based on ice concentration data for the 75×75 km grid monitored off both St. Lawrence Island communities, walrus were mostly taken when ice concentrations were above 0 and below 30% (Fig. 5.6). In both Savoonga and Gambell 88% of the harvests occurred for ice concentrations lower than 30% (Tables 5.9 and 5.10). These days account for 71 and 81% of all days on which walrus were harvested in Savoonga and Gambell, respectively.

While the spread of the data shown in Fig. 5.6 precludes major conclusions, the data do provide evidence that (1) in Savoonga disproportionately fewer walrus are caught at high (>60%) and very low (<10%) ice concentrations, with the opposite true for ice concentrations between 10 and 30%, and (2) in Gambell disproportionately fewer are harvested at ice concentrations <10 and >40%. For 36% of all days without a walrus take in Savoonga the sea ice concentration was below 10%. In Gambell the corresponding fraction was 47% of all days.

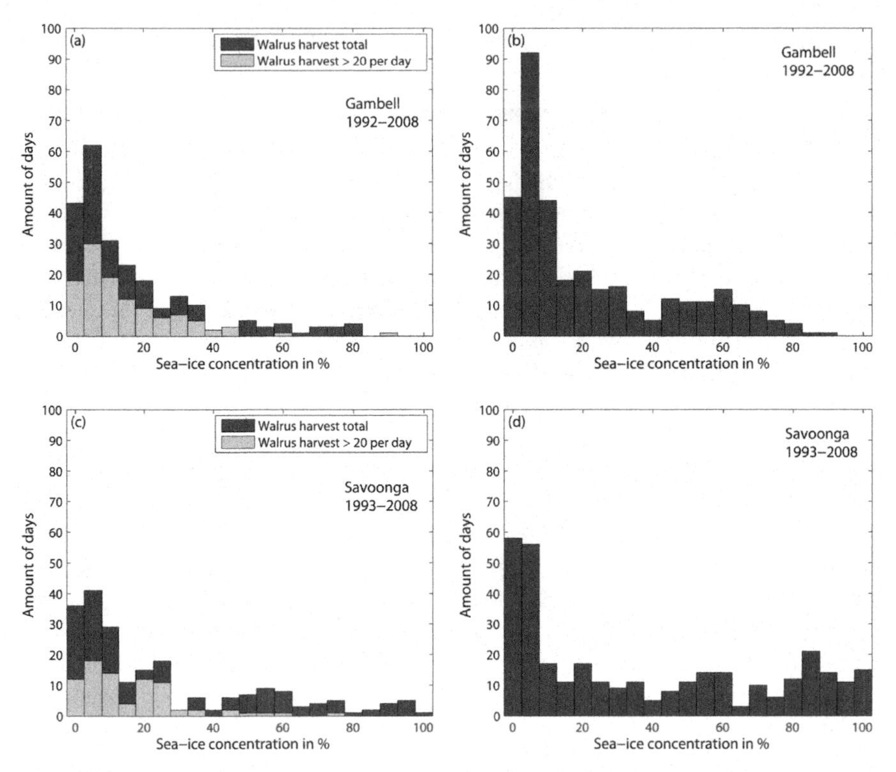

Fig. 5.6 Histogram of sea ice concentration days on which walrus were taken by (**a**) Savoonga and (**c**) Gambell, compared to ice concentration histograms for days (during the spring hunting season) on which no walrus were taken by (**b**) Savoonga and (**d**) Gambell

Table 5.9 Sea ice concentration and harvest data for Savoonga in spring of 1993–2008

SIC (%)	Number of walrus	Fraction of harvest (%)	SIC frequency (%)
0–10	2796	52	44
10–20	884	17	14
20–30	1019	19	13
30–40	182	3	3
40–50	98	2	5
50–60	162	3	7
60–70	70	1	5
70–80	83	2	3
80–90	4	0	1
90–100	36	1	4

Table 5.10 Sea ice concentration and harvest data for Gambell in spring of 1992–2008

SIC (%)	Number of walrus	Fraction of harvest (%)	SIC frequency (%)
0–10	3778	48	53
10–20	2165	28	18
20–30	944	12	11
30–40	626	8	8
40–50	166	2	2
50–60	18	0	3
60–70	55	1	2
70–80	23	0	4
80–90	2	0	0
90–100	22	0	0

Longer Term Change and Case Studies from the 1980s and 2000s

> The pattern [of walrus migration] is not the same today anymore like it used to be because of the climate change and the ice conditions have changed and the animals are affected by this global warming thing. That is sad to say. I think we are more adversely affected here because our walrus and whaling seasons are short, because of inclement weather . . . When I was growing up and later on as an adult hunting with my dad, we used to have good weather all the time (Leonard Apangalook, Sr., Gambell, St. Lawrence Island, in Metcalf 2003).

As expressed by Leonard Apangalook, Sr., and mentioned by George and Chester Noongwook (Noongwook 2000; Oozeva et al. 2004), hunters in Gambell have recognized changing sea ice, weather, and walrus migration patterns. These observations at the village scale coincide and as detailed by Huntington (2000) in many cases precede satellite observations of pan-Arctic sea ice retreat. This retreat has been most pronounced in the summer months, with the summer minimum ice extent declining by more than 10% per decade from 1979 to 2008, and record low extent in 2007 and 2008, almost 25% below the previous record minimum set in 2005 (Stroeve et al. 2008). Here, we examine how ice concentration has varied over those years in the ocean adjacent to Gambell and Savoonga (75 $\times$ 75 km as detailed in the methods section; Fig. 5.1), in areas important to walrus hunters.

Ice concentration anomalies based on annual averages for the ice year from September 1 through August 31 (Fig. 5.7a, c) and spring seasonal averages for March through June (Fig. 5.7b, d) for Savoonga and Gambell show substantially more inter-annual variation and a weaker trend than pan-Arctic data referenced above. Annual anomalies indicate a reduction in ice extent by 2%/decade for both Gambell and Savoonga, explaining 17 or 14% of the observed variance and are significant at the 95% level (t-test with the null hypothesis that the correlation coefficient is significant cannot be rejected, $p = 0.05$). Spring ice concentration anomalies are slightly higher, approaching 3%/decade. Since the mid-1990s, inter-annual variation in ice concentration appears to have increased significantly. The variance for spring concentration anomalies is generally higher than the annual anomalies.

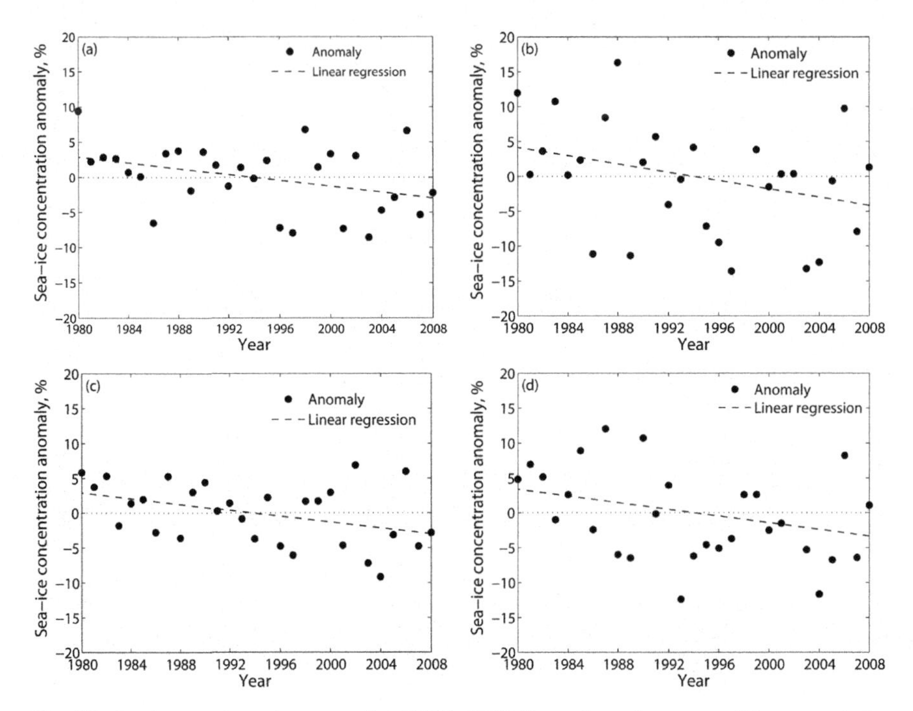

Fig. 5.7 Sea ice concentration anomalies (1980–2008) for each sea ice year and linear trends for entire time period in (**a**) Savoonga ($R^2 = 14\%$) and (**c**) Gambell ($R^2 = 17\%$). Sea ice concentration anomalies for the spring months (March through June) and linear trends are also shown for Savoonga ($R^2 = 9\%$) (**b**) and (**d**) Gambell ($R^2 = 9\%$). The anomalies are calculated by subtracting annual-averaged or spring-averaged values from the mean calculated for the entire time series. R^2 indicates the fraction of the variance explained by a linear regression

However, particularly noteworthy is the fact that none of the recent record minimum summer ice extent years register as record anomalies at the St. Lawrence Island locations.

As derived in previous sections of this contribution, hunters had the greatest success in taking walrus in the ice concentration range between above 0 and below 30% in Savoonga and Gambell. Taking these intervals as constraints for successful spring walrus hunts, we compared the years 1982–1984 as representative of the pre-decline Arctic ice extent with the years 2006–2008 as representative of reduced Arctic ice extent in recent years. In spring March–June of 1982–1984 ice concentrations were favorable for the hunt in Gambell (>0 to <30% ice concentration) on a total of 149 days. During the time period 2006–2008 (spring) there were only 92 days with such ice conditions, a reduction of 38% compared to the early 1980s. By contrast, in Savoonga no similar strong reduction was observed, with 100 favorable days in 1982–1984 and 94 favorable days in 2006–2008, a 6% reduction. In part, this is explained by the fact that in Savoonga disproportionately large numbers of walrus are taken on days with ice concentrations between 10 and 30%, which did not change significantly between the periods 1980s and 2000s. Gambell, on the other hand, experienced a reduction by 13% in the number of days amenable to a successful hunt (see also Fig. 5.6).

The hunting success during these observation periods in 1982–1984 was 116 walrus/week in Gambell and 60 walrus/week in Savoonga (weekly averages of the WHMP are calculated to reduce the error due to a decreased observation period). In the mid-2000s hunting success was slightly reduced by 3 walrus/week in Gambell, but had substantially increased by 36 walrus/week (38%) in Savoonga. Since the aggregated WHMP data (weekly totals) do not warrant a detailed analysis of hunting success on specific days (Benter, personal communication 2009), it is noteworthy that on highly successful days hunters brought in as many as 184 walrus in Gambell and 131 in Savoonga.

The time series of the number of days favorable to the walrus hunt determined for the satellite record (1979–2008) shows great inter-annual variability. While Gambell sees three fewer favorable hunting days (a reduction from 50 to 47 days) over the time period and Savoonga an increase of seven (37 days in 1980 and 44 days in 2008), these trends are not statistically significant (Fig. 5.8; *t*-test, $p = 0.05$). However, there is a clear link between the local ice conditions in spring (as determined for the study sites offshore from each village) and the number of favorable hunting days. Figure 5.9 illustrates how negative or positive sea ice anomalies in spring directly translate into more or fewer favorable hunting days. Spring ice concentration anomalies explain over 50% of the variance observed in the number of favorable hunting days (57% for Gambell and 63% for Savoonga, *t*-test, $p = 0.001$). Annual ice concentration anomalies account for 18 ($p = 0.05$) and 32% ($p = 0.001$) of the variance in Gambell and Savoonga.

Finally, we assessed whether the regional ice conditions in the Bering Sea might be a good predictor of the number of favorable days for the walrus hunt. While there is significant correlation between local ice conditions and ice concentration anomalies for the entire Bering Sea at the annual scale ($R^2 = 43\%$ for Gambell and

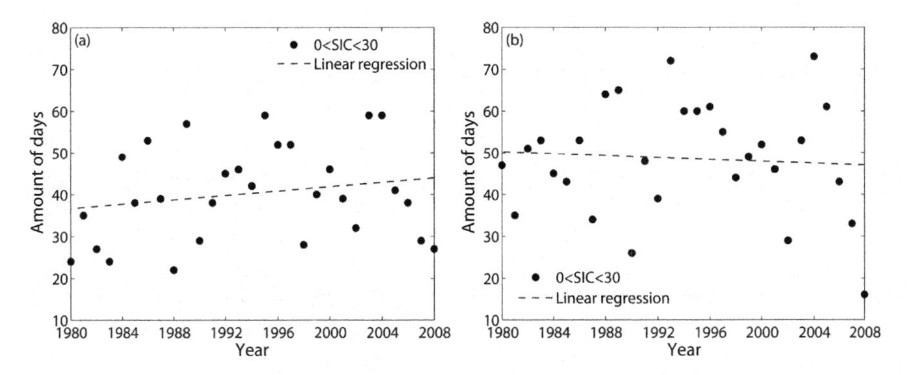

Fig. 5.8 Number of days when favorable sea ice conditions prevailed in (**a**) Savoonga and (**b**) Gambell

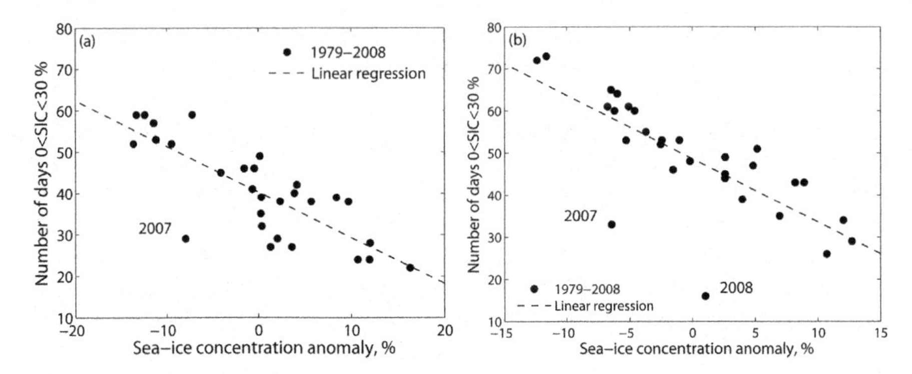

Fig. 5.9 Number of days on which the sea ice concentration exceeded 0% and was lower than 30% plotted relative to the sea ice concentration anomalies in spring for (**a**) Savoonga ($R^2 = 63\%$) and (**b**) Gambell ($R^2 = 57\%$)

$R^2 = 59\%$ for Savoonga; *t*-test, $p = 0.001$), this is less so for springtime anomalies ($R^2 = 12\%$ and $p = 0.05$ for Gambell; $R^2 = 27\%$ and $p = 0.01$ for Savoonga). However, most important is the finding that Bering Sea spring and annual ice concentration anomalies explain only 10% of the observed variance in the number of favorable hunting days.

Discussion

As evident from the results presented in the previous section, local, regional, and inter-annual variabilities and the shortcomings of coarse-resolution satellite data present a number of formidable challenges in reaching conclusions about the linkages between weather and ice conditions favorable to the walrus hunt on St. Lawrence Island. In part this is due to insufficient spatial and temporal resolutions, with respect to both harvest data (which are accurate to within a day, at least for the WHMP data) and weather records (which are daily averages of a limited set of observations). At the same time, the passive microwave satellite data analyzed here have significant limitations due to limited spatial resolution and errors in ice concentrations derived for the coastal zone (Massom 2009). To reduce the latter, we had to eliminate pixels that border on the coast from the analysis. While this does not affect overall trends, it may somewhat limit the perspective on hunting in close proximity to the village sites. We do note, however, that the study sites, 75 × 75 km in size, cover much of the region frequented by hunters, who on average travel between 13 and 146 km offshore during the season in Gambell (Benter and Robards 2009). Here, observations by Yupik sea ice experts, such as the detailed discussion of Gambell ice conditions provided by Leonard Apangalook, Sr., and Paul Apangalook (Chapter 14 by Krupnik et al. this volume), are invaluable in providing context and ensuring that satellite data are in fact representative of a specific site or ice-associated activity. We will discuss this aspect further below. However, despite such limitations, some tentative, broad conclusions with respect to the weather and ice conditions that favor spring walrus harvests can be drawn. Thus, Table 5.11 summarizes key findings and delineates the window of opportunity most favorable to hunting success in Savoonga and Gambell.

Table 5.11 reflects the fact that the most important role played by the weather is in controlling access to offshore sea ice that is bearing walrus while allowing for safe boating and hunting conditions among the ice pack or in open water. In particular the ice movement forced by winds and currents is of great interest to walrus hunters on St. Lawrence Island, evident in the description by George Noongwook (Savoonga) of a wind-driven ice phenomenon, referred to as *pequneq* in Yupik: "A coning of thin or freshly formed ice pushed by the storms before it crashes down [. . .] producing a small open lead for a very brief moment that closes almost instantly. If you get caught in one of these, you will not have time to escape! We have lost some of our people in these conditions all in the name of seeking food for survival" (Noongwook 2000). This description links the short-term variability of ice movement on a local scale with the ability of Savoonga hunters to interpret ice conditions. While the large-scale conditions prerequisite to *pequneq* can be extracted from satellite and

Table 5.11 Environmental factors relevant for local walrus hunting success off St. Lawrence Island, Alaska

Environmental factor or trend	Savoonga – favorable spring walrus harvest conditions	Gambell – favorable spring walrus harvest conditions
Sea ice concentration	Above 0% and below 30%; hunters need at least a few open leads that are wide enough to launch boats. Also, walrus prefer to rest on the ice, so a certain ice thickness (>60 cm; Fay, 1982) and floe size is necessary	
Wind speed	• 1–5 m s^{-1} Higher wind speeds present boating safety hazards	• 5–9 m s^{-1}
Wind direction	ENE, ESE, WSW; these are the main directions when hunting takes place. Northerly winds push ice against the shore and close off access to sea	NNE, ENE, SSW; northerly winds in Gambell do not have as much influence on the hunt as in Savoonga because Gambell has a beach on the north and west side to launch boats from
Visibility	>6 km; good visibility is a key factor for safety on the ice and important to spot walrus	
Air temperature	–5 to +5 °C due mostly to the time of the year (spring break-up); at low wind speeds and low temperatures (<–20 °C) leads tend to freeze (George et al. 2003)	
Cloud cover	Not important for the spring hunt; linked to temperature and indirectly to sea ice growth or melt and may hence on occasion correlate weakly with hunting success	
Sea ice concentration window favorable for spring hunt	1982–84: 100 days 2006–08: 94 days	1982–84: 149 days 2006–08: 92 days
Spring-hunt success	1982–84: 60 walrus/week 2006–08: 96 walrus/week	1982–84: 116 walrus/week 2006–08: 113 walrus/week

weather records, the temporal and spatial resolutions of such data are insufficient for deeper insight into such processes.

However, what this study does resolve quite well, as summarized in Table 5.11, are the differences in ice conditions (and to a lesser extent weather) between the villages of Savoonga and Gambell, as well as their impact on hunting success. Thus, small but significant differences in wind speed between villages on walrus harvest days with similar ice conditions and visibility suggest that hunting itself is not only dependent on the fact that sea ice is present. St. Lawrence hunters are very experienced and able to hunt and butcher walrus directly in the water (even if it is less efficient, inconvenient, and certainly the less preferred option from both, a hunter's or manager's perspective due to greater potential for loss) or to tow animals back toward shore for butchering (Benter, personal communication 2009). Hence, it is possibly also the behavior and distribution of walrus, as the animals are resting and nursing on the ice, that link ice conditions and hunting success. However, even if sea ice is present, sufficient open water is required for boat launching and navigation through leads or in loose pack ice. This circumstance is highlighted in the daily log of Leonard Apangalook on April 16, 2008: "No open water west side of Gambell where walrus, seals and whales pass by north on their northerly migration during this time of the season. Wide open water on the leeward side of the island, where Daniel Apassingok went out on boat from Kiyellek [Qayilleq] and got a bearded seal." With an ESE wind during that day ice was pushed against the coast east of Gambell while 5 km east from Gambell, at Qayilleq (a hunting camp, see Chapter 14 by Krupnik et al. this volume), ice got blown out and opened access to the sea. L. Apangalook's observation addresses three key points or prerequisites for access to ice and walrus:

1. There must be open water in form of large open areas or small open leads accessible, large enough to launch a boat
2. Wind stress can result in a wind-driven movement of sea ice that either packs ice against the coast or opens a coastal or flaw lead
3. The main advantage of Gambell compared to Savoonga is the position of the village, such that it has a north and a west beach and furthermore that the hunting spot Qayilleq is in close range around 6 km to the east, on the eastern side of the Gambell peninsula, allowing access to the lead system for different wind conditions. Savoonga hunters can compensate for more severe ice conditions and fewer options to launch boats by accessing other hunting spots on the south side of the island.

The first point is illustrated in the high-resolution Synthetic Aperture Radar (SAR) image of April 17, 2008 shown in Fig. 5.10, within less than one day of the ice conditions described by Mr. Apangalook and under comparable weather conditions. At this time, neither the west coast of the Gambell peninsula nor the north coast were ice free because of strong NNE winds on April 13 and strong SSW winds on April 16, 2008. Wind speeds of up to 9 m s^{-1}, possibly aided by currents, were able to move the ice up against the coast north and west of Gambell.

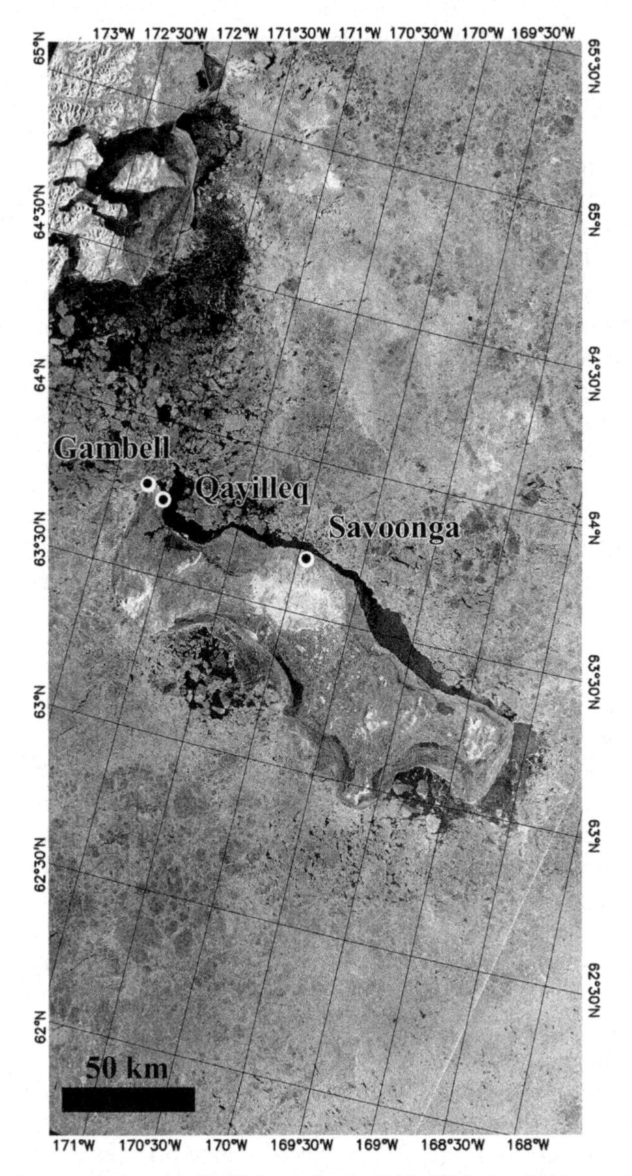

Fig. 5.10 Synthetic aperture radar (SAR) image for April 17, 2008, showing ice conditions around St. Lawrence Island, Alaska (image is 200 × 400 km in size, © Canadian Space Agency) and the locations Gambell, Qayilleq, and Savoonga

Therefore, hunters were not able to launch their boats from the beach at the community. However, the hunting camp Qayilleq had some open water extending well into the ice pack because of the southerly wind component on the 16th and 17th of April 2008 and wind speeds strong enough to move the ice. Hence, people were able to launch their boats at this site, in contrast with the main village of Gambell. Also at the coast of Savoonga a narrow lead opened due to the SSW winds, as seen in Fig. 5.10.

The second point emphasizes the importance of wind-driven sea ice motion, assuming that the under-ice currents are not strong enough to overwhelm the wind's forcing. The quote from G. Noongwook at the beginning of this chapter describes such a wind-driven sea ice phenomenon and its inherent danger. Winds control the opening and closing of leads under most conditions, but such processes are not directly observed at the relevant scale in passive microwave satellite imagery, precluding any analysis of linkages between wind speed and lead openings. However, since sea ice is much rougher than the open sea around the island and since ice impacts the larger scale surface wind patterns (Andreas 1998), sea ice performs an important service in support of harvest activities.

This is expressed by Leonard Apangalook, Sr.: "When there is no pack ice around Gambell anymore the winds are higher and people are not able to go out anymore [in spring]. [. . .] And as soon as the pack ice arrives [in fall] the wind calms down and makes weather conditions more reasonable" (personal communication 2009). This linkage between surface wind speed and ice concentration may also explain that Savoonga is slightly less windy (by 1 m s^{-1}) than Gambell in the spring with corresponding ice concentrations of 25 ± 27 and $17 \pm 20\%$. While the difference in wind speed is small but significant, Leonard Apangalook (personal communication 2009) confirms that Savoonga, due to its somewhat more sheltered position, experiences less windy conditions. In general, it appears that high wind speeds do not influence the hunt as long as they do not exceed a certain threshold. This threshold amounts to 9 m s^{-1} in Gambell and 5 m s^{-1} in Savoonga. Above these thresholds, very few walrus were taken. Similar observations by Lourie (1982) for the years 1981 and 1982 indicate that hunters in Gambell were most successful on days with average wind speeds of 3–5 m s^{-1}. Of course, as pointed out earlier, any factors such as personal judgment as to whether a given weather situation is good for hunting or not as well as other socio-economic factors that come into play here are ignored in our simplistic analysis. Similarly, current speeds and their impact on ice openings and boating safety are not considered and may explain some of the variability observed in hunting activities.

The third point addresses the importance of the local topography relative to prevailing winds and ice conditions. Gambell lies on a peninsula and has coastal access to the west and north, as well as at another location 6 km away, offering several options for boat launching and explaining the somewhat broader range in favorable wind conditions compared to Savoonga (Figs. 5.4a and 5.5a). In Savoonga, hunting primarily took place when winds were ENE, ESE, and WSW. Walrus hunters went out rarely during northerly winds, since such winds block access to open water north of town. These differences and the satellite image shown in Fig. 5.10 demonstrate how strongly conditions can vary between the two locations. A key conclusion from examining the summary provided by Table 5.11 is that despite the limitations discussed earlier, there is some value in this type of analysis in order to arrive at a rough delineation of the weather and ice conditions that favor or reduce the probability for a successful walrus hunt. This is a first step toward assessing how changes in large-scale surface climate and ice conditions have the potential to impact the walrus hunt in two Bering Sea communities.

With respect to long-term variability and change, analysis of the satellite data record from 1979 to 2008 leaves us with a number of points to ponder. First, as to be expected and demonstrated in studies off northern Alaska (Eicken et al. 2009), inter-annual variability in ice conditions is larger for the confined hunting grounds of Gambell and Savoonga as compared to the pan-Arctic trend toward reduced summer ice cover. This is in part explained by the fact that the winter and early spring sea ice extent in the Bering Sea is currently above normal, most likely in conjunction with the atmospheric circulation patterns associated with the cool phase of the North Pacific Decadal Oscillation (NPDO, Overland et al. 1999). As a result, even the record-low ice summer of 2007 registers with above-normal ice extent in the Bering Sea through mid-May. It is only in late May that the rapid retreat started which led up to the record minimum in the pan-Arctic and the Pacific Arctic sector (Eicken et al. 2010). However, since ice concentration anomalies capture deviations from mean extent as well as deviations from the average timing of the seasonal cycle, both 2007 and 2008 (with the second-lowest ice extent in summer on record) register as modestly negative anomalies.

Second, changes in the ice cover do impact access to and availability of ice-associated mammals such as walrus. We found that sea ice concentration anomalies in spring are directly connected to the number of favorable hunting days ($0 < \text{SIC} < 30\%$). This observation may seem trivial at first, i.e., if ice reduction were not shortening the ice season but simply decreasing the concentration on days with high ice cover. However, local ice experts' observations (see Krupnik et al. this volume) and our analysis indicate that in fact the picture is more complicated than that. Negative sea ice anomalies (reduced sea ice concentration) thus appear to increase the number of favorable opportunities for the hunt and hence provide the hunters with a longer time window during which they can successfully pursue walrus.

This can significantly reduce the risk associated with hunts on days where weather conditions or other environmental factors represent a hazard. Conversely, positive sea ice concentration anomalies, at least for the period studied, do impact the hunt in a negative fashion. The way in which the prevailing negative spring ice concentration anomalies in recent years have expressed themselves in the region is such that they resulted in improved conditions for the walrus hunt in spring. The spring ice conditions of 2007 and 2008 stand apart from this broad trend as evident in Fig. 5.9. In 2007, in particular, the number of favorable harvest days was anomalously small at Gambell. However, hunters in Gambell did not notice a decline in harvest numbers (Krupnik et al. this volume), in contrast with very poor harvests in communities further north, such as Wales in Bering Strait and communities on Alaska's North Slope (W. Weyapuk, Jr., personal communication 2007; J. Leavitt, personal communication 2007). The tight fit between favorable harvest days and local spring ice concentration anomalies evident from Fig. 5.9 suggests that this type of analysis can help in identifying extreme years – recognizing that hunters will find ways to address or adapt to such challenges. At the same time, the overall tight fit between favorable hunting days and local ice anomalies may provide an additional tool for walrus subsistence management that seeks to be responsive to environmental constraints on both animals and hunters (Robards 2008).

The comparison between the sea ice anomalies near the villages and for the whole Bering Sea has shown that even if the overall trend of sea ice on a scale of tens of kilometers to hundreds of kilometers in the Bering Sea is the same, the number of days with favorable ice conditions does not necessarily follow this trend at the regional level. This may be different for other subsistence communities, such as Barrow, where ice conditions are not as complex as in the vicinity of St. Lawrence Island. For example, an analysis of ice conditions in the eastern Chukchi and western Beaufort Sea suggests that local ice conditions track somewhat better with the regional ice anomaly patterns (Eicken et al. 2009) than observed here. Most important, however, this work illustrates a broader pattern of increasing decorrelation and decoupling between hemispheric scale processes that drive sea ice extent and concentration and those controlling the ice at the scale that matters to people hunting on and among sea ice (Gearheard et al. 2006).

Finally, in the past 4 years (2004–2008) a striking decline in the number of favorable harvest days is evident in Fig. 5.8. To assess to what an extent this trend mirrors potential changes in atmospheric forcing, we have examined data for the two most influential climate patterns in the region, the Arctic Oscillation (AO) and the North Pacific Decadal Oscillation (NPDO). By comparing surface ocean temperature and atmospheric pressure anomalies triggered by these patterns (annual averages of the sea surface temperature (SST) anomalies[1] triggered through NPDO and the leading mode of the 1,000 hPa height anomalies[2] for the AO, both poleward of 20°N, have been used) we found steadily decreasing SST anomalies starting in 2003 from +1 to –3 K in 2008. This significant correlation with the reduction in favorable ice conditions between 2004 and 2008 (Fig. 5.8) may be explained by the link between lower temperatures and above-normal ice concentrations, minimizing the number of days favorable to hunting from boats. The finding is also consistent with the fact that ice conditions were above normal during recent years in the Bering Sea during winter and early spring. By contrast, the AO anomalies did not show any significant correlation with any of the surface patterns and are hence deemed of lesser importance in explaining variations in hunting opportunities and success.

Hunters are well aware of changes in sea ice and climate and the associated evolution of the skill set inherited from past generations. This is echoed in George Noongwook's (Savoonga) comments: "The window of opportunity for conducting successful subsistence activities is going to become shorter because of warming climate. It is then up to [the hunters] to determine how, [they] can be more efficient and safety minded in terms of retrieving and stalking marine mammals" (Noongwook 2000). This comment reflects the fact that old hunting practices may have to change substantially in the future. It also speaks to the potential of younger hunters adapting their practices and improving hunting success, e.g., through the means of new technologies, more powerful boats, interpretation of satellite imagery, or the use of snowmobiles to access the south coast of the island.

Similarly, Leonard Apangalook, Sr., comments on less sea ice and higher winds at Gambell (personal communication 2009). Changes in sea ice conditions also have a direct influence on Pacific walrus (Robards 2008) and hence impact the subsistence hunt indirectly as well. These changes in turn bring about other impacts, such as shifts and potential conflicts with harvest times for other marine mammals, such as whales (Sease 1986; George et al. 2003).

Conclusions

This contribution explores the physical factors that are controlling the spring walrus hunt in Gambell and Savoonga. We were able to show that ice concentration, derived from satellite remote sensing at a scale of tens of kilometers, is significantly linked to the local walrus harvest success of the villages. At the same time, large-scale regional ice conditions in the Bering Sea scaled poorly with the number of favorable hunting days in each season, highlighting the need for downscaling approaches in linking sea ice use by coastal communities in this region to large-scale patterns of variability and change. The analysis of weather and ice data helped define the seasonal window of opportunity and the conditions under which hunters tend to be successful (summarized in Table 5.11). While this information is still limited in its scope and validity, it may be of value in developing scenarios or quantitative relationships between changes in weather and ice cover and potential hunting success that can be employed in model-based forecasts or studies of adaptation to climate variability and change.

Overall, the picture that emerges from this study for hunting success on St. Lawrence Island is complicated, as reflected by the nuanced and highly detailed statements hunters provide in this context. Overall, the research clearly demonstrated that for ice conditions at the local level off the village sites, the spring sea ice concentration anomalies are excellent predictors of favorable hunting conditions. Negative anomalies actually coincide with more favorable conditions, most likely because they correspond to a more accessible ice pack and potentially better access at the village site. The most recent years, however, with anomalously low arctic-wide summer ice extent clearly fall outside of this well-defined relationship since the rapid northward retreat of the ice edge in particular in 2007 greatly reduced the most favorable part of the spring shoulder season. Nevertheless, walrus hunters were able to cope and mostly by traveling longer distances (approaching 200 km away from the village for individual hunts, compared to studies by Lourie 1982, in the early 1980s when the maximum distance covered during the Gambell spring hunt was 113 km) the harvest season was deemed reasonably successful. Clearly, traveling longer distances increases vulnerability to storms and bad weather and greatly increases fuel costs, which already are at an all-time high. These potential challenges are exacerbated by the compression of the window of opportunity (Benter and Robards 2009).

As discussed in Chapter 4 by Krupnik et al. (this volume) and evident from the graphs and data shown here, comparatively low ice concentrations in the late winter and early spring do allow St. Lawrence Island hunters to venture out very early in the year. Here, the island is favored by its location that resides not too far within the maximum ice extent and hence results in loose ice pack conditions even in winter. At more northern coastal sites such winter hunts are not an option, also due to less favorable light and weather conditions. At the same time, at other villages the impact of more rapid ice retreat appears to be felt much more drastically (Eicken et al. 2010), requiring a more detailed region-wide analysis.

Another important conclusion to draw from this work is that there can be significant differences in weather and ice conditions between the two villages on St. Lawrence Island that impact access to walrus at any one time and also determine to what extent climate variability and change affect the villages. Hunters' insights in combination with high-resolution satellite imagery and weather records provide a very detailed picture of the conditions that favor access to walrus on sea ice (see also Table 5.11). Again, this information can help assess future developments and adaptation strategies if ice conditions were to continue to change. More important, as evident from the comparison between the window of opportunity in the early 1980s and spring of 2006–2008 (Table 5.11), the contrast between Savoonga and Gambell (the latter with a reduction in the number of favorable hunting days by 40%) can be substantial.

Given the importance of ice conditions and weather for a safe, successful hunt and recognizing that a shorter window of opportunity increases the risk to hunters (Noongwook 2000), it is noteworthy that currently forecasts for the region only cover the Alaska mainland and not conditions on St. Lawrence Island. Experienced hunters such as Leonard Apangalook report that these forecasts are significantly different from the conditions in Gambell (personal communication 2009). More analysis and further efforts are required to improve weather forecasts and potentially seasonal-scale forecasts of conditions that are applicable to the two villages' hunting grounds.

Acknowledgments We gratefully acknowledge the contributions by Leonard Apangalook, Sr., who shared his knowledge about weather and ice conditions around Gambell through regular ice observations and during a personal interview with one of us (M.-L. K.), and by all the other Yupik elders and hunters of St. Lawrence Island, who contributed indirectly to this chapter. We are also grateful to the native hunters and harvest monitors; their support was crucial in assembling the harvest-monitoring data set which made this research possible. Brad Benter (USFWS) supplied walrus harvest data from the USFWS and shared his personal insights into the walrus hunt on St. Lawrence Island with us. Steve Gaffigan (Alaska Ocean Observing System) provided access to and computational help with processing of satellite remote-sensing data. Martha Shulski (Geophysical Institute, University of Alaska Fairbanks) and Jim Ashby (Western Regional Climate Center) provided hourly weather data from stations in Gambell and Savoonga and Hyunjin Druckenmiller (Geophysical Institute, UAF) suitable SAR scenes. We appreciate the comments by Anthony Doulgeris, Brad Benter, and Igor Krupnik on earlier drafts of the chapter and Matt Druckenmiller's help with the map. This work was made possible through the National Science Foundation's support of the SIZONet project (0632398) and the SNAP project (0732758); opinions, findings, and conclusions are those of the authors and do not necessarily reflect any of the aforementioned individuals or organizations' perspective.

Notes

1. http://jisao.washington.edu/pdo/PDO.latest
2. http://www.cpc.noaa.gov/products/precip/CWlink/daily_ao_index/ao_index.html

References

ACIA (Arctic Climate Impact Assessment. Scientific Report). 2005. Cambridge: Cambridge University Press, 1042p.

Andreas, E.L. 1998. *The Atmospheric Boundary Layer Over Polar Marine Surfaces*. M. Leppäranta (ed.), Helsinki: Helsinki University Press, pp. 715–774.

Benson, A.J. and Trites, A.W. 2002. Ecological effects of regime shifts in the Bering Sea and Eastern North Pacific Ocean. *Fish and Fisheries* 3: 95–113.

Benter, R.B. and Robards, M., 2009. Subsistence walrus harvest trends in the Bering Strait. http://www.marinemammalscience.org/index.php?option=com_content&view=article&id=390&Itemid=214&abstractID=804.

Burns, J.J. 1970. Remarks on the distribution and natural history of pagophilic pinnipeds in the Bering and Chukchi Seas. *Journal of Mammalogy* 51(3): 445–454.

Eicken, H., Krupnik, I., Weyapuk, W., Jr., and Druckenmiller, M.L. 2010, in press. Ice seasons at Wales, 2006–2007. In *Kingikmi Sigum Qanuq Ilitaavut – Wales Inupiaq Sea Ice Dictionary*. I. Krupnik, H. Anungazuk, and M. Druckenmiller (eds.), Washington: Arctic Studies Center, Smithsonian Institution.

Eicken, H., Lovecraft, A.L., and Druckenmiller, M.L. 2009. sea ice system services: A framework to help identify and meet information needs relevant for Arctic observing networks. *Arctic* 62(2): 119–136.

Eppler, D.T., Farmer, L.D., Lohanick, A.W., Anderson, M.R., Cavalieri, D.J., Comiso, J.C., Gloersen, P., Garrity, C., Grenfell, T.C., Hallikainen, M., Maslanik, J.A., Mätzler, C., Melloh, R.A., Rubinstein, I., and Swift, C.T. 1992. Passive microwave signatures of sea ice. In *Microwave Remote Sensing of Sea Ice*. F.D. Carsey (ed.), Geophysical Monograph 68, Washington: American Geophysical Union, pp. 47–71.

Fay, F.H., 1982. Ecology and biology of the Pacific walrus, Odobenus rosmarus divergens Illiger. United States Department of the Interior, Fish and Wildlife Service, North American Fauna, Number 74.

Garlich-Miller, J.L. and Burn, D.M. 1999. Estimating the harvest of Pacific walrus, Odobenus rosmarus divergens in Alaska. *Fisheries Bulletin* 97(4): 1043–1046.

Gearheard, S., Matumeak, W., Angutikjuaq, I., Maslanik, J., Huntington, H.P., Leavitt, J., Kagak, D.M., Tigullaraq, G., and Barry, R.G. 2006. "It's not that simple": A collaborative comparison of sea ice environments, their uses, observed changes, and adaptations in Barrow, Alaska, USA, and Clyde River, Nunavut, Canada. *AMBIO* 35: 203–211.

George, J.C., Braund, S., Brower, H., Jr., Nicolson, C., and O'Hara, T. 2003. Some observations on the influence of environmental conditions on the success of hunting bowhead whales off Barrow, Alaska. In *Indigenous Ways to the Present: Native Whaling in the Western Arctic*. A.P. McCartney (eds.), Edmonton and Salt Lake City: Canadian Circumpolar Institute and University of Utah Press, pp. 255–275.

Gilbert, J.R. 1999. Review of previous Pacific walrus surveys to develop improved survey designs. In *Marine Mammal Survey and Assessment Methods*. G.W. Garner, S.C. Amstrup, J.L. Laake, B.F.J. Manly, L.L. McDonald, and D.G. Robertson (eds.), Rotterdam, The Netherlands: A.A. Balkema, pp. 75–84.

Grebmeier, J.M., Overland, J.E., Moore, S.E., Farley, E.V., Carmack, E.C., Cooper, L.W., Frey, K.E., Helle, J.H., McLaughlin, F.A., and McNutt, S.L. 2006. A major ecosystem shift in the northern Bering Sea. *Science* 311: 1461–1464.

Huntington, H.P. 2000. Using traditional ecological knowledge in science: Methods and applications. *Ecological Applications* 10(5): 1270–1274.

Huntington, H., Fox, S., Krupnik, I., and Berkes, F. 2005. The changing Arctic: Indigenous perspectives. In *Arctic Climate Impact Assessment*. Arctic Climate Impact Assessment (ed.), Cambridge: Cambridge University Press, pp. 61–98.

Krupnik, I. and Jolly, D. (eds.) 2002. *The Earth is Faster Now: Indigenous Observations of Arctic Environmental Change*. Fairbanks: ARCUS.

Krupnik, I. and Ray, G.C. 2007. Pacific walruses, indigenous hunters and climate change: Bridging scientific and indigenous knowledge. *Deep-Sea Research II* 54: 2946–2957.

Lourie, K.S. 1982. *The Eskimo Walrus Commission 1981–82 Walrus Data Collection Program*. Nome: Eskimo Walrus Commission, 233p.

Massom, R.A. 2009. Principal uses of remote sensing in sea ice field research. In *Sea Ice Field Research Techniques*. H. Eicken, R. Gradinger, M. Salganek, K. Shirasawa, D.K. Perovich, and M. Leppäranta (eds.), Fairbanks: University of Alaska Press, pp. 405–466.

Metcalf, V. and Robards, M.D. 2008. Sustaining a healthy human-walrus relationship in a dynamic environment: Challenges for comanagement. *Ecological Applications* 18(2): 148–156.

Noongwook, G. 2000. Native observations of local climate changes around St. Lawrence Island. In *Impacts of Changes in Sea Ice and Other Environmental Parameters in the Arctic*. H.P. Huntington (eds.), Marine Mammal Commission Workshop Report. Bethesda: Marine Mammal Commission, pp. 21–24.

Oozeva, C., Noongwook, C., Noongwook, G., Alowa, C., and Krupnik, I. 2004. *Watching Ice and Weather Our Way/ Akulki, Tapghaghmii, Mangtaaquli, Sunqaanga, Igor Krupnik. Sikumengllu Eslamengllu Esghapalleghput*. I. Krupnik, H. Huntington, C. Koonooka, and G. Noongwook (eds.), Washington: Arctic Studies Center, Smithsonian Institution, 208pp.

Overland, J.E., Adams, J.M., and Bond, N.A. 1999. Decadal variability of the Aleutian low and its relation to high latitude circulation. *Journal of Climate* 12(5): 1542–1548.

Ray, G.C., and Hufford, G.L. 1989. Relationships Among Beringian Marine Mammals and Sea Ice. *Rapports et Procès-Verbaux des Réunion Conseil International pour L'Exploration de la Mer* 188: 22–39.

Robards, M. 2008. Perspectives on the Dynamic Human-Walrus Relationship. Dissertation at the University of Alaska, Fairbanks.

Robards, M. and Joly, J. 2008. Interpretation of 'wasteful manner' within the Marine Mammal Protection Act and its role in management of the Pacific walrus. *Ocean and Coastal Law Journal* 13(2): 171–232.

Sease, J.L. 1986. Historical status and population dynamics of the Pacific Walrus. University of Alaska, Thesis, Fairbanks, 213 pp.

Stabeno, P.J., Bond, N.A., and Salo, S.A. 2007. On the recent warming of the Southeastern Bering Sea shelf. *Deep-Sea Research II* 54: 2599–2618.

Stabeno, P.J. and Hunt, G.L., Jr. 2002. Overview of the inner front and Southeast Bering Sea carrying capacity programs. *Deep-Sea Research II* 49: 6157–6168.

Stroeve, J., Serreze, M., Drobot, S., Gearheard, S., Holland, M., Maslanik, J., Meier, W., and Scambos, T. 2008. Arctic sea ice extent plummets in 2007. *Eos Transactions AGU* 89: 13– 14.

Chapter 6
Sila-Inuk: Study of the Impacts of Climate Change in Greenland

Lene Kielsen Holm

Contributing Author: Shari Gearheard.

When people talk about catastrophic climate change, there's a fair chance that Greenland is on their mind.
(Witze 2008:798)

Abstract Greenland is experiencing some of the most dramatic impacts of climate change in the Arctic. Much work has been done to study these changes through physical science, but little has been done to document the perspectives of local Kalaallit. In 2005, the Inuit Circumpolar Council-Greenland launched the Sila-Inuk project to do just this, interviewing local experts in 23 communities in south, west, and north Greenland. The analysis of this work is still underway and will be presented at the next ICC General Assembly in 2010. This chapter provides a sample of the work and an overview of the Sila-Inuk initiative.

Keywords Climate change · Inuit · Greenland · Sea ice · Weather · Indigenous knowledge

Introduction

A major news feature called "Climate Change: Losing Greenland" featured this opening line when it appeared in the journal *Nature* in 2008 (Witze 2008). Indeed, climate change in Greenland is creating some of the most rapid and dramatic environmental impacts of anywhere in the Arctic. The Greenland ice sheet, the main geographic feature covering some 80% of our island's territory of 2,166,086 km^2, has become a poster child for Arctic climate change, as now famous maps and graphics show how the maximum surface-melt area on the ice sheet has increased and the ice sheet itself has shrunk dramatically since the 1990s (Fig. 6.1).

Most of the attention on Greenland in terms of research has come from physical scientists, often with a focus on the ice sheet (e.g., Zwally et al. 2002, Luthcke et al. 2002, Hall et al. 2006). These studies and others have been immensely important

L.K. Holm (✉)
Inuit Circumpolar Council, Dronning Ingridsvej 1, P.O. Box 204, Nuuk 3900, Greenland
e-mail: lene@inuit.org

I. Krupnik et al. (eds.), *SIKU: Knowing Our Ice*,
DOI 10.1007/978-90-481-8587-0_6,

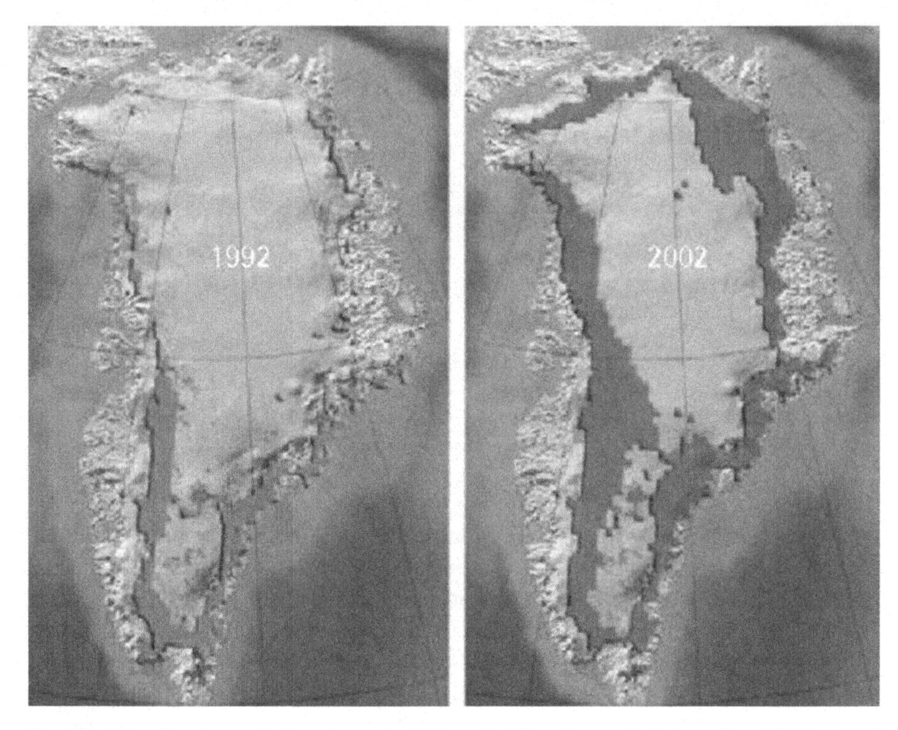

Fig. 6.1 Map showing surface-melt extent of Greenland ice sheet and changes from 1992 to 2002 (ACIA 2004:40)

for Greenland and for our understanding of how climate change is affecting and will affect our society. Studies that incorporate Greenland communities, in particular the knowledge of the Greenland Inuit (Greenlanders or Kalaallit), have been far fewer. Only a handful of studies have worked with the knowledge of Greenlanders to understand our environment and our experiences with climate change (e.g., Petersen and Siegstad 1993; Born et al. 2006). With an interest to add Greenlanders' observations and experiences to the collective knowledge on climate change, and inspired by similar work done in other countries such as Canada, the U.S. (Alaska), and Scandinavia (Krupnik and Jolly 2002; Huntington and Fox 2005; Helander and Mustonen 2004), the Inuit Circumpolar Council-Greenland in 2005 set out to design and conduct a study called "Sila-Inuk" that would engage Greenlanders in documenting their knowledge of climate change and its impacts.

Sila-Inuk Design and Methods

Sila-Inuk means literally "weather-human." Besides weather, *Sila* also refers to human intelligence. When the ending of *-suaq* is added to the word, *Silarsuaq*, it also means the universe. For those of us in charge of the project planning,

"Sila-Inuk" was a name that had the clearest expression of what we were trying to understand and document, namely the intimate and personal connection between people and their environment, their weather. Like other chapters in this book illustrate, Arctic indigenous communities and people have intimate links to and knowledge of the land, the sea, and the ice. The first-hand experiences of Greenlanders who live and work closely in and with their environment provide powerful and insightful knowledge about that environment, its processes, and its changes over time.

What is perhaps unique to Greenland, compared to the other Arctic communities and countries represented in this book, is that our knowledge holders include not only hunters and fishermen but also sheep farmers and agriculturalists. While ice sheets and sea ice certainly come to mind when we think about the Arctic, farms and vegetable gardens probably do not. But, farming and herding are traditional and important activities that define people's way of life in certain parts of Greenland and in particular southern Greenland. Those farmers from southern Greenland are some of the key experts that we have engaged within the Sila-Inuk project. They remind us that there are places in the Arctic where sea ice does not exist as a regular feature and that other pursuits, also vulnerable to climate change, are at the heart of local communities and cultures. They also embody the many diverse faces of our country – from our sheep farmers in the South to our sea ice hunters in the north.

With an ambitious goal to collect first-hand observations of climate and environmental change from experts in all Greenland communities, Sila-Inuk was developed by the Inuit Circumpolar Council-Greenland (ICC-Greenland) together with KNAPK (Kalaallit Nunaanni Aalisartut Piniartullu Kattuffiat, The Hunters and Fishermen's Organization in Greenland).[1] Our first community visits began in 2006 in the region of Kujataa (south Greenland). We then visited the regions of Avanersuaq and Kitaa (north and west Greenland) in 2007, 2008, and 2009.[2] Our main report and findings will be delivered at the next ICC General Assembly being held in Nuuk in 2010. This chapter provides initial findings of our project and a broad overview of shared themes among the communities we studied (Fig. 6.2).

In Sila-Inuk, expert observations are collected mainly through interviews conducted by myself in Greenlandic (Kalaallisut) and then transcribed by Anna Heilmann, Hanne Sørensen, and myself. These transcripts will later be translated to English during the making of the final report. The style of the interview used depended on the person and situation, and mostly semi-directed and open-ended interviews have been used (Huntington 1998). We interviewed mainly expert hunters, fishermen, sheep farmers, and others with a close connection to the land; those experts range in their characteristics from old to young and from men to women (see Table 6.1). Altogether, 55 people have been interviewed from 23 communities between 2006 and 2008. Therefore, the participants of Sila-Inuk offer a broad range of Greenland residents in terms of geography, livelihood, and experience. Often while interviewing we tour the expert's home, farm, hunting camp, or observe some of their daily activities (Fig. 6.3). This kind of interaction allows

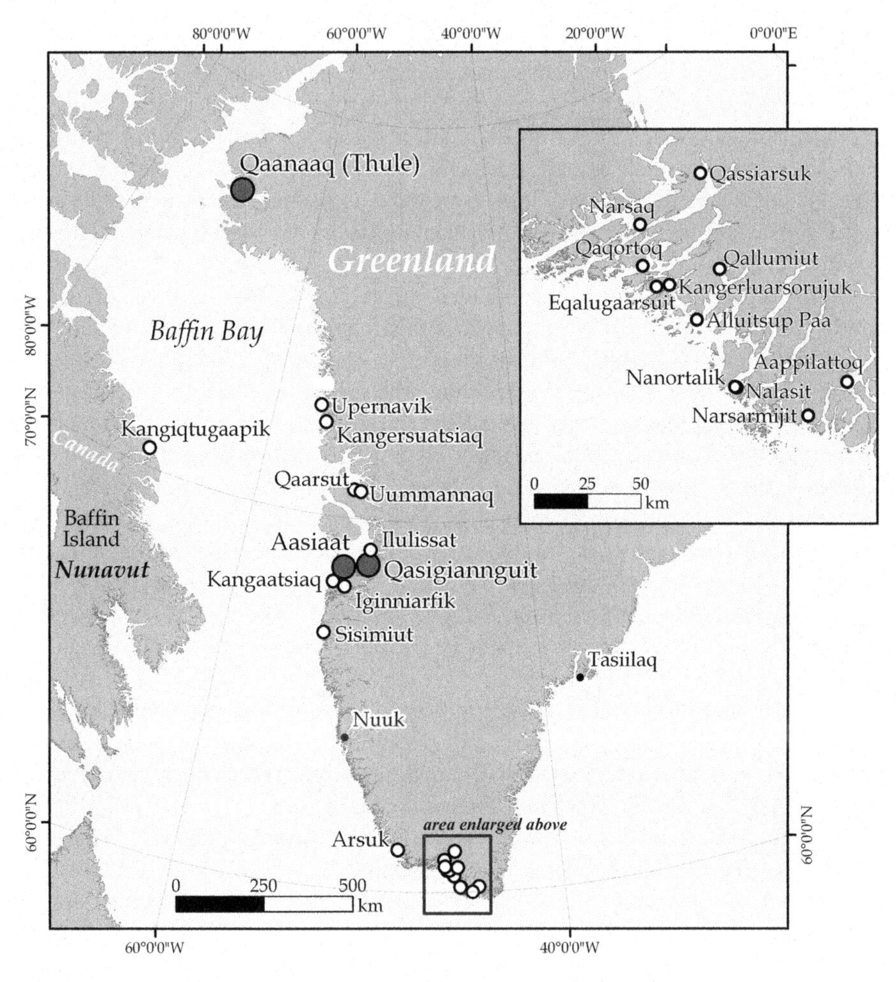

Fig. 6.2 Map of main communities of Greenland (courtesy of Naturinstituttet)

people to show us physical changes or speak specifically about how climate change is affecting their lives.

For example, when visiting Upernavik, I had a chance to sail with a local expert in his small boat to see some of the 18 (or more) active calving ice fjords, view bird nesting cliffs, and visit the two tiny communities of Aappilattoq and Kangersuatsiaq where I was able to interview several other people (Fig. 6.4). In the north, in Qaanaaq, I was able to travel the sea ice with expert hunters and collaborate with another joint project looking at sea ice, Siku-Inuit-Hila (see Chapter 11, this volume).

Interviewing was our main approach in collecting Greenland experts' knowledge of climate change but also we also used maps to help facilitate discussion

Table 6.1 Towns and settlements visited by Sila-Inuk including the number of interviews and year conducted. Centers of municipalities are given in bold; other settlements are positioned according to municipality (town) belong

Towns/*settlements*	Number of interviews	Year visited	Interviewer/s
Arsuk	5	2006	Lene Kielsen Holm
Nanortalik	1	2006	Lene Kielsen Holm
Aappilattoq	2	2006	Lene Kielsen Holm
Narsarmijit	3	2006	Lene Kielsen Holm
Nalasit	1	2006	Lene Kielsen Holm
Tasiusaq	1	2006	Lene Kielsen Holm
Alluitsup Paa	3	2006	Lene Kielsen Holm
Qallumiut	1	2006	Lene Kielsen Holm
Qaqortoq	3	2006	Lene Kielsen Holm
Kangerluarsorujuk	2	2006	Lene Kielsen Holm
Eqalugaarsuit	1	2006	Lene Kielsen Holm
Narsaq	6	2006	Lene Kielsen Holm
Qassiarsuk	1	2006	Lene Kielsen Holm
Sisimiut	3	2006	Lene Kielsen Holm
Aasiaat/ Kangaatsiaq	2	2007/2008	Lene Kielsen Holm and Aqqaluk Lynge
Iginniarfik	2	2007	Lene Kielsen Holm and Aqqaluk Lynge
Qasigiannguit	1	2007	Lene Kielsen Holm and Aqqaluk Lynge
Ilulissat	4	2008	Lene Kielsen Holm
Upernavik	3	2008	Lene Kielsen Holm
Aappilattoq	3	2008	Lene Kielsen Holm
Kangersuatsiaq	4	2008	Lene Kielsen Holm
Uummannaq	2	2008	Lene Kielsen Holm
Qaarsut	1	2008	Lene Kielsen Holm
Total	55		
		In collaboration with Siku-Inuit-Hila	
Qaanaaq	Workshops and traveling on sea ice	2007 and 2009	
Kangiqtugaapik, Nunavut	Workshops and traveling on sea ice	2008	

and document changes (Born et al. 2006). In addition, we used the radio to share what we were learning along the way. Radio is a powerful means of communication in Greenland and most citizens listen daily to our national radio programming in Greenlandic. As Sila-Inuk has traveled through communities collecting stories, we have encouraged the experts we met to share their observations over the radio. This means of sharing, along with radio interviews I have done to explain the Sila-Inuk project, have gone a long way to raising national awareness not only of climate change but also of the value of local knowledge in understanding climate change.

Fig. 6.3 Silas and Karen Bernhardsen grow turnips and other vegetables in Nalasit, South Greenland (Photo: Lene Kielsen Holm)

Climate Change in Greenland: Observations from Disko Bay

While our analysis is still underway and our major findings are expected to be delivered in 2010, the preliminary work of Sila-Inuk reveals some shared themes across different Greenland communities and regions. In this section, I present some of the initial findings from the Disko Bay area including the communities of Aasiaat, Iginniarfik, and Qasigiannguit. I focus on the interviews of three regional experts: Augo Davidsen (83 years old from Aasiaat), Kaspar Brandt (85 years old from Iginniarfik), and Tittus Magnussen (70 years old from Qasigiannguit). Though only samples from a much larger research effort, they are representative of some of the recurrent themes we have been hearing in this region of Greenland and in many other areas (see Table 6.2).

Changes in Weather Patterns and Seasonality

Residents of the Disko Bay area have observed that the weather is more unpredictable than in former times and that there has been a shift in the seasons. Winds

Fig. 6.4 Paulus Benjaminsen (*left*) of Aappilattoq, the southernmost inhabited place in Greenland, with Holm in 2006 (Photo: Nadia Holm)

in particular are unpredictable, and dependable winds like *avannaq* (the north wind) that used to be stable for long periods in the wintertime no longer follow the same patterns. Winds tend to shift direction more often and people can experience different winds even within the course of one day. Augo Davidsen highlighted the recent unpredictability of the winds by showing us his barometer that he had inherited from his father (Fig. 6.5). He has always been able to use it to help him determine the weather for the next day but "the brain of this barometer does not believe (or have faith) in the weather anymore" (Davidsen 2007).

Experts in these, as well as in other communities, have noticed a shift in the seasons, noting in particular changes in winter and spring. Winters are warmer. Tittus Magnussen (2007) noted that in former times common winter temperatures would be –30 or –40ºC, but today's winters barely reach –20ºC (Fig. 6.6). Instead, the experts observe that cold seems to have moved into the spring period. Winters are warmer, but when spring arrives and people expect warming and melting, they instead experience cold and high humidity. As Davidsen (2007) put it, "the warmth of the spring has shifted to winter and the cold of the winter has shifted to spring . . . In former times the cold period was October, November, December, January used to be very cold, but now it is February, March, April, May. So it is like the winter has shifted."

Table 6.2 Summary of observations from communities in South Greenland (Arsuk, Aappilattoq, Narsarmiut, Nanortalik, Nalasit, Tasiusaq, Qallumiut, Alluitsup Paa, Kangerluarsorujuk, Eqalugaarsuit, Qaqortoq, Narsaq, Qassiarsuk)

- *Sila Assallattoq* ("the weather has switched"). The weather is not what is expected for the season
- Directions of the winds are changing more often. Calm periods occur less and the speed of the winds is less than expected
- In summer it can get very warm during the day while in the night frost has been experienced
- More rain
- The currents are changing. During the 1990s the currents began to run from the Atlantic Ocean toward the west coast of Greenland and the cod disappeared. Now the currents flow out of the coast and the cod are coming back
- The levels of the tides are changing. High tide is higher and low tide is lower than in former times
- The sea ice that comes down the coast from east Greenland melts very fast. This melting ice creates meltwater that then refreezes, creating dangerous sea ice conditions for hunting
- Glaciers are receding
- The ice cover on the beaches (*qaanngoq)* has disappeared in the winter
- The winter ice cover of the fiords (*maneraq*) is not occurring any longer
- The behavior of birds and seals are changing
- *Aataat*, harp seals, are very high in numbers and are more stationary in the summer season
- *Natseq*, ringed seal, has moved
- Routes of minke whales have changed
- Arctic char are moving from the sea into the lakes earlier
- The molting period for seals is getting longer
- Algae is seen on the skins of the seals; something never seen before
- Algae also grows very fast on [the underside of] boats
- Snow cover is decreasing
- Less snow cover has resulted in poor crowberry seasons, but the blueberry season has not changed
- Humidity is higher

All experts mentioned as part of their assessment of this shift that the new cold, damp conditions of spring do not allow them to dry fish in the springtime, as the air is too humid. In the past, May would be a time to dry capelin, but cool and humid conditions, with more rain, prevent the meat from drying (Brandt 2007). Kaspar Brandt (2007) also mentions that when the spring does arrive, melt occurs very quickly, "the melting of the snow is as if the earth just swallowed it." For him the rapid melt is not only a sign of changing spring conditions but also the change in the permafrost conditions, as the rapid melt water would run off (not soak in), if the ground were frozen.

An interesting change in local weather was shared by Tittus Magnussen from Qasigiannguit. Magnussen explains that Qasigiannguit has a special storm called *Saqqarsarneq* ("the warm wind that melts all the snow in the winter time in no time"), which has been there for years. But *Saqqarsarneq's* duration and intensity have changed (decreased) in recent times. The new pattern of the *Saqqarsarneq* event in wintertime is another example of changing weather in the region.

Fig. 6.5 Augo Davidsen from Aassiaat explains how "the brain of his barometer does not believe (or have faith) in the weather anymore" (Photo: Lene Kielsen Holm)

Changes in Sea Ice

Over their lifetimes, our experts have seen varying sea ice conditions. The region has experienced periods of heavy ice but also light or even no ice during different periods in the past. For example, near Aasiaat in the 1940s, the sea ice was unstable, meaning people could not go dogsledding to villages to the south (Niaqornaarsuk and Iginniarfik). They could only get to those locations by boat. In the winter of 1946–1947, no sea ice was in the area at all. Davidsen explains that sea ice goes through cycles, peaks, and lows and that this period was the "lowest of the lows." However, years later, there were extreme low temperatures and the sea ice formed fully. Conditions were so severe that where there was a polynya each year, even the polynya was covered with ice. The polynyas are usually formed by strong current, so when they were traveling on top of the frozen polynya during these years, the dogs were very scared by the sound of the current under the ice.

Highs and lows in the sea ice were experienced in other communities as well. At Qasigiannguit, the sea ice did not form at all in the winter of 2006–2007 and boats were sailing year round. An example of a "high" in the cycle comes from Iginniarfik, where a place that never had sea ice in the past formed winter ice in 1993. The residents say that 1993 was a peak for winter ice in their region.

Fig. 6.6 Tittus Magnussen from Qasigiannguit (Photo: Lene Kielsen Holm)

Although the sea ice in the area experiences periods of highs and lows, the experts did indicate a trend overlapping this variability. For example, Davidsen observes

> When I became a trainee for hunting [in the 1930s] the sea ice would begin to form in October. Now, it doesn't come until January . . . Looking down at the place where they pull in the ships [the shipyards] in January and February it is amazing to see that they are able to pull ships into the shipyard [due to no ice].

Seal hunters who use nets at Qasigiannguit have noticed a change in sea ice within the last decade. During this time the sea ice has almost disappeared and they are not able to put nets in the ice at their usual hunting places. The hunters have their "own" places where they put in their nets to catch seals, but have not been able to do so since 2000. As well, Magnussen notes that a dogsledding route to a small settlement called Ikamiut has not been useable since the mid-1990s because of loss of sea ice and hunters must be very cautious when traveling on sea ice due to the ice being thin in places and having an uneven thickness. Again, there are connections to other regions (see also Chapter 2 by Taverniers, this volume). When Sila-Inuk visited north Greenland in 2007 and 2009, local hunters from Qaanaaq, Siorapaluk, and Savissivik also described the same thinning conditions, loss of sea ice, and loss of traditional sea ice travel routes (Oshima, Kristiansen, Simigaq, and Nielsen pers. comm. 2007 and 2009).

A very interesting sea ice change was shared by Augo Davidsen from Aasiaat. Born in 1926, Davidsen has observed that the number of *sassat* (beluga/narwhal entrapment in patches of open water amidst winter ice) has decreased.[3] *Sassat* did not occur every year, but over time they have been observed less and less in the area. Davidsen recalls the importance of *sassat* in former times:

> In 1951, I found a *sassat* which provided meat to many families from Qasigiannguit, Ilulissat, Qeqertarsuaq, Aasiaat, and other villages. I received 30 Kroners from the community for finding the *sassat*. My wife asked me to go down to the trade center to buy myself trousers, which, by coincidence, cost 30 Kroners. [Experiencing that *sassat*], it was like I was feeding all of Disko Bay.

Our experts report that changes in the sea ice are caused, at least in part, by a warming ocean. Davidsen explains

> It is obvious that the sea is warming up. Even if the air is very cold, you need cold water in order for it to freeze. In order for the sea ice to form, you need both cold air and cold water. But in [recent] years, even when the air is cold, the [temperature of the sea] must have risen because the ice still does not form.

One indication that the sea is warmer is changes in *qaanngoq* – ice that forms on the coast (land or shoreline) down to the waterline. When there is no *qaanngoq*, it means the water is warm. And, of course, when there is *qaanngoq*, it means the water is cold enough, so that it freezes at the shoreline.

Changes in snow can also have an impact on sea ice formation, processes, and characteristics. If there is snow on top of the sea ice, it is more difficult for the ice to thicken due to the snow's insulating properties. The sea ice can be used only after the sea ice has had a chance to freeze, free of new snowfall, for 3 or 4 days. In cold weather it only takes 3 or 4 days to form travelable sea ice; but this can only happen when there is not insulating snow on the ice surface.

Changes in Seals

Changes in seals have been observed not only in the Disko Bay area but also in south Greenland (Table 6.2). In both regions *natseq* (ringed seals) are disappearing and harp seals are on the rise. Magnussen notes that in the Qasigiannguit area, ringed seals do not even appear basking on ice in the spring sunshine anymore. The harp seal, on the other hand, stays year round because there is no longer sea ice. Greater numbers of harp seals are disturbing the fisheries and eating the cod stocks.

In south Greenland, the molting period of seals is getting longer and algae is being found for the first time on the skin of seals (Table 6.2). Seals with bald patches in their fur are also being found. Interestingly, when we met with Clyde River hunters from Nunavut, Canada, in Qaanaaq during our fieldwork for the Siku-Inuit-Hila project we learned that similar changes in ringed seals have also been observed in that area (Angutikjuak and Sanguya pers. comm. 2007). The Clyde River hunters also linked seal changes to sea ice changes. Since the sea ice season is also getting shorter in that region (3–4 weeks earlier break-up) seals are forced into the water before they have completely finished molting. Seals rely on sea ice for lounging and scratching off their old fur during the spring molting period. Since seals are in the water before molting is complete, hunters are harvesting seals with patchy (un-molted) fur in the summer time. These skins are of poor quality and are not useable for making clothing.

Discussion

Though I only present some samples of what we have learned from our work during the Sila-Inuk project to date, we can draw out some key points that warrant further discussion and exploration as we move into our analysis stage in the project. The first point is in regard to cycles and changes that people observed in the ice and weather conditions. The concept of "cycles" was explained many times during the discussions of sea ice in the Disko Bay area, where hunters have seen the highs and lows of sea ice extent and thickness over time and space. Cycles were also mentioned in conjunction with weather changes and weather variability. For example, in one of our interviews in south Greenland in 2006, sheep farmer Erik Rødе Frederiksen, when talking about changing wind conditions, said that the forefathers of his people would say, "*eqiterpaageeq anerlertarnissaminut silaannaap qatsingarujussuarlini,*" "the weather is collecting the future winds by being calm." In other words, if it is calm now, it will be windy in the future; and so the cycle is created.

The concept of "cycles" leads to the second area we can explore within the Sila-Inuk, namely the similarities of Greenlanders' observations and those of other Arctic residents. For example, many other Arctic communities have identified increasing weather variability (e.g., Krupnik and Jolly 2002; Huntington and Fox 2005; Helander and Mustonen 2005). The unpredictability of winds, later formation of sea ice in the fall, and weak sea ice during the wintertime are also common observations by local people that are shared across different regions (e.g., see Chapter 14 by Krupnik and Weyapuk, this volume).

The third point that Sila-Inuk helps demonstrate is that the observations and the depth of one's knowledge of climate change are intimately tied to how one uses the land. In Greenland, "people on the land" may have the most diverse range of occupations and expertise (such as fishermen, sheep-herders, small gardeners). For example, the indicators used, and knowledge gained, by fishermen, compared to sheep farmers or walrus hunters, are quite different. Thus, having a more diverse set of perspectives, due to many different occupations, can help us understand and articulate local and regional variations in climate change and the influence of observed change in weather and overall environment on various subsistence activities and livelihoods. What is more interesting is that, despite different engagement with the local environments, Greenland experts are seeing similar changes across many areas of the country. Changes like unpredictable weather and shifting seasonality are just one example.

Conclusion

In these years, Greenland is experiencing some of the most severe environmental impacts in the Arctic due to climate change. Sila-Inuk has been traveling from the southernmost region of Greenland to the northernmost region to document those changes and impacts from a local perspective. Fifty-five people have been interviewed from 23 communities between 2006 and 2008.

Since Greenland is the biggest island on earth one can imagine that we have a very diverse climate across our country, from the High Arctic to Sub-arctic climate zones. This is very different from how the public generally sees Greenland as one barren land "covered in ice." With diverse environments and livelihoods, Greenland is an excellent place to study the impacts of environmental change. For those of us in the Sila-Inuk project, asking the inhabitants of this big island about their observations of change is the best way to study and understand these changes. Greenlanders have been living here for millennia. The memory of the Inuit of Greenland has been handed over to us from our ancestors and is today a gift that we have to acknowledge. The appreciation of this kind of knowledge is growing and one can see how it complements science in better understanding how climate change is having an impact on Arctic environments, resources, and societies.

Take, for example, the observations of weather and animal shifts that we heard about in our project as we visited different regions. In south Greenland, people told us that the minke whale is moving north. In 2006, when we visited the southern community of Nanortalik, all of the people we were supposed to interview were out hunting for minke whale. They came back unsuccessful, saying that the whales had already moved north. In 2008, while in Sisimiut, we all heard a radio announcement that a minke whale had been caught in Siorapaluk, the most northerly community in Greenland (and in fact the world). No one had ever heard of a minke whale being caught so far north before (77° 47'N 70° 46'W). These and other observations of animals shifts and changes complement scientific research of past and present, such as the seminal work of Vibe (1967) who showed how climate fluctuations are connected to animal movements.

In another example, the weather shifts we heard about during our work add more detail to what are often generalized reports of "Arctic warming." Our experts help explain exactly how that warming is taking place, where, and with what local impacts. For example, Erik Røde Frederiksen, an 85-year-old sheep farmer from Qassiarsuk in south Greenland, explains in detail how warm temperatures and weather patterns have changed:

> One thing that also surprises me during the most recent years is that the weather patterns in Qaqortoq and Paamiut and for the east coast during the year, especially in the springtime and maybe also the warmer periods in the winter, have moved northwards. It is getting warmer north of us. For those of us who are living in the southern-most part of Greenland, it was [warmer] like that before. At these [former] times, this part of Greenland was the warmest part of the land throughout the year. I feel with my experience that the warmth that we used to experience solely here [at Qassiarsuk area] has moved to Paamiut. And this we can hear on the radio when they are forecasting the weather, for both the temperatures and the winds. And it seems that winds are also changing enormously in the northern part of Greenland.

Lastly, in terms of sea ice, the changes also vary across regions. Indeed, the very nature and importance of sea ice are different in various parts of Greenland. In south Greenland, when we speak of sea ice, we talk about the arrival of the pack ice from east Greenland. The main change observed in southern communities is less pack ice arriving due to changes in winds and currents in the east. In the north, the hunters depend on the formation of sea ice as a platform for their activities, and there the changes discussed are in terms of thinning sea ice, earlier break-up, and later freeze-up.

By traveling to the different regions of Greenland, Sila-Inuk has been able to document information about the use of and changes in the environment from some of the most knowledgeable people on these issues. In this chapter, I have shared some of the observations made by experts from mainly just one region (Disko Bay). These observations are diverse: from the winds, both in regard to its speed and direction, to the currents of the ocean, to animal behavior, and humidity of the air. Furthermore, we have documented observations of receding glaciers, thinning and disappearing sea ice, changes in multi-year ice coming from the eastern coast of Greenland, and how all these changes are having an impact on the resources that local people depend on from their environment. Some people are still dealing with the impacts of changing resources, such as the shift in the migration routes of hooded seals that follow the pack ice from east Greenland. Other changes, like the disappearance of cod, are shifting yet again and we see this resource coming back. The idea of climate and ice "cycles" brought out by some of our most knowledgeable experts in the interviews is a strong reminder that the processes on the land, ice, and in the sea are not always linear, so that experienced people always anticipate many different options ahead. The understanding and response to all these shifts are some of the challenges that we have to face in our communities and as a nation.

Inuit have faced changes to their environment and resources before, and even more so in Greenland, as documented by the rich history of our country. In the past our ancestors responded by developing new technologies and by moving to where the resources are. Today we are often able to do the same, but society has changed so that we are not as flexible as we used to be. We no longer have the mobility we once did, moving our camps and hunting grounds according to the seasons and our needs. We live in modern settlements and towns and depend on many modern conveniences. Since we are not as flexible as we used to be, it can seem that the changes are more visible and we feel them more profoundly. It also seems that climate change hits our livelihoods harder and it can be a risk to the security of those making a living from the land and the sea, and the living resources.

One of the goals for the Sila-Inuk study is to help in current and future policy-making by providing knowledge and insight collected from some of our most experienced people. By listening to those experts we may better address many urgent questions like: How are the changes impacting our environment? How can we best respond to these changes? And, what kind of environment are we to face and make a livelihood from in the near and farther future? We believe that the final report and the texts of the interviews we collected during the Sila-Inuk project will make us all better prepared to tackle these and other urgent questions that we face because of rapid climate change. One of the recommendations we can already identify is the need for local environmental monitoring around our country. Local monitoring that addresses the priorities of local people and contributes to the study of our environment and its changes is critical to acquiring detailed information. Our current climate research and weather forecasting is based on relatively few stations and are not representative of our complex environments. Local monitoring, including both quantitative measurements and systematic qualitative expert observations, will help to provide locally specific data that can improve local, regional, and national climate information. Local monitoring will provide more specific data for assessing the shifts like those mentioned above and the impacts that climate change is having on our communities and economies.

Another goal accomplished by the Sila-Inuk project is to have a contemporary study that can be compared to similar studies done in other parts of the Arctic and to earlier assessments of the historical climate changes in Greenland over the past centuries (e.g., Vibe 1967, Petersen and Siegstad 1993). We hope our project will help put the impacts of climate change in a broader perspective, with a holistic view rather than only by measuring certain environmental parameters, like the temperature, humidity, wind-speed, glacier melt, or the extent of sea ice. These parameters studied by scientists are of course of great importance in order to understand how climate change is affecting the environment. We believe that with a study like the Sila-Inuk we will have a more complete understanding of what we are to face in the near and farther future in Greenland.

Acknowledgments The Sila-Inuk Project is grateful to KNAPK, the Hunters and Fishermen's Organization in Greenland, for providing contacts to their knowledgeable members and to all the communities who participated in the project and shared their knowledge. I am grateful to ICC-Greenland for its support and belief in the project. Many thanks to NunaFONDEN for funding support and to Anna Heilmann and Hanne Sørensen for transcribing some of the interviews. Qujanaq to Shari Gearheard for being by my side and for her assistance in drafting this chapter, and to Igor Krupnik for helpful review and important comments.

Notes

1. Sila-Inuk is also collaborating with the SIKU project Siku-Inuit-Hila (Chapter 11).
2. Unfortunately, we were not able to visit east Greenland in the project. We hope to work in these communities in the future, as we understand that east Greenland has unique environmental conditions and changes that affect other regions.
3. The gathering of beluga and narwhal in swimming pool-sized openings in the sea ice to breath are called Sassat. The whales can become entrapped in those openings if the distance to the next breathing hole or open water is too far. When a group of whales is trapped in the openings of sea ice the result can be deadly for the animals, as too many of them are fighting for breathing room in a small space and many can drown. Hunting sassat is considered a humane response to these events. In the past and today, sassat provides a plentiful source of highly valuable food for local people.

References

Born, E.W., Heilmann, A., Holm, L.K., and Laidre, K. 2006. Isbjørne I Nordvestgrønland. En interviewundersøgelse om fangst og klima. Pinngortitaleriffik, teknisk rapport nr. 70.

Brandt, K. 2007. Kasper Brandt, Iginniarfik, July 2007. Interviewers: Aqqaluk Lynge and Lene K. Holm.

Davidsen, A. 2007. Augo Davidsen, Aasiaat, July 2007. Interviewers: Aqqaluk Lynge and Lene K. Holm.

Frederiksen, E. 2006. Erik Røde Frederiksen, Qassiarsuk, September 2006.

Hall, D.K., Williams, R.S., Jr., Casey, K.A., DiGirolamo, N.E., and Wan, Z. 2006. Satellite-derived, melt-season surface temperature of the Greenland ice sheet (2000–2005) and its relationship to mass balance. *Geophysical Research Letters* 33: L11501, doi:10.1029/2006GL026444.

Helander, E. and Mustonen, T. 2004. *Snowscapes, Dreamscapes: Snowchange Book on Community Voices of Change*. Tampere: Tampere Polytechnic Publications, Ser. C., 562pp.

Helander, E. and Mustonen, T. (eds.). 2005. *Snowscapes, Dreamscapes: Snowchange Book on Community Voices of Change*. Tampere: Tampere Polytechinic Publications. Ser. C., 562pp.

Huntington, H.P. 1998. Observations on the utility of the semi-directive interview for documenting traditional ecological knowledge. *Arctic* 51(3): 237–242.

Huntington, H.P. and Fox, S. 2005. The Changing Arctic: Indigenous perspectives. In *Arctic Climate Impact Assessment* (ACIA). C. Symon, L. Arris, and B. Heal (eds.), New York: Cambridge University Press, pp. 61–98.

Krupnik, I. and Jolly, D. (eds.). 2002. *The Earth Is Faster Now: Indigenous Observations of Arctic Environmental Change*. Fairbanks: ARCUS, xxviii+356pp.

Luthcke, S.B., Zwally, H.J., Abdalati, W., Rowlands, D.D., Ray, R.D., Nerem, R.S., Lemoine, F.G., McCarthy, J.J., and Chinn, D.S. 2002. Recent Greenland ice mass loss by drainage system from satellite gravity observations. *Science* 314(5803): 1286–1289.

Magnussen, T. 2008. Tittus Magnussen, July 2007. Interviewers: Aqqaluk Lynge and Lene K. Holm.

Petersen, H.C. and Siegstad, H. 1993. Registrering af levende ressourcer og naturværdier i Grønland. *Rapport no. 18*. Greenland Home Rule Administration, Direktoratet for sundhed miljø og forskning, Nuuk, Greenland.

Vibe, C. 1967. Arctic Animals in Relation to Climatic Fluctuation, *Meddelelser om Grønland* 170(5): 1–226.

Witze, A. 2008. Climate change: Losing Greenland. *Nature News* 452: 798–802, 16 April 2008.

Zwally, H.J., Abdalati, W., Herring, T., Larson, K., Saba, J., and Steffen, K. 2002. Surface melt acceleration of Greenland ice sheet flow. *Science* 297(5579): 218–222.

Part II
Using the Ice: Indigenous Knowledge and Modern Technologies

Chapter 7
The Sea, the Land, the Coast, and the Winds: Understanding Inuit Sea Ice Use in Context

Claudio Aporta

Abstract This chapter attempts to place Inuit sea ice knowledge in a broader context, first in connection to the knowledge of other environmental features and second within the practices of Inuit spatial orienteering and travel. The premise of this chapter is that any attempt to understand aspects of Inuit environmental knowledge without taking into account the context of mobility is limiting, as travel was an integral part of Inuit life before their establishment in permanent settlements. Inuit identities and environmental knowledge were historically connected not only to specific places (like a camp or the floe edge) but also, and significantly, to life on the move. The land, the sea, the floe edge, the shores, the sky, and the winds are all inseparable parts of the environment in which Inuit live. This chapter describes the two distinctive environments in which Inuit life takes place, namely the land and the sea, as well as the highly significant environment constituted by the shores, and how they all fit into a broader spatial framework constituted by the winds. The research for this chapter was undertaken in Igloolik, Nunavut.

Keywords Inuit · Sea ice · Wayfinding · Navigation · Spatial perception

Introduction

As the principal investigator for the Inuit Sea Ice Use and Occupancy Project (ISIUOP) and as one of the steering committee members of SIKU, I am part of a team that maps and documents aspects of Inuit knowledge that until only a few years ago were little known to non-Inuit. As scientists, we often focus on specific aspects of reality. Even as social scientists and anthropologists, aware as we are of the cultural restraints fixed within our own disciplines, we remain bound by forms of

C. Aporta (✉)
Department of Sociology and Anthropology, Carleton University,
Ottawa, ON, K1S 5B6, Canada
e-mail: Claudio_aporta@carleton.ca

I. Krupnik et al. (eds.), *SIKU: Knowing Our Ice*,
DOI 10.1007/978-90-481-8587-0_7,

knowing, learning, teaching, writing, and reporting that involve, to a certain degree, some sort of fragmentation of reality.

For many of the Inuit experts whom we have encountered while undergoing our research endeavors, this approach does not necessarily make sense. It is widely known that the way that Inuit knowledge and skills were traditionally shared was connected to a more comprehensive understanding of life, in which things like the weather, the animals, the winds, the sea, the land, the ice are all part of the same learning experience.

A study of Inuit knowledge and use of sea ice implies, inevitably, a simplification of the learning experience as it happens in its original context. The very definition of the research topic (e.g., writing a proposal) involves severing one particular aspect of Inuit knowledge (in this case, the sea ice) from the rest of a multidimensional approach to life and the world, where concepts as apparently diverse as social and environmental health are not completely understood without the other. This dilemma is part of the essence of doing ethnography, which involves the study of culture and cultural differences through varying frames of knowledge established within social science theory and methodology. However, once this fundamental problem is identified and once the limitations of documentation are acknowledged, the outcomes of an ethnographic enterprise may still be of significant value. Anthropologists such as Julie Cruickshank (2005) and Igor Krupnik (2002) have conducted research projects in Arctic and sub-Arctic North America where the final products were not simply conventional scholarly publications, but open spaces where the research participants could freely build their own narratives.

In multi-sited projects of the dimensions that most SIKU partners are conducting, such openness is difficult to attain. We create maps to document Inuit sea ice use, construct lexica of sea ice terms, and analyze how different groups deal with climate change, as ways to dissect the notable depth of Inuit understanding of the sea ice. Although fragmented, the compilation of studies in this volume offers detailed snapshots of what that knowledge is about and provides other scientists as well as the general public with an opportunity to learn from Inuit experiences. Furthermore, documentation of this type of highly technical knowledge may constitute a tool for younger generations of Inuit who do not have full access to ways of learning that were common in the past.

To compensate for the unavoidable fragmentation of knowledge, here I attempt to place Inuit sea ice knowledge in a broader context, first in connection to the knowledge of other environmental features and second within the practices of Inuit spatial orienteering and travel. The premise of this chapter is simple: any attempt to understand aspects of Inuit environmental knowledge without taking into account the context of mobility is limiting, as travel was an integral part of Inuit life before their establishment in permanent settlements. Inuit identities and environmental knowledge were historically connected not only to specific places (like a camp or the floe edge) but also, and significantly, to life on the move (Aporta 2009).

It is this relationship to movement that Knud Rasmussen, in his travels with the Fifth Thule Expedition, noted when he described the symbolic connections between a child's first journey and being introduced to the world (Rasmussen 1929:47). Anthropology theorist Tim Ingold proposed that wayfinding should be understood as a way of dwelling in the world, and that the answer to the question "where am I?" depends upon "situating [one's] position within the matrix of movement constitutive of a region" (2000:235) rather than narrowly defining a specific location. A region, in Ingold's terms, consists of the relationship between places, which "exist not in space but as nodes in a matrix of movement," and "wayfinding is a matter of moving from one place to another in a region" (2000:219).

It is within regions constituted by both land and sea that Inuit live and travel. The fact that the sea temporarily transforms into a landfast ice platform, supporting movement and life, makes the concept of "region" even more adequate, as the Inuit's well-established networks of trails are in constant transition between land and ice. When the landfast ice forms it connects shores seasonally separated by open water and simultaneously grants new resources to the hunters as marine mammals on and under the ice become available. The recurring topography of the ice (Aporta 2002) includes places of open water where seals, walrus, and polar bears are able to create their habitats and where Inuit hunters find their most important sources of food during most of the year. While Inuit continuously travel between land and ice, the presence and behavior of other environmental factors, such as the wind or the phases of the moon, are carefully considered, as they will determine the behavior of the moving ice. The land, the sea, the floe edge, the shores, the sky, and the winds are all inseparable parts of the environment in which Inuit live. Furthermore, travel and wayfinding are not separated from the daily business of living. A good hunter is always a good wayfinder because both hunting and wayfinding require a comprehensive engagement with the environment and because both activities are intrinsically connected to each other.

The following sections will describe the two distinctive environments in which Inuit life takes place, namely the land and the sea, as well as the highly significant environment constituted by the shores. This chapter will also place these regions within broader spatial frameworks constituted by the winds. The ethnographic research on which this chapter is based was undertaken in Igloolik between 1998 and 2008, but mostly between 2000 and 2002. All the interviews, conducted by the author or by others, are part of the Igloolik Oral History Project and can be found, following the reference codes, in the database hosted by the Igloolik Research Centre. Most of the interviews were conducted with elders that grew up on the land and that moved to the permanent settlement in the late 1960s. It should be noted that lifestyles, ways of learning, technologies of travel and hunting, and climate have all experienced transformation of various degrees. Therefore, the terms and processes described below should be placed in the context of change that is common to most regions and peoples in the Arctic.

The Region of the Igloolik Inuit

Igloolik is an island approximately 18 × 9 km in size, situated in northern Foxe Basin between Baffin Island and Melville Peninsula in the Canadian Eastern Arctic. Northern Foxe Basin and the island of Igloolik itself have been a center of Inuit and pre-Inuit cultures for more than 4,000 years as posited by archaeological evidence (Maxwell 1984). According to Crowe, throughout this long historical period "there has been a striking continuity in the cultural landscape, cultural history and cultural ecology of the region" (Crowe 1969:ix). Igloolik is situated in a biologically productive area, where cases of starvation and infanticide related to food scarcity have been rare (Mary-Rousselière 1984:436). The combined action of wind, the topography of the shore and seafloor, along with marine currents, create several polynyas in Fury and Hecla Strait (known in Igloolik as *Ikiq*). These features, including the productive floe edge southeast of the island, encourage the proliferation of microorganisms, larger invertebrates, and fish who in turn supply marine mammals with energy. Beluga, walrus, and seals are hunted in the summer and early fall, and caribou and polar bears are found throughout the year in different locations in Melville Peninsula and Baffin Island.

The patterns of travel and land use have changed throughout different historic periods. In the early 1820s Arctic explorer Parry pointed out that the most knowledgeable people of the Igloolik area were familiar with a territory which comprised "a distance of more than five hundred miles reckoned in a direct line [south-north], besides the numerous turnings and windings of the coast along which they are accustomed to travel" (1824:513). By the time members of Rasmussen's Fifth Thule expedition visited Igloolik in the 1920s, the patterns of travel included regular long journeys to the distant trading posts of Repulse Bay, Pond Inlet, and Arctic Bay (this last one was established in the 1930s). Sea ice travel was an integral part of the network of trails that connected this entire region, particularly because in pre-settlement, and pre-snowmobile times, Inuit routes often favored coastal and ice traveling (Aporta 2009).

As in most parts of the Canadian Arctic, the patterns of travel changed quite dramatically with sedentarization in the early 1960s and with the introduction of motorboats and snowmobiles. Part-time and full-time jobs along with formal education created new contexts for traveling, which is now frequently undertaken as a weekend activity. Despite all the changes,[1] traveling is still a very important part of people's lives (Aporta and Higgs 2005), and it takes place along routes that have been used for generations (Aporta 2004, 2009). Hunting and fishing are the main reasons for travel, but trips to visit relatives in the communities of Arctic Bay, Pond Inlet, Repulse Bay, Clyde River, and particularly Hall Beach are undertaken on a regular basis, with spring being the preferred season for long journeys. Hunting involves trips whose lengths range from a few hours (i.e., hunting on the floe edge) to several days (e.g., searching for caribou or polar bears). In Igloolik, practically all travel involves dealing with the sea ice.

The territory where all these activities take place is quite diverse. It includes deep valleys and fiords on Baffin Island and Northern Melville Peninsula, large expanses of flat tundra on the mainland across from Igloolik, and the crossing of long straits of frozen sea in northern Foxe Basin. The following three sections describe some aspects of how Inuit in Igloolik perceive the different elements of their environment. The mainland, the sea, and the coast play significant roles in the configuration of the Igloolik region. If language is a way of dwelling (Basso 1988), then naming can be conceived as a way of experiencing the environment. This is particularly true with Inuit cultures, where place names have enormous importance. Each section below includes a brief reference to place names collected by the author between 2000 and 2008. The names are associated with inland, coastal, and marine features. This paper will stress the terminology that Igloolik Inuit use when referring to traveling within their territory.[2] The descriptions below are by no means exhaustive, but they offer an important glimpse into the complex set of relationships that Inuit establish with their environment.

The Mainland

Inuit in Igloolik are familiar with the open space usually characterized by distant horizons.[3] Flat and mountainous landscapes as well as frozen seascapes are traversed by numerous routes that have been used for generations, are recreated year after year on trackless snow, and belong to the oral geographic knowledge of the Inuit of Igloolik. Judging from Inuit maps drawn for explorers, Spink and Moodie deduced that coastal Inuit knew little of the inland (1972). The authors argued that "the coastal Eskimo uses his rivers as reference points for aiding movement in coastal waters" (1972:15). This seems to be an overly simplistic conclusion, as rivers, creeks, and lakes are significant parts of Inuit land trails. Carpenter also pointed out that an old Igloolik hunter who was asked to draw a map "mentioned no names for most of the islands, though he did for salient points on their coastlines. In other words, he had no interest in land mass, only in geographical points" (1973:18). However, Inuit in Igloolik and in most of the Canadian Arctic have always practiced inland hunting and fishing, and some of the most important routes to distant places go through chains of lakes and rivers across large extensions of land (MacDonald 1998:162, Aporta 2004 and 2009, see also Boas 1888:450).

What counts as mainland? Although we may take for granted what the distinctions between islands and continental masses are, those notions are tied to cultural understandings and perceptions of the territory. The mainland in Igloolik is known as *iluiliq*, which can be defined as "a mass body of land without any islands" (Aqiaruq 1993a). *Iluiliq*[4] is applied to the mainland of Melville Peninsula, but elders in Igloolik remember that in the past the term *iluiliq* was also used when referring to Baffin Island, which was also considered to be mainland. Aqiaruq said that "if it was in the past [Baffin Island] definitely would be *iluiliq*, but now we tend to term it as Qikiqtaaluk ["big island"] ... this island is so huge [that] we used to refer to it as *iluiliq* because we did not know any better" (1993a). Igloolik was, therefore, located in the strait known as Ikiq between two large landmasses, each known as *iluiliq* (Fig. 7.1).

The definition of what distinguishes an island from a mainland is, of course, a matter of scale and perception. The Webster's Dictionary defines *island* as "a land mass, esp. one smaller than a continent, completely surrounded by water," and a *continent* as "one of the principal land masses of the world." For the Inuit, for whom space is mostly considered in terms of what is seen (or unseen) on the horizon (see

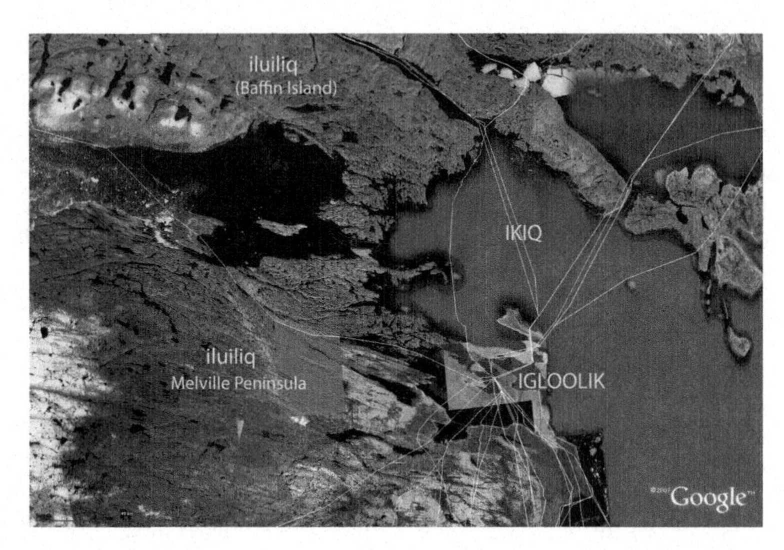

Fig. 7.1 Map of the two land masses (selected sea ice features – *darker lines* – adapted from Laidler and Ikummaq 2008; selected trails – in *white* – selected from Aporta 2009) (See also Color Plate 2 on page 471)

Aporta 2004), the criterion for the mainland is a large mass of land where the shores are not visible from the interior. The interior of the mainland is known as *nunavik*, which was formerly defined as "a place where the shore cannot be seen" (Kupaaq 1993). An *iluiliq*, therefore, contains a *nunavik*. Both Melville Peninsula and Baffin Island possess a *nunavik* and are thus considered by Inuit as mainland.

Although Baffin Island is no longer considered to be mainland, terms still used today in the community referring to travelers going back and forth between northern Baffin Island and Igloolik imply the consideration of Baffin Island as mainland. *Itijjaaq* ("gone overland"), for instance, is a term that refers to a traveler who has left Igloolik and is on his/her way to the Arctic Bay and Pond Inlet region (Amarualik 1994). Terms that refer to large, distant territories on the mainland are also used. The term *kiva*, for instance, means "where there appears to be nothing" and among the Inuit of Igloolik indicates the southern part of Melville Peninsula (the Kitikmeot area). This word is the root of words like *kivavvaat*, referring to travelers going to that region.[5] South of *kiva* there is *taungna*, which was known as "the land of the white man" (Kupaaq 1993).

Place names in Igloolik refer to a large variety of features. Of the 600 place names collected by the author in Igloolik, over 35% refer to land features, including lakes, rivers, hills, rocks, river bends, rock cairns, and portages.[6] The size or scale of the named features varies from large lakes, such as Tasiujaq (Hall Lake), to individual rocks (Iksivautaujaq). Names referring to different parts of the same geographic feature are also common. For instance, there is a river named Ajagutalik that has a bend named Sanguraq. Within this river bend, there is a precise turning point named Avalagiavvik. It is likely that such nesting of descriptions reflects a universal method to organize spatial memories (Kitchin and Blades 2002).

The Sea

Virtually all the surrounding environments the Inuit of Igloolik live in are highly dynamic. Raised beaches are a common feature around Igloolik and ancient campsites presently situated far inland depict dramatically different coastlines. Some old place names also remind people of the dynamics of the land. Qikiqtaarjuk, for instance, is a peninsula on the Island of Igloolik, but the name means "little island," which refers to a time when Qikiqtaarjuk was separated by the sea from Igloolik island proper.[7] The land also changes its appearance on a yearly basis and segments of routes are modified in relation to the snow coverage and the timing of ice formation and thaw within lakes and rivers.

The sea, however, remains the most dynamic of all environments, from the open water in the summer to the landfast ice and from boating to sledging.[8] The open sea has a certain type of topography that can be recognized and in some cases is geographically situated. Currents and swells are often predictably located and their occurrence is recognized by Inuit while they travel in their boats.[9] Interviews with elders in Igloolik (Kupaaq 1990; Qunnun 2002) reveal a wealth of technically precise terms. *Isuqtuq*, for instance, refers to turbid water created by mud or other dirt, *malliq* refers to rough sea, and *irringnangittuq* refers to turbid water created by fresh water flowing to salt water from the melting ice.[10]

In the sea environment islands are not merely positioned but understood in relation to their role in modifying currents and in creating the topography of the sea ice. Michel Kupaaq, for instance, explained that Simialuk ("the big plug") prevents the currents of Ikiq from flowing freely from the west and creates the conditions for the existence of three adjacent polynyas. The geologic characteristics of each island are well known to the knowledgeable hunter. Each island is recognized as having particular kinds of stones or gravel of different colors and shapes (Kupaaq 1987).

Twenty-three percent of the place names collected in Igloolik relate to the ocean, including islands, ice features, polynyas, and submarine terrain. The naming of islands does not follow a unique criterion. The island of Igloolik was only named as such by Parry's second expedition in search of the Northwest Passage. For the Inuit, the name Iglulik refers to a camp southeast of the island. There are 19 named features on the island, but the island itself has no Inuktitut name. Some other islands, however, are named as wholes (e.g., Saglarjuk). There are also names for clusters of small islands (e.g., Uqsuriattiangujaak, referring to three adjacent islands, or Uglit, referring to two islands). These kinds of naming patterns are similar in most of the Canadian Arctic.

The Sea Ice

Ice is part of the territory where Inuit live for most of the year and traveling on the ice may take up to 8 months every year in the Igloolik area.[11] Inuit hunters understand the codes of such a changing place and have discovered its predictability, to the extent that they can exploit the moving ice on a regular basis. In the past, they used to make the landfast ice their home for part of the spring (Aporta 2002). Places like Agiuppiniq (an ice ridge), Naggutialuk (an ice lead), Ivuniraarjuruluk (an ice build-up), and Aukkarnaarjuk (a polynya) recur every year at the same locations and are identified with names in a similar way as places on the land. The sea ice topography and processes are identified with complex terminologies, as several of the papers in this volume show (see also Laidler and Elee 2008; Laidler and Ikummaq 2008, and Laidler et al. 2008).

Once the sea ice attaches to the land, a significant new territory opens up for Inuit, where they can extend their travel routes to other coastal locations and to resources (particularly marine mammals) that become available on (or through) the ice. Because marine mammals (in Igloolik, mostly walrus and ringed and bearded seals) need both water and air to survive, it is not surprising that some of the most significant places on the ice are connected to open water, cracks, thin ice, or moving ice. Other features of this environment such as ice ridges are equally important as determining factors of the layouts of routes and as navigational aids and for locating seal dens. The sea ice, with its recurrent topography, offers, therefore, culturally and historically significant places for Inuit. In all the communities that are part of ISIUOP, Inuit collective and individual memories are often connected to some of these features and places on the sea ice, including, in some cases, the gathering in spring camps on the sea ice.

In Igloolik, and within the sea ice environment, the floe edge plays an important role, both for its productivity and as a frame of spatial reference. The environment that we generally refer to as sea ice is in fact constituted by several sub-environments, the most important of which are the landfast ice (*tuvaq*), the moving ice (*aulajuq*), the polynya (*aukkaniq*), and the floe edge (*sinaaq*). The floe edge plays a particularly important role in Igloolik, as compared to the other communities included in ISIUOP (Laidler et al. 2008). Therefore, the understanding of the timing of the moving ice in relation to the winds and tidal shifts is critical (Aporta 2002; Laidler and Ikummaq 2008). Ice cracks and leads are also identified, often in reference to whether they run parallel or perpendicular to the floe edge, as well as their physical characteristics or whether or not they refreeze (Laidler and Ikummaq 2008). Although the position of the floe edge fluctuates, it is always spatially situated in one general area which is visible in the horizon as a dark blue reflection of the water on the sky, a phenomenon known in Igloolik as *tunguniq* (MacDonald 1998:184). *Tunguniq* is sometimes strikingly visible as a thick line above the horizon and sometimes as a distant dark point. It becomes an important spatial reference when traveling across the flat tundra on the neighboring mainland.

Interestingly, the act of travel over open water and ice is sometimes distinguished by different terminology even when the terms refer to the same spatial action of going toward or away from the shore. Both *Kangivaq* and *tuvviaqttuq* refer to a return journey from the sea, but the former term is used when boating in open water while the latter when coming back from walrus hunting on the moving ice. *Sammuk* and *mauttut* are terms that refer to the same spatial action of "going out," the former referring to leaving the beach for the open water and the latter to leaving the landfast ice or the beach for moving ice.

Traveling on the sea ice is usually undertaken through trails that are often similar from year to year in their final layouts, but that vary throughout the year as the process of freeze-up takes place. The final layout of widely traveled ice trails will also depend on particular features of the ice connected to given years. As a general rule, however, the landing and launching places of trails are the same (named points, parts of bays, and other features on the shores) and are always traced on the same general locations in order to avoid recurrent dangerous ice conditions and to favor relatively recurrent favorable travel surfaces and efficient itineraries. When Inuit describe sea ice trails, they will describe particular ice features that travelers are going to find on the way, as well as place names of land features that can be seen in the horizon, and that can be seen and used as steering aids from the sea.

The Coast

Inuit in Igloolik interact with the marine environment throughout the year as most camps (past and present) are situated on the shores. The shores are also links between land and ice routes and they help determine good anchoring places or areas that are too shallow for boating. Rising beaches, deep cliffs, long fiords, broad bays, and low shores are all significant features that the traveler identifies and uses.

The coast plays a significant role as a framework of spatial orientation, as it does in other cultures where people's livelihoods are tied to the sea (Cablitz 2002). Fortescue noted that terms indicating *away from* and *down to* the shore and *left/right-along-shore* are an important part of coastal Inuit orientation systems across the Arctic (1988:25). The importance of the littoral is evident in old stories told by Inuit elders as the events are frequently placed in reference to the shores. The significance of the littoral as an important aspect of a spatial frame of reference is also illustrated by the terminology used to describe the shore and people's relative position to it. Noah Piugaattuk remembered how during the summer people would split between inland caribou hunting and staying on the littoral. "There was a time when certain individuals would be planning a trip for the inland for the summer... those that stayed on the littoral would hunt marine animals to store them for the winter. Those that stayed on the littoral would be in a place that was identified by the name of the land" (Piugaattuk 1989). In Igloolik, the term *ataartut* is used to refer to people going down to the littoral from the interior (Aqiaruq 1993b). People who remain on the coast while others went inland are known as *singmiujuq*.

Numerous features are defined in reference to their position relative to the shore. The terms *tilliq* (higher) and *salliarusiq* (the one further down), for instance, are sometimes used to refer to the relative position of mountainous ranges in reference to the shore (as seen from the sea) (Kupaaq 1993). Shore cracks (*qungiit*) are important, as they help to observe the tidal movements and tidal shifts. They are also named in relation to their relative position to the shore (here the observer is situated on the shore): *tilliqpaaq* ("the one that is higher than the rest") refers to the first shore crack, *akulliq* is the one in the middle and *salliq* the one further away (Imaruittuq 1990). A similar method is also used for naming the polynyas north of Igloolik, toward the Baffin Island shore. Other icecrack terminology describes other important spatial relations. For instance, *nagguti* is a crack that freezes and re-freezes and that goes from land to land, *napakkuti* is a crack that goes from land to the floe edge, and a *quppirniq* is a crack that forms outward from the floe edge, in moving ice (Laidler and Ikummaq 2008:131).

Almost 42% of the place names collected in and around Igloolik refer to coastal features. As with the names of land and sea features, place names that relate to the shores designate features of different scales. Many of them refer to points, but others refer to long stretches of shore. Names defining specific points within larger named features have frequently the same linguistic root as the name of the larger feature. Iqaluit Nuvua ("the point of Iqaluit"), for instance, refers to a point within the bay of Iqaluit. Other names cite fiords, cliffs, landing points, places with relatively high relief, low relief, and campsites. These place names infer that the naming of the shore is related not only to residence but also to travel. Both land and ice trails are fairly stable from year to year and they begin or finish at specific coastal places for landing and launching. While traveling across straits of open water or sea ice, named places represent recognizable points on a familiar horizon.

Horizons, Winds, and Spatial Frameworks

As stated at the beginning of this chapter, to better understand the way Inuit approach these diverse territorial entities (the sea, the land, and the shores), it is necessary to place them in the context of moving and traveling that were part of Inuit life before they moved to permanent settlements (and that, to certain degree, still are). In the context of traveling, the sea, the land, and the shores were closely intertwined in the way the environment was perceived. In the activity of wayfinding or orienting, these entities were part of a spatial framework, but were also placed in a larger spatial framework that will be described below.

Ingold defines wayfinding "as a skilled performance in which the traveler, whose powers of perception and action have been fine-tuned through previous experience, 'feels his way' towards his goal, continually adjusting his movements in response to an ongoing perceptual monitoring of his surroundings" (2000:220). Travel is not performed in abstract space, but through places of significance. Places, in turn, are nodes within a network of coming and going which Casey defines as a "region" (Casey 1996:24, cited by Ingold). In the context of Inuit culture, a region is constituted by the territory a person is familiar with either through his/her own travels or through somebody else's narratives. Wayfinding methods are used and understood in relation to a limited number of shared frameworks of spatial orientation known to all knowledgeable people in the community. In Igloolik (as in most other coastal Inuit communities), these frameworks are constituted by the direction of prevailing winds and by the position of the mainland, the shores, and the floe edge. All methods are understood in relation to such frameworks: animals and seaweed move in reference to the shore or the floe edge; sky features are situated in reference to wind bearings; and people move in and see the territory in terms of horizons where winds, shores, mainland, sea, floe edge, celestial marks, and familiar landmarks are situated, described, and experienced.

Horizons

The previous descriptions of land, coast, and sea are intended to stress the importance of looking at Inuit knowledge of any environmental feature as linked to a broader territory or region, an experience which comes into being through travel and movement (or dwelling). Traveling in Igloolik involves the transit from sea or sea ice to land on a regular basis, and spatial orientation is connected to how travelers situate themselves in that territory and on how they understand and describe the changing horizons that they encounter while they move. When using well-established Inuit methods of geographic representation (though narratives and without topographic maps, see Aporta 2004), the perception of what is seen or unseen on the horizon is crucial in order to situate, remember, and describe specific locations. Land and coastal features such as raised beaches or hills are often spotted from the sea during crossings and constitute a visual frame of reference for travelers (Fig. 7.2).

Inuit in most of the Canadian Arctic experience their territory in terms of vistas of the horizon even when talking about places that are far away, large extensions of land, small landmarks, or while describing a long journey or a route. This is why experienced hunters stress the importance of focusing on all directions which surround a traveler and not solely on the trail ahead. To be a knowledgeable, self-reliant traveler, they point out, one must be able to identify a place from a number of different angles. Although the slow motion of the dog team was better suited for this practice, hunters traveling by snowmobile still look around while taking tea breaks or when refuelling their machines. A common way of teaching younger or inexperienced people during these stops is to ask them to point at different places on the horizon.

Fig. 7.2 "Inuit horizon" from trail on the sea ice with a framework of winds (Nigiqpassik, Akinnaq, and Akinnaqpassik), place names (Qalirusik, Avvajja, and Qinngurmigarvigjuaq), and the main land (Iluiliq). (Photo: Claudio Aporta, content by Maurice Arnatsiaq)

For example, to remember or communicate where a broken sled is located, a hunter would describe what he/she "sees" on the horizon from that particular position, usually referring to those features by their names. The relative location of the place will then be described in terms of other spatial frames of reference, particularly the shore and floe edge.[12] Oral narratives tailored for people unfamiliar with a place may describe routes or specific locations (e.g., a turning point in a trail, a caribou cache, a fox trap, a broken sled) in relation to a landmark appearing on the horizon. For instance, when asked to describe what kind of information he would give to another hunter to describe the precise location of an object, Louis Alianakuluk explained

> If I left something behind at Ikiq... it might be that I left my machine behind, if I was to say that it is somewhere at Ikiq, no one would know where it is at. But if I was to say that it is just above the old floe edge, or if I was to say that it is close to the *nipititaaq* [moving ice that has stuck to and become part of the landfast ice], or just below it, at once someone would identify the location. Then I might say that I left my machine behind, where certain pressure ridges were in view at a certain direction. If I was to give this kind of information, even a person that did not leave the machine behind would now be able to go to the place. (2001)

Alianakuluk narrowed down the description of the location in different stages. First, he mentioned Ikiq, which defines a relatively large expanse of territory (the sea between Baffin Island and Melville Peninsula). Then he described some topographic features of the ice that people sharing his oral geographic knowledge would be able to identify and locate. To define a more precise location he mentioned that at the specific spot where the object is located there is a landmark visible on the horizon in a certain direction. His hierarchical descriptions of spatial relationships recapitulate what may be, in fact, a fundamental organizational property of human spatial cognition (Kitchin and Blades 2002). Sea ice, landmarks, and shores are all integrated in the same narrative.

Winds

As mentioned before, the most important frame of reference is provided by the winds. Winds occupy a central place in the lives of the Inuit of Igloolik, as knowledgeable hunters can divide the horizon into 16 directions or wind bearings (MacDonald 1998:181). The winds foretell weather changes, shape patterns on the snow and regulate – along with the tides – the behavior of the moving ice. They are by far the most discussed of all environmental phenomena (MacDonald 1998:182), they largely regulate hunting activities, and they play a fundamental role in spatial orientation. The Inuit of Igloolik recognize four primary winds: Uangnaq (WNW), Kanangnaq (NNE), Nigiq (ESE), and Akinnaq (SSW) (MacDonald 1998:181). Uangnaq and Nigiq are the two prevailing winds. Uangnaq produces snowdrifts that range in height from several centimeters to almost 1 m. These snowdrifts are named *uqalurait* (like a tongue) and are easy to recognize as their shape is distinctive and they harden to become permanent features of the snowscape. Hunters usually use *uqalurait* to set their bearings while traveling across large extensions of flat tundra or during periods of poor visibility due to weather or darkness.

The relative position of an individual or a place in relation to a landmark on the horizon is defined by the use of wind bearings. Theo Ikummaq remembered that on one occasion he was traveling to Repulse Bay when he became separated from the established and well-known trail only to be confused as he tried to locate it again. He used his shortwave radio to communicate with his uncle in Igloolik and ask for advice. Ikummaq explained how his uncle, after discussing the direction from which the wind was blowing, asked him several questions to understand where he was situated:

> Question: If you are facing down wind, what do you see?
> Answer: A couple of hills, a couple of large hills.
> Question: Facing toward the wind, what do you see?
> Answer: Some rocky outcrops.
> Question: To your left (that means towards Repulse) what do you see? Look to your left, what do you see?
> Answer: I see a narrow rocky outcrop, but it ends, and then it starts again a little further on and then it continues on.
> Question: That's the trail to take. You go between that and then you will find the main trail.
> And such was the case. He wasn't there. But he could determine where I was from what I described. (2000)

Ikummaq's story illustrates how the territory is seen and remembered as vistas oriented by an internalized spatial framework (in this case wind bearings). In a conceptual manner, this way of experiencing the territory is represented (and simplified) in Fig. 7.3.

A knowledgeable hunter, therefore, can move and talk about the land, using numerous spatial references in an environment that he can visualize from the perspective of the traveler: as vistas. Furthermore, the dynamics of the environments, coupled with the movements of animals and people, are conceived within the stability of fixed spatial frameworks provided by the shores and the winds and within familiar places and regions. A knowledgeable hunter is engaged in such a way with the environment that what may appear a homogenous landscape to an outsider is in fact dotted with spatial references.

In this environment, everything takes place and makes sense within several frames of spatial reference: fish swim up and down the shore, seals move against or toward the floe edge, birds fly toward and away from the shore, places are located in reference to the winds, and winds and celestial landmarks are identified with positions on familiar horizons. Descriptions and narratives make sense without the need to draw or point to places on a map. A clear example was provided by Siakuluk while he was telling a story of two people who committed murder. Siakuluk pointed out that after the murder was committed the murderers built an *inuksugak* (stone cairn) that was shaped like a human figure. The *inuksugak* "is facing the direction where the two fled. It is located some distance away from Ualinaaq towards the land away from the littoral" (1996). Siakuluk situated the story by naming a place (Ualinaaq)

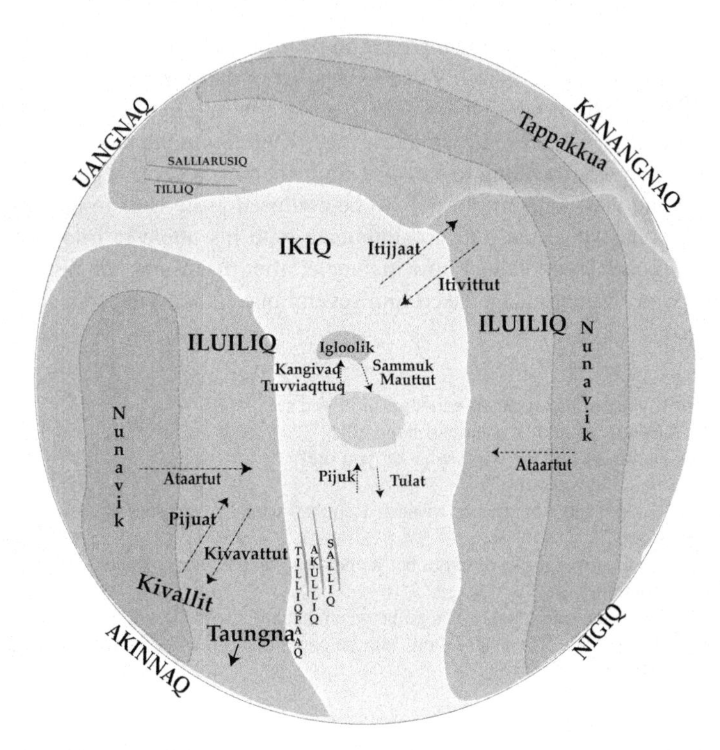

Fig. 7.3 Schematic view of Inuit representation of space based on winds, shores, and floe edge

and using a spatial referent (the coast) for direction. Siakuluk's narrative was, therefore, situated in a frame of spatial reference without the need of pointing at a map. It is within this frame of reference that horizons are updated from the perspective of the traveler. It is also within this type of geographic representation and narrative that the three territorial entities (land, sea, and shores) are constantly related to each other, as a traveler will always find his/her own present position is relative and, to some degree, transitional, and in relation to broader frames of reference.

Both social and environmental changes are challenging this well-established way of understanding and representing space. The younger generations, born and raised in town, are less attuned with the environmental factors that were observed and experienced on a continuous basis since an early age. New technologies and representations of space (e.g., GPS and maps) are making this type of spatial perception less essential. Despite relying more on new methods of orienting, however, the main aspects of these spatial frameworks are still known and used by most active hunters in Igloolik. Environmental changes, however, can represent a bigger challenge. Inuit in most of the communities involved in the ISIUOP project report changes in the direction and patterns of the winds and different degrees of changes in formation and dynamics of the sea ice. It remains to be seen how people will adapt their established spatial frameworks and environmental observations to account for the ongoing changes.

On the other hand, this well-established approach to space also presents a challenge to the researcher who is using mapping as a tool of representation of Inuit land and sea use. Older Inuit, who rely on this oral, horizon-oriented approach to space, are forced to translate their own knowledge into the two-dimensional, bird's eye perspective of the map. Current and historical evidence, however, indicate that such translation is less dramatic than what one might anticipate as Inuit were quickly able to grasp cartographic representation since their first contact with Europeans (see, for instance, Bravo 1996).

Conclusion

Perhaps one of the main reasons why many of us remain fascinated by how Inuit relate to the sea ice has to do with how foreign this experience is to our own, Western, perception of the environment. The fact that ice can be interpreted as "home" and that it can be a place of culture and history, with a known, recurrent topography and features associated with toponyms, is quite far from the way southerners (including Arctic explorers) have approached and understood this environment.

The sea ice is still part of the daily environment of most Inuit communities, even in the face of climate and social changes. For Inuit who lived on the land before moving to permanent settlements, the sea ice was part of a travel surface where life took place.

Documenting the knowledge and skills associated with traveling and hunting on the sea ice fills up a gap in our understanding of Inuit environmental knowledge. The technical detail and precision of this knowledge were developed over generations of experience interacting with this environment. In practical terms, however, the knowledge and skills to travel, live, and hunt on the ice were not – in the learning experience – dramatically separated from other spheres of knowledge and experience. In placing Inuit sea ice knowledge and use in the context of traveling, this paper has attempted to overcome some aspects of the inevitable fragmentation of reality that comes with documentation. When sea ice is considered in this context, relations between the sea and the land, as well as the significance of the coast, become clear. The sea ice remains the most dynamic of all the environments that Inuit face, and its dynamics closely tied to tidal shifts, topographic features of the coast, and the effects of temperature and, principally, the winds. The winds are the broadest spatial reference within which the development and interaction of these territorial entities take place and are understood by Inuit.

The International Polar Year, through our projects ISIUOP and SIKU, has made possible the first systematic study of Inuit and sea ice across regions. The documenting and understanding of this topic will be greatly improved as a result of these projects. For younger generations of Inuit, however, the key to this knowledge still seems to be the direct experience of the sea ice within broader environmental and spatial contexts, through the experience of travel. Environmental, technological, social, and language changes, however, cannot be ignored as we continue to try to make sense of this fascinating aspect of Inuit knowledge.

In understanding the way older generations understood their relation to the environment, it is perhaps useful to cite the late George Kappianaq, a knowledgeable elder of Igloolik. Kappianaq made it clear that places are not to be separated from past experiences, including such emotions as sadness, happiness, or love:

> My memories would return to the times we all lived in big families. There are so many things that remind me when I go back to the locations where we used to live; today memories will return to me as if things were there; with the passage of time I no longer feel too much emotion but memories still come back to the time when we were whole. [What] I remember most was the time both of my parents were alive and well. Even the thought of the late wife will come... thinking she too was alive and well. I would imagine others too have fond memories. (1990)

Projects like the ones conducted in the context of ISIUOP and SIKU may sometimes reflect the constraints of academic research, where such comprehensive experiences of the environment can be overlooked. It is, perhaps, in the narratives that we document that such wholeness can still be detected, preserved, and transmitted and, hence, the importance of making this documentation available to the communities beyond the limits of academic publications or scientific reports.

Acknowledgments This chapter could not have been written without the support and help of Igloolik hunters, elders, and friends. Beyond the experts cited in the reference list, Louis Tapardjuk, Maurice Arnatsiaq, and Theo Ikummaq were key in helping with trips, translations, and interpretations. John MacDonald provided feedback and insight on the early stages of this research. Funding for this research was provided by an IPY Canada research grant and by earlier SSHRC and Wenner-Gren grants. I would also like to thank Igor Krupnik, Gita Laidler, Chase Morrison, and Tim De Leo Browne for their comments and edits on earlier drafts. All mistakes and inaccuracies are my sole responsibility.

Notes

1. For a detailed description of social and demographic changes in contemporary Igloolik, see Rasing (1994).
2. The terminology cited in the chapter was developed through consultation with several Inuit experts and also extracted from the Igloolik Oral History Project. The terms and the meanings were checked for accuracy with several elders and interpreters, and in collaboration with John MacDonald (former coordinator of the Igloolik Research Centre) while in the field and through interpreters. Any mistakes and omissions are the author's sole responsibility.
3. This varies significantly in other ISIUOP communities such us Clyde River and Pangnirtung, where traveling through very intricate land features is part of the daily travel experience. Place naming and route tracing, however, are remarkably similar.
4. For clarity, Inuktitut terms will be printed in italics, while Inuktitut place names will be written in normal font.
5. Fortescue has suggested that the word Keewatin can be an anglicized form of *kivallin* (southerners) from the directional stem *kivat*, which in turn refers to the spatial organization of the *igloo* (1988:10).

6. Percentages are given here only to provide a sense of proportion. Categorizations of names by geographic location or by topic do not offer any significant value, as hills may be orienteering features when traveling by sea, rocks may be situated on the shores, and ice features may be connected to topography of the land.
7. It has been estimated that Qikiqtaarjuk became a peninsula about 300 years ago, when the sea levels receded (John MacDonald, personal communication 2008).
8. The complexity and dynamics of the sea in relation to travel are also seen in the practice of boating in the floe edge, in which boats are pulled on sleds by snowmobiles from the land to the open water.
9. The recognition of water features is, of course, not unique to Inuit, but common to all marine peoples (see, for instance, Lewis 1994, for a description of how Puluwat navigators recognize swells).
10. "Once the fresh water had sank to the heavy or salt water and mixes, the water will clear again" (Kupaaq 1990).
11. Climatic changes are making the sea ice season shorter in most communities involved in the ISIUOP study.
12. Having a spatial framework is not seen here as imposing an abstract grid onto the world, but as a way of experiencing or perceiving the environment through the engaging process of moving (literally or figuratively) in it. These frameworks are shared by the members of the community and have been developed through generations.

References

Alianakuluk, L. 2001. Interview for Igloolik Oral History Project. Igloolik: Archives of the Inullariit Society, Igloolik Research Centre, Igloolik, Nunavut (IE-481).

Amarualik, H. 1994. Interview for Igloolik Oral History Project. Igloolik: Archives of the Inullariit Society, Igloolik Research Centre, Igloolik Research Centre, Igloolik, Nunavut (IE-314).

Aporta, C. 2002. Life on the ice: Understanding the codes of a changing environment. *Polar Record* 38(207): 341–354.

Aporta, C. 2004. Routes, trails and tracks: Trail-breaking among the Inuit of Igloolik. *Études Inuit Studies* 28(2): 9–38.

Aporta, C. 2009. The trail as home: Inuit and their pan-Arctic network of routes. *Human Ecology* 37(2): 131–146.

Aporta, C. and Higgs, E. 2005. Satellite culture: Global positioning systems, Inuit wayfinding, and the need for a new account of technology. *Current Anthropology* 46(5): 729–754.

Aqiaruq, Z.U. 1993a. Interview for Igloolik Oral History Project. Igloolik: Archives of the Inullariit Society, Igloolik Research Centre, Igloolik Research Centre, Igloolik, Nunavut (IE-272).

Aqiaruq, Z.U. 1993b. Interview for Igloolik Oral History Project. Igloolik: Archives of the Inullariit Society, Igloolik Research Centre, Igloolik Research Centre, Igloolik, Nunavut (IE-269).

Basso, K.H. 1988. Speaking with names: Language and landscape among the Western Apache. *Cultural Anthropology* 3(2): 99–133.

Boas, F. 1888. *The Central Eskimo*. Toronto: Coles Publishing Company, 1974.

Bravo, M.T. 1996. The Accuracy of Ethnoscience: A Study of Inuit Cartography and Cross-Cultural Commensurability. *Manchester University Monographs in Social Anthropology* 2. Manchester.

Cablitz, G.H. 2002. The acquisition of an absolute system: Learning to talk about space in Marquesan (Oceanic, French Polynesia). In Meetings of the Child Language Research Forum, April 2002, Stanford, California.

Carpenter, E. 1973. *Eskimo Realities*. New York: Holt, Rinehart and Winston.

Casey, E.S. 1996. How to get from space to place in a fairly short stretch of time: Phenomenological prolegomena. In *Senses of place*. S. Feld, and K.H. Basso (eds.), Santa Fe: School of American Research Press.

Crowe, K.J. 1969. *A Cultural Geography of Northern Foxe Basin*. Department of Indian Affairs and Northern Development, Ottawa, Series 69–2.
Cruickshank, J. 2005. *Do Glaciers Listen? Local Knowledge, Colonial Encounters, and Social Imagination*. Vancouver and Toronto: UBC Press.
Fortescue, M. 1988. *Eskimo Orientation Systems*. Meddelelser om Grønland, Man and Society, No. 11, Copenhagen.
Ikummaq, T. 2000. Interview for Igloolik Oral History Project. Igloolik: Archives of the Inullariit Society, Igloolik Research Centre, Igloolik Research Centre, Igloolik, Nunavut (IE-466).
Imaruittuq, E. 1990. Interview for Igloolik Oral History Project. Igloolik: Archives of the Inullariit Society, Igloolik Research Centre, Igloolik Research Centre, Igloolik, Nunavut (IE-101).
Ingold, T. 2000. *The Perception of the Environment: Essays in Livelihood, Dwelling, and Skill*. London: Routledge.
Kappianaq, G. 1990. Interview for Igloolik Oral History Project. Igloolik: Archives of the Inullariit Society, Igloolik Research Centre, Igloolik Research Centre, Igloolik, Nunavut (IE-069).
Kitchin, R. and Blades, M. 2002. *The Cognition of Geographic Space*. London: I. B. Tauris.
Krupnik, I. 2002. Watching ice and weather our way: Some lessons from Yupik observations of sea ice and weather on St. Lawrence Island, Alaska. In *The Earth Is Faster Now: Indigenous Observations of Arctic Environmental Change*. I. Krupnik and D. Jolly (eds.), Fairbanks: ARCUS, pp. 156 –197.
Kupaaq, M. 1987. Interview for Igloolik Oral History Project. Igloolik: Archives of the Inullariit Society, Igloolik Research Centre, Igloolik Research Centre, Igloolik, Nunavut (IE-017).
Kupaaq, M. 1990. Interview for Igloolik Oral History Project. Igloolik: Archives of the Inullariit Society, Igloolik Research Centre, Igloolik Research Centre, Igloolik, Nunavut (IE-098).
Kupaaq, M. 1993. Interview for Igloolik Oral History Project. Igloolik: Archives of the Inullariit Society, Igloolik Research Centre, Igloolik Research Centre, Igloolik, Nunavut (IE-272).
Laidler, G.J., Dialla, A., and Joamie, E. 2008. Human geographies of sea ice: Freeze/thaw processes around Pangnirtung, Nunavut, Canada. *Polar Record* 44(231): 335–361.
Laidler, G.J. and Elee, P. 2008. Human geographies of sea ice: Freeze/thaw processes around Cape Dorset, Nunavut, Canada. *Polar Record* 44(228): 51–76.
Laidler, G.J. and Ikummaq, T. 2008. Human geographies of sea ice: Freeze/thaw processes around Igloolik, Nunavut, Canada. *Polar Record*. 44(229): 127–153.
Lewis, D. 1994. *We, the Navigators: The Ancient Art of Landfinding in the Pacific*. Honolulu: University of Hawaii Press, 1972.
MacDonald, J. 1998. *The Arctic Sky: Inuit Astronomy, Star Lore, and Legend*. Toronto: Royal Ontario Museum and Nunavut Research Institute.
Mary-Rousselière, G. 1984. Iglulik. In *Handbook of North American Indians*. D. Damas (ed.), Washington: Smithsonian Institution, vol. 5, pp. 431–447.
Maxwell, M.S. 1984. Pre-Dorset and Dorset prehistory of Canada. In *Handbook of North American Indians*. D. Damas (ed.), Washington: Smithsonian Institution, vol. 5, pp. 369–376.
Parry, W.E. 1824. *Journal of a Second Voyage for the Discovery of a Northwest Passage From the Atlantic to the Pacific*. London: J. Murray, 1969.
Piugattuk, N. 1989. Interview for Igloolik Oral History Project. Igloolik: Archives of the Inullariit Society, Igloolik Research Centre, Igloolik Research Centre, Igloolik, Nunavut (IE-031).
Qunnun, A. 2002. Interview for Igloolik Oral History Project. Igloolik: Archives of the Inullariit Society, Igloolik Research Centre, Igloolik Research Centre, Igloolik, Nunavut (IE-505).
Rasing, W.C.E. 1994. *Too Many People: Order and Nonconformity in Iglulingmiut Social Process*. Nijmegen: Rish & Samenleving.
Rasmussen, K. 1929. *Intellectual Culture of the Iglulik Eskimos*. Report of the Fifth Thule Expedition. vol. 7 (1). Gyldendalske Boghandel, Copenhagen, 1976.
Siakuluk, N. 1996. Interview for Igloolik Oral History Project. Igloolik: Archives of the Inullariit Society, Igloolik Research Centre, Igloolik Research Centre, Igloolik, Nunavut (IE-384).
Spink, J. and Moodie, D.W. 1972. *Eskimo Maps From the Canadian Eastern Arctic*. Cartographica, Monograph No. 5, Toronto: B.V. Gutsell.

Chapter 8
The Igliniit Project: Combining Inuit Knowledge and Geomatics Engineering to Develop a New Observation Tool for Hunters

Shari Gearheard, Gary Aipellee, and Kyle O'Keefe

Contributing Authors: Apiusie Apak, Jayko Enuaraq, David Iqaqrialu, Laimikie Palluq, Jacopie Panipak, Amosie Sivugat, Desmond Chiu, Brandon Culling, Sheldon Lam, Josiah Lau, Andrew Levson, Tina Mosstajiri, Jeremy Park, Trevor Phillips, Michael Brand, Ryan Enns, Edward Wingate, Peter Pulsifer, and Christine Homuth.

Abstract This chapter provides an overview of the Igliniit project, an International Polar Year (IPY) project that took place in Clyde River, Nunavut, from 2006 to 2010. As part of the larger IPY projects, SIKU and ISIUOP, the Igliniit project brought Inuit hunters and geomatics engineering students together to design, build, and test a tool to assist hunters in documenting their observations of the environment. By combining a global positioning system (GPS) receiver, a mobile weather station, a personal digital assistant (PDA), and a digital camera, the hunters and engineering students in Igliniit co-developed and piloted a system that allows hunters to contribute to environmental research in an active way, through the regular *use* of their environment, documenting observations and experiences in context, as they happen. Despite hardware problems and the challenges of using such technology in Arctic winter, the data collected by hunters provide detailed, dynamic, geo-referenced information about the environment that could otherwise not be collected. With continued development, this technology could be useful in many different regions and applications for understanding the environment and human–environment relationships over time and space. The approach, of supporting local people in their own activities year-round and outfitting them with a simple but powerful tool to document their environmental observations, proves a promising method in future community-based environmental research and monitoring, with applications as well in land use planning, resource management, hazards mapping, wildlife and harvest studies, and search and rescue operations.

S. Gearheard (✉)
National Snow and Ice Data Center, University of Colorado at Boulder, Clyde River, NU X0A 0E0, Canada
e-mail: shari.gearheard@nsidc.org

I. Krupnik et al. (eds.), *SIKU: Knowing Our Ice*,
DOI 10.1007/978-90-481-8587-0_8,

Keywords Inuit · Trails · GPS · Collaborative research · Geomatics · Traditional knowledge · Sea ice · Environmental monitoring · Nunavut

Introduction

In Inuktitut, *igliniit* refers to trails routinely traveled. Countless trails are known and used by Inuit; these trails join to create a vast network across the Canadian North and the Arctic (Aporta 2009). The location, use, condition, and changes in *igliniit* over space and time can help reveal a great deal about the environment and human–environment relationships. In our project, the use of *igliniit* provided inspiration for the creation of a tool to help Inuit hunters document observations of their environment as they travel. This tool, in turn, provides a potential means for Inuit to stay active on the land, get younger Inuit involved in land activities, and both contribute to and lead environmental research and monitoring efforts.

The development of the Igliniit project (or simply "Igliniit," as its team members refer to it) happened in Clyde River (Kangiqtugaapik), Nunavut, in 2005. At that time Gearheard, a Clyde River resident, had been working with the community for 5 years documenting Inuit knowledge of climate and environmental change (Fox 2002, 2003, 2004). In anticipation of the 2007–2008 International Polar Year (IPY) (proposals were due in 2006), Gearheard and her local research partners began discussing ideas for local IPY projects. A common theme in those discussions was a desire for research and activities that involved documenting and communicating knowledge through *practice*. Hunters and elders were most interested in projects that would allow them to spend time on the land, in sharing knowledge with each other, youth, and others, as they practiced travel, hunting, and other skills. Gearheard and local hunters discussed the idea of using handheld global positioning system (GPS) receivers that were becoming more and more popular in the community, in way that would allow hunters to document their environmental observations as they traveled, in context, as opposed to the more common methods of having people recall information during interviews, focus groups, or mapping sessions. The idea was refined through discussions with the local Nammautaq Hunters and Trappers Association (HTA) in late 2005; the general outline for what would become the Igliniit project was approved by the HTA and Hamlet Council in early 2006. Around this time, Gearheard was contacted by the Inuit Sea Ice Use and Occupancy Project (ISIUOP) to submit a research idea and join their collaborative effort as a sub-project (see Chapter 1, Introduction for background on ISIUOP). When ISIUOP was successful in obtaining funding from the Canadian IPY Program, Igliniit was ready to get underway.

From Soft Snow to Software

The idea of using a GPS unit to document environmental observations is not new (e.g., Aporta 2003). The use of GPS by local citizens to make observations is also not new and has been applied in several contexts, perhaps most notably and successfully in South Africa, at Kruger and Karoo National Parks where local expert animal trackers use handheld field computers/GPS and CyberTracker software to monitor wildlife and habitat (Liebenberg et al. 1999, Blake et al. 2000). GPS-based ability to facilitate the so-called citizen science in these areas has been well illustrated and has been credited with achievements such as large-scale environmental monitoring (MacFadyen 2005) and the discovery and analysis of various specific phenomena, like the impact of the Ebola virus on lowland gorillas in the Congo (Leroy et al. 2004).

The development of a field computer (better known as a personal digital assistant (PDA)) and GPS system for Inuit hunters in the Arctic, however, was new. Extreme environmental conditions and very demanding travel conditions (i.e., snowmobiles over rough sea ice) present a number of serious challenges. An exploratory study in Alaska in 2005–2006 attempted to develop a system there, but initial consultations and testing indicated that while there was keen interest in the concept, both the available hardware and software that were tested were not sufficiently robust or adaptable to support potential uses without an unacceptably high probability of failure (H. Huntington, pers. comm. 2009).

The special considerations of working in the Arctic environment, as well as access to technical support, were early considerations in the Igliniit project as well. To meet these challenges, we first had to build our team and our tool.

Building a Team

We knew at the outset that our project depended on a strong, diverse, and well-linked team. It has become apparent over the years that the role of our project manager and interpreter Gary Aipellee (Fig. 8.1) was essential in the creation and functioning of our team. Aipellee's skills as a leader, researcher, and interpreter allowed team members with very different backgrounds, experiences, and skills to talk to one another and work together. Facilitating such a diverse group is no easy task and even more so in a project where there are diverse cultures and languages working together, not to mention thousands of kilometers of distance. There is no question that without Aipellee, Igliniit could not have succeeded.

With Aipellee on board, the other project members came into place. Gearheard, during the proposal phase, had contacted Dr. Kyle O'Keefe at the Department of Geomatics Engineering at the University of Calgary Schulich School of Engineering to establish a collaboration for developing the technical equipment in the project. O'Keefe felt the opportunity would fit well with a senior undergraduate course (ENGO 500) where students completed an independent design project. O'Keefe initiated a selection process that led to identifying the handful of students with the right mix of skills and interests to undertake the project. Three different student groups over three consecutive years worked on Igliniit (11 students in total) with five students visiting Clyde River in two separate visits to work directly with the hunters.[1]

The project had funds for two full years of development and testing. Four hunters participated in the first year and six participated in the second year. In the first year, the HTA suggested a list of hunters they thought would be interested in the project and could provide the most appropriate skills and expertise. The HTA also recommended working with a range of ages, from elders to hunters in their twenties, so that the younger participants could learn from older hunters and younger hunters could help the elders with the technology. Gearheard and Aipellee contacted the HTA's list of eight potential hunters and identified four that were available and most interested. In the second year, the original four decided which two hunters would be most appropriate to invite to join them for year 2.[2]

Iterative Engineering

With a team in place, we set to work on developing a handheld computer/GPS system that could be used by the hunters to record their observations of the environment. First, we worked with the hunters to identify the most basic requirements that would drive the choices for the most appropriate hardware. Those basic requirements included GPS tracking capability, ability for hunters to log observations quickly on a touch screen, touch screen text in Inuktitut, ruggedness, reliable power source, and the ability to record the weather. The students, after considering our project budget and our interest in keeping the system affordable, decided that a combination of some off-the-shelf products would be the best option. For the PDA they chose a TDS (Trimble) Recon outfitted with an XC Pathfinder GPS receiver (also from Trimble). For the mobile weather station (connected to the PDA by serial cable) they chose a Kestrel 4,000 pocket weather meter from Nielsen-Kellerman. The hunters approved of the size and weight of the instruments, noting that they had to be lightweight and easy to move, store, and operate.

With the basic hardware identified, we spent the better part of fall 2007 working together to create the most important aspect of the system – the interface that would allow the hunters to record their observations. Facilitated by Aipellee and Gearheard, the group of hunters in Clyde River met regularly to discuss their ideas. After each session, Gearheard compiled meeting notes and shared them by e-mail with the engineering students in Calgary. The students in turn would respond with questions, answers to questions, and photographs and drafts of potential designs.

The hunters decided that in the first year of testing it would be more manageable for everyone if they chose a short list of observations to be available in the first iteration of the PDA program. The hunters developed a list of 20 key observations and organized them according to five categories: animals, sea ice features, land features, hunters' list, and other (Table 8.1). The "animals" category consists of the most hunted species and the sea ice list consists of those ice features that the hunters deemed most important in terms of hunting, travel, and safety. The land features list consists of "lake" and "river"; both are key travel routes, landmarks, and fishing places, as well as potential hazards. *Inuksuk*, a rock cairn used as a marker, is also included as a land feature. The "hunter's list" includes an icon to mark the network of cabins hunters routinely use and an icon to mark the caches of meat left after a harvest. The final "other" category provides a list to help obtain more information for other observations. For example "danger" and "catch" would be

Fig. 8.1 Gary Aipellee leads a project discussion (Photo: Shari Gearheard)

used along with observations like "crack" and "seal." A hunter may observe a crack in the sea ice, so he would log "crack." If it was a hazard, however, he would press "crack" and then "danger," so that the information was logged and geo-referenced together. Likewise a hunter may observe a seal, so he would press "seal," but if he caught the seal he would press "seal" and "catch." One can imagine the various combinations as hunters highlight different land or sea ice features, travel different routes, and harvest different animals. Last, the list includes an icon for "photo" to geo-reference a photograph taken with a digital camera (time synched with the GPS units so photographs could be matched to maps later), an "overnight" icon to mark an overnight camp, and a suggested "remark" icon, for those observations that could not be accounted for using the list.

The hunters' complete list was sent to the engineering students at the University of Calgary where they began the task of writing the software for the PDA – creating the touch screen interface on the handheld computer that the hunters would use to

Table 8.1 Original list of observations identified by Clyde River hunters for the Igliniit PDA/GPS

Animals	ᐅᒪᔪᐃᑦ
Seal	ᓇᑦᑎᖅ
Polar bear	ᓇᓄᖅ
Fish	ᐃᖃᓗᒃ
Caribou	ᑐᒃᑐ
Narwhal	ᕿᓚᓗᒐᖅ
Fox	ᑎᕆᒐᓂᐊᖅ
Ptarmigan	ᐊᕿᒡᒋᖅ
Carcass	ᐅᒪᔪᒥᓂᖅ ᑐᖁᙵᔪᖅ
Sea Ice Features	ᓯᑯᒧᙵᔪᐃᑦ
Seal hole	ᓇᑦᑎᐅᑉ ᐊᒡᓗᖓ
Floe edge	ᓯᓈᖓ
Crack	ᓇᒡᒐᑎ
Thin ice	ᓴᒃᑐᖅ ᓯᑯᒃ
Rotten ice	ᓯᑯᑐᖃᓗᒃ
Iceberg	ᐱᖃᓗᔾ
Very rough ice	ᒪᓂᓚᕋᓗᒃ
Land Features	ᓄᓇᒧᙵᔪᐃᑦ
River	ᑰᒃ
Lake	ᑕᓯᖅ
Inuksuk	ᐃᓄᒃᓱᒃ
Hunters' List	ᐊᖑᓇᓱᒃᑎᓄᙵᔪᐃᑦ
Cabin	ᐃᒡᓗᕋᓛᑦ
Cache	ᐅᒪᔪᖅᑕᒥᓂᑦ
Other	
Danger	ᐊᑦᑕᕐᓇᖅᑐᖅ
Catch	ᐅᒪᔪᖅᑕᒥᓂᑦ
Photo	ᐊᒡᔭᙳᐊᑦ
Overnight	ᓯᓂᑦᑕᕐᕕᑦ
Remark	ᓇᓗᓇᐃᕆᐊᕈᑎᒃ

enter their observations as they traveled (Fig. 8.2a–d). The basis of the interface was visual icons that represented the different animals and environmental features identified by the hunters. This made locating the various observations quick and easy, but the hunters also requested the written names of each to be included, in both Inuktitut and English.

The students thus incorporated the syllabic writing system of Inuktitut into the program. After presenting various options to the hunters, the students designed a screen where the program could be viewed in all Inuktitut or all English depending on preference, with the possibility to toggle back and forth between the two languages if desired. Since younger hunters use both Inuktitut and English, the hunters wanted to be able to access both languages quickly (Fig. 8.2).

The First Year of Testing

With the hardware selected, the software written, and the hunters' observations incorporated into an icon-based interface, the units were ready for their first year of testing. In January 2008, two of the engineering students, Tina Mosstajiri and Brandon Culling, visited Clyde River and brought all the equipment to launch the Igliniit system. The team spent an intense week together – the Clyde group getting

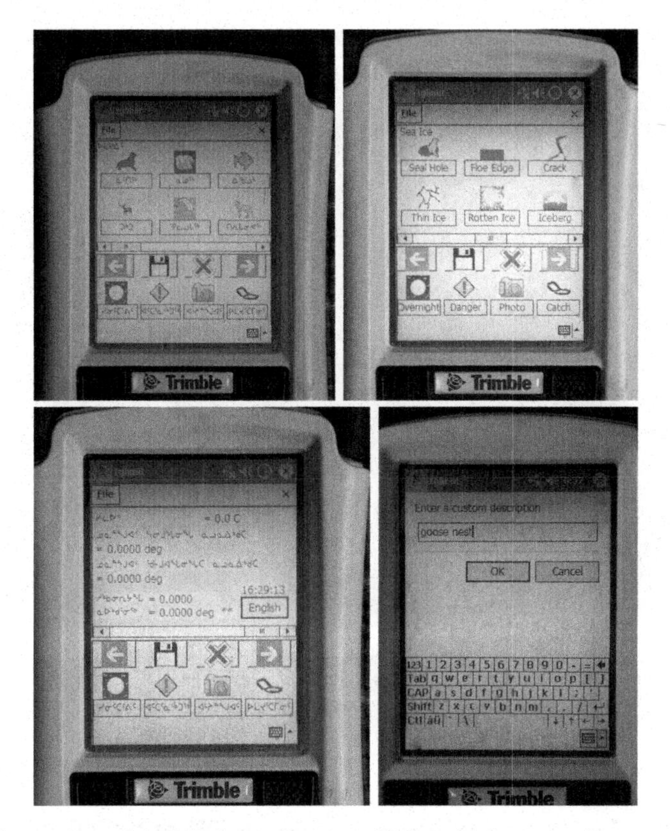

Fig. 8.2 Screen shots of the Igliniit user interface. *Right* and *left arrows* allow a user to scroll through the various screens of icons (Table 8.1). Located between the *arrows*: computer disk ("save" (to save an observation)) and *X* to cancel (to delete the last observation). The icons for "overnight," "danger," "photo," and "catch" also remain available (the icons shown at the *bottom* of the screen). *Top left*: One of the animal pages in Inuktitut. *Top right*: One of the sea ice features pages in English. *Bottom left*: Last page of the system in Inuktitut shows real-time temperature, latitude and longitude, speed (of travel), bearing, and time. Pressing "English" would allow a user to scroll through the pages in English. *Bottom right*: Screen that appears when a user selects the "remark" icon. A keyboard appears and can be used to make notes

familiar with the new technology and the students getting an introduction to the Arctic climate and a taste for the environment that the system was up against.

The team spent 3 days indoors reviewing the functions of the PDA and weather station and practiced entering observations into the touch screen (Fig. 8.3).

We immediately saw how having the system in Inuktitut broke down some barriers to using computers. The three unilingual hunters in our group were able to pick up the unit and start working with the pages and entering observations right away. The icons and Inuktitut made it simple to navigate through pages and to choose observations quickly. The engineering students had also included a surprise for the hunters – a selection of games (Fig. 8.4). Computer solitaire is a very popular game

Fig. 8.3 Brandon Culling (*standing*) introduces the software interface he and his fellow students custom programmed for the hunters. Sitting, from *right*: David Iqaqrialu, Laimikie Palluq, Jacopie Panipak, and Apiusie Apak (Photo: Shari Gearheard)

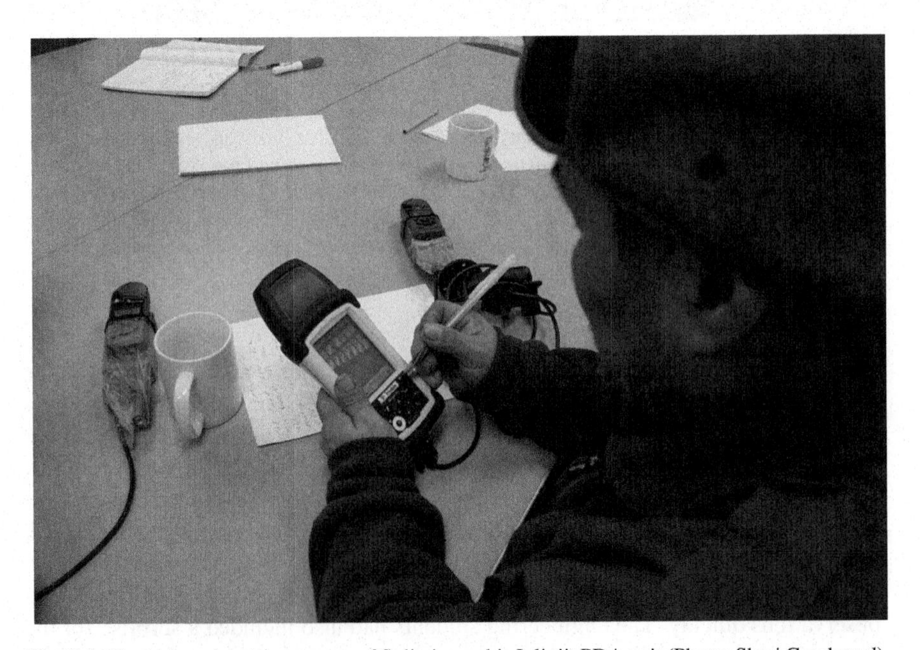

Fig. 8.4 David Iqaqrialu tries a game of Solitaire on his Igliniit PDA unit (Photo: Shari Gearheard)

in the community already, so the hunters were happy about the addition. The games ended up having a very useful and practical purpose during field testing as well; they provided welcome entertainment when the hunters were waiting out bad weather or when the kids were getting bored at camp.

Installation

Once the hunters were familiar with the hardware and using the interface, it was time to install the units onto the hunters' snowmobiles. The hunters led this portion of the (mechanical) engineering since they are the most familiar with the machines. Most Inuit hunters are skilled mechanics by necessity, as they are frequently making repairs and adjustments at home and on the trail. Installation involved (i) mounting the PDA units and weather stations on the snowmobiles where they are easily accessible to hunters using manufactured ruggedized mounts, (ii) hooking the units up to the snowmobile battery supply (the units themselves have several hours battery life, but are charged and can run on the snowmobile's battery power with minimal draw), and (iii) connecting the weather station device to the PDA unit. To ensure accurate temperature readings, the weather station was positioned at the back of the snowmobile away from any engine or exhaust heat and connected to the PDA by a long cable that several hunters tucked under the upholstery of their seat so it was out of the way. The final setup is shown in Fig. 8.5.

Fig. 8.5 Jacopie Panipak with the Igliniit system installed on his snowmobile. The PDA is mounted with ruggedized mounting hardware on his handlebars. The weather station is mounted at the back of the snowmobile. A serial cable connects the two units and weather data are stored automatically on the PDA (Photo: Shari Gearheard)

Hitting the Trail

With the students still in Clyde River, the Igliniit team started its first season of field testing. The group spent 2 days traveling the sea ice near the community, practicing entering observations, asking questions, and identifying any problems. We stayed relatively close to the community both days, traveling approximately 60 km each round trip. Since Igliniit is meant as a tool for hunters to use during their regular travel, the trips entailed accompanying them to check on seal nets and visit some nearby cabins, common activities for the hunters. We met several other hunters on the trail who then traveled with us and had an opportunity to view the system as well. Three dog teams were part of the travel during the first day, giving the students an opportunity to experience dog sledding and discuss with hunters possibilities for eventually deploying the system on dog sleds (see year 2 in following sections). As we traveled, the hunters were encouraged to enter in any observations that they felt were important, again as part of their routine travel, as the objective of the project is to not monitor anything specific or send them on predetermined observation missions.

During the 2 days of testing we recorded few problems with the units. Two of the four systems worked well, but one failed due to a loose battery and the other due to a loose connection in the GPS receiver. Both problems were fixed with no difficulty. At the time of our testing the weather was unusually mild for January in Clyde River. Temperatures during that week were around –15°C (5°F), warmer than the students' home city of Calgary at the same time. As we learned in coming weeks, the mild temperatures hid some of the key problems that we would have with the units in coming months.

Troubleshooting

Once temperatures returned to normal (–30°C and below), we started to notice the PDA units freezing. This was not unexpected, but the units are rated to –30°C and we experienced the screens freezing at –27°C. Most of the time, the units would still operate (the computer), but the screen was completely frozen and the hunters could not see the program. The students tried to solve this problem after discovering an article on the Internet posted by someone from Iqaluit, Nunavut's capital, who was troubleshooting a frozen GPS unit. They suggested placing an electric snowmobile handlebar warmer between the GPS and the mounting cradle. The students made the necessary hardware purchases to set up the warmers, but by the time they arrived in Clyde River by mail from Calgary the temperatures had warmed enough so that the warmers were no longer needed. Also, hunters were skeptical about drawing more power for the hand warmers from their snowmobile batteries and decided they did not want to implement this solution. No access to computer support services or computer hardware in Clyde River, combined with slow postal service, attributed to the delay in fixing another minor problem, the cracking and failure of some of the serial cables connecting weather stations to PDAs. Replacing the cables would be a simple repair, but caused several weeks delay in the remote community.

After the coldest months of January and February, we no longer had trouble with units freezing, but we then discovered another problem. The many weeks of hard traveling had taken its toll on the mounting hardware. By May, all the hunters had broken mounting brackets for either their PDAs or their weather stations. Fortunately, we did have extra mounting systems on hand, so these were easily replaced. Spring presented a new problem with wet conditions and the improvised plastic wrap used to protect serial cable connections did not hold up to the harsh environmental and travel conditions. By June, the hunters were still logging observations, but several had broken or missing mounts and in one case the hunter had to stop using the unit when the cable connections were no longer working and we had run out of replacements.

Data, Discussions, and Identifying Refinements

Through the entire test season, Gearheard, Aipellee, and the hunters met every Friday for 1 h to download data, photographs, and discuss any questions and problems. These meetings were extremely important for keeping up momentum in the project and providing a designated time for hunters to share their experiences. The team members bonded through these meetings and supported each other in matters not only related to the project, but often in terms of sharing advice for travel, scheduling trips together, or sharing equipment and other resources. Often hunters helped each other troubleshoot problems with their units (during meetings and outside of meetings) and helped each other learn the various features of the units. Sometimes hunters would come with new discoveries about the PDAs, such as one hunter's daughter showing him how to download music onto the unit, something quickly adopted by the other members of the team.

Through the weekly meetings we twice identified additional observations that the hunters wanted to add to their PDA list and these were sent to the students who immediately added new icons and e-mailed us an updated program that was easily (re)installed onto each hunter's PDA (the additional observations were "island," "polar bear tracks," "dog team," "government trash," and "community trash"). The meetings also provided dedicated time to talk about hunting that had occurred over the past week, weather conditions, sea ice conditions, hazards, and more. These discussions helped to provide additional, contextual information to the specific data that was collected on the units. For example, if someone had identified a crack that was dangerous, they could elaborate on why it was dangerous and how they were able to find a safe crossing point.

The photographs collected by the hunters became a very important part of the project. The hunters carried pocket-size digital cameras (Canon PowerShot SD1000s) and they became extremely skilled good at taking quality photographs and small videos. The resulting collection of several thousand photographs is unique because the hunters travel constantly and to places and in conditions that most other people rarely experience, such as the extremely thin sea ice of freeze-up, polar bear hunts by dog team, and to the remote reaches of central Baffin Island and the Barnes Ice Cap. The photographs the hunters took were often of their activities, including hunting, traveling, camping, preparing meat, and working on tools. The photographs also show many of the observations that the hunters recorded on their PDAs about animals, successful catches, floe edge position, dangerous ice conditions, and cabins. We always viewed each person's photographs at our regular meetings and that was another opportunity to share information, experiences, and stories. Some of the photographs are easily geo-referenced because the time/date stamp on the camera is synced to the PDA/GPS. However, if the PDA or camera was reset for any reason (e.g., loss of battery power) the times would become un-synced. Most of the hunters can identify where they took the photo so the locations can be recreated. For now, the Igliniit maps that have been produced (see below) do not include the hunters' photographs, but future work on creative products for displaying their photographs in a useful way (e.g., books or web sites that teach about sea ice conditions, travel routes, place names, and landmarks) is planned.

By the end of the first year, the Clyde River team had tested the units and met weekly for almost 6 months. By that time we identified a wish list for the engineering students for year 2 and refinements were made over the summer and fall before the new test season that started in January 2009 (see Table 8.2).

A new student group (Michael Brand, Ryan Enns, and Edward Wingate) tackled the list and was successful with fulfilling almost all requests. They simplified the PDA interface so that starting up the Igliniit program was a one-touch icon (before it had to be selected through a menu). Several of the favorite games were also restructured into the one-touch format. This "one-touch" refinement alone greatly simplified PDA operation for the hunters. In addition, all extra features of the PDA (Internet access, e-mail, documents folders, additional programs) were removed since they were not used and the hunters could no longer accidentally get lost in unwanted menus or applications. The Inuktitut font was stabilized and the additional

Table 8.2 List of requested changes sent by the Igliniit hunters to the engineering students after the first year of testing in 2008

1. Simplify the PDA operating system for launching the Igliniit program
2. Make the system deployable on a dog team
3. Solve the PDA screen freezing problem
4. Make sure the user-entered data are recorded (there were problems with some observations not being saved)
5. Make the Inuktitut font stable (two of the units experienced periodic problems displaying Inuktitut)
6. Add a map display (i.e., active GPS map)
7. Provide more rugged cables and mounting hardware
8. Add new observation icons ("duck," "goose," "Arctic hare," "glacier," "glacier runoff," "soft ice," "*aliuqqaniiq*" (hollow snow))

observations (icons) were added. The students researched and had new military-grade serial cables custom fabricated to replace the previous versions and these proved to stand up much better to the cold and wear and tear over year 2.

The map display that was included in the wish list was one request that the students could not fulfill. The complexity of programming was beyond the scope and time frame of the project and the team decided that this aspect would have to be postponed for now. The students noted that this screen option (a functioning GPS map) was possible, but could not be accomplished within the constraints of the project. This was acceptable to the hunters who were pleased that the rest of their requests were met. The refinements that took the most research and effort were solving the freezing issue and preparing the system so it could be used on a dog team. The freezing problem (specifically the screen freezing) was ultimately solved by a custom-insulated bag and electric heater system that the students researched, designed, and fabricated themselves (Fig. 8.6).

The bag is made up of three layers: a ballistic nylon shell, layer of microfiber insulation, and an internal layer of nylon. A fold-back flap provides access to the screen and a pocket in the screen flap was designed to hold a flat handlebar heater (Fig. 8.6). The bag works in two ways. First, the handlebar heater can be activated to provide heat directly to the PDA screen when the flap is closed (and held firmly in place with Velcro). Controlled condition tests by the students in the lab showed that with heating, a constant screen temperature of –10°C was maintained for more

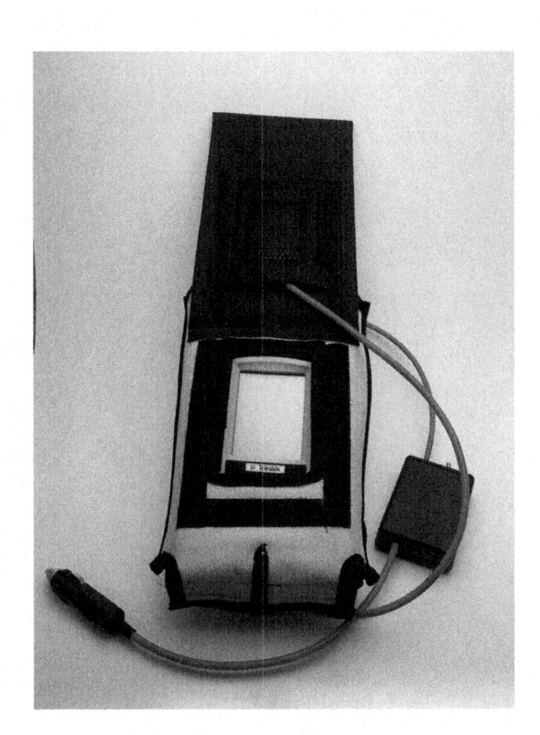

Fig. 8.6 New insulated bag and heating system custom-designed and manufactured by engineering students Edward Wingate and Michael Brand (Photo: Edward Wingate)

than 8 h in an external temperature of –40°C (recall the screen starts to freeze at –27°C, so –10°C is more than adequate to prevent freezing and also has negligible draw on a snowmobile battery). Second, the bag is an effective insulator on its own, without active heat. Lab tests showed that the bag alone can keep the PDA screen above –10°C for several hours.

The Second Year of Testing

When Brand, Enns, and Wingate arrived in Clyde River in January 2009 to deliver the insulation innovation and introduce other new changes, the hunters put the bag to the test. In the field, the hunters found that the bag alone was enough to prevent their units from freezing. Several hunters noted that in extremely cold conditions they would turn on the handlebar heaters (using the quick on/off switch included) for 30 min or so to warm up the bag, but they did not have to leave it on. Since the hunters were still wary of additional draws on their snowmobile batteries, they were pleased that the bags were enough to keep their units working. The flap design made access to the screen easy (Fig. 8.7). Also, the bag could remain on the unit even when it was moved on or off the snowmobile mount and the students designed

Fig. 8.7 David Iqaqrialu logs sea ice information with his Igliniit system (Photo: Edward Wingate)

a convenient holder on the flap for the PDA's stylus[3] (plastic stick used for pressing icons on the touch screen).

The week that Brand, Enns, and Wingate spent in Clyde River in January 2009 was also used to introduce two new hunters to the team, Jayko Enuaraq and Amosie Sivugat. The two new hunters spent a few days getting familiar with the Igliniit system and learning about the new changes that the students had introduced. Two things were very apparent during these meetings. The first was that the original four hunters from the first year were clearly invested in the project and still had tremendous motivation. They were excited about launching a second year and they took the lead in training the two new hunters on how to operate the system. It was clear that, once a final design was in place and operational, the hunters would be best suited to train other hunters and observers about the technology. Second, the hunters were clearly impressed that the students had listened to their concerns from year 1 and had systematically addressed each one, providing either a solution or an explanation about what they had researched to solve the issue. This effort on both sides (to test and identify issues on the part of the hunters, and to listen and respond on the part of the students) added to the team building that took place over the project's life span. The mutual interest and good communication between hunters and students (facilitated by Aipellee and Gearheard) made everyone feel supported and contributed to strong project momentum.

With the six hunters familiar with the new system, the group started working to set the units up for use on a dog team. Two of the hunters (Laimikie Palluq and David Iqaqrialu) who had dog teams would test the set-up in year 2. Since the system was designed to be compact and easily moved, the hunters could simply take the units on their snowmobile or dog team depending on what mode of transportation they were using each trip.

The dog team setup was exactly the same as that for snowmobiles, except that the power source was a 12 V battery that the hunters carried on their *qamutiit* (sleds). Each hunter found a space on the sled to store the battery, PDA unit, and weather station that was convenient for them. The grub box that most *qimuksiqtiit* (dog teamers) carry on their *qamutiik* was a useful place to store the PDA as it is easily accessible even on the move (Fig. 8.8).

Overall, the results of year 2 testing were positive, except in terms of hardware. Several of the units continued to suffer from the rough travel over sea ice. Cable connections failed (and were replaced and failed again) and GPS receivers became loose. Though the freezing problem had been solved, the so-called ruggedized PDAs were simply not rugged enough for use by Inuit hunters. The weather stations, on the other hand, did stand up very well, never freezing and operating reliably on lithium batteries even in the coldest months. The mounting hardware, even with modifications attempted by both the engineering students and the hunters, also did not stand up to the thousands of kilometers over bumpy sea ice and land. The team was constantly replacing and repairing mounts. And while the hunters were happy with the system design for the dog team, they found themselves not using it as much as on snowmobile, where it was more easily accessible and hunters did not have the

Fig. 8.8 Laimikie Palluq and Ryan Enns take the system for a test drive on Palluq's dog team in January 2009 (Photo: Shari Gearheard)

other work that comes with driving dogs. The assessment at the end of year 2 was straightforward; the concept was exciting, the bilingual software worked well, and the results were promising (see next section). The hardware, on the other hand, would eventually have to be re-researched and replaced.

Results: Mapping Observations

While the hunters and engineering students were working hard on the design and equipment testing, Aipellee and Gearheard also collaborated with experts at the Geomatics and Cartographic Research Centre (GCRC) at Carleton University to produce maps using the Igliniit output. The ultimate goal is to get Igliniit working as a complete "plug-and-play" package where users can download data and print maps immediately. However, data integration and display for the current Igliniit system requires working with several standalone, independent programs until the technology (especially the hardware) is finalized. The biggest challenge in displaying the data is finding a way to represent all of the information in a manner that is easy and quick to read (in both English and Inuktitut). Following the success of using icons in the software (easily identified by users), we used the same approach in producing maps. Figure 8.9 shows one of the early attempts at creating a map product using data collected from Igliniit hunter Apiusie Apak.

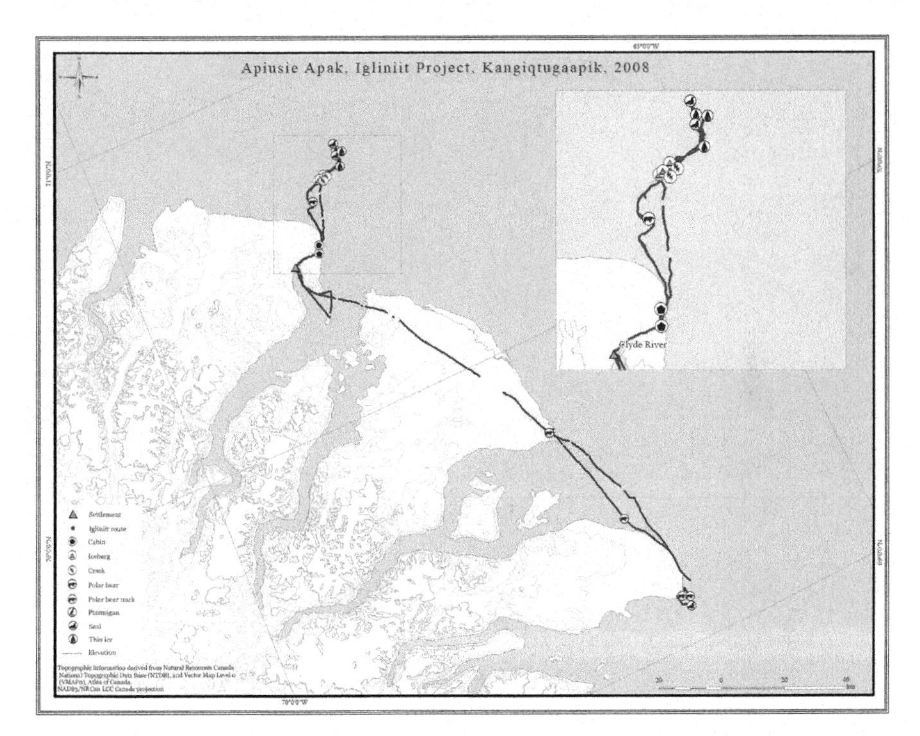

Fig. 8.9 Example map using Igliniit test data. This version uses easily identifiable icons and a legend. The blank spaces in the trail are due to loss of GPS signal while travelling

The map shows Apak's trails and the inset map helps to zoom in on an area with multiple and overlapping observations (hunters may log a number of features at one location, in this case several seals and thin ice, and icebergs and cracks in the sea ice). After several iterations with the hunters, we were able to identify some key features that were important to include in the example maps: (i) who had collected the data, (ii) data should be date/time stamped, (iii) maps should be available in English and Inuktitut, and (iv) base maps should show topography and include realistic colors. The first item is addressed in the example map shown in Fig. 8.10 by including the hunter's name in the map title. Since all the Igliniit hunters agreed to be credited by name in the Igliniit project and since the data are only sample data, including a hunter's name was considered acceptable and important. The hunters did discuss the possibility of needing to code names in the future when data are actually being used either by the hunters themselves or as part of other projects or monitoring efforts. Some hunters may not want to be identified or some of the data may be sensitive. Coding to protect data and the identity of data collectors is certainly a possibility and would be explored in any future applications. However, whether identified or coded, the hunters agreed that knowing who collected the data was important so that the observer could be consulted for more information if needed. For example, if a hazard was mapped, the observer could be contacted for more details on the nature of the hazard, how it can be negotiated, or how it may be avoided.

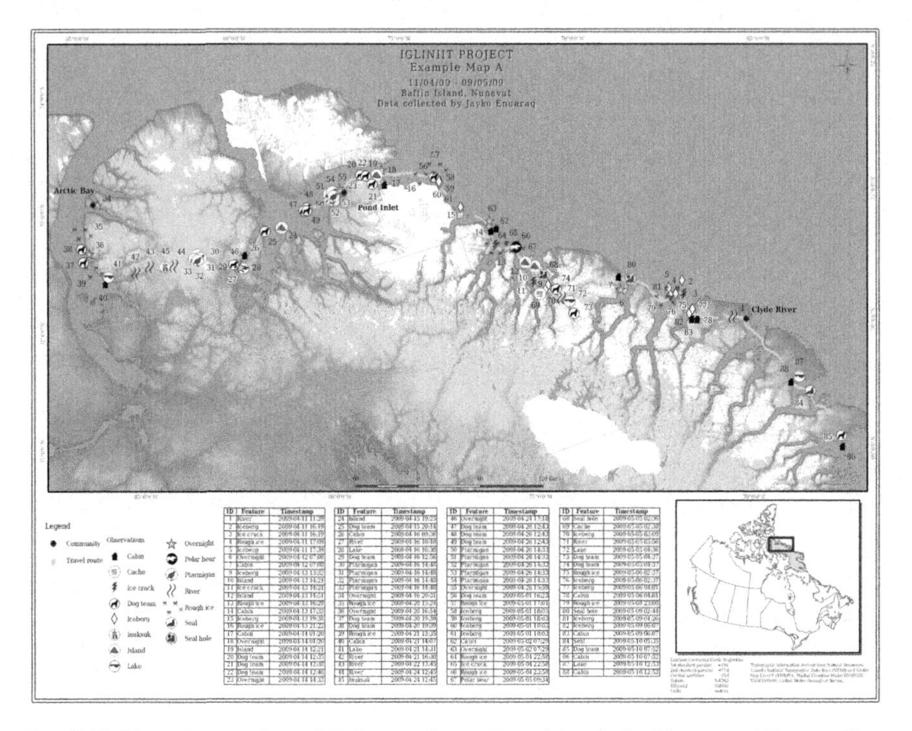

Fig. 8.10 Example map from Igliniit test data collected by Jayko Enuaraq while travelling between Clyde River and Arctic Bay in 2009 (see also Color Plate 3 on page 472)

It was also important to the hunters that the observations be notated with the date and time. This information is key for a variety of reasons including identifying any patterns in animal sightings or environmental changes. The hunters also valued having a record of the dates and times of different trips, to see when people traveled, how long trips took, and what was observed along the way. Displaying time-stamped information was first attempted by including it alongside icons, but soon the map was too crowded with information. More recent iterations have tried numbering observations and including information in chart form (Fig. 8.10).

In terms of the hunter's need for maps in both languages, Fig. 8.10 is shown in English but is readily available in Inuktitut as well or can be displayed in both languages together. The use of symbols helps to cut down on the need for text and in future iterations we may consider trying to reduce the text even further (for example, the "feature" descriptor in the map table may not be needed).

Lastly, Fig. 8.10 also shows the changes made to address the hunters' request for more realistic coloring. Originally, GCRC cartographers were interested in a more artistic/aesthetic representation of the land and water using unconventional coloring (Fig. 8.9). While the hunters thought the maps looked attractive, they felt that they were not intuitive and did not represent the topography clearly. They opted for a more "brown for land/blue for water" look that also captured elevation. The result,

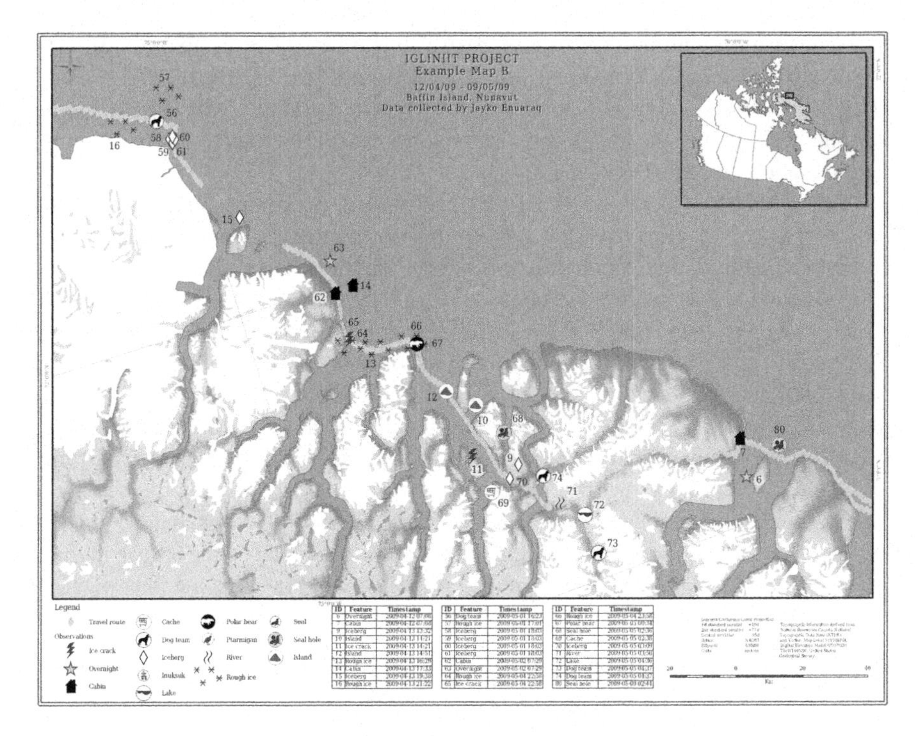

Fig. 8.11 Zoomed-in image of example map A (see also Color Plate 4 on page 473)

they felt, could be more quickly understood and interpreted, especially for those not familiar with the landscape.

As Figs. 8.9 and 8.10 demonstrate, one of the challenges of mapping such dense collections of observations is creating a clear map in a small space. Locally, most people only have access to printers that can process 8.5″ × 11″ paper, so maps must be easy to read in that size (and in black and white). One option that may help to address this is providing zooming capability. For example Fig. 8.11 shows a zoomed-in view of a portion of example map A (Fig. 8.10). In this view, the icons are more clearly visible, as well as the timestamp information. The goal is to include this type of zooming capability in future software developments that will allow hunters/observers to download their own data, zoom as needed, and print their own maps.

Conclusion

The Igliniit project is an example of multi-disciplinary and cross-cultural collaboration in the deepest sense. Engineering, technology, and Inuit knowledge came together to produce a new tool that provides tangible, useful information with broad application. It highlights the results that can be achieved when academic research is blended with community-based expertise toward a common goal. It also provides a case study for creating a practical, project-based experience for students. The "client-based" approach of Igliniit for the engineering students gave them a specific task with a specific group to satisfy, but most importantly, in the context of collaboration and co-development.

The ultimate goal of the Igliniit technology is to provide hunters and other observers with a user-friendly and Arctic-worthy method for recording georeferenced observations of their environment. These observations, when mapped, can be used locally and in other contexts (e.g., land use planning, wildlife management, protected area planning) for data collection, analysis, and decision making. Over space and time, the maps can be used to show patterns in trail usage, environmental changes, harvest activities, animal populations, and more. As well, the technology could be augmented with other features such as GMRS (general mobile radio service), FRS (family radio service), EPIRB (emergency position indicating radio beacon) or ELT (emergency locator transmitter) devices, or satellite telephone transmitters, allowing the hunters to not only log their observations but also to send out a signal so they themselves can be located. Live-tracking or the ability to transmit a locator distress signal was identified by the Igliniit hunters as another potential feature that would be very useful in the Arctic context. In addition to the up-to-date hazard maps they can already produce, this technology could add another benefit to the search and rescue applications of the Igliniit system.

The project focused on the development and testing of a concept and was met with success and challenges. As mentioned previously, the hunters agree that the idea of logging observations using such a system is very exciting and they found the program developed by the engineering students to be easy to use and modify according to their changing interests (i.e., adding or deleting observations). The hardware, on the other hand, presented serious challenges in the most extreme conditions. As the hunters became more familiar with the technology and more comfortable incorporating this type of observation into their regular activities, the hardware malfunctions and breakdowns became increasingly frustrating. The "ruggedized" units may stand up to other field applications, but not to the demands of the Arctic or the Inuit hunter. On the positive side, it seems plausible that these problems will eventually be solved through using the right hardware or engineering a model that can endure normal Arctic travel conditions. The challenge will be to find these alternative solutions in a system that is affordable to individual users.

While this chapter focused more on the "nuts and bolts" of the project, other complex issues such as data sensitivity, ownership, access, and archiving have been discussed. During the weekly meetings held over the 2 years of testing, these subjects were also explored. For example, during one of our very first project meetings, the hunters requested a private meeting amongst themselves. After the meeting, the hunters explained to Aipellee and Gearheard that they wanted to review in private whether or not to include polar bear dens as part of the list of observations to be recorded. The hunters explained that they considered this information particularly sensitive and decided that at this time they would remove polar bear dens from the list. The discussion and decision illustrate that community members are well aware of the sensitivity and potentially political nature of these types of data and therefore caution must be taken in deciding how to share them and how to protect them.

While recognizing the potential sensitivity of the data, the hunters also recognized the power of collecting such data. Many times the group discussed the potential power of the technology, once finalized, if it could be implemented in a large-scale, systematic manner. Examples that were brought up include how data over time and space could influence decisions regarding shipping lanes, protected areas, development areas, and even hunting quotas. The technology could be an especially powerful tool for monitoring the environment and wildlife, a topic of great interest to hunters in Clyde River and other communities. In August 2009, after a few months off since the end of the second year of testing, the Igliniit team in Clyde River met again to view more map iterations and discuss ideas for future work. After seeing the latest map versions and discussing the potential the technology could have for environmental monitoring, David Iqaqrialu explained the following:

> We already know about the animals. We know where they are and when they will be in certain areas. We also know about climate change and where certain places are changing, and what is changing. This [tool], it helps us. It can be combined with our knowledge to document it and share it with others.

Iqaqrialu raises one of the most important lessons learned in the entire Igliniit experience. The Igliniit technology cannot, and never will, replace Inuit knowledge. Just like a snowmobile, a gun, or a motorboat, the system is just a tool and is useless on its own. What the computer can do is assist and empower users, enabling them to extend their abilities (Blake et al. 2000). The Igliniit system cannot recognize what needs to be recorded. More importantly, it cannot *interpret* the observation (is it dangerous? Should it be characterized as a hazard?). It cannot do what the hunters can do. Technology like the Igliniit system simply provides a quick but effective means to help the hunter log his/her observations in a specific time and place and then communicate that information to others.

Whether or not the Igliniit project will find support to continue its refinements remains to be seen. The partnership with ISIUOP and SIKU, and funding from the Canadian International Polar Year Program, made it possible for initial advances and partnerships to be made. This project also raised awareness about the potential for PDA/GPS technology in the Arctic. In time we may yet see Inuit and other Arctic communities collecting data using this technology and contributing information to local, territorial, national, and even international research and monitoring efforts. This potential for local contributions to large-scale science and monitoring is real and it is exciting. The Igliniit team looks forward to continuing efforts to reach this goal and to share our knowledge and experiences with other interested Arctic communities.

Acknowledgments We would like to thank the community of Clyde River and the Nammautaq Hunters and Trappers Association for their collaboration and support on this project. We are extremely grateful to our primary funder, the Government of Canada International Polar Year Program for making this project possible. Thank you to the Clyde River RCMP for providing warm workspace and to the Nunavut Arctic College for much needed meeting space. Thank you to Igor Krupnik and Claudio Aporta for useful comments on earlier versions of this chapter.

Notes

1. 2006–2007: Desmond Chiu, Sheldon Lam, Andrew Levson, and Jeremy Park. 2007–2008: Brandon Culling, Josiah Lau, Tina Mosstajiri, and Trevor Phillips (Mosstajiri and Culling visited Clyde River in January 2008). 2008–2009: Michael Brand, Ryan Enns, and Edward Wingate (all three visited Clyde River in January 2009).
2. The original four hunters were Apiusie Apak, David Iqaqrialu, Laimikie Palluq, and Jacopie Panipak. Five hunters actually started in year 1, but the fifth hunter, Jaykurassie Palluq, had to withdraw early on after breaking his foot and left unable to travel. The two hunters who joined the project for the second year of testing were Jayko Enuaraq and Amosie Sivugat.
3. Fitting with their good senses of humor, several of the hunters took to using bullets instead of a stylus for their touch screens.

References

Aporta, C. 2003. Using GPS mapping software to map Inuit place names and trails. *Arctic* 56(4): 321–327.

Aporta, C. 2009. The trail as home: Inuit and their pan-Arctic network of routes. *Human Ecology* 37: 131–146.

Blake, E.H., Steventon, L., Edge, J., and Foster, A. 2001. A Field Computer for Animal Trackers. Report from ACM SIGCHI South Africa Chapter. 2000. 9 pps.edwin@cs.uct.ac.za.

Fox, S. 2002. "These are things that are really happening": Inuit perspectives on the evidence and impacts of climate change in Nunavut. In *The Earth Is Faster Now: Indigenous Observations of Arctic Environmental Change*. I. Krupnik, and D. Jolly, (eds.), Fairbanks: Arctic Research Consortium of the United States, pp. 12–53.

Fox, S. 2003. When the Weather Is Uggianaqtuq: Inuit Observations of Environmental Change. Multi-media interactive CD-ROM. National Snow and Ice Data Center (NSIDC) and Arctic System Sciences (ARCSS), National Science Foundation.

Fox, S. 2004. When the Weather Is Uggianaqtuq: Linking Inuit and Scientific Observations of Recent Environmental Change in Nunavut. Unpublished Dissertation. Department of Geography, University of Colorado at Boulder.

Leroy, E.M., Rouquet, P., Formenty, P., Souquiere, S., Kilbourne, A., Forment, J., Bermejo, M., Smit, S., Karesh, W., Swanepoel, R., Zaki, S., and Rollin, P.E. 2004. Multiple Ebola virus transmission events and rapid decline of Central African wildlife. *Science* 303(5656): 387–390.

Liebenberg, L., Stevenson, L., Benadie, K., and Minye, J. 1999. Rhino tracking with the CyberTracker field computer. *Pachyderm* 27: 59–61.

MacFadyen, S. 2005. Electronic ranger diaries: The Kruger National Park CyberTracker Programme. http://www.sanparks.org/parks/kruger/conservation/scientific/gis/cybertracker.php. Accessed August 23, 2009.

Chapter 9
Assessing the Shorefast Ice: Iñupiat Whaling Trails off Barrow, Alaska

Matthew L. Druckenmiller, Hajo Eicken, John C. George, and Lewis Brower

Abstract At Barrow, Alaska, local Iñupiat whaling crews annually construct a network of seasonal trails through the shorefast ice during the traditional spring hunting season. These trails originate at locations along the coast and pass through diverse ice features, including ridged and rubbled ice, new and potentially flooded ice, and tidal cracks, before terminating at the shorefast ice edge where camps are established. The safety of this hunt relies on the careful observation of evolving ice characteristics from freeze-up onward and the understanding of how the interplay between ice dynamics, ice thermal evolution, and ocean and atmospheric processes leads to both stable and dangerous conditions. Partnering with Barrow whalers, a multi-year documentation of whaling trails, alongside a geophysical record of shorefast ice conditions, provides insight into how Iñupiat hunters monitor the development of the shorefast ice throughout winter and spring and how individual and community assessments of ice conditions and associated risks, traditions and knowledge, and personal preference determine trail placement. This contribution also discusses how the documentation of human use of the ice environment contributes to integrated observations of Arctic change and adaptation.

Keywords Barrow · Alaska · Iñupiat · Local knowledge · Shorefast sea ice · Whaling

Introduction

Along a 35-km stretch of coastline in northernmost Alaska, the Iñupiat Eskimos of Barrow have hunted the bowhead whale for centuries (Stoker and Krupnik 1993). As the whales migrate northward in spring toward summer feeding waters in the Beaufort Sea, the ocean is covered with sea ice that is continuously responding to

M.L. Druckenmiller (✉)
Geophysical Institute, University of Alaska Fairbanks, Fairbanks, AK 99775-7320, USA
e-mail: mldruckenmiller@alaska.edu

I. Krupnik et al. (eds.), *SIKU: Knowing Our Ice*,
DOI 10.1007/978-90-481-8587-0_9,

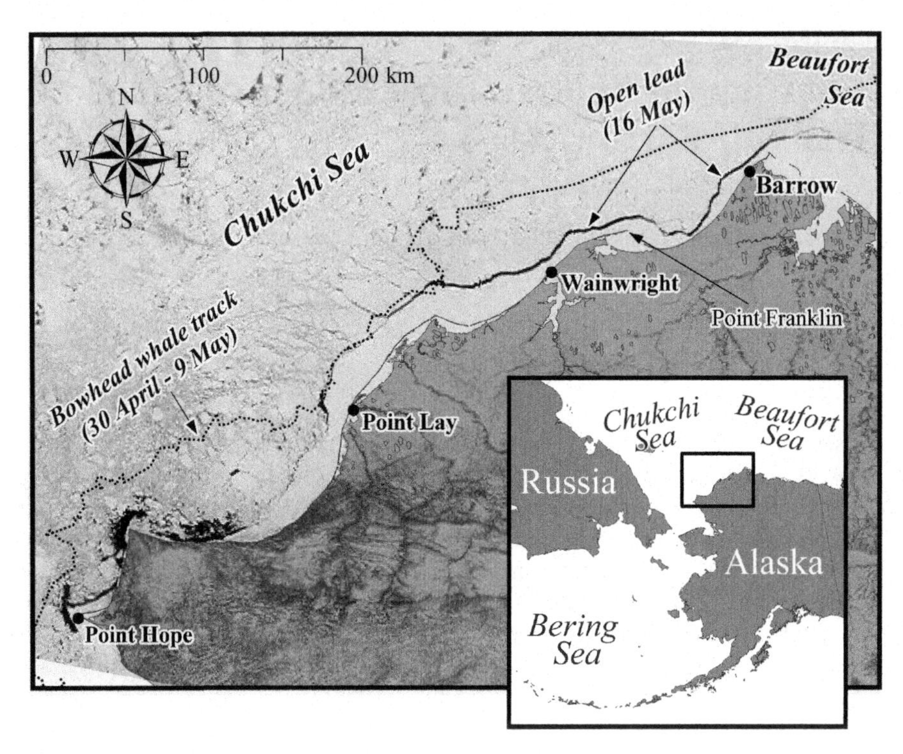

Fig. 9.1 Map of the western Arctic coastline of Alaska. Barrow, Wainwright, Point Lay, and Point Hope are whaling communities officially recognized by the Alaska Eskimo Whaling Commission and hunt in a similar fashion from the edge of the shorefast ice in spring. The *dotted line* shows the migration path of a satellite-tracked bowhead whale during the spring of 2009 (unpublished data provided by Alaska Department of Fish and Game). The whales typically migrate along this coastline between late March and early June and regularly surface in open water of the persistent coastal lead system. The background of this map, a Terra MODIS (Moderate Resolution Imaging Spectroradiometer) satellite image from May 16, 2009, shows a lead opening in a pattern that mirrors the shape of the coastline

the forces of winds and currents (see Fig. 9.1). A narrow shelf of coastal shorefast ice extends out from the land into potentially dangerous waters, shaping the environment that local hunters have come to understand. Whaling crews base their hunt from a network of trails that traverse the shorefast ice (*tuvaq*), often leading them as many as 16 km offshore. While hunters may not travel far, the ice conditions they experience are always changing and each year brings new challenges. Hunting efficiently, safely, and respectfully according to Iñupiat customs requires careful observation, years of experience, and the accumulated knowledge passed down from earlier generations.

In March the shorefast ice off Barrow has been shaped by several months' history of ice growth and dynamics – events that have anchored the ice to the sea floor and the coast. A short traverse of only a few kilometers from the village reveals a vast assortment of ice types, thicknesses, and morphological features (cracks, rafted ice, etc.). Interpreting the make-up of the ice in terms of safety and ease of travel partially determines where the hunters will establish their camps. Understanding whale behavior and recalling past ice conditions additionally informs the hunters' strategy. In late March whaling crews begin to move out onto the ice in great numbers. At this time, the shorefast ice is still evolving, and over the course of the whaling season (mid-April to late May) it deteriorates from its cold winter state. Hunters carefully assess the evolving conditions in relation to safety, on-ice travel, and successful hunting.

We observed the location of these trails and spoke with hunters about how ice conditions informed and shaped their hunting and travel decisions during three consecutive springs, 2007, 2008, and 2009. During this same time various components of a geophysical-based ice monitoring system were recording information on shorefast ice thickness, growth, decay, and deformation. Relating these observations to those of the hunters has stimulated conversations about the specific ice features and processes that the whalers consider important, led to interesting generalizations about shorefast ice variability, and provided important considerations for how to move forward in making scientific observations of sea ice useful to the community. It is our hope that this chapter sheds light on how hunters understand and interact with sea ice under current climate conditions. Our research also examined the relationships between physical environmental processes, such as those during freeze-up and break-up, and ice characteristics that can be monitored on scales relevant to the community's use of the ice (Druckenmiller, n.d.).

Throughout this chapter, Iñupiaq terms for sea ice are used to illustrate the diversity of the Barrow whalers' ice terminology and the complexity of their knowledge. Whenever brief definitions (explanations) are offered, they may not capture the full meaning of the term attributed by Iñupiat experts.

The Shorefast Ice Environment

Shorefast sea ice is present off Barrow for much of the year, typically between November and July. Recent research has shown that the ice is forming later in fall and breaking up earlier in late spring (Mahoney et al. 2007a) and that multi-year ice is becoming less abundant (Drobot and Maslanik 2003). Scientific predictions for a warmer Arctic and further reductions in summer minimum ice extent, and hence multi-year ice area, raise additional concerns for whether or not the "familiar" shorefast ice environment prior to the 1990s will persist. Community observations also indicate that sea ice is changing, but these observations are made against a different "baseline condition" than the one scientists often use. Hunters understand and observe ice conditions largely in relation to how they and their ancestors have used the local ice cover for travel and hunting. Each hunter comes to understand the local environment based on his personal experiences and those of the elders that taught him. While accounts expectedly vary between men, a basic understanding of the primary factors that shape the local shorefast ice environment is shared by

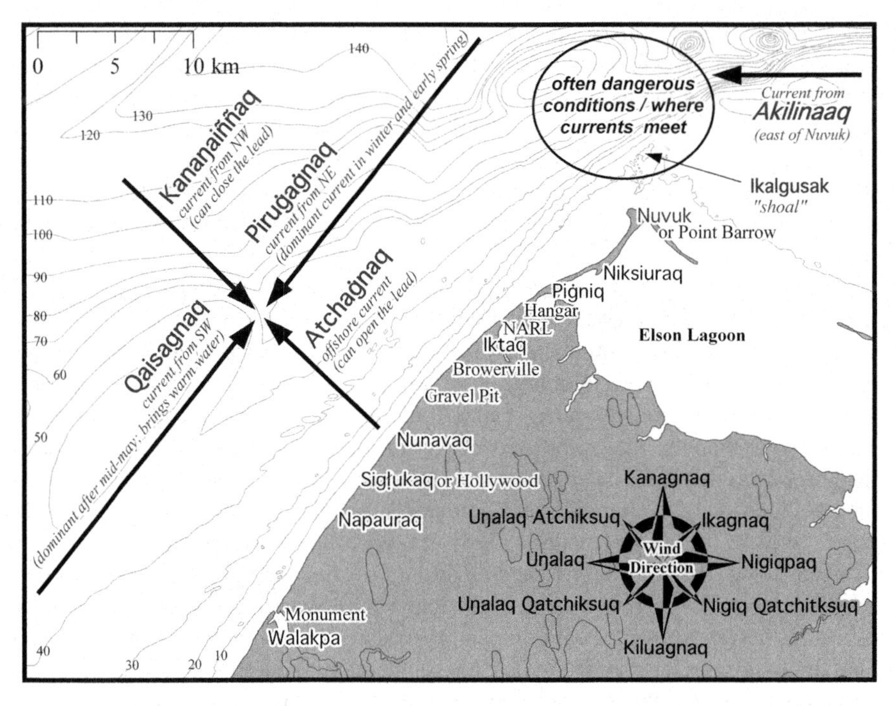

Fig. 9.2 Barrow's traditional knowledge of currents, winds, and ice drift directions. A subset of the many traditional place names and other commonly used terms unique to Barrow are shown at locations along the coast. "NARL" refers to the sight that was previously the Naval Arctic Research Laboratory. Bathymetry is shown as 10 m contours

the community. Figure 9.2 presents a map that shows a few of the key factors that various hunters have mentioned as important.

Point Barrow, located only several kilometers north of the present village, extends into the waters where the Chukchi Sea meets the Beaufort. While Barrow resides on the Chukchi Coast and the community bases their spring whale hunt on that side of the Point, it is not uncommon for hunters to also travel north of the Point. A regional perspective is important as hunters attribute ice dynamics to winds and currents and understand that stresses in ice are often transmitted over great distances, such as from the Beaufort side of the Point to the Chukchi side.

The dominant current is from the northeast (*piruġaġnaq*) during most of the year. However, in mid-to-late May, there is a shift in the major current direction to that from the southwest (*qaisagnaq*) and also an increase in current speed. *Qaisagnaq* is known to bring warm water that accelerates the melt and break-up of shorefast ice. In 2007, while discussing ice conditions near the Point, the late Arnold Brower, Sr., a Barrow elder and whaling crew captain, explained that there is more than one current that parallels Barrow's coastline. Often pieces of drifting ice of similar size can be seen side by side, yet traveling at different speeds. Somewhere near the Point these currents converge with each other and also with the current that comes from east of the Point. During drift ice conditions this meeting of currents can be observed by looking at ice floes turning in circles. As recalled throughout our interviews with hunters, old stories in the community tell of strong currents and ice conditions near the Point that are very dangerous for spring whaling in comparison to ice conditions to the southwest. Traveling north of the Point during an east wind is particularly dangerous. Today's whalers often describe past experiences north of the Point as defining moments for when they began to fully understand the risk of hunting on ice.

The dynamic conditions north of Point Barrow can lead to the formation of massive ridges, often through shear. These ridges are believed to ground in this area due to the presence of a shoal or *ikalgusak* (shown in Fig. 9.2). Here, large ridges serve as a point of deflection for drift ice coming from the east that could potentially impact and destabilize the shorefast ice off Barrow's Chukchi coast.

Winds play a major role in the drift of pack ice and in determining whether or not the lead along the shorefast ice is open. Onshore winds from the north to southwest may bring in pack ice to close the lead, while offshore winds from the northeast to south may open the lead. When an offshore wind is strong enough, it can locally depress sea level by developing an offshore current, which can cause certain areas of the shorefast ice to detach when cracks form around grounded ridges (George et al. unpublished; observations on landfast ice break-off events "Uisauniq" near Point Barrow, Alaska).

There are general patterns in how the shorefast ice develops along Barrow's coastline. On the regional scale, Barrow whaling captain Eugene Brower explained in 2009 that the shape of the coastline between Wainwright and Barrow can influence local ice conditions. With Point Franklin providing a deflection point (see Fig. 9.1), Barrow's coastline north of *Nunavaq* bears the brunt of the pack ice moving in from the southwest in comparison to that south of Nunavaq. For this same reason, and perhaps also due to the intricacies of coastal currents, the pack ice typically approaches the lead edge slower and with less force south of Nunavaq than it does further north. This leads to the ice south of *Nunavaq* and *Sigḷukaq* being flatter and less rough than the ice further north. Also in this region the shorefast ice typically extends out further than it does to the North. Once again, these are only general patterns and hunters clearly state that ice conditions are different each year.

Springtime Whaling: A Sequence of Observations

Barrow has been the location of continuous settlements for at least 1,300 years, with periodic settlements traced back as far as 5,000 years. In the mid-1800s Yankee whalers began regular contact with the settlement at Point Barrow, and by 1890 multiple commercial whaling operations were in full swing employing hundreds of Iñupiat whalers (Braund and Moorehead 1995). When 1908 brought the end to commercial whaling, subsistence whaling continued according to tradition, yet infused with technological advances, such as the bomb lance. Today, there are approximately 50 licensed whaling crew captains in Barrow that are responsible for supplying both the immediate community and their extended families across Alaska with food from the bowhead whale. Barrow whalers still use skin boats – *umiat* (wooden frames covered with bearded seal skins) – and thrive as expert hunters by applying knowledge and skills that have been transmitted across generations for centuries. The success of the hunt relies on assessing the shorefast ice – one of the more complex, ephemeral terrains on earth.

Evaluating the Ice in Preparation for the Hunt

Even though shorefast ice begins to appear off Barrow in November, January through March are the major ice-building months. This is the time when hunters count on heavy pack ice coming in to create ridges. Careful attention is given to how the different regions of the shorefast ice develop (see Fig. 9.3). First there is the flat ice zone (*igniġnaq*), which is typically either floating or bottom-fast, between the shore and ridges (*ivuniq*). Second, there is the zone where grounded pressure ridges (*kisitchat*) develop and provide the anchoring strength. These ridges typically form in shear. Multiple rows of ridges often exist between the *igniġnaq* and any extended floating shorefast ice (*iiguaq*), which is vulnerable to impact by drifting pack ice (George et al. 2004).

Whaling captain Crawford Patkotak explained that you have to observe the ice while thinking about what happened before. Hunters look for sediment entrained in ridges for clues that the ice scraped and grounded to the seabed as it formed. They

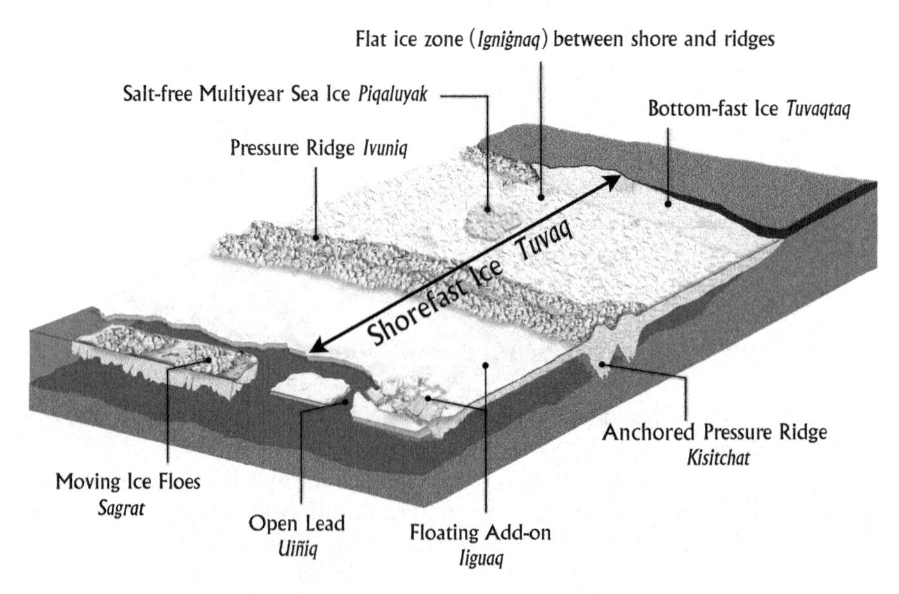

Fig. 9.3 Schematic representation of coastal sea ice in the Chukchi Sea off Barrow. Both English and Iñupiaq terms for ice structures are given (George et al., 2004). Modified with permission from an illustration by Deb Coccia

examine the ice makeup and question whether the ice near pressure ridges was there when the ridge formed or rather came in later. Every attachment represents a point of fusion where the ice may break out later in the season. Low winter temperatures are important for fusing the ice together. Winter is the time to watch the ice and decide where one needs to closely monitor throughout the whaling season. In addition to safety, hunters also examine the ice to determine the layout of snowmobile trails to be made from the beach to the open lead (*uiñiq*).

Building the Ice Trails

The physical process of building trails begins in late March. Both experienced and young members of the whaling crews use snowmobiles and ice picks to blaze and cut their trails across the shorefast ice according to a strategy that varies from crew to crew. A range of considerations exists for a whaling captain deciding where to place his crew's trail.

Safety and stability. The most important consideration for any whaling captain is the safety of his crew, which includes everyone from the hunters who will camp on the ice and pursue the whale to those who will come out to help with butchering and hauling meat. While every captain will agree that a successful hunt is not worth the loss of human life or of vital whaling equipment, different hunters have varying perceptions of risk. However, in general, a trail is chosen such that it traverses ice that is well grounded or securely attached to stable ice. Therefore, knowledge of the locations of cracks and points of attachment is important. It is quite common that winter seal hunters, who often travel on foot, provide detailed initial assessments of ice conditions.

Construction effort. Hunters must consider how much work it will take to build a trail. While a trail of several kilometers that connects many flat pans of ice may take only a few days work for a few men, a trail of similar length that traverses extremely rough ice and multiple rows of large ridges may take several weeks. Those crews taught to go where the ice is rough, thick, and well grounded inevitably accept that they will work harder for their trail. It is common for several crews to work together on the same trail near the shore. Once the trail nears the edge, however, the crews will split their efforts to build individual trails that will branch off from the main one.

Navigability and potential for evacuation. It is important to be able to drive a snowmobile quickly along a trail, especially when in need of swift evacuation off the ice. In consequence, trails are made as straight and as smooth as possible, utilizing large interconnected pans of flat ice, and are built wide enough to allow two snowmobiles to pass each other. When describing these strategies for Wainwright in the 1960s, Richard Nelson (Nelson 1969) discussed the use of refrozen cracks as a way to efficiently travel through areas of highly deformed sea ice. Trails nearly always approach the lead edge perpendicularly to the coast since this represents the shorter distance to land. Alternative evacuation routes are often considered, resulting in more than one trail leading to the beach or to safer ice. It is in this region of safe ice that a crew will often place their *naŋiaqtuġvik*, which is a place where they store their whaling equipment and camp when waiting for the lead to open or for other favorable conditions.

Ice edge conditions. Conditions at the edge are also critical for a successful hunt. Hunters prefer to find thick heavy ice (or rafted thinner ice) where they can place their camp, build a boat launch, and pull up a whale. Hunters indicate that ice thicker than 1.5 m is needed to haul up a large whale more than 16 m (53 ft) in length. Some prefer to find ridges near the lead that can be used as a perch to watch the water. Trail building when the lead is closed requires hunters to utilize observations made earlier in the season to make predictions for where the edge will be when the lead eventually opens. Even at times when the lead is open, ice conditions are not always ideal due to unstable or thin add ons (*iiguat*; plural form for *iiguaq*), leaving the crews in wait for more suitable ice edge conditions to develop.

Proximity to other crews and distance from town. Some crews prefer to hunt in places far removed from others as they prefer solitude and because they believe it promotes self-sufficient hunting practices. When a crew is on their own, they must focus on killing the whale with the first strike, as they cannot rely on help from other crews. Conversely, some crews prefer to remain close to assist each other when needed or to share favorable ice and trail conditions. The price of fuel and the time it takes to get to a hunting location also play a role in trail placement.

Forecast of late spring conditions. Hunters must consider both the conditions at the time they build their trail and those that will be encountered toward the end of the season. A trail that crosses large flat pans of thinner ice is at greater risk of having the ice wear dangerously thin once air temperatures warm, snow melts, and the warm current from the southwest arrives. Some may build trails on top of ridges and keep on higher elevation ice for as long as possible. The advantage is not only that they can see greater distances to landmarks and open water but also because it reduces the likelihood of the trail being eaten away by warm water or snowmobile traffic. In contrast, other crews may decide to place their trail in the lower elevation ice between and throughout ridges since the ridge walls serve as side ramps to the trail and prevent heavily loaded sleds from tipping.

Bowhead whale behavior. Understanding how the whales behave as they migrate along the ice edge also advises the hunter where to place his camp. When predicting where a whale will surface, hunters employ different strategies. Barrow elder Warren Matumeak explained that whales will swim beneath young thin ice, avoiding large ridge keels, and will surface in embayments along the edge (*kaŋikłuk*) (see Fig. 9.4). "Camping on the north side of these embayments and facing south" (*manilinaaq*) provides a good place to watch whales coming toward you and a good place to launch a skin boat. In turn, iluliaq refers to a location where you have only a view of whales traveling away. There are also hunters that prefer to place their camps at points along the ice edge (*nuvuġaq*) since these tend to provide good visibility and access to whales that swim from point to point and bypass embayments. Some hunters have also been taught that the whales are attracted to thick multi-year ice because it is shiny and may also provide feeding advantages. It is believed that ice with deep keels (thick multi-year ice or ridges) causes the water to churn and stir up krill.

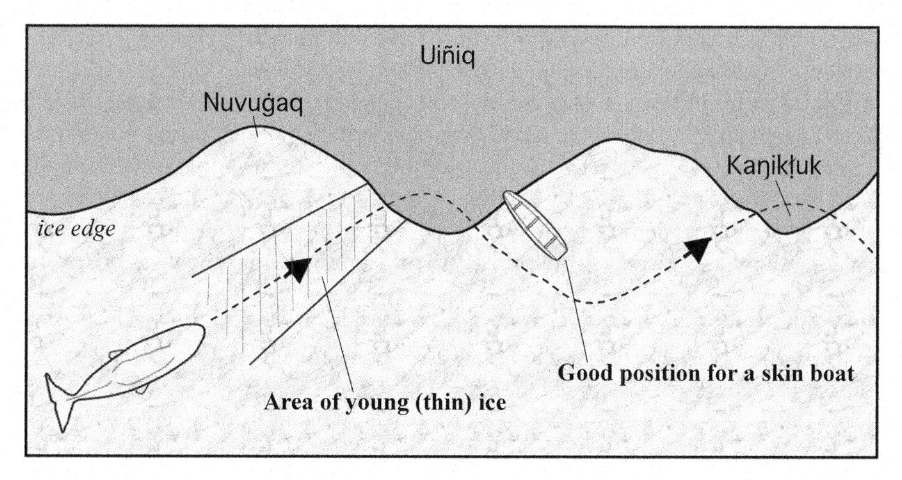

Fig. 9.4 Likely path of a bowhead whale as it swims along the ice edge. *Kaŋikļuk* is an embayment along the ice edge. *Nuvuġaq* is a point of ice extending out from the lead edge. Adapted with permission from a sketch by Warren Matumeak

Elders' knowledge. In the end, the decision on where to place an ice trail for spring whaling may most strongly be influenced by tradition and what was taught by elders. Arnold Brower, Sr., noted that he learned from his elders to hunt in the north early in the season before the current from the southwest strengthened, thus minimizing the risk of losing a struck whale that is carried under the ice by the current. When the current intensified in mid-to-late May, he would move his crew to the south. Whaling Captain Nate Olemaun discussed how he was taught that the waters off *Sigļukaq* are rich feeding waters and are a good place to see whales. As noted earlier, many captains prefer to hunt south of the dangerous and unpredictable conditions north of the Point, despite acknowledging that this is a good place to see whales.

The trail network built by the Barrow whaling community evolves throughout the season as ice conditions continuously change and crews move locations. To assist in navigation most crews use distinct markers for their trails, such as painted wooden stakes or flags. Markers often note the crews' names. Trails are typically referred to by the name of the captain or crew or by the trail's point of origin, using the place-names and landmarks shown in Fig. 9.2.

Observations at the Ice Edge

When crews are "along the edge of the ice observing the environment and looking for whales" (*nipaaq*) they must continuously monitor the ice on which they are camped and the pack ice beyond, both of which are influenced by wind and current. To monitor the currents, hunters typically drop a sounding line into the water. Barrow whaler Joe Leavitt explained how an increase in a current's strength starts at the bottom and develops upward over the course of a few days, providing advanced notice of potentially precarious ice conditions. In particular, ice moving against the wind is an indication that the current is moving with considerable strength. Currents, especially when bringing in warmer water, can lead to the break-up of ungrounded ridge keels near the edge resulting in the "throwing-up of ice" into the lead (*muġaala*; see Fig. 9.5), presenting a danger to boats. When the lead is closed, these broken pieces can remain under the ice, only to emerge when the lead reopens.

A "water sky" (a dark band along the horizon that indicates open water; see Fig. 9.6) serves as a way to monitor for incoming pack ice that may present a threat to those at the ice edge. If the dark band begins to disappear, the pack ice is approaching. This is of particular concern when camped on *iiguaq*. In these conditions a whaling crew is forced to retreat to safer ice. When camped on multi-year ice at the lead, encroaching pack ice presents less of a hazard.

It is also important to monitor the current's strength to avoid striking a whale when conditions may prevent the crew from being able to haul it to the ice edge for butchering. A strong current, especially near the Point, has been known to defeat the efforts of several boats attempting to haul a single whale to stable shorefast ice. A decision "to launch a boat from the ice edge to go to the whale's path" (*pamiuqtak*) must be done only when conditions present an acceptable risk for the entirety of the hunt, which ends when the meat, *muktuk*, equipment, and people are on safe ice.

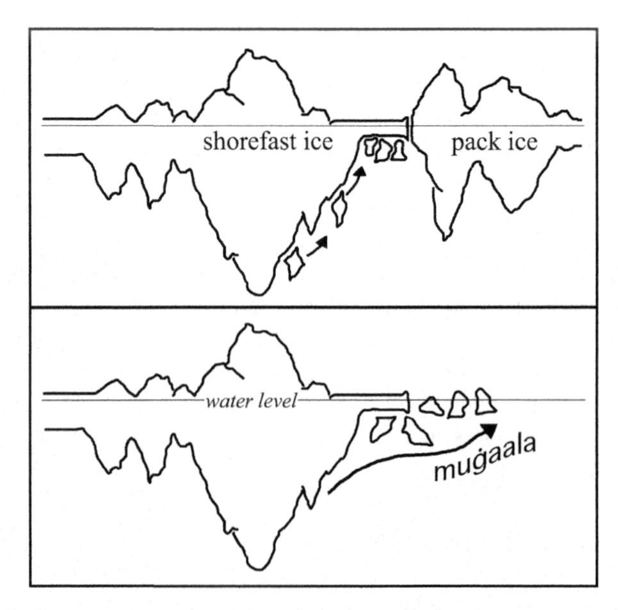

Fig. 9.5 *Muġaala* ("throwing-up of ice") at the ice edge. Following the opening of the lead as pack ice drifts away, loose pieces of ice detach from ridge keels (or from the bottom of rafted ice) and float up beneath the level ice or into the open water. When such pieces hit the level ice, they can produce a loud sound that is often misinterpreted by a hunter as a crack forming in the ice. Based on a description provided by Lewis Brower

Fig. 9.6 Dangerous open water on a whaling trail. Such "holes" may be attributed to heavy snowmobile traffic and warm water melting the ice from beneath. A "water sky" can be seen on the horizon, indicating an open lead. (Photo: M.L. Druckenmiller)

Monitoring the Shorefast Ice

Hunters carefully watch the shorefast ice along their trail throughout the season. There are several features that they pay particular attention to, such as previously identified cracks, newly formed cracks in the flat thin ice near grounded ridges, and areas where slush ice has been incorporated into the shorefast ice. Barrow whaler Lewis Brower told of how his father Arnold Brower, Sr., had taught him to build small handmade rows of compacted snow to perpendicularly extend across cracks so that fracturing or disturbances to the snow piles would serve to monitor the cracks' activity. Cracks or weak points where new ice has been added to the shorefast ice become particularly important in determining where the shorefast ice may break-out. *Katak*, which means "to fall," is the Iñupiaq term used to describe a sudden drop in sea level where the flat ice near grounded ridges cracks and may lead to a break-out (George et al. unpublished; Norton 2002).

Another feature that must be monitored is *muġaliq*, which is slush ice that forms through shear and the incorporation of snow. This ice can be found anywhere throughout the shorefast ice zone since it can freeze in place as the shorefast ice develops and evolves throughout the year. These areas are observed closely since they represent a particular danger as spring progresses. When frozen *muġaliq* warms it rapidly loses its integrity and acquires a quicksand-like consistency, breaking in a quiet manner. A quiet break-up process is particularly disconcerting to hunters since they often rely on sounds to warn of potentially threatening conditions, such as cracking and ridging.

By mid-to-late May, warmer air temperatures and the arrival of the warm current from the southwest escalate the transition of shorefast ice toward increasingly unsafe conditions. The "glue" that is holding the weak areas together begins to release. Old cracks melt out, and newly formed "cracks open up, never to refreeze" (*nutaqqutaq*). After the snow melts, trails develop dark areas of water or extremely thin ice, where snowmobiles can easily fall through (see Fig. 9.6). Also by this time, the majority of passing whales become increasingly large and difficult to pull up onto thin ice at the edge. *Kasruq* ("when one is done with whaling and pulls their gear off the ice") takes place either when Barrow has reached its quota of strikes or when ice conditions are no longer suitable for whaling.

Looking for Old Ice

The retreat and thinning of the Arctic's perennial ice, observed each September as the ice extent is at its annual minimum, is a clear indication that conditions in the Arctic have changed over the last 40 years. Since 1979 when satellites first began monitoring Arctic ice, the extent has declined as much as 10.2% per decade (Comiso et al. 2008). After the mid-1970s, hunters along Alaska's Chukchi coast also began observing that ice conditions, in particular shorefast ice morphology and stability, began to deviate from what was considered normal for prior decades, as reflected in direct observations and elders' teachings (Norton 2002). In large part, these observations note that multi-year ice, which here refers to ice that has survived at least one summer's melt season, is becoming less abundant over the long term. Figure 9.7 shows multi-year ice near Point Barrow.

Both hunters and scientists view the presence or absence of multi-year ice as an indicator of change and as proxy for a range of processes related to stability and decay of coastal and offshore ice. Scientists view multi-year ice as important for regulating the amount of solar energy that enters the ocean over the course of the summer and early fall, thus partially controlling the growth conditions for new ice in late fall. In the coastal environment, multi-year ice assists in the formation of shorefast ice by providing anchoring points. When winter approaches and the prevailing clockwise circulation pattern in the Beaufort Sea brings multi-year ice south- and westward, multi-year floes enter the coastal region during a time when ice dynamics and the growth of new ice build shorefast ice. The degree to which these processes coincide determines the amount of multi-year ice entrained into the shorefast ice zone.

With perennial ice retreating further to the north and less multi-year ice present during fall freeze-up (Maslanik et al. 2007; Nghiem et al. 2007), the period of stable shorefast ice has grown shorter as well. A widespread concern of the Barrow community is that with the loss of multi-year ice, ice conditions will become increasingly unfamiliar and the hunting season will shorten. In the past, the whaling

Fig. 9.7 Multi-year ice near Barrow. Whaler Roy Ahmaogak is shown standing in front of *piqaluyuk* that grounded near the shoal north of Point Barrow in 2009. Roy stated that "Ten years ago, the reduction in multi-year ice was not so noticeable. But in recent years we have seen a large disappearance of multi-year ice. I was surprised to see two-story *piqaluyuk* north of Nuvuk this year." The *lower panel* of images presents three QuikSCAT satellite scenes from December 1 of 2006, 2007, and 2008 – prior to the whaling seasons discussed later in this chapter. The regions of ice appearing bright – corresponding to higher radar backscatter – north of Alaska can be interpreted as multi-year ice. In general, the amount of multi-year ice drifting near Barrow at this time of year is related to the amount of multi-year ice incorporated into the shorefast ice environment, which forms around this time. (Photo: M.L. Druckenmiller)

season often extended into the month of June, while in recent years the hunt has concluded around the third week of May. Crawford Patkotak, for instance, recalled that in 1987, a year with a lot of heavy multi-year ice off Barrow, his father Simeon Patkotak, Sr., landed a 16 m (52 ft) whale on June 15.

Whalers consider the advantages and disadvantages of multi-year ice. Hunters, similar to Arctic engineers, appreciate the physical properties of multi-year ice and understand that it possesses greater strength than more saline first-year sea ice, but that it is also much more brittle and can shatter upon impact or as a result of a build-up of internal stresses due to surface cooling or heating. When hunters discuss multi-year ice, it is often noted that it is dangerous to camp on for this reason. However, it is also often referred to as a stable platform to base a hunt from at the lead edge. This apparent discrepancy comes from the fact that whalers do not group all multi-year ice into one class. Ice is not simply first-year ice or multi-year ice. In fact, "multi-year" ice is not a term commonly used by Barrow hunters. *Piqaluyuk* is the term used to refer to multi-year ice that is salt-free and serves as a preferred source of drinking water. Large pans of this type of salt-free ice may shatter upon impact. *Tuvaġruaq* is a large region of old ice (perhaps often "old" first-year ice or second-year ice, and younger than *piqaluyuk*) that is stable and will not shatter. Some hunters describe *tuvaġruaq* as not only a single type of ice but rather as a stable conglomerate of different types, potentially even of *piqaluyuk* and younger thin ice. This type of ice, when found along the edge, is resistant to break-out and is suitable for pulling up a heavy whale. However, because of its thickness and associated freeboard, a ramp (*amuaq*) must be cut at the edge in order to pull the whale from the water.

Monitoring and Mapping the Ice Trails

The research presented in this contribution is part of a broader effort to put in place a coastal sea ice observatory at Barrow that addresses both scientific research questions and the information needs of the community and other stakeholders that conduct activities on sea ice (Druckenmiller et al. 2009; Eicken et al. 2009). A key aspect of the observatory is to examine how geophysically derived ice thickness measurements and the monitoring of near-shore ice movement and deformation are relevant to the whaling community's springtime assessments of ice stability and safety. A coastal radar mounted on a building that overlooks the shorefast ice where many of these trails are located monitors the movement and stabilization of ice throughout the year. Collaboration with hunters and the community has enabled data collection during a time when they are most active on the ice.

Between 2001 and 2006, the North Slope Borough Department of Wildlife Management maintained periodic records of where and over what types of ice (e.g., grounded ridges, rubble, flat pans of ice) the community placed ice trails during spring whaling. After a suggestion that a more thorough and complete mapping of the trails take place each spring, we began this effort by mapping the ice trails during 2007, 2008, and 2009. The trails were traveled by snowmobile with a handheld Garmin GPS (geographic information system) device. Using ArcGIS, a collection of GIS software products, the tracks were plotted and placed on recent SAR (synthetic aperture radar) satellite images to produce maps for the community. With input from the community and iterative improvements, these maps have evolved into a product that is useful for on-ice navigation, general ice-type discrimination (flat ice versus rough ice), and as a reference for Barrow's Search and Rescue operations.

With permission from the individual whaling crews, continuous ice thickness measurements were made along most trails using an electromagnetic induction device (Geonics EM31 conductivity meter), which estimates ice thickness by detecting the distance between the surface of the ice and the sea water below. This device was placed on a large wooden sled and hauled along the trails to provide quick indirect measurements (see Fig. 9.8). Measurements are most accurate (to within a few percent of total thickness) over un-deformed ice less than 3 m thick as compared to thicker, rough ice, such as ridges, but still provide detailed information about ice

Fig. 9.8 Snowmobile hauling a sled with the ice survey instruments. Shown here is the EM31 conductivity meter that measures ice thickness, a highly accurate differential GPS, and a radar-reflector mast, which allows the measurements to be located in the imagery collected by the coastal radar in downtown Barrow (see Fig. 9.9). The skyline of Barrow can be seen in the distant background. (Photo: M.L. Druckenmiller)

thickness variations across the entire extent of shorefast ice (Haas et al. 1997). While this chapter presents an overview of the data, a specific discussion of how these measurements relate to changing ice conditions and the responses of the hunting community will be discussed in a later contribution.

This is not the first such project to map sea ice travel by high Arctic communities. Other studies have done so (Aporta 2004; Tremblay et al. 2006) and likewise describe trail breaking and navigation of these temporary landscapes as requiring an experienced ability to discern reoccurring environmental patterns.

A Brief Survey of Weather and Ice Conditions During 3 Years of Whaling

Each year brings new and unique ice conditions to Barrow, and with each year a story can be told about how the community interpreted these conditions and responded during the spring whale hunt. From 2007 to 2009, we visited Barrow each spring to investigate ice conditions, map the ice trails, and speak with hunters. While these years may be characterized as "typical low multi-year ice years," the information presented here places many of the observations made by the community into a framework for year-to-year comparison.

2007: Successful Whaling on Thin Ice Following a Break-Out

The 2007 whaling season was very successful with Barrow landing 13 whales, including a record number of small juvenile whales, known as *ingutuks*. The locations where many of the crews chose to hunt demonstrated two important points. First, hunters tolerate ice conditions that may first appear unsafe if other conditions – the wind, currents, and tides – are favorable. The risk associated with specific ice conditions clearly relates to the length of time a hunter may decide to stay on the ice in that area. Second, hunters choose their camp locations based on not only ice conditions but also whale behavior.

On March 31, 1 week before crews began constructing their trails, a break-out event occurred in the shorefast ice off Barrow (see Fig. 9.9). Immediately following this event, adjacent first-year ice from south of the location piled up and replaced

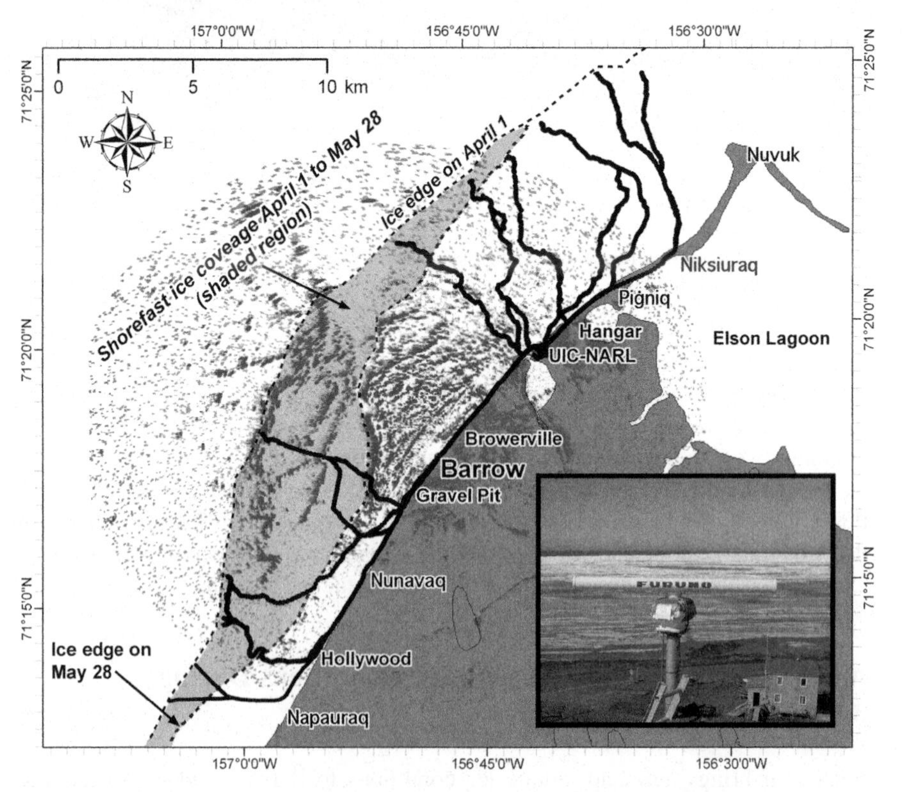

Fig. 9.9 Map of the 2007 whaling trails. Many of the trails shown here traversed the region that existed in the shorefast ice between break-out events on March 31 and May 28. The background in this image shows a sample radar backscatter image (*dark speckles* represent ice features) as recorded during the break-out on May 28. The location of the main trail off Napauraq was hand-drawn after the whaling season ended based on the input from members of the community. The 10 kW X-band Furuno marine radar in downtown Barrow is shown in the *lower right* photo. (See also Color Plate 5 on page 474)

the ice that broke out. This ice, despite being quite thin relative to the shorefast ice to the north and possessing few grounded ridges, remained in place throughout the entire whaling season and provided the location where most of Barrow's whales were landed (Druckenmiller et al. 2009). This circumstance may be in part due to the observation of one hunter that the whales were following the edge of the southern lead and overshooting the crews camped at the lead edge further north. Barrow reached its quota on May 25 and the ice broke out again on May 28 at approximately the same location as on March 31. Figure 9.9 shows the area of shorefast ice present between these break-out events. Also shown in this figure are the trails that traversed this region and a radar image from the March 31 break-out as recorded by the Observatory's coastal radar.

Barrow whaler Joe Leavitt, along with elders Arnold Brower, Sr., and Wesley Aiken, observed that this first-year ice was held in place by only a few "key" ridges and that favorable conditions allowed the community to successfully whale in this area. Except for between May 7 and 13, the wind throughout the season (see Fig. 9.10) allowed the lead to remain open and prevented pack ice from colliding with the shorefast ice. Prior to the May 28 break-out, however, the trails in

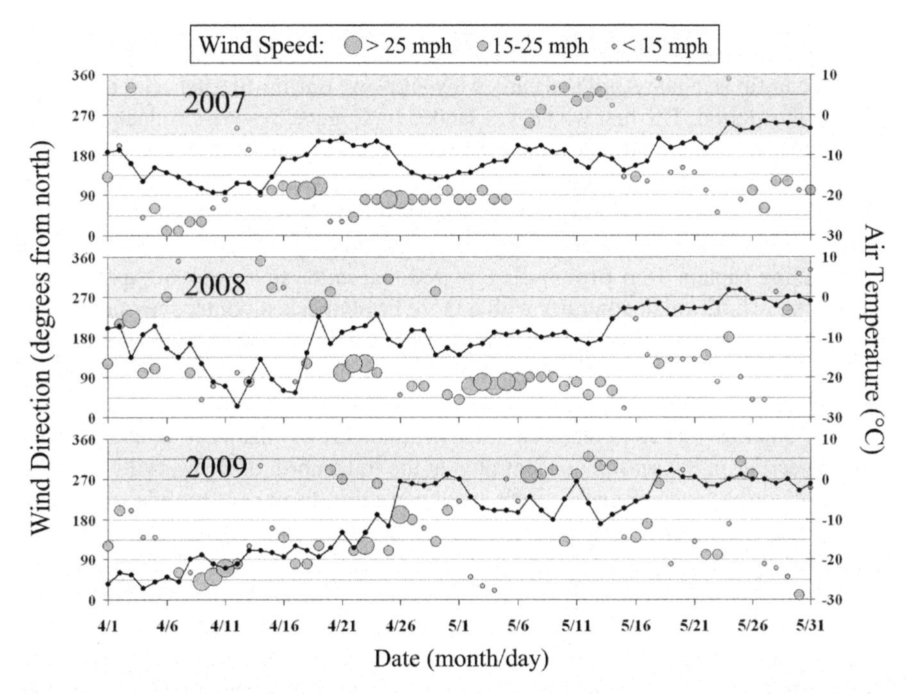

Fig. 9.10 Winds and air temperature during the 2007, 2008, and 2009 whaling seasons. Wind direction and speed (maximum 2-min readings) and air temperature (daily averages) are denoted by the *gray circles* and *solid lines*, respectively. The *shaded* regions span the wind directions (southwest to north) that tend to close the lead. Data were recorded at the Post-Rogers Memorial Airport and accessed from the National Climate Data Center

the south were worn dangerously thin by large amounts of snowmobile traffic and previously refrozen cracks began to open, which may have significantly contributed to the second break-out (Druckenmiller et al. 2009). After the trails in the south deteriorated some crews moved to the trails in the north to take advantage of safe ice conditions persisting later into May.

2008: Whaling in the North Long After Southern Trails Deteriorate

During the 2007–2008 ice year, stormy conditions during the period when ice moved in and stabilized along the coast contributed to a rough shorefast ice cover composed of highly deformed thin first-year ice. In some areas, ridges were exceptionally close to the beach due to the high winds driving these ridges near shore. This was particularly evident off Nunavaq, where some ice had even blown up onto the beach. Whaling captain Harry Brower, Jr., explained that there was a repeated sequence of ridge building followed by ice coming in to add on that contributed to a rough but stable ice cover. Brower decided to place his trail off Nunavaq (see Fig. 9.11) because the ice off NARL was too rough.

Cold conditions in early April helped to provide stable shorefast ice at the start of whaling. Similar to 2007, but in stark contrast to 2009, 2008 experienced a dominating east wind that kept the lead open (see Fig. 9.10). Most crews encountered good conditions in late April and early May allowing Barrow to catch a lot of whales during this period. The first whale was landed by Eugene Brower's Aalaak Crew on April 26. However, while the east wind tended to keep the lead open, it also presented a hazard – crews pulled off the ice when winds approached 25 mph believing that such an offshore wind can drop the water level and lead to a break-out as floating ice cracks away from grounded ridges.

Whaling captain Tom Brower III reported that in the first week of April a late-season snowfall, which contrasts with a more firmly packed winter snowfall, led to hazardous conditions. First, the fresh snow served as an insulating layer allowing the warm currents to more efficiently melt the thin ice from below. Later in May, when air temperatures increased, the snow quickly melted, which then in turn accelerated surface ablation through enhanced solar heating. Crews that were unable to land whales earlier in the season concentrated at the trails north of Browerville as those to the south became dangerous with areas worn thin from snowmobile traffic and warm water. Brower reported having to abandon their trail off Napauraq in early May only after a few days of heavy use.

Figure 9.11 shows the 2008 trails and where ice thickness measurements were made during the season. While these data are useful from the standpoint of tracking long-term trends in the thickness distribution of shorefast ice, it also assists in understanding how different types of ice are used by the community. For example, Fig. 9.12 shows the cross-sectional thickness profiles from two trails. The thin ice at the end of Jacob Adams Crew's trail was chosen for a camp since it was identified as flat ice where whales would be swimming beneath (see Fig. 9.4 and related discussion) and surfacing at the edge. Their crew, among many others, decided this

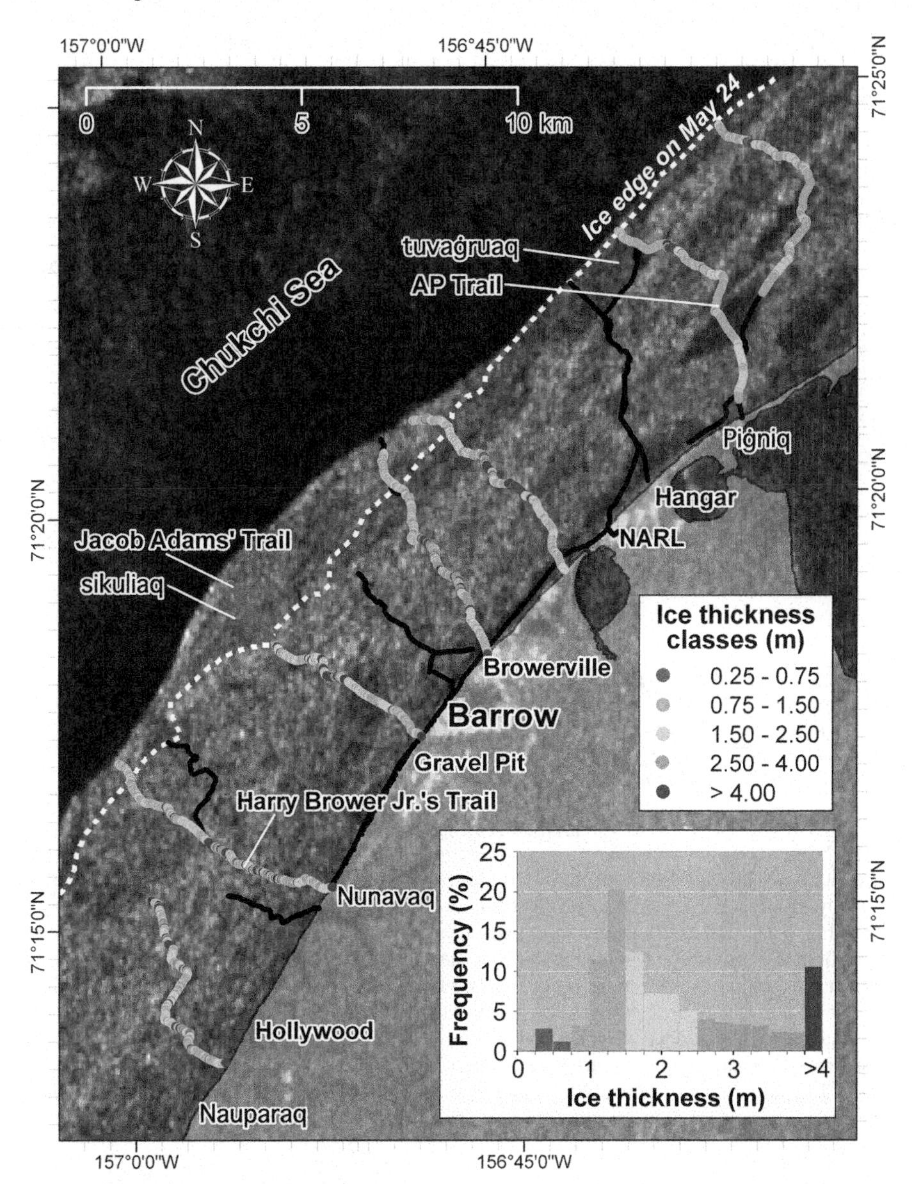

Fig. 9.11 Map of the 2008 whaling trails. Trails are shown here with ice thickness data overlaid on select trails where measurements were made. The two trails south of Nunavaq were not fully mapped since they were incomplete at the time of mapping in early to mid-April. The trail off Barrow was abandoned before making it to the ice edge. The SAR image, acquired by the RADARSAT-1 satellite and provided by the Canadian Space Agency and C.E. Tweedie and A.G. Gaylord, is from April 5, 2008. (See also Color Plate 6 on page 475)

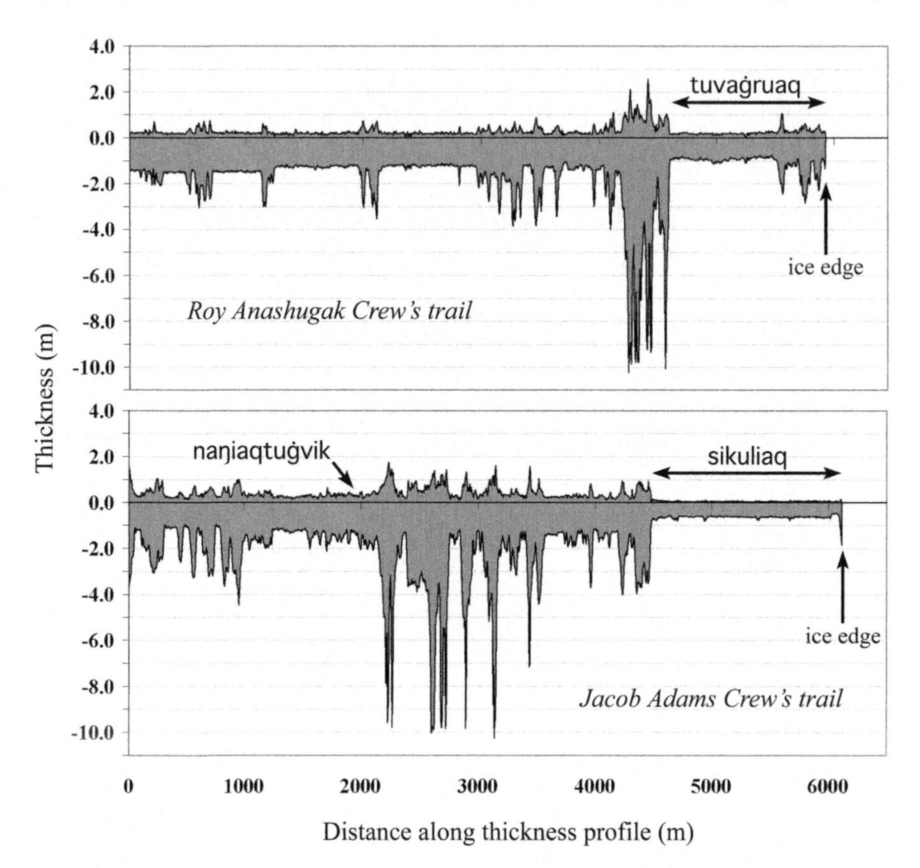

Fig. 9.12 Cross-sectional ice thickness profiles along two different whaling trails. The 2008 trails of Roy Anashugak Crew and Jacob Adams Crew were measured on April 5 and 7, respectively (see Fig. 9.11). Labeled features are based on interviews with Jacob Adams, Herman Ahsoak, and Gordon Brower. The level ice in the zones labeled "*tuvaġruaq*" and "*sikuliaq*" had average thicknesses of 1.0 and 0.5 m, respectively. The location of the *naŋiaqtuġvik*, or "safe camp," is shown for Adams' trail but was not documented for Anashugak's trail. Differential GPS was used to survey the surface elevation and an EM-31 conductivity meter was used to measure ice thickness. True thickness is the total thickness of ice above and below the water line, which is at zero. The proportionality between the *x* and *y* axes is such that the thickness is emphasized. Ridge thicknesses over 4 m are underestimated by up to 30% due to instrument limitations (Haas 2003)

sikuliaq (young ice formed along the edge of solid ice) was a good place for a camp. However, they reported having to retreat to their *naŋiaqtuġvik* (safe camp; labeled in Fig. 9.12) multiple times when the west wind brought in the pack ice and when the strong east wind threatened to drop sea level and break the extended floating ice from the grounded ice. Significant portions of this *sikuliaq* broke off following impact with pack ice brought in with the west wind on April 25. However, the ice remained safe and allowed several crews to stay camped there. Adams landed a 9 m (30 ft) whale on May 7, just before the remainder of the *sikuliaq* broke out. Adams noted that such ice is suitable for pulling up a whale approaching 12 m (40 ft) in length, but would not be sufficient for a whale of 15 m (50 ft).

By mid-May, crews abandoned Adams' trail since newly deformed rough ice at the edge prevented easy access to open water. As previously mentioned, many crews moved to the trails north of Browerville. Older, thicker, and more stable flat ice (*tuvaġrauq*) near the edge allowed crews to easily connect these trails together near the lead (beyond the last row of ridges) with secondary trails (not shown in Fig. 9.11). This enabled hunters to travel between camps without the need to come a long way back toward shore in order to get on another trail. Also, connected trails always provide more numerous escape options in the case of dangerous conditions. Unlike the *sikuliaq* further to the South this *tuvaġrauq* remained in place into late May, beyond the end of the whaling season.

2009: West Wind Leads to Unsuitable Ice Edge Conditions

The shorefast ice of 2009 was representative of typical ice conditions in recent years with a few noteworthy differences, which are discussed later in this section. Off NARL and Browerville the shorefast ice was heavily ridged and deformed with few areas of level ice. Despite the near-absence of larger pieces of old ice, it was very stable all the way to the last major row of ridges at about 3 km offshore. The few scattered pieces of *piqaluyuk* were landward of already well-grounded ice, providing little service to the crews, other than as a source of drinking water. The last row of grounded ridges was separated by a system of cracks from the outermost floating ice. The ice off Gravel Pit had formed in place and was very flat and thin yet with no noticeable cracks. Due to the lack of anchored ice, except for a few ridges close to shore, the crews in this area were extremely cautious of any drift ice that approached. The conditions off Hollywood were similar to those off Gravel Pit – flat ice that had mostly formed in place – although many hunters indicated that it was more firmly grounded. In this area, notable cracks developed later in the season. The ice off Monument formed a large promontory of shorefast ice (*nuvuġaqpuk*) that extended approximately 11 km offshore (see Fig. 9.13). The distance required to reach the edge was one reason crews may have decided against hunting in this area, but perhaps the more important reason is that most believed this promontory of ice would eventually collide with pack ice and break away. However, surprisingly, the *nuvuġaqpuk* remained throughout the entire whaling season.

Despite stable conditions along the general extent of the shorefast ice, the pack ice, winds, and currents never cooperated to make the ice edge suitable for whaling. Tuuq is when the pack ice collides with the shorefast ice edge and acts as a chisel (George et al. 2004). While such events surely present danger to crews camped at the edge, they are also relied on by hunters to "fix-up" the ice – to thicken thin ice through deformation and to rid the edge of dangerous attachments (*iiguat*). An ideal sequence of events would involve heavy pack ice coming in to "fix-up" the ice, driven by the wind and/or current in such a manner that hunters are able to foresee

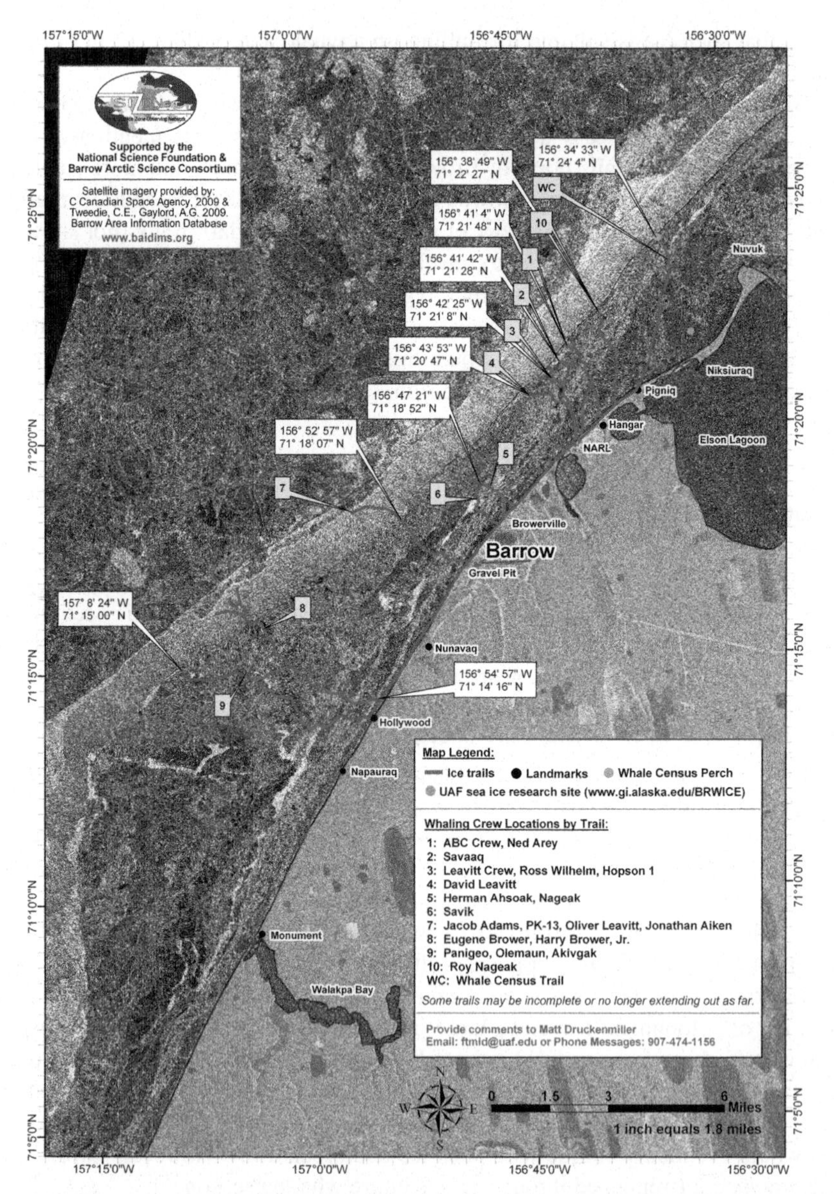

Fig. 9.13 Map of the 2009 whaling trails. This exact map was provided to the community during the whaling season. The SAR image, acquired by the European Remote Sensing satellite ERS-2 and provided by the Canadian Space Agency and C.E. Tweedie and A.G. Gaylord, is from May 16, 2009, just prior to the opening of the lead shown in Fig. 9.1. Various GPS locations are labeled to assist with navigation. Locations are also shown for the camp of the 2009 bowhead whale census orchestrated by the North Slope Borough's Department of Wildlife Management and of our sea ice mass balance site that measured level ice growth and other variables of interest. (See also Color Plate 7 on page 476)

the event and pull off the ice. Next, the lead would open to reveal ice edge conditions suitable for safely hauling up a whale.

Beginning on April 20, just as most crews were finishing their trails, the west wind arrived and dominated throughout the remainder of the whaling season (see Fig. 9.10). The wind-driven pack ice either formed *iiguat* or on occasion built up a moving one-story high wall of slush ice (*muġaliq*) along the edge. Especially for the crews off NARL, *iiguat* persisted and when one broke off another formed. Gordon Brower recalled that in late April his crew was fortunate to be camped on *tuvaġruaq* for a few days but that the area was still considered unsafe since it was only connected to the grounded ice by thin young ice. Many hunters described 2009 as a "waiting game" – waiting for the lead to open or for edge conditions to improve. Most were only camped at the edge for 1 or 2 days and were prepared to run at a moment's notice. Some hunters never even brought their boats onto the ice.

The year 2009 was also difficult since the *muġaliq* incorporated within the shorefast ice never froze solid. Potentially thawing temperatures arrived on April 26, followed by a refreeze and then another thaw later on May 18. Warm weather and the arrival of warm water (as suggested from interviews with the hunters) led to a quick deterioration of trails and cracks, and in particular to those south of *Nunavaq*. This sequence of events made the conditions in the South very unsafe. Some crews pulled off the ice as early as May 12 due to these unsafe conditions, but also because the larger whales were beginning to move through. Similar to 2007 and 2008, the trails off NARL and northward remained intact longer than those to the south.

On May 16 a southeast wind opened the lead for a short time (see Figs. 9.1 and 9.10) and in the early hours of May 17, ABC Crew (Arnold Brower, Sr., Crew) landed an 8 m *ingutuk* from trail number one (see Fig. 9.13) and was able to find a pan of *tuvaġruaq* to successfully haul up the whale. Three other crews caught whales before May 23 but experienced great difficulty in finding a suitable place to butcher because of *muġaliq* at the edge. One whale was struck and butchered at trail one and two were struck from the trails off Hollywood. Of these latter two, one was hauled to trail seven and the other to trail four in hopes of finding ice that would support the weight of the whales and also because the trails off Hollywood were not safe enough to permit safe passage for the large number of people required to butcher a whale. Each attempt failed and as the whales were pulled onto the ice, they would immediately break through. In all three cases they had to cut off the heads of the whales (1/3 of the whale's body) in the water and anchor it to the ice edge. These poor butchering conditions unfortunately did not allow the crews to retrieve the entirety of the whale meat, and in one case they were only able to collect the skin and blubber (*muktuk*). Joe Leavitt stated that if heavier ice conditions had existed in 2009 butchering all four whales would not have been a problem.

To some members of the community the success of the whale hunt is more than just about climate and ice conditions; it is connected to the well-being of the people. Roy Ahmaogak, for instance, said, "One of the most heartbreaking things about this year was that we weren't given the opportunity to practice traditional whaling because of the ice. Barrow and its people have been feuding and bickering at each other all this last winter. This is the reason we think the ice didn't go out this year and it stayed closed. This will make us think this coming year that we have to watch our tongue and to watch what we say to people. We are lucky to have two blanket tosses this year. It will teach Barrow and people like us." In the end, Barrow joyously celebrated the four caught whales during two *Nalukatak*. Barrow then set their sights on the non-traditional fall bowhead hunt, which is done in open water with outboard engines and aluminum boats.

Discussion

The initial placement of ice trails is largely in response to ice conditions, traditional practices, and crew preference; however, there seems to be a pattern of the trails to the south having to be abandoned earlier in the season due to the ice deteriorating, causing the crews to concentrate at the trails north of town. This is in contrast to that which Arnold Brower, Sr., explained regarding how crews in the past typically moved south later in the season as currents intensified and conditions in the north became dangerous. This raises the question of whether climate and environmental change are impacting how the community uses the ice cover. The seasonal summaries presented here span a 3-year period and accordingly can only present a brief look at how present climate and ice conditions impact spring whaling.

Changes in shorefast ice characteristics are much more complicated than the obvious reduction in the presence of multi-year ice. During our conversations, Arnold Brower, Sr., and Tom Brower III both noted that shorefast ice prior to the 1980s extended much further out, was flatter, and was composed of thicker level ice than today. While detailed analysis of how shorefast ice characteristics have changed over time is beyond the scope of this chapter, it is clear that changes are taking place that present a new assortment of challenges for the whaling community. If hunters continue struggling to find sufficiently grounded and stable ice, such as the *tuvaġruaq* experienced in 2008, they may increasingly have to deal with the problems encountered in 2007 and 2009 – early spring break-out events close to shore and ice edge conditions that are not suitable for pulling up a whale.

The presence of *muġaliq* dominated the observations of hunters in 2009. This slush ice, which forms through shear at any time throughout winter or spring, represents a type of ice that lacks the drainage of salt water that typical thermodynamic ice production promotes, thus rendering it potentially unstable and responsive to slight changes in temperature. Hunters acknowledged that this ice is common but that 2009 was remarkable because it was so widespread and air temperatures did not allow for this ice to retain its integrity late into the season. Coupling this phenomenon with an understanding that advection of warm water can lead to the destabilization of shorefast ice by melting grounded ridge keels (Mahoney et al. 2007b) and refrozen cracks reveals that shorefast ice as a stable platform for hunting and travel is closely linked not only to climate change but also to weather and oceanographic variability.

Lastly, the summary of how the crews responded to ice conditions during these 3 years begs the important question of whether there is a clear and distinguishable local zonation of ice conditions along Barrow's coastline. This topic, worthy of further investigation, may underscore the fact that Barrow's ice environment allows for understanding not only how ice conditions respond to climate but also to subtleties in local and regional conditions, such as bathymetry and coastal currents. This may present an opportunity for scientists to further discover the local expert sea ice knowledge found in Barrow, which likely possesses an intricate understanding of the processes that may govern a local zonation of conditions and further lead to improved scientific monitoring that is relevant to the community's activities on ice.

Conclusions

The Barrow community continues to practice successful traditional spring whaling from shorefast ice while making observations that lend a new perspective to understanding processes that dominate the present-day coastal sea ice environment. Hunters assess shorefast ice in a highly specialized manner as they consider safety, navigation, hunting strategies, and traditional knowledge and practices. Detailed characteristics of year-to-year ice conditions, which are unobservable by standard scientific monitoring programs, manifest in impacts to the whaling community. Utilizing the collaborative and experiential (as opposed to experimental) approach presented here – a type of ethnoglaciology – we are working toward an improved understanding of how to observe the local environment in a manner to track changes important to both climate study and the community. This research may ideally begin to illustrate how strategic adaptations in the way the community uses the shorefast ice are indicative of and responsive to environmental change.

Mapping Barrow's ice trails allows us to piece together how ice characteristics spatially and temporally relate to the community's use of the ice. These maps are providing a valuable product to the community while also serving as a useful reference tool for scientists and hunters to communicate across barriers of culture and experience. It is our hope that this project continues as a long-term monitoring effort to unite advanced scientific instrumentation and expertise, traditional knowledge, and ice use by a modern arctic whaling community.

Acknowledgments This research was made possible with the assistance of several experienced Iñupiat whalers: Billy Adams, Jacob Adams, Roy Ahmaogak, Herman Ahsoak, Arnold Brower, Sr., Eugene Brower, Gordon Brower, Harry Brower, Jr., Lewis Brower, Tom Brower III, Jeffrey Leavitt, Joe Leavitt, Warren Matumeak, Ben Nageak, Nate Olemaun, and Crawford Patkotak. Thanks to Ronald Brower Sr., for reveiewing the use of Iñupiaq terminology. We would like to also thank the North Slope Borough Department of Wildlife Management, Barrow Arctic Science Consortium, Barrow Whaling Captains Association, and Allison Gaylord and the Barrow Area Information Database. This publication is the result in part of research conducted as part of the Seasonal Ice Zone Observing Network with financial support from the National Science Foundation (OPP-0632398), the Oil Spill Recovery Institute, and the Cooperative Institute for Arctic Research (Project CIPY-34) with funds from the National Oceanic and Atmospheric Administration (cooperative agreement NA17RJ1224 with the University of Alaska).

References

Aporta, C. 2004. Routes, trails and tracks: Trail-breaking among the Inuit of Igloolik. *Etudes Inuit* 28(2): 9–38.

Braund, S.R. and Moorehead, E.L. 1995. Contemporary Alaska Eskimo bowhead whaling villages. In *Hunting the Largest Animals: Native Whaling in the Western Arctic and Subarctic*. A.P. McCartney (ed.), Edmonton: The Canadian Circumpolar Institute, University of Alberta, pp. 253–279.

Comiso, J.C., Parkinson, C.L., Gersten, R., and Stock, L. 2008. Accelerated decline in the Arctic sea ice cover. *Geophysical Research Letters* 35: L01703, Doi:10.1029/2007GL031972.

Drobot, S.D. and Maslanik, J.A. 2003. Interannual variability in summer Beaufort Sea ice conditions: Relationship to winter and summer surface and atmospheric variability. *Geophysical Research Letters* 108: C7.

Druckenmiller, M.L. (in preparation) Observing Alaska shorefast sea ice from an ice-user's perspective: Integrating geophysical and Iñupiat knowledge. Doctoral Thesis, University of Alaska Fairbanks, Fairbanks, Alaska.

Druckenmiller, M.L., Eicken, H., Johnson, M.A., Pringle, D.J., and Williams, C.C. 2009. Toward an integrated coastal sea-ice observatory: System components and a case study at Barrow, Alaska. *Cold Regions Science and Technology* 56(2–3): 61–72.

Eicken, H., Lovecraft, A.L., and Druckenmiller, M.L. 2009. Sea-ice system services: A framework to help identify and meet information needs relevant for Arctic observing networks. *Arctic* 62(2): 119–136.

George, J.C., Huntington, H.P., Brewster, K., Eicken, H., Norton, D.W., and Glenn, R. 2004. Observations on shorefast ice dynamics in Arctic Alaska and the responses of the Iñupiat Hunting Community. *Arctic* 57(4): 363–374.

Haas, C. 2003. Dynamics versus thermodynamics: The sea-ice thickness distribution. In *Sea Ice—An Introduction to Its Physics, Biology, Chemistry and Geology*. D.N. Thomas and G.S. Dieckmann (eds.), Oxford: Wiley-Blackwell, pp. 82–111.

Haas, C., Gerland, S., Eicken, H., and Miller, H. 1997. Comparison of sea-ice thickness measurements under summer and winter conditions in the Arctic using a small electromagnetic induction device. *Geophysics* 62(3): 749–757.

Mahoney, A., Eicken, H., Gaylord, A.G., and Shapiro, L. 2007a. Alaska landfast sea ice: Links with bathymetry and atmospheric circulation. *Journal of Geophysical Research* 12: C02001.

Mahoney, A., Eicken, H., and Shapiro, L. 2007b. How fast is landfast ice? A study of the attachment and detachment of nearshore ice at Barrow, Alaska. *Cold Regions Science and Technology* 47: 233–255.

Maslanik, J.A., Fowler, C., Stroeve, J., Drobot, S., Zwally, J., Yi, D., and Emery, W. 2007. A younger, thinner Arctic ice cover: Increased potential for rapid, extensive sea-ice loss. *Geophysical Research Letters* 34: L24501, Doi:10.1029/2007GL032043.

Nelson, R.K. 1969. *Hunters of the Northern Ice*. Chicago and London: The University of Chicago Press, 429pp.

Nghiem, S.V., Rigor, I.G., Perovich, D.K., Clemente-Colón, P., Weatherly, J.W., and Neumann, G. 2007. Rapid reduction of Arctic perennial sea ice. *Geophysical Research Letters* 34: L19504, Doi:10.1029/2007GL031138.

Norton, D.W. 2002. Coastal sea ice watch: Private confessions of a convert to indigenous knowledge. In *The Earth Is Faster Now: Indigenous Observations of Arctic Environmental Change*. I. Krupnik and D. Jolly (eds.), Fairbanks: Arctic Research Consortium of the United States, pp. 127–155.

Stoker, S.W. and Krupnik, I. 1993. Subsistence whaling. In *The Bowhead Whale*. J.J. Burns, J.J. Montague, and C.J. Cowles (eds.), Lawrence: Allen Press, pp. 787.

Tremblay, M., Furgal, C., Lafortune, V., Larrivée, C., Savard, J., Barrett, M., Annanack, T., Enish, N., Tookalook, P., and Etidloie, B. 2006. Communities and ice: Bringing together traditional and scientific knowledge. In *Climate Change: Linking Traditional and Scientific Knowledge*. R. Riewe and J. Oakes (eds.), Winnipeg: Aboriginal Issues Press, University of Manitoba, pp. 289.

Chapter 10
Creating an Online Cybercartographic Atlas of Inuit Sea Ice Knowledge and Use

Peter L. Pulsifer, Gita J. Laidler, D.R. Fraser Taylor, and Amos Hayes

Abstract A team of community and university researchers, Inuit experts, Inuit organizations, and software developers are developing a Cybercartographic Atlas of Inuit Sea Ice Knowledge and Use. In keeping with a cybercartographic approach, the Atlas combines maps with text and multimedia representations including images, sound, video, and visualizations. Ultimately, members of the communities involved in the Inuit Sea Ice Use and Occupancy Project are interested in evaluating the utility of such approaches for their educational potential as classroom tools, as well as to ensure more dynamic forms of knowledge documentation that can be easily updated and accessed over time. At the user interface level, the Atlas presents documented Inuit knowledge in new and innovative ways. The ability to support innovative representations is underpinned by a flexible data model that is populated with knowledge documented through a participatory mapping process. The Atlas presents a variety of topics including "Our Partner Communities," "Our Contributors," and Inuit knowledge of "Ice Conditions" and "Uses." Future iterations of the Atlas will see a restructured and greatly expanded table of contents and potentially the addition of user-contributed content functionality.

Keywords Inuit knowledge · Web mapping · Participatory mapping · Interoperability · Interactive Atlas

Introduction

This chapter provides an overview of past and current efforts toward the development of a Cybercartographic Atlas of Inuit Sea Ice Knowledge and Use, henceforth referred to as the Atlas. As one of the four sub-projects within the Inuit Sea Ice Use

---

P.L. Pulsifer (✉)
Department of Geography and Environmental Studies, Carleton University, Ottawa, ON K1S 5B6, Canada
e-mail: ppulsife@connect.carleton.ca

I. Krupnik et al. (eds.), *SIKU: Knowing Our Ice*,
DOI 10.1007/978-90-481-8587-0_10, © Springer Science+Business Media B.V. 2010

and Occupancy Project (ISIUOP) (see Chapter 1 Introduction, this volume), the Atlas is being developed using a cybercartographic approach that combines maps with multimedia representations and involves a team of partners (i.e., community and university researchers, Inuit experts, and Inuit organizations) and developers (university researchers, technical staff, and programmers). Ultimately, cybercartography is not a technology or a methodology per se, but a conceptual framework that both guides development and reflects on and discusses the implications of new methods of cartographic production. Therefore, our efforts to create this sea ice Atlas grew from local interest in having accessible, dynamic, and educational ways of using the documented information shared by Inuit experts, as well as evolving technological and cybercartographic tools that could facilitate such development.

This chapter begins with a brief overview of the importance of mapping and representing change with respect to sea ice knowledge and use. To provide context for our presentation of results and the discussion of our methods, the concept of cybercartography is defined and elaborated through a review of recent and emerging trends in the domain of cartography and geographic information. This chapter is followed by a description of the conceptual framework supporting Atlas development. The results of the first iteration of Atlas development are then presented along with an overview of design strategies used. The participatory mapping approach used to document Inuit knowledge in the form of digital geographic information and multimedia content is discussed. In the context of this chapter, an important component of Inuit knowledge documentation is the establishment of a data model that can effectively and appropriately represent and store the information collected as part of the documentation process. An examination of the current design of the ISIUOP data model is followed by an account of how the Atlas interface and functional elements have been implemented. In highlighting the processes undertaken to date in Atlas design and creation, we outline our future plans for ongoing improvements and refinements of Atlas structure and functionality.

Mapping and Representing Change

We use "sea ice" as a general term referring to the frozen ocean surface, whereby more specialized terminology is used to characterize the various states, descriptors, and uses of sea ice that reflect a complex set of interrelated, dynamic social – ecological features. Central to sea ice phenomena is the concept of change (i.e. seasonal evolution of freezing and decay, annual variations, or long-term climate-induced changes). Indeed, much of the current debate and discussion associated with sea ice revolves around change. Change, however, is difficult to document. No sooner do we log an observation, or digitize a line on a map, than the condition (or local perspectives of it) has already changed. As such, we acknowledge the limitations of efforts to document change, while also arguing for the importance of such activities to help improve our understanding of the interconnections between changing environments, uses, habitats, and positive and negative implications (even if the changing of conditions can never be fully represented). Therefore, while developing the Atlas, we explored – and continue to explore – differing means of documenting or representing change to evaluate their utility in conveying the unique environmental and cultural contexts of change in each partner community. Some important components of this process of documentation include iterative participatory mapping carried out during interview and focus group sessions, and exploring different representational forms of results. For early iterations of Atlas design and development we worked predominantly with information previously documented as part of Laidler's thesis research (Laidler 2007) with community researchers and experts in Cape Dorset, Igloolik, and Pangnirtung, Nunavut (see Chapter 3 by Laidler et al. this volume). Our exploratory approach to developing representations of Inuit sea ice knowledge operates within a conceptual framework we refer to as cybercartography.

Cybercartography

The concept of "cybercartography" was proposed by D. R. Fraser Taylor in his keynote address entitled "Maps and Mapping in the Information Era," presented to the 1997 International Cartographic Conference in Sweden. The concept has been expanded through a number of publications (Taylor 2003, 2005; Taylor and Caquard 2006). Cybercartography is a broad conceptual framework that describes and theorizes on movements related to cartography and geographic information processing. While a detailed discussion of cybercartography can be found in the previously cited works, key elements of the framework are presented here to provide context for our discussion of the Atlas. Cybercartography is defined as

> ...the organization, presentation, analysis and communication of spatially referenced information on a wide variety of topics of interest and use to society in an interactive, dynamic, multimedia, multi-sensory format with the use of...multi-modal interfaces. (Taylor 2003:404).

According to Taylor, cybercartography

1. is multisensory using vision, hearing, touch, and eventually smell and taste;
2. uses multimedia formats and new information and communications technologies such as the World Wide Web;
3. is highly interactive and engages the user in new ways;
4. is applied to a wide range of topics of interest to society, not only to location finding and the physical environment;
5. is not a stand-alone product like the traditional map but part of an information/analytical package;
6. is compiled by teams of individuals from different disciplines; and,
7. involves new research partnerships and the private and civil society sectors. (Taylor 2003:407)

An important creation that emerges from a cybercartographic process is a cybercartographic Atlas. A cybercartographic Atlas is a metaphor for many kinds of qualitative and quantitative information linked through location and the concepts embedded within geographically situated knowledge.

The numerous themes comprising cybercartography have been the subject of academic discussion for some time, with some research focusing on the multimedia aspects of contemporary cartography (Cartwright et al. 1999, 2007; Peterson 2008), while others examine collaborative components of cartographic production (Lauriault and Taylor 2005). Additional studies examine the link between general knowledge modeling and cartographic visualization (Skupin and Fabrikant 2003). Taylor's cybercartography aims to examine these developments in an integrated manner and so corresponds well with our effort to understand and represent Inuit sea ice knowledge and use in a more comprehensive way. Furthermore, there are a number of recent and emerging trends in cartography and geographic information that are of interest from a cybercartographic perspective.

Trends in Cartography and Geographic Information

In recent years, the domain of cartography and geographic information has changed dramatically with the introduction of online mapping services and virtual globe technologies. Online mapping services (e.g., Microsoft® Bing™ Maps) provide detailed mapping and satellite image coverage coupled with high-performance user interfaces. Virtual globe technologies (e.g., Google™ Earth) allow you to view and "fly over" a spherical representation of the Earth that appears to be three dimensional. These tools allow non-experts to add their own spatial content and then share those additions with others. With some software development expertise and a (typically) free license, users can integrate content from a number of sites and services to create a "mashup." This may include linking photos from a photo-sharing site with shared video from another site and news feeds from a favorite blog. The previously described services and mashups can be described as Web 2.0[1] applications. In practical terms, developing such geospatially enabled applications was not possible for most people until the aforementioned services were released early in the twenty-first century. This is a fundamental shift from a model that, until very recently, saw government-based agencies as the dominant and sometimes legislated producers and stewards of mapping information. This provides an unprecedented opportunity for communities, including Inuit communities, to create geographies that are representative of their needs, knowledge, and worldviews. This empowerment of communities in creating their own geographic "stories" establishes an important conceptual foundation for the Atlas.

At the base of many Web 2.0 applications is the concept of interoperability. In simple terms, interoperability is the ability of information systems to exchange information and to effectively use the exchanged information. In contemporary information systems, interoperability is increasingly supported by the use of standards, and in particular, broadly used open standards. These include standard Web application architectures and standard data exchange formats. While a detailed discussion of interoperability and open standards is not provided here (cf. Kresse and Fadaie 2004; Goodchild et al. 1999), interoperability and related standards and practices are an important part of recent trends in cartography and geographic information processing and are critical considerations in the Atlas development process.

The open source software movement has also played an important role in recent development trends. Open source is a type of software licensing that provides free access to source code (the programming code created when developing software), free redistribution, and non-discriminatory/non-restrictive license terms (see www.opensource.org/docs/definition.php). Members of the geospatial software development community have embraced open source licensing and the result has been a set of high-quality, powerful software applications ranging from Web mapping toolkits to full featured, analytical Geographic Information Systems. Coordination of the most mature and popular applications is now overseen by the Open Source Geospatial Foundation (www.osgeo.org). The Atlas has been developed entirely on open source software developed by the Geomatics and Cartographic Research Centre and other projects. In our experience, open source geospatial software projects tend to be early adopters of open standards and provide us with the flexibility required to work in a research environment. This includes allowing our researchers to modify existing functionality and add new functionality as needed, and then readily sharing the results of these innovations with our research partners and others in the research community.

While Web 2.0 applications focus on facilitating relationship building and content generation among other things, the Web 3.0 movement (as some people are calling it) is focused on knowledge management and advanced content integration and processing models. Many believe that Web 3.0 will include a "semantic Web" that allows computer systems to interpret and process the meaning of the information available on the Web. While the Atlas is not a Web 3.0 application, it is being designed with these forward-looking developments in mind. Although Web 3.0 presents exciting visions for the next-generation Web, such developments must be critically examined to understand the implications for broader society, local communities – including Inuit communities – and individuals. These issues have been discussed in previous work (Pulsifer 2008).

While many trends in cartography and geographic information are primarily technical in nature, there are also methodological developments that are important with respect to developing the Atlas. Participatory mapping methods established through community-based research and development projects have played an important role in the ISIUOP. Participatory mapping is, "in its broadest sense, the creation of maps by local communities – often with the involvement of supporting organizations including governments (at various levels), non-governmental organizations (NGOs), universities and other actors engaged in development and land-related planning" (International Fund for Agricultural Development 2009:1). As discussed in subsequent sections, participatory mapping is a fundamental component of the information collection process used in the ISIUOP as a whole and in the creation of the Atlas more specifically. Moreover, the knowledge shared and understanding gained through the participatory process is driving the design of information models, representations, interface functionality, and the Atlas as a whole.

All of the trends discussed here are, to some extent, being considered in the design and development of the interactive, multimedia Atlas that is being built to represent Inuit sea ice knowledge and use as contributed by our partner communities. The idea of the Atlas has been supported by the local communities involved in ISIUOP, as most people are interested in exploring the utility of such technologies for their educational potential as classroom tools, as well as to ensure more dynamic forms of knowledge documentation that can be easily updated and accessed over time.

Conceptual Framework of the Cybercartographic Atlas of Inuit Sea Ice Knowledge and Use

Although much technical effort has been expended in developing the Atlas, the process has been carried out in the context of a larger conceptual framework that goes beyond the technical. Through a collaborative research and learning process, ISIUOP researchers undertook extensive documentation of local observations of sea ice conditions, dynamics, uses, and changes with several Inuit communities (see Chapter 3 by Laidler et al. this volume for a description of methods employed). Throughout this process, contributors made it clear that the best way to learn – and pass on – their knowledge was through experience in the context of use (i.e., traveling or hunting on the ice). It is therefore apparent that the Atlas can only provide a tool toward learning in a broader context. The results of our research (and the information displayed in the Atlas) can – at best – provide a partial, fragmented, indication of the depth, and complexity of Inuit knowledge of the marine environment. Nevertheless, this research developed as a result of strong local interest in each community to document their sea ice knowledge and observations to ensure that their perspectives can be represented – and taken more seriously – in broader efforts to monitor and combat climate change, and also for the benefit of younger generations less familiar with sea ice conditions. Local perspectives often get overlooked in climate change research, and thus contributors saw this project as an opportunity to share their expertise and to reach a broader audience. As highlighted earlier, the learning experiences summarized in this chapter reflect our initial efforts to create an Atlas prototype based on C-SIKU results, but we are in the process of expanding to include the other ISIUOP partner communities while tailoring various components to their specific sub-project interests and goals.

In addition to incorporating elements of the cybercartographic conceptual framework and development paradigm, the Atlas has been and continues to be built on a set of research and development principles as follows:

Community partnership. Research questions, results, and dissemination strategies are developed in partnership with community organizations and experts.

Preservation. Ensuring that the documented knowledge is available now and for many generations to come.

Appropriate use of qualitative methods. This includes the use of participatory mapping and a variety of qualitative analytical techniques.

Information content over presentation. Interface level display techniques and functionality can change and rapidly become obsolete. Priority is placed on establishing rich data, information, and knowledge structures that are quite independent of the display interface used. This provides the best possibility for long-term preservation.

Interoperability. The use of standards and development techniques to promote flexible and widespread use of information in a variety of technical contexts. This supports the objective of making the information available to a broader audience and increases the chances of preservation.

Iterative design and development. Rather than trying to establish a static set of application requirements at the outset of the project, design and development are carried out in an iterative fashion with discussion among partners taking place between and during development cycles. Theory and practice (Graham 2003) have established that an iterative approach typically results in more useful and usable software.

Use of open source technology. This limits the potential for partners to be excluded now and in the future due to prohibitive software licensing costs or inflexible licensing conditions. This does not imply that open source does not carry costs, however, the costs can be more appropriately allocated among partners.

Capacity building. Wherever possible, supporting knowledge sharing and skills development among partners so that the capacity exists to create the most appropriate knowledge representation, whether that be a paper map, online Atlas, or podcast.

Constant dialogue. To the extent possible, promote a constant and constructive dialogue among partners to ensure that research outputs meet the needs of all involved.

A Cybercartographic Atlas of Inuit Sea Ice Knowledge and Use

This section presents an overview of selected content and functionality from the current iteration of the Atlas at the time of drafting this chapter. This preliminary iteration organizes the Atlas into seven broad topics including: Welcome, Background, Inuit Knowledge, Changes, Technologies, Monitoring, and Learn More (Fig. 10.1). Each topic then contains more detailed content modules. The iteration of the Atlas reported here was used to facilitate discussion with project partners and elicit feedback from the broader community. The results of this chapter are being incorporated into the design of future iterations of the Atlas. The reader is encouraged to experience the Atlas to fully appreciate the

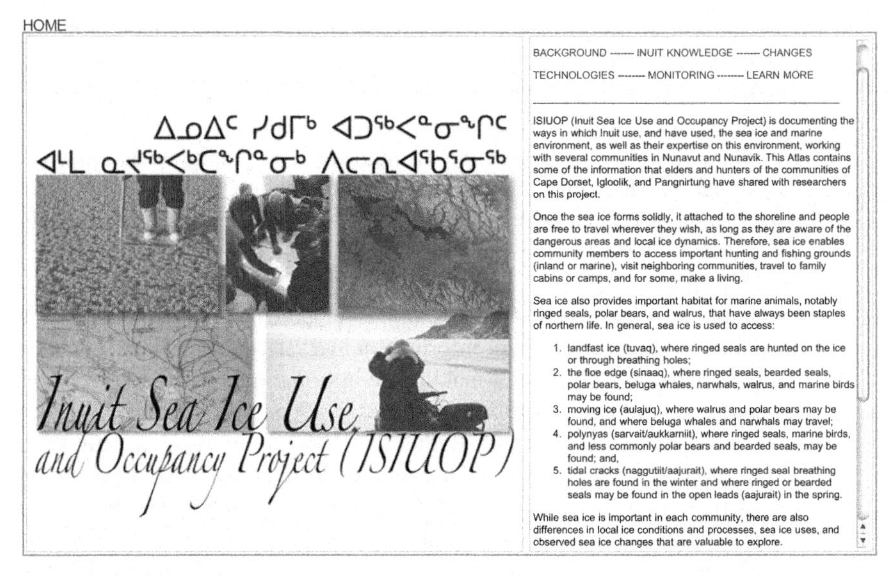

Fig. 10.1 The Atlas Introduction page provides an overview of the ISIUOP. The *top right* of the screen provides links to other atlas topics

descriptions and potential shared here, with the latest version of the Atlas accessible at gcrc.carleton.ca/isiuop-atlas/

Atlas Welcome and Background

The Welcome topic of the Atlas provides a general overview of the project and presents the user with an overview of the project in written and pictorial form (Fig. 10.1). The Background topic (not shown) documents the history of the ISIUOP, introduces project contributors (including researchers and local experts) and supporters, outlines general research methods used, and highlights our partner communities, along with providing important contact information. Much of the Welcome and Background material presented in this iteration was drawn from existing publications and communications documents. At this point, therefore, the content is too detailed and/or is not written with a general audience in mind. Future iterations of the Atlas will contain modified Welcome and Background content, and a few particular modules are described further below.

Our Partner Communities

Included in the Background topic is a series of content modules about the partner communities involved in the research (Fig. 10.2). An interactive map is used to access more detailed information about each community including textual

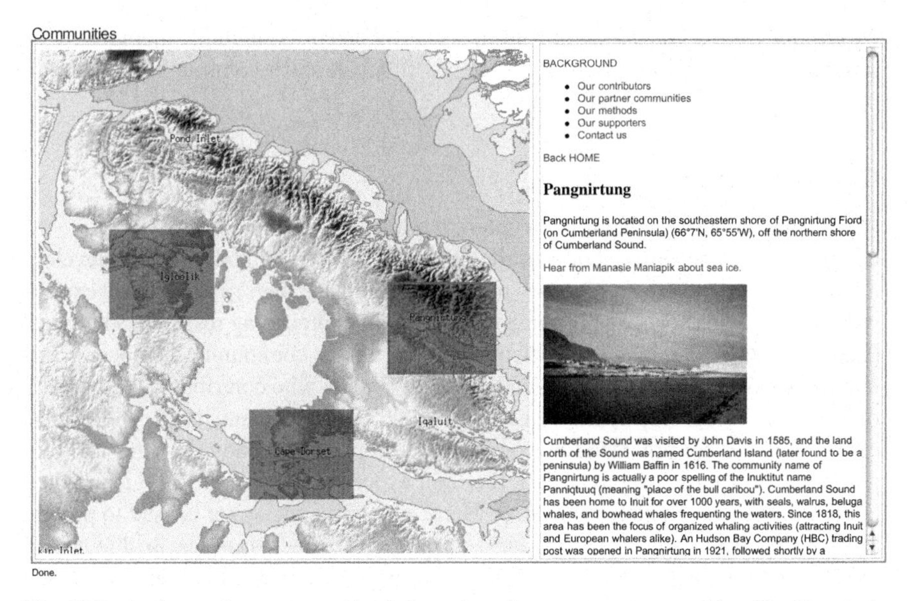

Fig. 10.2 An interactive map provides information about partner communities. The historical, geographical, and sociological aspects of the communities are described using text, photographs, sound, and potentially video (See also Color Plate 8 on page 477)

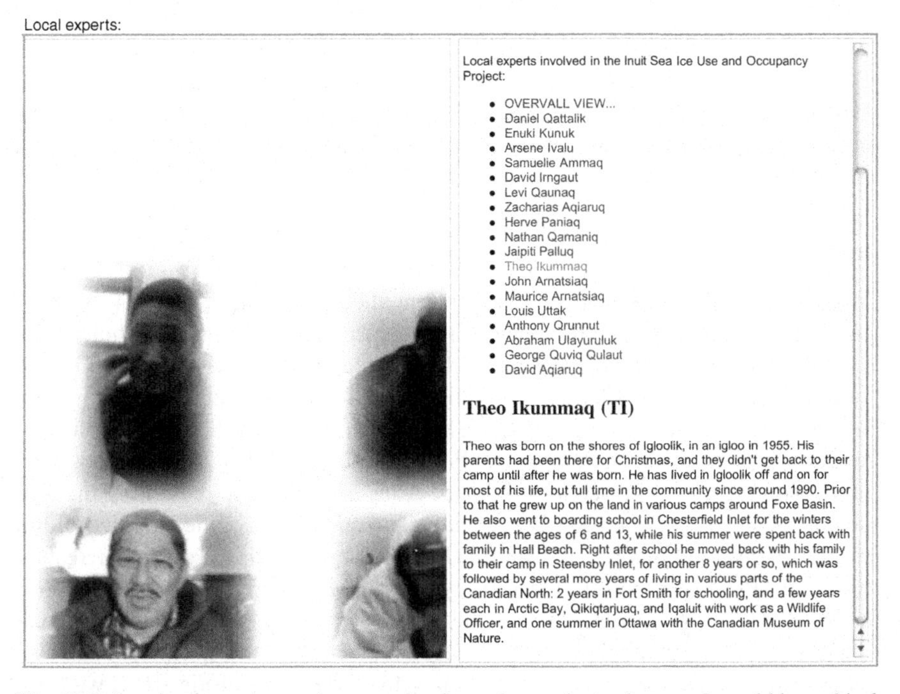

Fig. 10.3 Local experts. A mapping metaphor is used to navigate photographs and biographical information

description, photographs, and oral narratives provided by community members. This provides users with information that helps to situate each community geographically, historically, and socially.

Our Contributors

As part of the Background topic, content modules for each community provide the user with a virtual introduction to the contributors who helped with the development of content (whether university or community researchers) and the local experts who shared their knowledge through the ISIUOP and preceding research. The Local Experts (Fig. 10.3) module is particularly important as community members gravitate toward exploring the backgrounds of local experts who contributed their wealth of expertise to develop Atlas content, as much as the content itself. These local expert modules are organized by community and provide a photograph of and biographical information about each of the Inuit experts who consented to having their contributions represented in the Atlas. Here, a mapping metaphor is used to present the content. The photo "map" is used to provide interactive access to biographical information. Future iterations of the Atlas will see a closer integration of Local Experts information with modules that document ice conditions and travel.

Inuit Knowledge

The Inuit Knowledge topic is a collection of content modules focused on Ice Conditions, Uses, Hazards, and Terminology, all from the perspectives of Inuit experts who shared their knowledge with ISIUOP researchers. At the time of drafting this chapter, the Hazards modules were still under development and are not discussed here.

Ice Conditions

The Ice Conditions content modules present the bulk of the participatory mapping data collected through the ISUOP. The sea ice feature data are combined with data about the people who contributed the knowledge and other details about the interview process (Fig. 10.4). One could attempt to plot the data on a paper map, however, with more than 1,300 features documented across three geographically disparate communities, designing a useful map would be difficult if not impossible. In this situation, the interactive web-based Atlas is effective in presenting the data in more manageable "chunks." The menu system on the base of the interface allows users to reduce the number of features displayed by choosing a particular feature type (e.g., Floe Edge, Crack, Open [polynya], or Melt [early melt area]) and/or a knowledge contributor (e.g., based on the person's initials [also detailed in the Local Experts module]). By clicking on a map feature, the interactive query function also allows users to learn more about the feature. In the example in Fig. 10.4, we learn that the selected floe edge feature was established during an interview on May 6, 2004, with Mosesee Nuvaqiq in Pangnirtung. The feature information displayed in

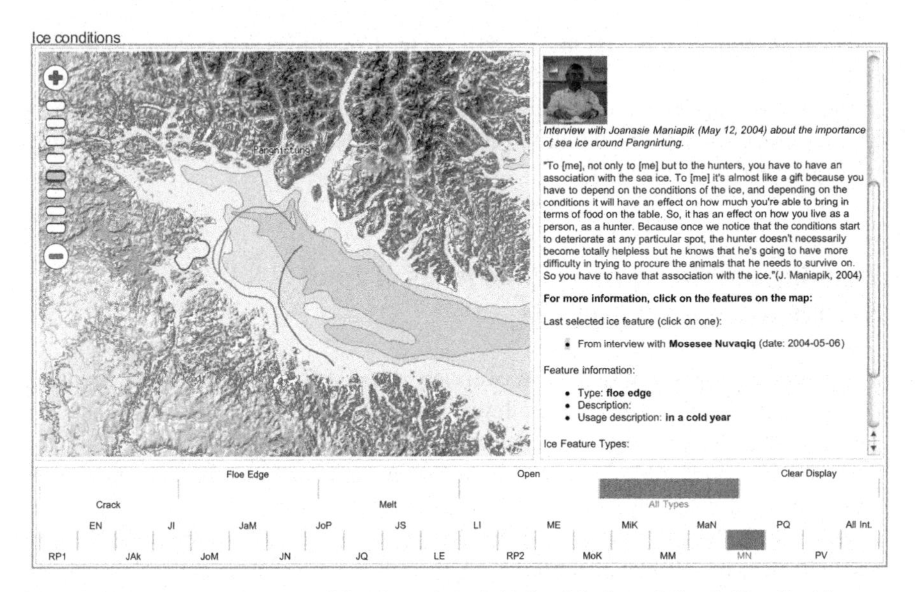

Fig. 10.4 Users select the type of feature and the initials of the knowledge holder (in this case MN [Mosesee Nuvaqiq]) using the selection bars. Clicking on a feature displays more information about that feature in the *right* frame of the window. This information may include multimedia content such as digital photographs, audio clips, or video clips. In this figure an audio clip (*top* of information box) is provided along with a photograph of the knowledge contributor (See also Color Plate 9 on page 477)

the right-hand text pane can be anything stored in the database (e.g., description, related dates, stories) or a value computed from database values (e.g., the duration of elapsed time as calculated from a start date and end date). This function also has the potential to display information through links to other Web sites or online data services. This allows for the creation of a "mashup" as previously discussed in this chapter.

At the time of this writing, significant amounts of multimedia content have yet to be associated with map features. The process of documenting these associations is underway, and future iterations of the Atlas will include more extensive multimedia content that will include audio recordings of interviews, videos of narratives, photographs of landscapes and seascapes, textual transcripts of interviews, illustrations, and visualizations. This will provide valuable context to both local expert contributions (e.g., descriptions of ice features or uses in the words of the individual, audio descriptions or video demonstrations) and the ice feature depictions (e.g., pictures and video to represent "on-ice" conditions).

Uses

The Uses sub-topic modules present additional data related to sea ice usage collected through the ISIUOP. Like the Ice Conditions content modules, the user can access data related to ice use in more manageable chunks (i.e., by travel route [dog team, snowmobile, or boat] and/or established camp [traditional or contemporary]). The interactivity model draws on that used in the Ice Conditions component of the Atlas in that a user can zoom-in and out and pan across the map, filter by type of feature (travel route or camp), and then interact with the map by clicking to find out more about a given feature (including the contributor, relevant dates, and particular uses or importance to contributor).

Building a Cybercartographic Atlas of Inuit Sea Ice Knowledge and Use

The preceding section presented an overview of the Atlas content and functionality as experienced at the user interface level. Underlying the Atlas user experience is a series of three iterative and interrelated development steps: (i) Data Collection and Conversion; (ii) Data Modeling and Publication; and, (iii) Atlas Implementation. A relatively brief discussion of each phase is presented here. While every attempt is made to limit the use of jargon, some topics require the use of specialized terms and will thus be defined accordingly. Although we attempt to make the material accessible to a broad audience, there are some sections that are necessarily technical in nature. These details are retained to provide a comprehensive overview of the Atlas and to serve interested readers.

Data Collection and Conversion

The ultimate objective of the Atlas is to build a system to facilitate documentation, representation, access, and preservation of the information recorded about sea ice knowledge and use. Knowledge is constructed from observations, use, and experience of the environment. However, the documentation process results in a form of data that attempt to represent this knowledge, while the actual knowledge remains within the minds of the individual contributors. We use the term "data" throughout this chapter for lack of a better descriptor of documented information, but in using it we recognize that we are documenting more than simple observations and measurements typically referred to as data. Furthermore, while the data collected provide us with only a partial representation of the knowledge held by community members, it does provide an important link from the knowledge-holder, to representations of their documented knowledge, and importantly, to the end user of the Atlas. The following section provides a description and discussion of how the information was collected for use in the Atlas.

Participatory Mapping and Maps

Information was collected through interviews, focus groups, and sea ice trips in the communities of Cape Dorset, Igloolik, and Pangnirtung, Nunavut, Canada (see Chapter 3 by Laidler et al. this volume). This qualitative method was augmented

Fig. 10.5 Eric Joamie and Gita Laidler work with Pangnirtung Elders and hunters to verify sea ice features mapped as part of earlier work that contributed to ISIUOP and the Atlas (May 2005; Photo: Ame Papatsie)

by the use of a participatory mapping approach where paper topographic maps of the research participant's area of experience were used to facilitate discussion. Incorporating a mapping element in interviews and focus groups (see Fig. 10.5) had a number of advantages beyond the goal of documenting/representing Inuit knowledge of sea ice processes, use, and change, such as

1. the visual element provided by incorporating maps proved to be valuable in triggering memories of sea ice conditions or travels, stories of sea ice travel or hunting, and descriptions of ice conditions or terminology;
2. contributors frequently incorporated the map into their descriptions to help focus on a specific geographic area, explain a particular ice condition or localized process, and/or demonstrate stories or specific travel routes; and,
3. the use of maps as a visual aid helped to improve communication between the interviewer and the contributor by providing a useful discussion piece that helped to emphasize interactive conversation and minimize the tensions associated with more structured question and answer sessions.

Therefore, the incorporation of maps, as initiated by the contributor, was effective in leading to richer interview responses, as well as more ice features/ice uses/indicators of change actually being drawn on the maps. Thus, the participatory mapping component provided a medium through which people were able to bridge abstract descriptions with their experiences and knowledge of the complexity of the sea ice environment.

While recognizing the advantages of using maps as part of the interview process, there are also some important practical considerations to highlight when attempting to incorporate this form of local and regional landscape/ocean expanse representations around each community, including issues of map scale, and coordinate system and projection.

Map scale, coverage, and size are important considerations depending on the features of interest and the requirements for large- or small-scale coverage. We used a 1:250,000 map scale as a viable middle ground between the coarse-scale maps employed in regional land use and occupancy studies (e.g., 1:500,000 in Freeman (1976)) and fine-scale maps employed in some place-name and local harvesting studies (e.g., 1:50,000 in Brice-Bennett (1977)). It should be also noted that 1:250,000 maps are normally used by hunters in everyday situations or while traveling. While this medium scale provided a flexible compromise, this scale lacked the detail necessary to indicate fine-scale features (e.g., tidal cracks, travel routes) in some instances, or in other cases, did not provide adequate coverage of the full extent of areas that people had covered in their travels, or that they wished to describe, in relation to ice conditions or uses (e.g., travel routes, hunting/harvesting destinations). Practically speaking, it would not have been feasible to create large composites to increase coverage as map sheets already exceeded most regular table sizes (i.e., would have been unwieldy in small interview spaces). Thus, there is an inevitable tradeoff between map detail, coverage, availability, and size that must be considered when establishing the interview methodology.

The topographic map sheets used are projected and referenced using the common Universal Transverse Mercator (UTM) coordinate reference system. This system comprises a series of zones that prevents the simple mosaicking of map sheets where an area of interest is large enough to cross between UTM zones. Because many areas of interest crossed UTM zones, map sheets had to be digitized separately, followed by data re-projection into a common coordinate system (typically a Lambert Conformal Conic projection), and subsequent editing to join features at the zone boundaries. These necessary steps did not alter the data itself, but it can be a challenge to ensure map accuracy and precision after several rounds of data processing.

Conversion of Mapped Data

Although paper maps were used in the participatory mapping process, the maps typically acted as a reference base mounted below a sheet of transparent or translucent polyester film. The documented data were transcribed on the polyester film to reduce the cost of map production while ensuring that each contributor had their own blank map and dedicated recording of their individual-mapped features to work with. Once complete, the polyester film sheets were scanned using a large format scanner. Using desktop mapping software, the image files resulting from the scanning process were registered to the original coordinate system (UTM in this case). The scans were then traced using a heads-up digitizing process to produce discrete vector representations of points, lines, and polygons. The vector data were stored in ESRI Shapefile format, a popular GIS data format that links geographic location with a table of attribute values. In addition to being simple and efficient, this method has the added benefit that the scanned polyester film sheets and the subsequent vectors act as a backup of the original film sheets. This facilitates long-term preservation of the original data. The Shapefiles resulting from the digitizing process became the source data for construction of the ISIUOP database.

GPS Data

GPS receivers were used to record travel routes and key features/places during sea ice trips. These receivers establish the position of the handheld unit to an accuracy of approximately 10 m or fewer. The GPS units used were capable of storing trip log points that documented the route traveled. Points of interest were also collected as way-points. Date and time information was stored in the log along with positional information. The date and time information is important as it allows researchers and community members to associate a logged location with other time-stamped material such as digital photographs, digital audio files, and video footage. The GPS data were downloaded from the units in the GPS Exchange Format (GPX). The GPX data were then converted to a format that could be used in the ISIUOP database.

Audio Recording, Video Recording, Photographs

In addition to producing map data, the community-based interviews, focus groups, and sea ice trips resulted in the production of other data in the form of audio recordings, video recordings, and photographs. Audio recordings were transcribed to produce a textual record of the contributed knowledge.

This digital media provided an additional way to include more visual references for context, as well as different means of learning, interpreting, and interacting with the knowledge and observations shared as part of this project. The incorporation of such diverse media could be considered more in tune with Inuit oral culture and the emphasis on learning through watching (and doing). Animations and the ease of updating digital map representations are also helpful as a means of overcoming some of the limitations of hard copy maps and their static representations. There is also strong local interest in moving more toward multimedia representations of Inuit knowledge, for potential use in educational settings for youth (Inuit or not). This is a different form of conceptual mapping if you will, necessitating the description and application of complex relations between varying media types (including maps) to tell the interrelated parts of "the story" with the differing representations most suited to particular aspects (i.e., visual, audio, video, interactive, spatial) (Aporta 2009).

In keeping with the cybercartographic development framework, the use of multimedia representation plays an important role in the ongoing development of the

Atlas. Media elements are typically, but not always, organized by location through association with map features. As such, they are an integral element of the data model discussed in the next section.

Data Modeling and Publishing

Data Modeling Concepts, Design, and Challenges

We have previously described how map data were converted to a GIS-compatible format (i.e., Shapefiles). For some applications, the Shapefile format is perfectly adequate as the production level format for spatial data. However, we have already highlighted the complex nature of Inuit sea ice knowledge and use, whereby the mapped data are only one component of efforts to represent a constellation of related data. Thus, using a simple file format like the Shapefile was not sufficient for our needs. The following sections outline the approach taken to address the representational requirements of the Atlas.

As part of the Atlas development process, map data collected in the communities were converted from the digitized Shapefiles to a centralized database. While Shapefiles and similar file-based geographic information formats are simple and powerful due to their capacity to associate a geographic location (as a point, line, or polygon) and the information attributes used to describe the location (e.g., place name, elevation, type of use), these formats are best suited for representing one-to-one relationships (Fig. 10.6).

Early in the Atlas design process, it quickly became clear that the different aspects of sea ice knowledge and use data and associated media involved complex relationships that could not adequately be described through one-to-one relationships (Fig. 10.7). For example, one location may have more than one place name associated with it, resulting in a one-to-many relationship. Conversely, a given sea ice feature may be associated with more than one knowledge contributor, resulting in a many-to-one relationship. In the case of organizing multimedia content, this was evident in the situations that commonly arose where a photograph or video was related to several sea ice features (e.g., one icescape photo) while at the same time, a single sea ice feature could be related to many photographs (e.g., multiple photos of the same feature). This situation represents a many-to-many relationship. While these complex relationships can be managed using simple file-based formats, doing so requires duplication of data and potentially irregular data structures that result in increased maintenance effort and a significant increase in the risk of data entry errors and corruption.

To adequately represent the data collected, a relational data model was developed and implemented using a relational database management system that supports the representation and processing of geographically referenced information. An overview of the current model is presented in Fig. 10.8 as a simplified view of the database comprising a set of data packages that may contain one or more related tables. In this model, a set of data classes was created and established in the database

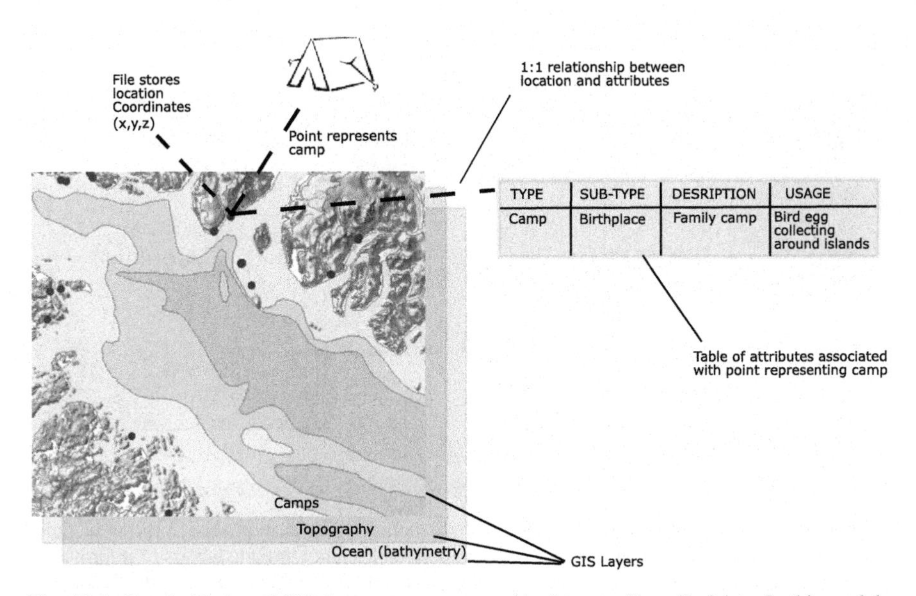

Fig. 10.6 Simple file-based GIS formats store geographic data as a "layer" of data. In this model, a one-to-one relationship is represented, where one sea ice usage feature is associated with one set of attributes called a record. This model does not readily support one to many relationships (e.g., one feature associated with many records)

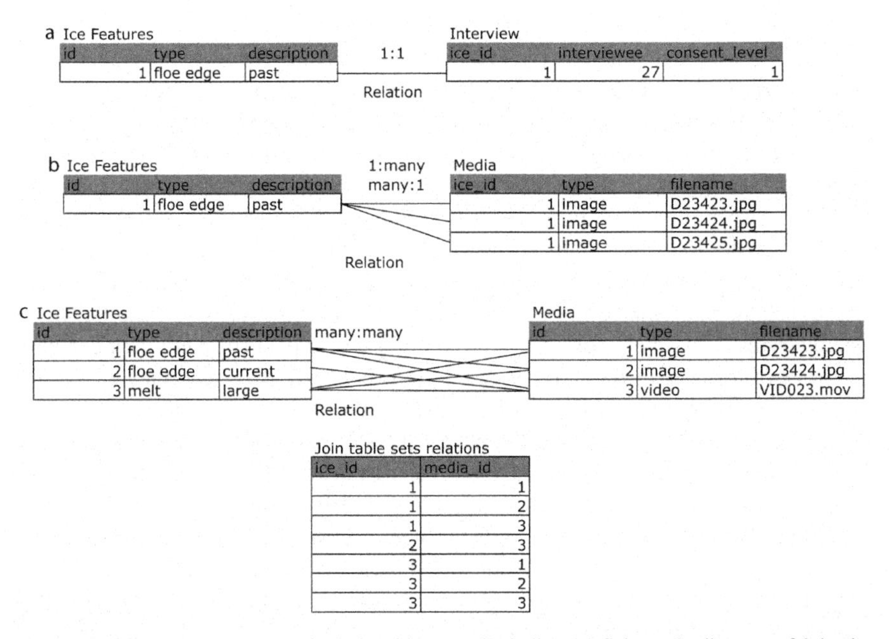

Fig. 10.7 Schematic depiction of relationship types including (a) Schematic diagram of 1:1 relationship; (b) Schematic diagram 1:many relationship and many:1 relationships; and (c) Schematic diagram many:many relationships

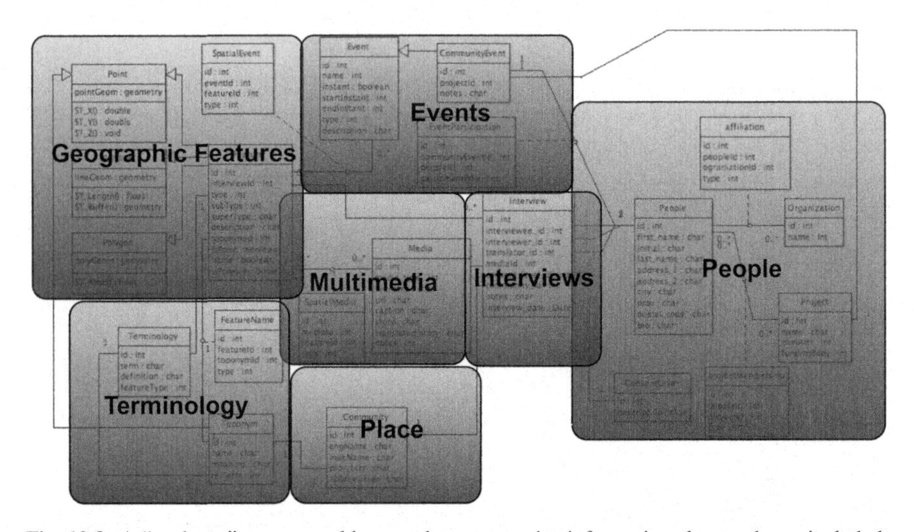

Fig. 10.8 A "package," represented by *gray* boxes, contains information about a theme included in the model. Each package may contain one or more tables that represent a class of information. Packages can be associated through the establishment of relations between tables in the package. In the ISIUOP model, the Interviews package is used to associate many other packages (Reproduced with the permission of the Geomatics and Cartographic Research Centre, Carleton University.)

as tables of rows and columns. For example, People, Interviews, Events, and Media were established as data tables. The People table contains data columns such as first name, last name, community membership that describe people involved in the project (e.g., knowledge contributors, researchers, translators). The Interviews table contains data about an interview such as date, location, people involved, and their respective roles. As previously explained, different types of relationships can be used to associate tables. Thus, the People table can be associated with the Interviews table to indicate, for example, that two contributors, two researchers, and one translator participated in one particular interview (one-to-many relationship – one interview to many people). Developing this sort of data structure then becomes the basis of forming different types of relations between tables and their respective attributes to support the depiction of complex descriptions (i.e., stories).

The ability to represent complex relationships is also useful for representing the landscape and icescape as dynamic and evolving phenomena. For example, the type of sea ice use may be different from season to season. Using the model described, more than one use type can be associated with a location. So, in conjunction with stored temporal information, the change of usage over time can be derived and displayed. Similarly, the model supports the association of multiple geometries with a feature. A given feature may be perceived as a line or a polygon or both, depending on the context (e.g., state of feature, scale of representation). The model used allows the feature to be represented as either a polygon or a line or both.

Although the database structure used is relatively complex and took significant effort over several iterations to develop, Atlas developers and end users do not necessarily see the underlying complexity. Using the established relationships, a data "view" is created. This view presents a selection of records and combination of tables as needed for application development and the desired representations of stories. For example, although geographic features, people, and interviews are separate tables, a view is used to create a tabular representation that combines mapped feature data with information about the interview within which the data were collected, and who contributed the information. Although the view may contain duplicate information (e.g., the same researcher involved in many interviewers), the underlying source maintains only one record of the data, which helps to maximize efficiency and data integrity. Because the database is designed to work with geographic information, this view can be used to create a visual representation in the form of a map.

At the time of drafting this chapter, the ISIUOP data model was one of the most significant results and time-consuming efforts of this particular sub-project. The model is still being refined to ensure stability, however, like in all sub-projects the model development is iterative and will change and adapt to meet emerging needs.

Data Publication

Raw data (i.e., not a "view") in the ISIUOP database can only be accessed via the local network or Internet by trusted users through a number of database management tools including command line interfaces, graphical tools, and GIS applications (e.g., QGIS www.qgis.org). This type of access requires that the user have an account on the system, and available functions (e.g., deleting records) are limited by user group with various levels of permissions. Data accessed over the Internet are encrypted for security so that in the unlikely event that a data stream is intercepted en-route, the contents of the data would not be compromised. This ensures that, in part, the confidentiality conditions stipulated in the research consent form are met.

Under the terms of consent to participate in the research project, much of the original contributions shared can be made public.[2] To do this, a more interoperable method of data publication is used in the form of Web Services. A user or application can make a request to a service and the service will provide a suitable response. For example, a user requests a map constructed from several layers, covering a particular geographic area, presented using a particular color scheme, and referenced in a particular map projection. This request is made to a Web Map Service (WMS) that provides a graphic map image (e.g., JPG) in response. If the user needs data in raw tabular form, a Web Feature Service (WFS) is used where the consent constraints established in the underlying database are met. Alternatively, data that are best represented as regularly spaced grids such as Digital Elevation Models can be retrieved using a Web Coverage Service (WCS).

Figure 10.9 provides a schematic depiction of the Atlas architecture, including a representation of the data publisher layer. These data publication services are based on open standards and support interoperability by providing data access through many different tools including modern GIS, desktop mapping programs,

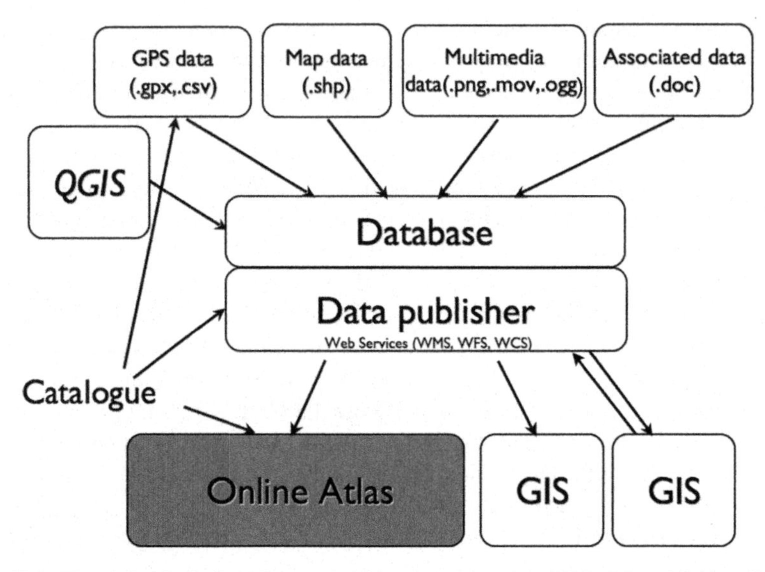

Fig. 10.9 The Atlas is built using a multi-level architecture. GPS, Map, Multimedia, and Associated data are loaded into the Database. Only trusted users can access the Database directly using appropriate tools (e.g., QGIS). To simplify the data structure, views are created and published using the Data publisher layer. Published data can be accessed using a variety of tools including atlas development frameworks such Nunaliit, resulting in an Online Atlas. A catalogue is used to document various aspects of the system (Reproduced with the permission of the Geomatics and Cartographic Research Centre, Carleton University.)

and application development toolkits such as the Nunaliit framework (see Atlas Implementation section).

In addition to promoting interoperability, the use of Web Services results in very practical benefits for the Atlas. Other organizations are also publishing their data using these services. This eliminates the need for our team to download, style, and host tens of digital 1:250,000 basemaps. Instead, we connect to a Web Map Service hosted by Natural Resources Canada that provides topographic basemap layers seamlessly and "on the fly" over the Internet. Using this service, we create customized Atlas basemaps using the published data layers (e.g., water courses, contours, shaded relief).

Atlas Implementation

The Atlas has been implemented as a Web-based application. Due to the complex programming typically associated with creating a cybercartographic Atlas, implementation may require significant programming resources. While this was the case for this Atlas, the level of effort was significantly reduced by using the Nunaliit (www.nunaliit.org)

The Nunaliit framework aims to make it easier for developers and researchers to build a cybercartographic Atlas. Currently, Nunaliit uses a markup language for organizing and connecting content into a meaningful state, and a compiler program to render that information out to an interactive Web interface. Although the authors of the Atlas do need to be comfortable with using a markup language, the level of knowledge and experience required to develop an Atlas is significantly less than that needed to develop a similar Web-based application using typical Web programming languages (e.g., JavaScript). A detailed description of Nunaliit can be found at the cited Web site and in previous publications (Pulsifer et al. 2008). For our purposes here, it is sufficient to point out the features of Nunaliit that made it suitable for developing this iteration of the Atlas. In this regard, the key features of Nunaliit include support for:

- linking of geographic features with other features and multimedia objects (including text);
- timelines and other alternative types of menus;
- a rich set of features related to the use of sound;
- producing a dynamic and interactive browser model – interacting with one interface element (a piece of text for example) will trigger a behavior in another element;
- connection to standard publication services such as Web Map Service and Web Feature Service;
- different types of deployment. One method is used to publish the Atlas to a Web server, while another is used to produce a local copy of the Atlas that is suitable for use from a data key or a CD ROM

Figure 10.10 outlines a typical Nunaliit workflow. It is important to note that Nunaliit is neither a Geographic Information System nor a database management system. If data are to be used in Nunaliit, they must already be published in a suitable form (e.g., Shapefile, Web Services), and thus the Atlas accesses the ISIUOP database views through a published Web Service rather than directly. Nunaliit acts as a mediator software that compiles data resources and established relationships into a particular end product – a Web application in the case of the Atlas. Thus, Nunaliit was used to create the interactive, Web-based component of the Atlas. In Fig. 10.9, the application produced by Nunaliit is represented by the box-labeled "Online Atlas." This Web application accesses the aforementioned data publication services. These publication services in turn draw information from the ISIUOP database, which is constructed from the information resources collected during the community-based research process. Although our primary use of the supporting infrastructure illustrated in Fig. 10.9 is currently the Atlas, the system is quite flexible and can provide layers of information, multimedia content, etc., to other systems such as desktop mapping or GIS software. Thus, the publication approach used supports data sharing outside of the Atlas context, as well as supporting applications such as large format printed map production.

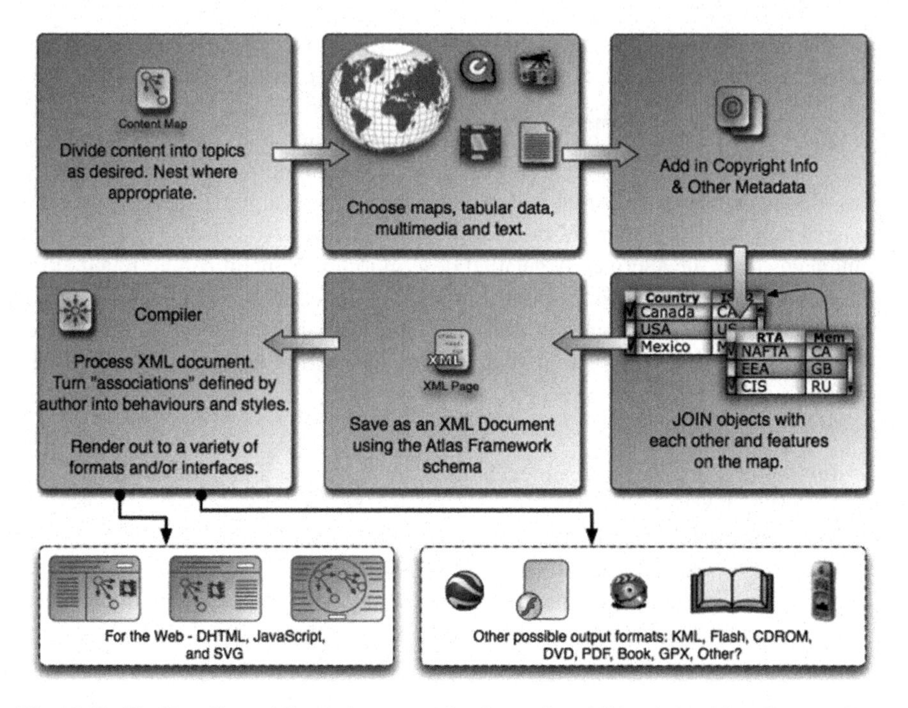

Fig. 10.10 The Nunaliit workflow helps to organize data and establish relationships. The compiler is then used to generate a Web-based atlas or other potential outputs (some currently available and others remain under development) (Reproduced with the permission of the Geomatics and Cartographic Research Centre, Carleton University.)

Effectively managing source data, databases, and publication services with interoperability and preservation in mind means that we must document the information contained in the system. A Web-based catalogue has been established for this purpose. This catalogue uses standard documentation protocols to record key information about the system. This information can then be used by community members, researchers, application developers, members of the science community, the general public, and others to discover the information that exists in the Atlas and underlying database.

Future Work and Conclusion

In addition to the intended developments stated in previous sections, there are a number of future developments already underway or in early design phases. Two are discussed here. First, we have focused on the addition of new content to populate an expanded information structure established in the second iteration of Atlas development. Figure 10.11 is a visualization of the table of contents for the Atlas at the time of writing. New content modules (as indicated in the diagram) will include

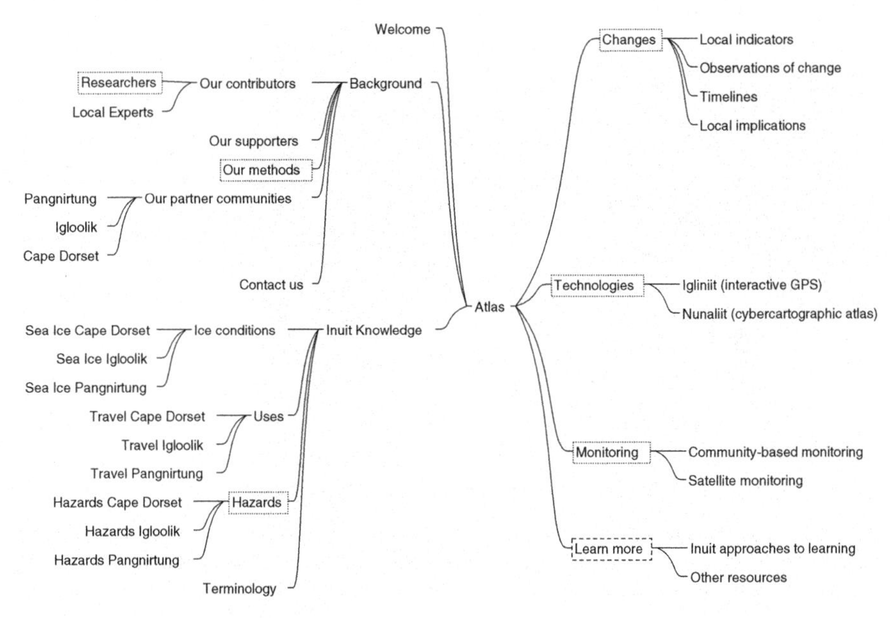

Fig. 10.11 Second-generation table of contents for the Atlas represented by a tree diagram. This information structure uses first-generation content within a significantly revised hierarchy and organizational scheme. *Dotted boxes* indicate new content modules that have been added in the second iteration of Atlas development. The structure better reflects the overall organization of the ISIUOP and the SIKU projects

results of other ISIUOP sub-projects. Content from the first-generation Atlas is still being used, however, where content is sparse, new material is being added. The primary focus of this effort is the addition of new multimedia content.

Second, the most significant development being considered is the implementation of a Web 2.0 application approach that supports user-contributed content. Through other projects, the Geomatics and Cartographic Research Centre has extended the functionality of existing open source Web mapping tools to support user-contributed content. Current applications (Fig. 10.12) use Google™ satellite basemapping as a backdrop (although other Web Mapping Services could be used). It is currently applied within the Kitikmeot Place Name Atlas (atlas.kitikmeotheritage.ca), whereby a user can register and then log on to the system. Once logged on, the user can add his/her own spatial information by providing a feature (point, line, polygon) on the map as reference. Once the location is established, a form is presented that allows the user to enter data about the location. Once the form is saved, the new location is plotted on the map. The contributor or other users can then comment on the feature, including the addition of media (sound, photo, video). Comments are displayed below the attributes provided by the original contributor (Fig. 10.12).

We aim to adapt this technology for use in the sea ice Atlas. We are currently considering the technical, structural, and cost implications of adopting this new

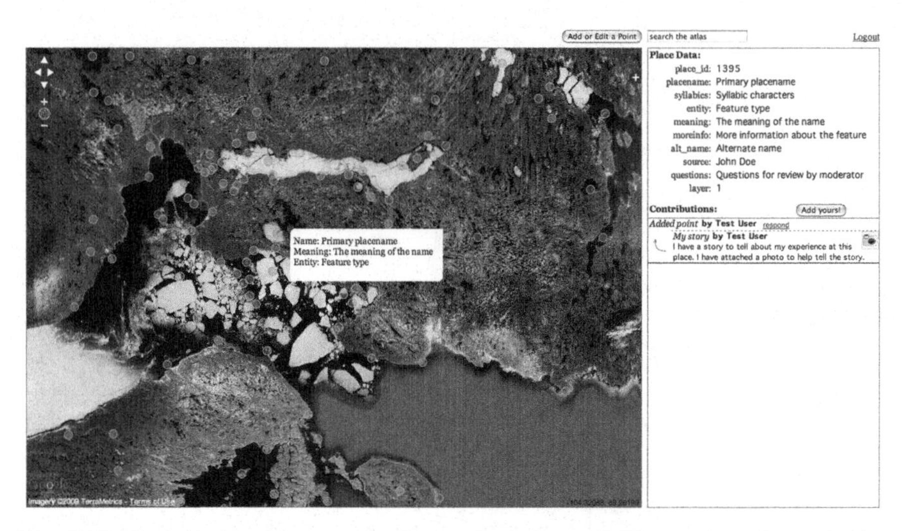

Fig. 10.12 User-contributed content. The displayed point was added by a registered user. In this atlas, anyone may contribute comments and/or multi-media to existing features as a "Guest" user. Registered users may also add, edit, move, or delete features on a "user-contributed" layer of the map. Administrators may transfer user-contributed features to other layers (e.g., moderated layers with more restrictive security policies.) The security framework is flexible and supports layer and attribute level restrictions by user groups

technology within a broader geographic context and within a sea ice framework (i.e., would need ability to represent feature dynamics and support additional associations between people, events, features, media, etc.). This chapter presented a high-level overview of the Cybercartographic Atlas of Inuit Sea Ice Knowledge and Use being developed as part of the ISIUOP. The Atlas development process is built on the visions and objectives of Inuit community members around knowledge preservation and sharing, and establishing equitable consideration of their knowledge in current affairs. Community-based visions, objectives, and documented forms of Inuit knowledge are an integral part of the Atlas development process, however, as we move forward, it is clear that great care must be taken when representing Inuit knowledge using emerging representation technologies, methods, and theories. Participatory mapping supplemented by GIS and Web-based mapping techniques, combined with the use of multimedia and user-contributed content mechanisms, present a powerful approach to representing the complex and nuanced knowledge held by Inuit. To take full advantage of the opportunities afforded by new technologies, methods, and theories, requires a full understanding of Inuit sea ice knowledge and use. Thus, partnership with communities is, and will continue to be, a central component of the ISIUOP and its successor projects. The Cybercartographic Atlas of Inuit Sea Ice Knowledge and Use will be developed in partnership with Inuit communities to support the goals of the community members and researchers participating in the partnership.

Acknowledgments The authors are grateful for the support and collaboration of community members in Cape Dorset, Igloolik, and Pangnirtung who have shared their knowledge with us and who are willing to share it with the world. We gratefully acknowledge the support of the Government of Canada International Polar Year Programme, the Social Sciences and Humanities Research Council of Canada, and the Natural Sciences and Engineering Research Council of Canada. Specific thanks is also extended to Glenn Brauen, Sébastién Caquard, Christine Homuth, Kelly Karpala, and Karen Kelley for their project and data management assistance. A special thanks to the book and section reviewers for their constructive comments, which led to considerable paper improvements.

Notes

1. Web 2.0 is the term used to describe current movements in Web application development that are characterized by Web-based communities who use a variety of tools (wikis, blogs, social networking sites, etc.) to facilitate relationship-building and/or generate content.
2. Using the view methodology previously described, selection criteria embedded in the database prevent data that lack necessary consent from being exposed by the public Web Services.

References

Aporta, C. 2009. The trail as home: Inuit and their pan-arctic network of routes. *Human Ecology* 37: 131–146.

Brice-Bennett, C. (ed.) 1977. *Our footprints are everywhere: Inuit land use and occupancy in Labrador*. Nain, Labrador: Labrador Inuit Association.

Cartwright W., Peterson M.P., and Gartner G. (eds.) 1999. *Multimedia Cartography*. Berlin: Springer.

Cartwright W., Peterson M.P., Gartner G. (eds.) 2007. *Multimedia Cartography*. Berlin; Heidelberg: Springer.

Freeman, M.M.R. 1976. Inuit Land Use and Occupancy Project (3 volumes). Ottawa: Department of Indian and Northern Affairs.

Goodchild M., Egenhofer M., Fegeas R., Kottman C. (eds.) 1999. *Interoperating Geographic Information Systems*. Norwell: Kluwer Academic Publishers.

Graham, I. 2003. *A Pattern Language for Web Usability*. Toronto: Addison-Wesley.

International Fund for Agricultural Development (IFAD) 2009. *Good Practices in Participatory Mapping*. Rome: IFAD.

Kresse, W. and Fadaie, K. 2004. *ISO Standards for Geographic Information*. Berlin: Springer.

Laidler, G.J. 2007. Ice, through Inuit eyes: Characterizing the importance of sea ice processes, use, and change around three Nunavut communities. Dissertation, University of Toronto

Lauriault, T.P. and Taylor, D.R.F. 2005. Cybercartography and the new economy: Collaborative research in action. In *Cybercartography: Theory and Practice*. D.R.F. Taylor (ed.), Amsterdam: Elsevier.

Peterson, M.P. (ed.) 2008. *International Perspectives on Maps and the Internet*. Berlin: Springer.

Pulsifer, P.L., Hayes, A., Fiset, J.P., and Taylor, D.R.F. 2008. An open source development framework in support of cartographic integration. In *International Perspectives on Maps and the Internet*. M. Peterson, (ed.), Berlin: Springer, pp. 165–185.

Pulsifer, P.L. 2008. An ontological exploration of Antarctic environmental governance: Towards a model for geographic information mediation. Dissertation, Carleton University

Skupin, A. and Fabrikant, S.I. 2003. Spatialization methods: A cartographic research agenda for non-geographic information visualization. *Cartography And Geographic Information Science* 30(2): 99–115.

Taylor, D.R.F. 2003. The concept of cybercartography. In *Maps and the Internet*. M. Peterson, (ed.), Cambridge: Elsevier.

Taylor, D.R.F. 2005. Introduction: The theory and practice of cybercartography. In *Cybercartography: Theory and Practice*. D.R.F. Taylor (ed.), Amsterdam: Elsevier.

Taylor D.R.F. and Caquard S. (eds.) 2006. Cartographica Special Issue On Cybercartography 41(1): 1–5.

Part III
Learning, Knowing, and Preserving the Knowledge

Chapter 11
The Power of Multiple Perspectives: Behind the Scenes of the Siku–Inuit–Hila Project

Henry P. Huntington, Shari Gearheard, and Lene Kielsen Holm

Abstract The Siku–Inuit–Hila (Sea ice–people–weather) project presents a new approach for collaborative research in the Arctic that links Inuit and scientific knowledge. For perhaps the first time, Inuit have undertaken comparative environmental research in a formal structure: not only comparative across Inuit knowledge and science but also comparative across time and place. By involving local research team members in community knowledge exchanges, we blurred the distinctions between "researchers" and "participants," giving each team member a variety of roles during the project, including host, visitor, teacher, and student. The exchanges were complemented by quantitative sea ice measurements taken from specially designed local monitoring stations and information gathered during regular sea ice expert group meetings held in each community. Our experiences illustrate that this approach to collaborative research can yield new insights into sea ice processes, changes, and impacts at the local and regional scales.

Keywords Sea ice · Inuit · Traditional knowledge · Collaborative research · Research methods · Nunavut · Alaska · Greenland

Introduction

The involvement of, and indeed reliance on, local residents in field studies has a long and rich history in both the social sciences and the natural sciences. From the first visitor who asked a local, "what is that?" travelers and researchers have long depended on local knowledge and expertise for navigation, safety, and at least an introduction to the flora, fauna, and physical environment of a region (e.g., Simpson 1855; Amundsen 1908; Irving 1976). Giving credit to local, often indigenous, knowledge and working in partnership with local people has generally

---

H.P. Huntington (✉)
Pew Environment Group, Eagle River, AK 99577, USA
e-mail: hhuntington@pewtrusts.org

I. Krupnik et al. (eds.), *SIKU: Knowing Our Ice*,
DOI 10.1007/978-90-481-8587-0_11,

come more slowly, but as many recent studies and publications – not least those in this volume – demonstrate, a recognition of the value of local expertise is blossoming in many ways (e.g., Salenave 1994; Brewster 1997; Wolfe et al. 2007; Pearce et al. 2009). Such expertise is also increasingly recognized as a source of valuable insight and interpretation, and not simply as raw material for the academically trained to analyze (Huntington et al. 2004).

The Siku–Inuit–Hila[1] ("Sea ice–people–weather") project extends this movement in Arctic research, engaging local residents not only as guides and explicators of their own areas but as co-researchers studying other locations where they, too, are visitors. Bringing together a team of hunters, elders, and researchers, Siku–Inuit–Hila set out to investigate sea ice and sea ice use in three Arctic communities, where the sea ice acted as the common denominator for a wide range of experience and knowledge. For the Inuit, Iñupiat, and Inughuit (Clyde River, Barrow, and Qaanaaq) members of our team we sought not only insight into how they see the sea ice environment in which they travel, hunt, and live, but also how they apply their understanding when they visit other locales where conditions are both familiar and different. In addition, discussion among the local team members allowed them to raise topics they thought important, rather than only to respond to the interests or ideas of the academically trained members of the group. Our team is currently working together on a book about our research, containing the results of our work and aimed at a broad audience. In this chapter, writing for an audience of research practitioners, we focus mainly on the local partnership component of our efforts: *how* we organized and conducted the project, *what* we learned from this approach, and *why* we think it is a fruitful approach for certain research goals.

How the Project Came Together

The Siku–Inuit–Hila project is a successor to an earlier project, funded by the U.S. National Science Foundation and conducted in Clyde River, Nunavut, and Barrow, Alaska, in the spring of 2004 (Gearheard et al. 2006). The pilot study focused on creating a small team of researchers and Inuit sea ice experts to study sea ice and sea ice use at the two communities. A central component was the exchange of hunters from the two locations, testing the idea that such an exchange could lead to new insights into sea ice conditions and human use of sea ice by comparing Inuit and Iñupiat perspectives based on first-hand experience in both locales (see Fig. 11.1). The results included specific observations and comparisons made by the hunters (e.g., differences in sea ice dynamics and processes that create rough conditions in Barrow and relatively smooth conditions in Clyde River, along with local adaptations to using each environment) and a more general confirmation that the hunter-exchange approach was a useful way to document knowledge and create strong partnerships. All team members felt that a more detailed and thorough study, using the same approach in conjunction with other methods, would be a huge benefit to learning more about sea ice and sea ice use at the community scale.

Fig. 11.1 Researchers (from *left*) Ilkoo Angutikjuak, Joe Leavitt, Shari Gearheard, Craig George, and Geela Tigullaraq spend time learning about sea ice off Barrow, Alaska, in 2004 (Photo: Henry Huntington)

Accordingly, we proposed the larger study to the Human and Social Dynamics program at the U.S. National Science Foundation in 2005 and were awarded a grant in 2006. We were thus slightly in advance of the International Polar Year (IPY) 2007–2008 program planning, but became an "affiliated initiative" of SIKU, and thus were connected to the overall IPY effort. Our NSF funding takes Siku–Inuit–Hila into 2010 and we received additional support from Health Canada for preparation of our book and related materials.

In planning the larger study that was to become Siku–Inuit–Hila, we sought to build on the successes of the earlier study by retaining the community exchanges as the central piece. We expanded by adding Qaanaaq[2] in Northern Greenland, to Barrow and Clyde River as host communities and research partners, and by including two new components. First, we worked with our local partners to establish a "local sea ice experts working group" in each of the three communities. These working groups organized sea ice trips and other activities when the rest of the team was visiting. In addition, the groups met regularly throughout the sea ice seasons (normally once per month) to discuss and document sea ice conditions and changes, and work on activities such as mapping. The working groups often invite other local experts to participate in the meetings and share their knowledge, and the meetings provide a mechanism for the local team members to contribute to our book by documenting each meeting and collecting writing contributions that individual members worked on at home. Second, we instituted a program of sea ice measurements to gather quantitative, localized data in each area. These measurements, including sea ice and snow thickness and sea ice and snow temperature, are taken by trained local observers and complement the qualitative observations gathered by the expert working groups.

Logistically, the community exchanges (Table 11.1) were by far the most complicated part of the project. Travel of course faced the usual complications of Arctic weather during spring, the season in which our team wanted to visit each place. In addition, Arctic travel routes tend to go north and south, not east and west, so the route to get from Qaanaaq to Clyde River (which are within range of each other on the VHF radio) goes from Qaanaaq to Thule to Baltimore to Ottawa to Iqaluit to Clyde River. That routing was in fact a blessing as our National Science Foundation funding allowed us to travel with the military directly to the U.S., otherwise an even more circuitous route via Copenhagen would be involved. Of course all the international travel meant that passports were required. In Barrow, an electronic sign outside City Hall advertises passport services, so the procedure there was relatively straightforward. In Clyde River, however, several team members did not have basic documentation such as a birth certificate, leading to various paperwork and bureaucratic obstacles. Even obtaining an acceptable passport photo can be difficult in a remote village and it took many months to ensure all the necessary identification was in place, in particular for Qaanaaq and Clyde participants (during the 2004 study, passports were not yet required for travel between the U.S. and Canada, but Ilkoo Angutikjuak, an elder from Clyde, found that returning to Canada with only his firearms permit was not so simple and had to answer a number of questions at the border.)

Table 11.1 Participant list and dates of community sea ice knowledge exchanges

Community	Dates	Visiting team members
Qaanaaq	March 2007	*From Clyde River*: Ilkoo Angutikjuak, Joelie Sanguya, Igah Sanguya, Shari Gearheard *From Savissivik*: Qaerngaaq Nielsen *From Barrow*: Joe Leavitt, Nancy Leavitt *Researchers*: Andy Mahoney (Boulder, Colorado), Lene Kielsen Holm (Nuuk), Yvon Csonka (Nuuk)
Barrow	May 2007	*From Clyde River*: Ilkoo Angutikjuak, Joelie Sanguya, Igah Sanguya, Geela Tigullaraq, Shari Gearheard *From Qaanaaq*: Toku Oshima, Mamarut Kristiansen *From Savissivik*: Qaerngaaq Nielsen *Researchers*: Andy Mahoney (Boulder, Colorado), Henry Huntington (Eagle River, Alaska)
Clyde river	April 2008	*From Barrow*: Warren Matumeak, Joe Leavitt, Nancy Leavitt *From Qaanaaq*: Toku Oshima, Mamarut Kristiansen *From Savissivik*: Qaerngaaq Nielsen *Researchers*: Andy Mahoney (Boulder, Colorado), Lene Kielsen Holm (Nuuk), Yvon Csonka (Nuuk), Henry Huntington (Eagle River, Alaska).

In addition to those who traveled, many people in each place contributed to hosting the visitors and sharing knowledge about local ice conditions and other features of the area.

Getting people to each place, however, was just the beginning. We then had to arrange lodging, food, and local travel so that we could actually go out on the ice we were there to study. Again, Barrow was the simplest by far, thanks to the services of the Barrow Arctic Science Consortium (BASC), which is funded by the U.S. National Science Foundation to provide exactly those kinds of services to visiting researchers. In Qaanaaq, we worked closely with the Kommunia (municipal office) who helped us reserve the local hotel and find an additional empty house we could use for our stay. In Clyde River, where Gearheard lives full time, we were able to use one empty house in the town, another person rented us their home, and a few other team members found space on Gearheard's floor.

The elaborate travel and logistical adventures actually helped to build the interpersonal relationships that made us truly a team. The traveling team members spent time getting to know one another in airports, hotels, and restaurants. In each community, we helped cook for each other, took turns as host and visitor, and joined local festivities. When we arrived in Clyde River in April 2008, the third and last of the community visits, a group of skiers on the same flight were bemused by the shouts of joy and the hugs of our reunion at the airport, as we greeted our friends once again.

Traveling on the sea ice in each place allowed our team to visit sea ice features and locations that have importance for the resident team members. The time on the ice also let us join in local activities (e.g., hunting, fishing, ice travel) and acquire a set of shared experiences for further discussion. Traveling by dog sled in Qaanaaq over thin ice in March drove home the extent of environmental change occurring there and its life-and-death importance to local hunters. Watching a bowhead whale hunt in Barrow and taking part in butchering the whale provided a vivid example of the significance of sea ice hunting, not only for food but for a sense of identity and shared and individual fulfillment, a shared trait in all three places. A weekend trip through the fjords of Baffin Island to a fishing lake allowed the Clyde River team to explain how they had lived in camps spread across the land, and how the arrival of sea ice each year provided the means to join other camps, visit family, and catch up on the news from other places. We were reminded through that journey we took together that the land and ice are "home" far beyond the limits of today's communities.

In contrast to the logistical challenges of the exchanges, the establishment of local sea ice expert groups was relatively straightforward thanks to team members in each community who took leadership of the groups. Joe Leavitt in Barrow, Joelie Sanguya in Clyde River, and Toku Oshima in Qaanaaq were team members that had participated in the community exchanges, and thus knew the full scope of the project and could share what they had experienced in the other places. They led a group of local sea ice experts in the regular meetings over the course of the project; some of the experts had also participated in the community exchanges and others were involved when the research team visited. The expert groups each have a core of four to six members, with additional participants attending the meetings at times. At first, the expert discussions proceeded of their own accord, with the groups recording what they found important, such as terminology for sea ice and a general description of the seasonal round of sea ice use. Later, when the community exchanges were complete and we had a better sense of what information was of most interest to the research team as a whole, we prepared more specific agendas for each group to follow so that they could produce materials that could be compared across communities and included in the book we outlined and planned as a group during our last exchange in Clyde River.

Another driving force behind the success of the local experts groups was the recognition that their purpose was to document *their* perspectives and ideas, not just to have them answer questions from the academically trained researchers. (Igor Krupnik and Henry Huntington had found a similar "Aha!" moment in the St. Lawrence Island project "Watching Weather and Ice Our Way" (Oozeva et al. 2004). That project did not blossom until Igor visited the islanders and explained that the goal was for them to document the way *they* saw the ice, not to try to anticipate what the outside scientists might be interested in having written down). In one sea ice experts' meeting in Barrow, attended by Henry Huntington in February 2009, elder Wesley Aiken responded to the question, "How do you see the sea ice?" by saying, "The sea ice is a beautiful garden." The image as he described it in detail, with agricultural as well as Biblical overtones, is in remarkable contrast to the usual descriptions in Arctic documentaries and the like, showing the Arctic Ocean as a hostile and forbidding place dominated by storms, cold, and danger. With the groups providing a comfortable place for sea ice experts to document their own knowledge in their own way, this and other creative expressions were freely shared and recorded and include stories, illustrations, maps, family photos, and artwork.

The work of the expert groups provides the most material for the project and the book we are currently preparing, fleshing out in detail some of the concepts we had first discussed during the community exchanges. We had noticed, for example, during the 2004 study that whenever the local team members took the lead in the discussions we had, the conversation turned immediately to food. "What kinds of fish do you have?" "How do you catch seals"? Sea ice, it became clear, was interesting and important, but it was also only a means to provide for one's family and community. The real interest was in what one was able to *do* by means of knowing and using sea ice. Accordingly, the experts groups prepared material on food associated with sea ice, along with other sections on tools and clothing, travel and safety, and sea ice as home and belonging. This is what sea ice means to them.

The third component of the project, the installation of sea ice stations and quantitative measurement of sea ice near each community, added another dimension to our collective ways of looking at sea ice and our collective understanding of sea ice dynamics in the three locations. While remote sensing can provide a great deal of information about larger scale sea ice dynamics and processes, some data simply cannot be obtained except locally. Andy Mahoney (the project sea ice physicist) developed a simple but robust method for taking measurements including sea ice thickness, snow depth, and sea ice and snow temperature using wooden stakes, metal cables, a generator, and only a few other items (Mahoney and Gearheard 2008; Mahoney et al. 2009). The method was refined over the project with feedback from local sea ice monitors and a manual was produced in English and Inuktitut (Mahoney and Gearheard 2008). The cost-effective measuring stations can be set up quickly using almost all locally available materials and monitored regularly, allowing local residents to participate and provide data that could not have been obtained in any other way. Trained local monitors were installing, maintaining, and removing stations independently by the second year. In Clyde River and Qaanaaq, the communities have been able to take over the measurements and will continue them after Siku–Inuit–Hila has ended.

The measurements serve two purposes. First, they allow Andy to use the quantitative data to help assess the processes of freeze-up and melt in the three locations. One discovery from the station measurements helps support what Qaanaaq residents had suspected, that the ice in the fjord melts from the bottom in spring due to influx of heat from warm currents (Mahoney et al. 2009). This process is in contrast to Clyde River, where the dominant cold water currents from the north create minimal ice-bottom melting. Second, the measurements provide a basis for discussions both during the community exchanges and for the experts groups. For example experts groups have discussed the timing of freeze up and progression of sea ice thickness, considering if the measurements and their own observations were consistent with past conditions or if current conditions represented a change. An encouraging sign of the value of the methods for measurement we implemented is that other communities in Nunavut have requested copies of the training manual, and in Nunavik (northern Quebec) they have switched current ice monitoring methodology to the one we have been using (Chris Furgal personal communication 2009). With a standardized methodology in place that works for local communities, the possibility to start a sea ice observing network in the North has great potential.

What We Learned

As with many other similar projects with strong local involvement, the partnership between local residents, local researchers, and visiting researchers gave us access to a great depth of information and insight. The title of the paper describing our 2004 study quoted Joelie Sanguya, "It's not that simple" (Gearheard et al. 2006). He was responding at the time to the promptings of visiting researchers, who were trying to boil down the impacts of a shorter sea ice season on local residents. Joelie took great pains in disabusing us of the idea that impacts could be measured on a simple, linear scale from "more seals" to "fewer seals." Instead, he said that the hunters were still able to get seals, but they were out in boats at a time when they felt they should be traveling over sea ice. It just felt odd. Only a long and close relationship with the environment can produce such a feeling of dislocation as a result of change. These feelings are an integral part of change and impacts. Reducing sea ice knowledge and assessments of change to neat timelines, flowcharts, or tables of cause and effect do not provide a complete picture or address what impacts *mean*. The precise, local meaning of change is what is most relevant to the people experiencing it, even if that meaning can be difficult to pin down and explain to outsiders.

Discussions like the one with Joelie helped broaden and deepen the appreciation of the visitors for the role of sea ice in local life. Of course, similar insights have been gained from many projects, such as the ones included in this volume. A difference for Siku–Inuit–Hila is that the category of "visitors" blended with the category of "locals" when we had our sea ice exchanges. Those with local expertise in their home area nonetheless became visitors when they went to another community. It quickly became clear that the basic dynamics of sea ice were well understood by these visitors even in the new locations, but they also had great respect for the differences large and small that can turn success into failure, or safety into disaster.

None of the visitors felt comfortable traveling on sea ice alone in the new locations. Instead, they relied (as did all of us) on local expertise and guidance. Understanding sea ice in general is not the same as understanding sea ice in a particular area, under particular conditions, with full knowledge of the development of that ice through the current season. Arriving in spring to a new area, regardless of one's knowledge from home, does not allow one to understand the hazards of a new area. This degree of humility is hardly surprising in cultures where humility is valued and its lack can have fatal consequences. Nonetheless it was instructive to those of us who thought we were learning something about sea ice during this and other research efforts – even those whom we regard as masters of sea ice insist on treading carefully when away from home.

The exchanges provided common ground in our discussions for comparisons and reference to shared experiences in other places. "Do you remember the pressure ridges in Barrow?" "The open water on the way to Siorapaluk was scarier than this." Even the preparation of maps and other graphics for our book has made us consider the significant differences among the communities in sea ice and in the ways people use it. Clyde River and Qaanaaq have prepared detailed maps showing the growth of sea ice during fall and winter and its breakup during spring, and Qaanaaq used maps to compare present and past conditions. In Barrow, by contrast, the ice forms along the coast and is liable to shift throughout the winter, but does not display the same consistent pattern of growth, most likely because Barrow lacks the fjord coastline of northern Greenland and Baffin Island. In consequence, currents and wind rather than coves and bays dominate ice formation and movement in Barrow, making communication via maps more difficult.

Currents were another common thread in all communities. Everyone agreed that currents were crucial for understanding sea ice, and that monitoring currents was essential for safety near the ice edge. It also became apparent that neither local residents nor scientists have much detailed understanding of currents in any of the locations nor does anyone have an inexpensive way to monitor currents. Early on, elder Warren Matumeak from Barrow asked Andy Mahoney what happened to all the water when the current runs straight under the shorefast ice toward shore. Obviously, the water has to go somewhere – it cannot simply disappear. Nonetheless, no one knows what happens. Warren at first enjoyed teasing the academically trained for their inability to answer apparently simple questions, but as the project and relationships developed, we all began to appreciate our common ignorance as well as our common understanding.

The presence of a sea ice physicist on the team added another comparative dimension to the study, which was invaluable in trying to make sense of the observations and interpretations offered by the rest of the team. Studies of this kind are hybrids in that they often employ social science methods to examine natural science topics (e.g., Huntington et al. 2009). Without some expertise in both areas, there are many potential pitfalls for researchers. Social science training can help ensure effective cross-cultural communication and avoid misinterpretations. Natural science training can help recognize key observations and avoid being misled into thinking an unfamiliar statement is unknown to science. In addition to helping the social scientists understand what the locals were saying at times, Andy was also able to share considerable scientific information with local team members (e.g., remote sensing data such as satellite imagery, scientific perspectives, and explanations on various sea ice dynamics), who greatly appreciated the chance to learn more about sea ice from another point of view. The measurement program was simply the contact point between local and scientific views of sea ice. One result was that the local partners also had the opportunity to learn another way to observe and understand sea ice, and were not simply sources of information. The exchange of skills went in the opposite direction as well, as local experts shared with the whole team their methods for testing ice thickness and evaluating hazards as we traveled the sea ice together.

The tangible products from Siku–Inuit–Hila center on the book described above. In addition, we have so far written a paper on the ice measurements (Mahoney et al. 2009), completed the manual for taking measurements (Mahoney and Gearheard 2008), worked with Kalaallit Nunaata Radioa (KNR) TV to produce a documentary on our Qaanaaq work for Greenland television,[3] and contributed to a compilation of sea ice research methods (Huntington et al. 2009). All the raw data from our project will be housed in each community at organizations they choose. For example Qaanaaq is arranging for its raw data to be housed at the local school and in Clyde

River at the local Ittaq Heritage and Research Center. Barrow will decide the best place for their data. Copies of the data may be shared more broadly with other communities, researchers, and educators, through a service such as the Exchange for Local Observations and Knowledge of the Arctic (ELOKA[4]), but our team will make these decisions once the project and book are complete.

Why It Is Valuable

While it would be wonderful to claim brilliant new insights into sea ice and the human dimensions thereof, in fact it is not simple to point to single new facts or brief conclusions that might demonstrate why the Siku–Inuit–Hila approach is worthwhile. Perhaps the simplest lesson we all learned was humility. We may be experts in our field or in our local area, but we also all have a great deal to learn, especially when it comes to new territory. Studying one area can make extrapolation to other places tempting. As the saying goes, a person with one watch knows what time it is, whereas a person with two watches is never sure. After visiting Qaanaaq, Barrow, and Clyde River, all members of the research team have in some senses three watches. If we add scientific understanding, we may even have four watches each. Joelie was right to say of Clyde River that "it's not that simple." His words are even more relevant if we consider the Arctic as a whole.

Having four watches does not, however, mean that we are left with nothing but uncertainty. To the contrary, we have a deeper and better understanding of what sea ice means, how it relates to local cultures, and why sea ice change is likely to be the dominant environmental factor for Inughuit, Inuit, and Inupiaq communities in the coming decades. There is simply no escaping the multitude of ways that sea ice matters to people and their environment. Here is where the real uncertainty lies. As people face such a future, the more resources they have at hand, the better. Projects like Siku–Inuit–Hila will not solve every challenge, but they can help build relationships such as those between scientists and locals that will be valuable. Even if the individuals involved are different, a degree of mutual respect and understanding is a solid foundation for communication and collaboration. Similarly, awareness that others are facing similar challenges can also be helpful, both for simply knowing that one is not alone and for learning from others' experiences.

In a small way, Siku–Inuit–Hila benefited greatly from exactly this kind of intercommunity encouragement. Sharing the stories that were compiled, the information that was documented, and the lists of sea ice terms from one community to the next helped sustain the energy and excitement that drove the experts groups in their work through the winters. When Barrow people read hunting stories from Clyde River or heard how Qaerngaaq Nielsen,[5] a highly respected Greenland elder and a member of the research team, had written many pages about his experiences, they too were inspired to tell their stories. When the Clyde River experts group read what the Barrow visitors had written about their experiences visiting Clyde, the group became very emotional and took the time to read each page carefully. When hearing about the interest that all team members had in their community and culture,

Qaanaaq team members provided hundreds of photos from their family collections to help share more about their way of life on the ice. Encouragement from the academics alone is less likely to have spurred the local teams to give full rein to their creativity.

In a larger way, Siku–Inuit–Hila has shown the power of comparative research that includes local residents among those making the comparisons. Whereas many studies have used local residents as experts and guides in their home areas, our approach gave our community partners a new role, one that they relished and one through which we gained considerable depth of insight. Furthermore, the connections among the three communities produced information and insight far beyond that captured in publications, because visitors and hosts learned from and exchanged with one another in many ways. On our last day in Clyde River, at our wrap-up meeting before everyone left, Qaerngaaq Nielsen expressed to the group how grateful he was to have shared a tent with Clyde River elder Jacopie Panipak during our extended sea ice trip. He was grateful that he had the opportunity to get to know an elder just like himself, but who was from a different place and had some different knowledge than he had. If there was anything he had learned about sea ice it was not so much from us looking at it, photographing it, mapping it, or talking about it, it was from making a new friend and talking about a shared *life with sea ice.* They exchanged stories of their families, skills on how to make various tools, and stories of travel and hunting. This is how they exchanged sea ice knowledge. Berkes (2002) writes of fostering inter-community and inter-regional links to share information and increase resilience to climate change and other challenges. The Siku–Inuit–Hila approach provides opportunities for direct learning and exchange, without the filter of a small (largely exogenous) research team.

Finally, one aspect of the project and especially the community exchanges that cannot be omitted is that they were *fun.* As the scene at the Clyde River airport indicates, we were genuinely happy to see one another when we got together. We looked forward to spending time in the communities, on the sea ice and land, and yes, even at the airports en route. The experience was special and the feeling was infectious. While laughter and friendship may not be sufficient goals for scientific research, they are still valuable outcomes and perhaps necessary for building the strong and lasting relationships upon which mutual understanding depends.

Conclusion

Siku–Inuit–Hila, building on a much smaller study, further developed a new approach to collaboration between communities and visiting researchers. By involving local research team members in the community exchanges, we blurred the distinctions between “researchers” and “participants,” giving each team member a variety of roles during the project, including host, visitor, teacher, student, and not least, friend. For perhaps the first time, indigenous peoples have undertaken comparative environmental research in a formal structure, though it should be noted that exchanges and migrations have occurred as long as there have been people in the

Arctic and have inevitably required some degree of comparison of one environment with another. One result of our effort to do so systematically was the documentation of a great range and wealth of diverse information, which is proving challenging to compile into a single book.

Another result was some genuinely new insights into sea ice dynamics, such as the importance and lack of understanding of the role of currents, as well as a greater awareness of the limits of knowledge and its applicability in different settings. The measurement program and its unique methodology offers a way for Arctic communities to generate important information that can be of use to them as well as to a wider scientific audience, and as such constitutes a major contribution to Arctic science. Sustaining the link between measurement efforts and that wider scientific audience will take some degree of nurturing, perhaps through the establishment of formal connections between communities taking measurements and projects or programs that will benefit from the resulting data. Some encouraging advances have already been made with CliC, the Climate and Cryosphere project established by the World Climate Research Programme (WCRP) and with the Arctic Observing Network (AON). CliC works to assess and quantify the impacts of climate variability and change on the cryosphere and climate system. In recent years they have included indigenous knowledge in their discussions and have expressed interest in how their scientific program can make links to locally collected data. The AON has specifically included local and traditional knowledge in its definition of what must be incorporated into any Arctic observing network (National Research Council 2006) and the National Science Foundation has funded community-based observation and monitoring projects to help begin to develop community-based monitoring for the AON. Further discussions and future collaborations along these lines can help bring local observations and measurements into broader research campaigns.

Organizing and running a project like Siku–Inuit–Hila is not a trivial undertaking. From the complexities of travel logistics to the burdens on the host community, to the intellectual challenges of interpreting the wide range of results, there were many opportunities for mix-ups and even catastrophe. The former were inevitable and fortunately did not develop into the latter. Nonetheless, patience and tempers were frayed from time to time. These stages are perhaps common to most if not all field projects, collaborative and otherwise, but we mention them as a caution that anyone considering community exchanges think carefully about what they are getting into.

The outcomes have, in our minds at least, more than justified the difficulties and it is significant that everyone on the research team has stuck with the effort throughout the entire course of our effort. This is probably due to many factors, among them the way that the project evolved to reflect the interests and talents of those involved so that everyone had a strong intellectual engagement, and also the development of equally strong social connections that created a sense of shared endeavor and mutual dependence for success. Laughter was but the superficial indication of the degree of commitment and enthusiasm that characterized the research team and allowed us to quickly overcome the comparatively trivial obstacles along the way (Figs. 11.2, 11.3, 11.4, 11.5, 11.6, 11.7, 11.8, 11.9, and 11.10).

Fig. 11.2 Traveling as a team, via seven dog teams, from Qaanaaq to Siorapaluk, 2007 (Photo: Shari Gearheard)

Fig. 11.3 Mamarut Kristiansen, a Qaanaaq hunter, teaches Ilkoo Angutikjuak, a Clyde River elder and hunter, about fishing for *qaleralik* (Greenland halibut) through sea ice Qaanaaq style in 2007 (Photo: Lene Kielsen Holm)

Fig. 11.4 Joe Leavitt, a Barrow whaler and hunter, shows Clyde River hunters Joelie Sanguya and Ilkoo Angutikjuak different characteristics of the Barrow floe edge in 2007 (Photo: Shari Gearheard)

Fig. 11.5 The Siku–Inuit–Hila research team travels the sea ice together in Sam Ford Fjord on Baffin Island in 2008 (Photo: Henry Huntington)

Fig. 11.6 Elder Qaerngaaq Nielsen, from Savissivik, Greenland, shows off his catch of Arctic char during travels with the Siku–Inuit–Hila research team near Clyde River in 2008 (Photo: Shari Gearheard)

Fig. 11.7 The Qaanaaq sea ice experts working group at work in early 2009. From *left*: Uusaqqak Qujaukitsoq, Mamarut Kristiansen, Taliilannguaq Peary, Otto Simigaq, Ilannguaq Qaerngaaq, and Toku Oshima (Photo: Shari Gearheard)

Fig. 11.8 Clyde River sea ice monitor Teema Qillaq installs the first ice monitoring station of the 2008–2009 sea ice season as his son, Ken, and Lasalie Joanasie look on (Photo: Shari Gearheard)

Fig. 11.9 Siku–Inuit–Hila glaciologist Andy Mahoney (*left*) and Clyde River elder Ilkoo Angutikjuak became close friends over the course of the project after rooming together and through their shared interest in playing string games (Photo: Shari Gearheard)

Fig. 11.10 Laughing during a quick break from sea ice travel, Siku–Inuit–Hila researchers and friends, from *top left* clockwise: Toku Oshima, Lene Kielsen Holm, Igah Sanguya, and Shari Gearheard (Photo: Henry Huntington)

Acknowledgments Siku–Inuit–Hila would not have been possible without the help and support of many people in the communities of Barrow, Clyde River, and Qaanaaq. There are too many people and organizations to name here, but to all of you – quyanaqpak, qujannamiik, qujanaq!! Joe Leavitt, Joelie Sanguya, and Toku Oshima, our team leaders in each of the communities, deserve very special thanks. The project would not be possible without generous funding and encouragement from the U.S. National Science Foundation (HSD 0624344), Health Canada's Climate Change and Health Adaptation in the North Program, First Nations and Inuit Health Branch, and the Inuit Circumpolar Council-Greenland. We are grateful to our SIKU colleagues, Igor Krupnik and Gita Laidler, for many valuable comments to the first draft of this chapter.

Notes

1. For those familiar with Inuktitut and related languages, there might be confusion about the title "Siku–Inuit–Hila," as opposed to Siku–Inuit–Sila or Hiku–Inuit–Hila as would be the proper spelling if we adhered to one dialect or another. Since pronunciations vary among the regions involved, we decided to share the spellings, Siku being common to both Barrow and Clyde River, and *hila* following the usage in Qaanaaq.
2. Qaanaaq joined after we partnered with the Inuit Circumpolar Council-Greenland on the project. We presented our project to ICC and asked if they would like to work together and if there would be interest in conducting the work in Greenland. They responded positively and recommended Qaanaaq as the community to invite into the research. We, along with ICC,

invited Qaanaaq via their Kommunia council and they were enthusiastic to join. The local government then had meetings to decide which local hunters and elders would be most appropriate to invite as researchers to the team (see Note 5).

3. "Inuit Isaannit Silaannaq" http://www.knr.gl/index.php?id=2022
4. See www.eloka-arctic.org
5. Qaerngaaq is considered part of the Qaanaaq team, but lives in the neighboring community of Savissivik. Qaerngaaq is considered one of the most experienced and knowledgable hunters and elders of the region and well known in Qaanaaq and other communities. When Qaanaaq joined the project and the Kommunia had meetings to decide which local hunters would be asked to participate in the project, Qaerngaaq was immediately identified.

References

Amundsen, R. 1908. *The Northwest Passage*. London: Archibald Constable, 2 vols.

Berkes, F. 2002. Epilogue: Making sense of arctic environmental change? In *The Earth is Faster Now*. I. Krupnik and D. Jolly, (eds.), Fairbanks, AK: Arctic Research Consortium of the United States, pp. 335–349.

Brewster, K. 1997. Native contributions to arctic science at Barrow, Alaska. *Arctic* 50(3): 277–288.

Gearheard, S., Matumeak, W., Angutikjuaq, I., Maslanik, J., Huntington, H.P., Leavitt, J., Matumeak Kagak, D., Tigullaraq, G., and Barry, R.G. 2006. "It's not that simple": A collaborative comparison of sea ice environments, their uses, observed changes, and adaptations in Barrow, Alaska, USA, and Clyde River, Nunavut, Canada. *Ambio* 35(4): 203–211.

Huntington, H.P., Callaghan, T., Fox, S., and Krupnik, I. 2004 Matching traditional and scientific observations to detect environmental change: A discussion on Arctic terrestrial ecosystems. *Ambio* 33(7): 20–25.

Huntington, H.P., Gearheard, S., Druckenmiller, M., and Mahoney, A. 2009. Community-based observation programs and indigenous and local sea ice knowledge. In *Handbook on Field Techniques in Sea Ice Research (A Sea Ice System Services Approach)*. H. Eicken, R. Gradinger, M. Salganek, K. Shirasawa, D. Perovich, and M. Leppäranta (ed.), Fairbanks: University of Alaska Press. pp. 345–364.

Irving, L. 1976. Simon Paneak. *Arctic* 29(1): 58–59.

Mahoney, A. and Gearheard, S. 2008. *Handbook for Community-Based Sea Ice Monitoring*. NSIDC special report 14. Boulder, CO: National Snow and Ice Data Center. http://nsidc.org/pubs/special/nsidc_special_report_14.pdf

Mahoney, A., Gearheard, S., Oshima, T., and Qillaq, T. 2009. Ice thickness measurements from a community-based observing network. *Bulletin of the American Meteorological Society* March 90(3): 370–377.

National Research Council. 2006. *Toward an Integrated Arctic Observing Network*. Washington: The National Academies Press, 115 pp.

Oozeva, C., Noongwook, C., Noongwook, G., Alowa, C., and Krupnik., I. 2004. *Watching Ice and Weather Our Way*. Washington: Arctic Studies Center, Smithsonian Institution, 280 pp.

Pearce, T.D. and 19 others. 2009. Community collaboration and climate change research in the Canadian Arctic. *Polar Research* 28: 10–27.

Salenave, J. 1994. Giving traditional ecological knowledge its rightful place in environmental impact assessment. *Northern Perspectives* 22: 16–19.

Simpson, J. 1855. Observations on the western Esquimaux and the country they inhabit. Reprinted in *The Journal of Rochefort Maguire 1852–1854*. J. Bockstoce (ed.), 1988. London: Hakluyt Society, pp. 501–550.

Wolfe, B., Armitage, D., Wesche, S., Brock, B., Sokal, M., Clogg-Write, K., Mongeon, C., Adam, M., Hall, R., and Edwards, T. 2007. From isotopes to TK interviews: Towards interdisciplinary research in fort resolution and the Slave River Delta, Northwest Territories. *Arctic* 60(1): 75–87.

Chapter 12
Knowings About Sigu: Kigiqtaamiut Hunting as an Experiential Pedagogy

Josh Wisniewski

Abstract Complex and ever-changing sea ice coverage results in often highly differentiated spring hunting conditions on an annual basis in eastern Bering Strait. *Kigiqtaamiut* (Shishmaref) hunters' way of knowing about biophysical phenomena reflects this variability. A continuously advancing, experientially informed analysis, the hunters' way of knowing is an ongoing processual engagement in the world. This chapter explores what Kigiqtaamiut hunters know about local sea ice conditions and how they come to make authoritative claims about what they know. A comprehensive analysis of Kigiqtaamiut hunters' ice knowledge includes an equally critical engagement with the sociocultural mechanisms of local knowledge construction. Drawing upon an experientially driven ethnography of spring-bearded seal hunting conducted over three seasons, this chapter examines how Kigiqtaamiut hunters' understanding of sea ice and their way of learning are mutually constitutive and inseparable components of Kigiqtaamiut ways of knowing about sea ice in an environment of extreme temporal variability and annual fluctuations.

Keywords Shishmaref · Alaska · Inupiaq · Sea ice · Ice knowledge · Indigenous pedagogy

Introduction

Late one evening during the spring breakup in 2008, Clifford and I were returning to Shishmaref from reindeer herding camp on snow machines. We crossed the Shishmaref lagoon and skirted along the edge of a long stretch of "funny ice" where the *issuruaq* (deep water current) at *Sinŋaazruaq* (West Channel) was beginning to break up and move the ice, making travel from the island to the mainland more

J. Wisniewski (✉)
Department of Anthropology, University of Alaska Fairbanks, Fairbanks, AK 99775, USA
e-mail: jwisniewski@alaska.edu

I. Krupnik et al. (eds.), *SIKU: Knowing Our Ice*,
DOI 10.1007/978-90-481-8587-0_12, © Springer Science+Business Media B.V. 2010

precarious. Later, Clifford and I discussed how quickly the lagoon ice conditions changed over a couple of days. Clifford is a Kigiqtaamiu[1] hunter in his mid-1960s and is respected in Shishmaref as a highly knowledgeable elder. He has held many roles in my life over the past 6 years: as my primary instructor in the field, adoptive father,[2] and hunting and boat building partner. He has been one of the biggest supporters of my research and persistently my most vocal and constructive critic.

As Clifford and I discussed if the ice was safe enough to cross to go back to West Camp the next day, a biologist visiting the community and staying at his house asked, "How do you read the ice, how do you know it's safe?" "Well," Clifford responded, in his characteristic laid-back style, a Winston cigarette balanced on his lip. "I can't really explain it. I just know the ice, I can just tell, I watch it, you know, when it freezes up, so I know what it's doing. I don't know; I just know the ice."

Clifford's comments highlight something of the nature and structure of hunters' ways of knowing about sea ice in Shishmaref, Alaska, an Inupiaq hunting village in Bering Strait (Fig. 12.1). Their way of knowing is personal, intuitive, and experiential, a continuous process of coming to know. This active view of knowledge in motion speaks to Ingold's (2000:21) notion of *knowledgeability* as " . . . The capacity to situate such information and understand its meaning within the context of direct perceptual engagement within our environments." This form of knowing and of coming to know does not differentiate between these two aspects of knowledge. Rather they are intrinsically inseparable dimensions of Kigiqtaamiut hunters' ways of understanding that are collapsed into each other, where knowledge of and learning about are integrated continuous active processes realized through the practical activities of hunting and traveling through the sea ice environment.

Building upon the foundations of Ingold's "knowledgability," which speaks to knowledge *in action*, Kigiqtaamiut hunters' expertise can perhaps be best understood as knowledge that is not knowledge at all. At least not in the classical Cartesian epistemological duality, built on a separation of *res cogitas* from *res extensa* and the direct link between the thinker-being-mind and the substance of the world. It is this separation of the thinker-self from the materiality of the world that allows for an objective means of understanding the world in order to form a cohesive body of generalized, transmittable information that can be in turn applied to other phenomena observed.

In contrast, knowledge in the Kigiqtaamiut hunting context can be better conceptualized as *knowing*. Knowing emphasizes both the active application of knowledge and processual learning without delineating any differentiation between the two. Moreover going a step further, we can understand it as *knowings*, in plural. Kigiqtaamiut hunters place great emphasis on individual personal perspectives and experiences as the basis for having developed an understanding of the world. Conceptualizing hunters' knowledges as "knowings" allows for both shared common understandings and personal engagements with commonalities. This speaks to Casey's (1996:45) conceptualization of local knowledge as "an intimate understanding of what is true in the locally obvious." Thus understanding Kigiqtaamiut "knowings" of the sea ice environment requires equally attentive inquiry into the processes through which it has emerged.

Fig. 12.1 Map of the study area (Wisniewski 2005)

Hallowell (1976[1960]:358) argues that a "higher order of objectivity" can only be achieved through adopting a perspective that focuses on considering the sociocultural outlook and understandings situated within their particular local ontological context. This suggests an "objective" analysis of local understandings emphasize locally meaningful contexts and the sociocultural processes through which localized understandings are derived.

To expand on these comments and to consider Kigiqtaamiut knowing about sea ice and the role and meaning that the sea ice terminology plays in contemporary hunting practices, the following discussion examines the marine mammal hunting complex in Shishmaref and how it has changed in recent years in response to local sea ice coverage and quality. Concurrently it attends to experiential understandings articulated in local hunting stories that serve as both a fundamental means of sharing information and a context for learning.

Hunting in a Sea Ice Environment, Past and Present

Engagement with Kigiqtaamiut hunter's understandings of sea ice requires consideration of the role sea ice and marine mammal hunting play in everyday life. Bearded seals (*Erignanthus barbatus*) are of primary significance in both historic and contemporary marine mammal hunting economies throughout western Alaska (Burch 1998; Burns 1967; Fienup-Riordan 1983, 2007; Nelson 1969). The relationship between people and bearded seals is further articulated through diverse regional social and ritualized milieus (Fienup-Riordan 1983, 1994). Bearded seals (*ugzruk*[3]) or simply "*ugs*" are the most important sociocultural and economic wild resource in contemporary Shishmaref and throughout the Bering Strait region, Shishmaref is identified as a prime ugzruk hunting community. The annual seasonal round of subsistence activities begins with the spring ugzruk hunting, social distribution, and processing. At this time of the year, hunting families look forward to hunting with the greatest degree of anticipation.

Bearded seals have a clear and unique social position in relation to other seals. This hierarchy between bearded seals and other seal species is outlined in the Yup'ik oral tradition "The Boy Who Lived with the Seals" (Fienup-Riordan 1983, 1994), wherein seals sit in the *qasgiq*[4] according to rank with the fully mature bearded seals at the top of the hierarchy. The social hierarchy of seals and the importance of ugzruk in Shishmaref are revealed through casual conversation. The hunters differentiate ugzruk from ringed seals (common), spotted seals, and ribbon seals. If one asks a Shishmaref hunter after a fall seal hunt, "how many seals did you get?" a typical answer could be "I caught three common and two spotted." A closer look in his boat or sled might inform you that he also caught two young ugzruk (*anmiak*[5]). That these seals were not included in the count is not absent-mindedness, but rather an example of the conceptual differentiation between ugzruk and (other) seals. Had the question been phrased "what did you get?" the answer would include the two *anmiat* as well. Within local discourse, ugzruk are not seals but hold an exceptional and distinguished status above seals. Shishmaref hunters regularly speak to biologists about ugzruk as bearded seals and are keenly interested in biological information related to population dynamics, diminishing ice coverage, or migration patterns.[6] Hunters oscillate between local ugzruk hunting discourse and speaking to outsiders and scientists in more abstracted language of bearded seals.

The social position of ugzruk in community life is further demonstrated through the ritual distribution of the first ugzruk to be caught during spring hunting. By distributing the first-killed ugzruk of the year to village elders, hunters seek to promote individual and community luck in future hunts by demonstrating of respect to both elders and ugzruit. In addition, most familial hunting crews also distribute their first ugzruk of the spring hunt so people "get a taste."[7] Though less formally structured than seal parties described on Nelson Island (Fienup-Riordan 1983), the specific butchering techniques and ritual distribution of the "firsts" is locally identified as an important sociocultural dimension of Kigiqtaamiut spring hunting traditions.

The importance of bearded seals in daily life speaks to the local importance and valuation of the sea ice that bearded seals are dependent on. Though the role of ugzruk hunting has remained central to Shishmaref hunters, hunting practices and sea ice conditions have undergone marked changes over the last 100 years and even within the past 20 years. A school was founded in Shishmaref in 1906 and the government teachers began keeping records of hunting activities. While the records are limited and sporadic, they show that school would close down by May 1, when families would leave for spring hunting camps using dog teams to transport large skin hunting boats (umiat).[8] The boats would be used for hunting following the breakup of the shore ice, as well as to transport the hunters' families, their seal oil and meat back to Shishmaref later in the summer. During early spring hunting, before the pack ice separated from the shore ice and drifted northward, hunters traveled on to pack ice with dog teams to find open leads that ugzruit might be traveling through. Another strategy would be to find seals resting on the ice when they came through a hole that had rotted in the ice. This form of hunting is today referred to as "potholing" as hunters often refer to small bodies of open water as potholes.

The stability of the pack ice is crucial to this mode of hunting where elders have reported that within living memory they had to travel for 2 days to reach open water. Shishmaref is situated on the northwest-facing littoral of the Seward Peninsula. A light northerly wind would hold the drifting pack ice close against the shore ice, allowing hunters to cross onto the pack ice for pothole hunting. If the wind switched to a southerly direction, hunters have to hurry back to the shore ice to avoid being drifted out with the pack ice. Getting caught on the drifting pack ice was not an uncommon occurrence and some elders have spent more than a week out on the pack ice before making it back to the shorefast ice. This hunting format would continue until a strong southerly wind brought high water and broke up the shore ice, or until the shore fast ice became too rotten to travel across (see also Burch 1998, 2006). After this time hunters would use *umiat* to hunt among the scattered flows.

This general pattern of hunting did not change dramatically until the early 1990s. After the snow machines replaced dog teams in the early 1970s, hunters continued to travel out potholing in the early spring, often spending the night on the pack ice sleeping in their basket sleds. Once the shore ice broke up, hunters would switch to using boats. By the early 1990s, this general pattern also changed. The quality and character of the ice had changed, becoming less reliable, and to venture far out onto the sea ice with snow machines became increasingly dangerous. Hunting practices adapted to a new set of the ice dynamics. Rather than relying on a north wind to hold the pack ice in place, which was now not safe to travel onto, hunters instead waited for a south wind to spread out the pack ice. Hunters then used their snow machines to drag their boats across the shore ice to hunt in leads and on large floating ice pans called *iluqnaut*. This is the general spring hunting system in Shishmaref today.[9]

Using boats to hunt among scattered ice flows in open water following the breakup of shorefast ice has always been a part of the spring marine mammal hunting complex in Shishmaref. But the timing and dynamics surrounding boat use have changed dramatically in recent years. Whereas groups of hunters used to travel together or independently with a minimal amount of individual gear (snow machine, sled, rifle, ice tester, retrieval hooks, and a sleeping bag), hunting today requires a greater financial commitment and is more labor intensive. Hauling boats and outboard motors across the jumbled shore ice is hard on equipment. Broken snow machines and sleds and excessive wear and tear on boats are the expected norms. The same environmental factors that made potholing an effective hunting practice now works against hunting with boats. Kigiqtaamiut hunters note that the thickness of the ice and the extent of the ice coverage in Bering Strait have declined. Historically, a light north wind would hold the pack ice against the shore ice on the northwestern coast of the Seward Peninsula. Recently, the pack ice has retreated northward much faster in the spring, while a persistent north wind will pile the remnants of the drifting pack ice against the shore, slowing down the breakup of the shorefast ice and extending the shorefast ice apron. This extended ice apron is too thin, unstable, and rough to cross to get to open water, especially while dragging a 26-ft plywood boat and if open water is 20 miles out. At the same time as the north wind piles ice against the shoreline, the long warm days of spring melt the ice, making it even more dangerous to cross, until a south wind scatters the ice enough to permit hunting. Active hunters will hunt on as many days as possible when they can get out. Nonetheless, the financial costs of hunting in unstable ice conditions have contributed to a dramatic reduction in the number of days hunters can go out.

Older hunters who grew up hunting ugzruit primarily with dog teams speak nostalgically about potholing, describing it as a pleasurable way to hunt that was relatively simple and stress-free, particularly when compared to the logistical, physical, and financial stresses associated with today's hauling of large boats across unstable ice, in order to reach open water. This is demonstrated in one of Clifford's story that I recorded during the spring hunting season in 2008:

> There are changes in the currents out here too, we don't see those big ice bergs in December, that use to be good drinking water alright, never mind lakes . . . I sure missing sitting on edge of the ice with a sled, hook and a good rifle, we use to do that a lot . . . Dad, he says "sonny one time to get to the lead it took me two days to get out the lead to get ugzruk" And then, ah, his hunting partner was Fred Avasuk, you know Georgie Ann's father. And so he (Fred) woke up first and he saw ugzruk on top the ice he got up from his sleeping bag took his rifle in bare feet, just to make stories, he got it all right. *Suksuk* was his name (Clifford speaking Inupiaq) and he use to laugh like this "he he he he" "with my bare feet I killed it.
>
> [T]hat ocean sure change always north wind now, I don't know why . . . It never use to be like that. You know Daniel and I sure miss being on the edge of ice. You know, like with Ben, when we use to sleep on sleds. I don't know what happened, I don't know what happened. It use to be so much easier just to get ugzruk. Currents must of change or . . . I don't know. (Josh Wisniewski, notebook/recording, June 14 2008)

Over the course of my fieldwork, the timing of hunting has fluctuated dramatically. The date when the first ugzruk is caught in Shishmaref generally marks the beginning of the hunting season and indicates how different the ice conditions along the northwestern Seward Peninsula can be annually. In 2006, hunting began in May and the first ugzruk was caught on May 15. That year the ice was thick and solid. Boats were pulled less than 5 miles across the shorefast ice to reach open water, where there were lots of large floating ice pans. In 2007, no ugzruk were caught in Shishmaref until June 17. In spring 2007, there was a persistent north wind piling remnant pack ice against the shorefast ice for much of the season. Open water was more than 20 miles away, but the ice near open water was too thin and jumbled to permit hauling boats across. Additionally, as the spring progressed, the shorefast ice became increasingly rotten, so that hunting could not start until the shorefast ice broke up and began to spread out. Most hunting took place 50 miles southwest of Shishmaref among the scattered flows of shorefast ice as the pack ice retreated northward. The 2008 hunting and ice conditions were similar to the 2007 season and it was not until June 20 that hunters were able to catch an ugzruk. Again a strong north wind prevented earlier hunting while the main pack ice retreated northward. A small body of open water opened up a mile offshore from Shishmaref in early spring. However, no leads connected it to open water further out. As a result no bearded seals were able to come in closer to shore. As in 2007, the ice had to deteriorate enough to permit launching boats from the shore for crews to begin hunting. Crews traveled northeast of Shishmaref "chasing the ice" toward Cape Espenberg, where there is less current and the ice lingers longer.

Unusually cool and wet weather followed late hunting in 2008. Persistent rainfall throughout most of July forced families to attempt to dry meat and make seal oil indoors for the first time in living memory. The process of splitting, air drying meat, and rendering seal oil generally takes 4–6 weeks. Kigiqtaamiut families have historically counted on dry spring weather and have meat dried and oil rendered by early July, so that they may start other subsistence activities. Summer 2008 represented a dramatic departure from the previously experienced conditions. Yet on May 13, 2009, a text message from Clifford's daughter stated "Ugs on the beach, butchering tomorrow." She said that some families were able to hunt with snow machines early on, and they got everything they needed before "boating." These examples highlight the high degree of variability hunters currently experience.

Yet, elders' hunting stories also imply dramatic variability in the timing of spring hunting annually. Clifford's example of how his father had to travel for 2 days to reach open water is supplemented by his own accounts of the relative ease of ugzruk hunting, as he remembers it compared to contemporary hunting. On another occasion, Clifford described a spring when ice conditions were so poor that Shishmaref hunting crews could not hunt at all. He recalled how after the ice broke up enough for boat travel, he traveled with his grandfather to trade with and buy seal oil from families living at Cape Espenberg. This high degree of variability speaks to both the need for and role of understanding the unpredictability as a constant. It is here that we need to consider how Kigiqtaamiut hunters' experiences frame local vision of the phenomenal world.

Aŋizuġaksrat iniqtiġutait: The Rules of Old Folks

Examination of the language of experience and the meanings inherent in the varied Inupiaq and English descriptions of sea ice phenomena are good openings to considering the experiential template that influence hunters' understandings. These underpinnings shape local contextual meaning of sea ice terms and hunting stories that connect personal experiences to specific ice forms. Much as Hallowell (1976[1960]:358) suggests, adopting a locally meaningful analytical framework is crucial to examine hunting stories that hunters tell to describe their actions in response to specific ice conditions.

When elder Harvey Pootoogooluk described boating though a crack in shorefast ice that opened perpendicular to the shoreline, he began his story by stating "I broke the rule that time." On another occasion, Harvey told me how after falling through thin ice he was scolded by elders, for "play taking a short cut" to get across thin ice.[10] When telling this story of his experiences on sea ice, he again suggested that he broke a rule. Within Kigiqtaamiut hunting practices, there is a wide range of shifting, relational, and circumstantial modalities in response to specific sets of conditions in a given moment. Younger hunters for whom Inupiaq is not their first or primary language conversationally refer to these rules as Eskimo Law[11] or simply as "the rules," that which you are supposed to do. For elders they are *Aŋizuġaksrat iniqtiġutait* or "the rules of old folks." *Aŋizuġaksrat* translates as "elders" or "older people" while *iniqtiġutait* is "warnings," admonishments or rules.

Several considerations must be taken into account as we consider this particular usage of rules and what adherence to rules is implied in Kigiqtaamiut hunting practices. Rules do not state what people do. The conceptualization of rules suggests what people *should* do in different circumstantial contexts. *Aŋizuġaksrat iniqtiġutait* does not represent an a-historical, timeless, and idealized version of the Kigiqtaamiut society. Rather, when historical narratives are used to stress a point, more often than not they are applied to underline the ramifications people experienced for not adhering to them. In suggesting what people should do, *aŋizuġaksrat iniqtiġutait* implies that failure to follow them can potentially bring harmful or dangerous ramifications. Just as failing to check the oil in one's car can lead to a seized engine and considerable financial expenditure, failure to acquiesce to the Kigiqtaamiut rules can result in injury, loss of life, or one's inability to manifest luck[12] in catching animals. The following example from the 1930s provides an experiential basis for how *Aŋizuġaksrat iniqtiġutait* is considered in contemporary life.

> Johnny: I guess he must of stay at *Ipnauraq*, 18 mile and they hunt owl by putting pole and trap. He must of caught that one and instead of killing that owl he must of cut the sinew of that owl and let it go. And later on, I don't know how late or how many months or weeks, they saw Clifford[13] go qayaq on a real calm day and someone that was staying at *Iglut* saw *utuqtaq* (Clifford) tip over in his qayaq and they saw the owl was pestering him or bothering him and let him turn over and let him drown. They say he drown on a calm day. And they must of go rescue him because someone from Iglut saw by Ipnuaraq. And the only thing I remember I don't know how, but we go by skin boat from Ipnauraq there was Eddie Tocktoo, Grace Tocktoo, my mother myself, and one old lady, the way I remember I think she is *Kiqsiuq*, that old lady that stay with us and we bring the body, real calm day from Ipnauruq to here (Shishmaref), and we park the boat south side of church and I can't remember where they bury him or if they bury him here. So they always tell us not to play with the animals. They say it always go back to you or harm you. (Josh Wisniewski notebook/recording March 21 2008)

"Relationality" is the understanding that persons are in the world, and subsequently that human actions and intentions both influence and are influenced by the phenomenal world. Hunters' "relationality" forms a central and a priori assumption that shapes their individualized ways of knowing. Part of Kigiqtaamiut relationality in hunting is the recognition that biophysical, spiritual, and not completely known forces shape and are concomitantly shaped by human behavior. These can never be fully known. Briggs (1991:262) provides a similar characterization of the Inuit experiences – " Inuit regard the world as a place where little can be taken for granted, where answers are not fixed and nothing is ever permanently knowable." Briggs' statement suggests, like Kigiqtaamiut hunters, many Inuit directly experience and interact with the world from a relational perspective. Rather than viewing the phenomenal world as composed of fixed forms, they engage it as non-permanent and shifting entity based on the specific circumstances of a particular engagement.

Recognition of the environment as unfixed, not fully knowable, and full of potential and intentional factors, alive to personal experience is implicated in the Inuit/Inupiaq notion of *sila*. *Sila* is a term variously construed across the broad spectrum of Inuit research. Broadly viewed the word *sila* refers to the environment, the organization of the world, consciousness, as well as the weather, without implying any categorical differentiation between these diverse dimensions of the world. In Kigiqtaamiut practice, *aŋizuġaksrat iniqtiġutait* speaks to the sentience implied in *sila*, suggesting that not being aware of factors that cannot be fully known or ruled out is potentially dangerous.[14]

The phenomenal world, including the range of factors contributing to the sea ice dynamics, cannot ever be fully known or predicted. Even claiming to fully know or understand phenomena or to be able to predict the outcome of events can be disastrous. Therefore, learning the Kigiqtaamiut hunting rules should not be interpreted as the memorization of a set of prescribed actions or steps, such as in a formal dance. Rather, following rules suggests continuous cautions, careful, and deliberate actions in response the circumstances of the moment.

As opposed to a generalized set of ideals, Kigiqtaamiut rules are personal, continuous, and flexible adjustments to the changing circumstances of the moment, with an awareness of how ones' actions contribute to shaping these circumstances. The significance of Kigqtaamiut hunting rules is application. To apply rules is to know them. Indeed there is no differentiation between knowing and applying, knowing in this hunting context *is* being. Hunters actively live this knowing. We can examine local Kigiqtaamiut terminology of the sea ice within this framework.

Kigiqtaamiut Sea Ice Terminology

The concept of *aŋizuġaksrat iniqtiġutait* and how it contributes to the experiential framework of the Kigiqtaamiut hunting pedagogy may be applied toward the examination of both Inupiaq and English terminology and understanding the contemporary Kigiqtaamiut hunting lexicon used to describe various ice features. Over the course of three seasons of hunting and group discussions with the Shishmaref elders, I was able to document 65 Inupiaq terms for phenomena related to the sea ice dynamics.[15]

Though Shishmaref is historically a community in which Inupiaq was the primary language, its usage has been largely eclipsed by English, particularly among hunters under 50 years, the majority of active hunters today. Among this group approximately 20 Inupiaq terms are consistently and regularly used to describe ice conditions. These terms are often used in sentence structure composed primarily of English words among both fluent Inupiaq speakers and younger hunters. The grammatical rules of both languages are not used with much concern in daily conversations and are often applied interchangeably. As Rosaldo (1986:108) points out, linguistic exchanges among hunters are articulated through "telegraphic shorthand," as "speakers can safely assume their listeners depth of knowledge about the landscape, hunting practices, the huntsmen's abilities previous hunts in the area and elsewhere."

During my fieldwork, an elderly woman and fluent Inupiaq speaker offered to help me with the translation of hunters' stories about hunting on the sea ice. She was unable to translate the materials. Minor dialectal differences notwithstanding, the reason she could not understand what hunters were talking about, as she herself suggested, was because they spoke in their own language, used their own words, and made up terms directly connected to their personal experience. She found hunters' discourse largely impenetrable. Therefore, to understand the meaning imbedded in hunters' ice terminology we must focus on the manner in which different terms are used. The examples below demonstrate how certain terms used in the context of Kigiqtaamiut telegraphic shorthand describe a range of conditions and personal understandings that speak to the range of dimensions that Kigiqtaamiut hunters associate with ice phenomena.

A term frequently used by Shishmaref hunters to refer to large pieces of ice is *iluqnauq*. During 2007–2008 group discussions elderly hunters defined *iluqnauq* in the following way:

> An *iluqnauq* is a large piece of floating ice that ugzruit like to rest upon. This is particularly true when *aupkanit* (holes that go all the way through the ice) form which allow ugzruit a way to get on and off the ice while staying in the middle of a large pan. *Iluqnauit* (plural) are formed when large ice pans break off of the pack ice that has moved south during the winter. An *iluqnauq* can be flat or rough ice. It is not topography that defines an *iluqnauq*, but its overall size. Their main defining feature is that they are large free-floating ice platforms. Depending on the wind conditions large *iluqnauit* can block the animals from traveling closer to the shore ice edge from 'big water' further west. A large *iluqnauq* that is held against the shore ice by the north wind can also slow down the breakup of the shore ice by holding it in place. This makes hunting more difficult as the shore ice rots in place it becomes dangerous to travel on. Yet if the ice doesn't move hunters are forced to find ways to get to open water. An *iluqnauq* serves in one capacity as good habitat for bearded seals. While in another capacity under specific environmental constraints it can serve as an impediment to hunting.

Coming from a long and unsuccessful hunting trip in May 2008, I walked to Tony's house to share my experience with him. The following brief dialogue highlights a typical example of local "Shishmaref-style" telegraphic shorthand in exchange of information between me and another hunter as we discussed an Iluqnauq.

Josh: Real big *iluqnuaq* man, that open water is good alright, and we see lots of seals, no ugs though. But we couldn't get to that "big water"

Tony: Oh, *iluqnauq*?"

Josh: Iyah, *iluqnauq*. Big one.

Examined in relation to a literal translation of the term, *iluqnauq* would seem to imply that they are good places to look for animals. Stories told by elders and active hunters confirm this. These are important ice habitats that ugzruk seek out, particularly when holes rot through them away from their edges. They provide places where ugzruk can rest away from the ice edge where they are more susceptible to attack from polar bears. While hunting in boats, hunters look for pressure ridges along the sides of *iluqnauit*. Hunters can then climb them, concealed by wearing white parka covers and look for resting animals (Fig. 12.2). Yet, what I knew from having been out on the ice that day and what Tony assumed through his own life history of experiences was that despite there being open water less than 1 mile offshore, persistent north winds were causing large *iluqnauit* to come and rest against

Fig. 12.2 Clifford Weyiouanna and his grandson Tyler look for ugzruk, while concealing themselves behind an *iunŋiq* (ice pressure ridge), June 18, 2007. (Photo: Josh Wisniewski)

each other, blocking ugzruit from coming closer to shore or hunters from being able to go further out to look for them. *Iluqnauit* provide good ice habitat for seals and a good place to look for ugzruit in specific circumstances; in other circumstances they can be highly disruptive to hunting.

These examples show some of the ways *iluqnauit* are experienced and referred to. Yet the local meanings and associations connected to an *iluqnauq* are equally connected to local historical events. In the following (partially abridged) discussion between myself and elder Arthur Tocktoo about a 1948 ugzruk hunt, Arthur, who was 16 at the time, described an encounter with a *kununigaq* (mermaid) (Fig. 12.3). This experience is now central to today's understandings of *iluqnauq* in Shishmaref, among elderly and active hunters alike. Arthur's story tells about hunting and how one should behave in the event one encounters such or another non-human being:[16]

> That (i.e., painting – refers to the image featured in Fig. 12.3) describes a female; but it was a male what we been see. Me and him (Davy Ningeulook) run off the boat to see if we could see that thing, you know? While were gone, they call us, they want us to go back right away ... Yeah we didn't even see it when we were boating, but when we stop (to) get up the ice we saw it. (It) was sitting like this on the ice, part in the water. Real clear, real white ice, like what ugzruks always be on, that real clean ice. Yeah we see one. They say there is someone sitting over there, and that the older guy said it's not a person, that's a mermaid, but never say mermaid he say *kununigaq*. It got one long leg, not like person, like a seal. We never see, but we ask that old guy, he said they got no legs, just a tail. But to the people they say if you see that kind, not to bother that kind, that's how come that elder with us, he called "come on, come on we gotta go, before something happens to us. . .Then when we go out

Fig. 12.3 "Eskimo Mermaid" by James Moses. The digital image of the Eskimo Mermaid was provided by Molly Lee and is used here with permission

> again Jake Minguna said "I hope we won't see nothing like that again". We gotta hunt what were hunting. But I think they (animals, mermaids) always hear us. Well good thing I tell somebody before I die!! Yeah, I am the last one alive now who was there. (Josh Wisniewski Recording w/Arthur Tocktoo April 3, 2008)

Removed from the context of its narrative, Arthur's story is difficult to follow. It assumes a general shared understanding of hunting activities, places, persons, and genealogies, that most if not all hunters in Shishmaref would know. The purpose of Arthur's story is to inform. He does not try to explain the occurrence of a *kununigaq*. His goal is more pragmatic and immediate, as he stresses that these beings are out in the world, and that they may be encountered while hunting on the sea ice. He also highlights that behaving correctly when seeing a *kununigaq*, in essence, by leaving him/her alone is an important "rule" and a part of knowing about *iluqnauit*. When elders were asked about *iluqnauit*, Arthur's hunting experience was central to their discussion. In recounting Arthur's story, local historical and practical dimensions were presented interchangeably and as inseparable dimensions of what knowing about *iluqnauit* means in Kigiqtaamiut hunting context. As a term and as a feature of the frozen seascape, *iluqnauq* embodies a range of relational and contextual meanings about the experienced and known world.

The range of associations embodied in *iluqnauq* offer an example of how the Kigiqtaamiut view and know the ice (Figs. 12.4 and 12.5). The early fall slush

Fig. 12.4 Butchering ugzruit on an *iluqnauq*, 2006. (Photo: Josh Wisniewski)

Fig. 12.5 Fall seal hunting at west channel near Shishmaref, *qinu* solidifying, November 2007. (Photo: Josh Wisniewski)

ice condition called *qinu* is another example (Fig. 12.6). Beginning once again by defining the condition, elders discussed *qinu* as

> *Qinu* is not considered ice as a form of *sigu*. It is a form of snow mixed with seawater that consolidates and freezes. Slush ice is considered very dangerous, because it is not a solid platform like ice, it is not strong, and cannot support weight. It is often compared to quick sand, because one who falls through can be stuck. One can also fall all the way to the bottom of *qinu* where it meets the ocean floor. *Qinu* is very white. When hunting in the spring, it is important to be able to distinguish between *qinu* and ice when traveling out to the ice edge. This is one of the reasons travel to edge of shoe ice can proceed along a long circuitous path as an experienced hunter goes through great pains to avoid *qinu*.

Another term that speaks to variations in *qinu* is *qaimut* (plural), slush ice formed by water waves that forms snow burms. *Qaimut* can be used to navigate in marginal weather. However, when *qinu* was brought up in a discussion with a group of elders, the discussion quickly turned to the story of *Tulimaq*. Tulimaq was *Qulliaġzruit*[17] (a Siberian Native) who once led a group of warriors across Bering Strait to Shishmaref during early fall, when freezing conditions and early stages of ice and slush formation made it a notably dangerous time to travel. Due to his resourcefulness and understanding of ice conditions, Tulimaq was able to cross a large stretch of not yet solidified *qinu*. Fabricating snowshoes out of materials in their skin boat, he walked ashore and subsequently pulled his boat through *qinu* to shore and led an attack against the Kigiqtaamiut. Tulimaq was subsequently killed in

Fig. 12.6 Resting on an iluqnauq, waiting for ice conditions to change. Umiaq hunting, June 2008. (Photo: Josh Wisniewski)

an ensuing epic battle by *Iġizrgaiyuk*. Like Arthur Tocktoo's account with a *kununigaq* on an *iluqnauq*, the story of Tulimaq is central to local knowing about *qinu*. It suggests the danger of fall travel and of *qinu*; it also speaks to the importance of improvisation and individual resourcefulness in different contexts. Though Tulimaq arrived to inflict violence against Shishmaref people, his knowledge and ingenuity in crossing *qinu* render his story historically significant toward local knowledge of sea ice (Fig. 12.7).

These and other recorded examples suggest that the meaning inherent in Kigiqtaamiut terminology goes far beyond translation and categorical organization. Instead, emphasis is placed upon the importance of an individual term's specific contextual usage and the role of shared activity for fostering understanding. This leads to another important consideration. Locally meaningful information about ice conditions is not bound to the descriptive qualities of Inupiaq terms but takes on meaning through shared experience and local history. Therefore, non-Inupiaq language terms that are actively used by hunters also warrant careful consideration.

Since Inupiaq is not the primary spoken language among the majority of Shishmaref hunters today (though many of them know and use Inupiaq terms), most of the Inupiaq words for ice are not in active use. Instead, I recorded 20 English terms that feature prominently in the contemporary hunting lexicon. They include terms such as "funny ice"; "bum trail"; "pothole"; "pond"; "river"; "trail"; "drymouth"; "big water"; and "white ice". Examined in relation to shared experiences,

Fig. 12.7 Checking ice conditions, before crossing a crack while "potholing," May 2006. (Photo: Josh Wisniewski)

contextual usages, and personal understandings, these terms are equally important as their Inupiaq counterparts, suggested by the following exchange that took a place a few days following our *iluqnauq* discussion, as Tony returned from a long snow machine ride down the coast.

Josh: Which way did you guys go?
Tony: *Pingupaq* we tried to go pot holing, I guess. There was game in that one river near *pingupaq* but the current changed. I thought I saw an ugzruk but, maybe, I was just getting excited. We tried to go Ikpik. There was big water down there. Yeah, lots of water down there.

A "pothole" is typically a small body of open water surrounded by ice that ugzruit and seals will travel through. A "river" or *kuuk* can be a pothole or a wide crack or lead connecting potholes, or leading out to open water, or something entirely different depending on the circumstances of the use. "Big water" can mean the ocean, any large body of water surrounded by ice or has other meanings depending on circumstances of the moment. These terms like their Inupiaq equivalents appear to only provide a generic description of a set of ice conditions. Yet in the context of their specific usage to portray the subtlety of phenomena among other knowledgeable practitioners these terms are highly useful in expressing Kigiqtaamiut hunters' experiences on the ice.

Conclusion

More than 50 years ago, the journal *Polar Record* published an illustrated ice glossary of 68 terms used by Arctic meteorologists and mariners (Armstrong and Roberts 1956). Juxtaposed with the 65 Kigiqtaamiut sea ice hunting terms documented thus far, the terms and definitions recorded in the *Polar Record* glossary (such as, ice pan, flow edge, brash ice, polynya) offer many parallel descriptions of sea ice phenomena. The two lists are also close regarding the number of terms.[18] Yet there is a notable difference in these lists. As opposed to the general description of sea ice conditions encountered in Arctic and sub-Arctic oceanic conditions offered by the *Polar Record* glossary, Kigiqtaamiut ice terminology emphasizes a highly local vocabulary used to describe sea ice forms as experienced in context of hunting. The translation and classification of the Kigiqtaamiut sea ice terminology provides a set of descriptions of ice phenomena; yet, this does not speak to the depth of hunters' understanding and ways of knowing. Kigiqtaamiut hunters' ways of knowing, like the Inupiaq language structure, are flexible and context-specific. "Knowings" are rendered significant through activity. Just as being in the flow of specific circumstances requiring meaningful actions, the significances embodied in hunter's sea ice lexicon are continuously and creatively constructed in relation to ongoing activities.

Relations with bearded seals, changing hunting practices, ongoing and historic responses to variations in sea ice, conceptions of rules, and highly contextualized local associations with sea ice terminology connect in intricate and individualistically divergent ways. Ambiguously and continuously emergent local understandings of sea ice in Kigiqtaamiut hunting experience are encapsulated and embodied in diverse shared and personal associations, connections and understandings articulated through the contextual usage of sea ice terminology.

Sartre (1965:87–91) captures the essence of this writing that "The man can only mean what he knows" and to that end "things can reflect for individuals only their knowledge of them." This speaks to a dual role that experience as pedagogy plays toward both informing and engaging hunters' ways of knowing. Hunters learn through their own experiences on the sea ice, the hunting stories of elders, and the constant exchange of information between other hunters. Likewise the ethnography of hunting and of hunters' ways of knowing about sea ice develops through being on the ice with other hunters, sharing stories and experiences with others, and discussions with elders. Here we see that the processes of coming to know in hunting life and coming to know through the ethnography of hunting life are two intertwined and mutually informing projects. Just as hunters' way of knowing is processual, incomplete, and ambiguous, the anthropological understanding of local perspectives is equally an incomplete process. It develops and changes through time in relation to experience. To that end, experience as pedagogy in hunting and in ethnography suggests how hunting and the ethnography of hunting are largely inseparable.

The examination of dimensions of Kigiqtaamiut hunters' ways of knowing about sea ice demonstrate that hunters have much to teach us about marine mammal behavior, historical ecology, local resource management, and the pragmatic skills of a community of hunters. Equally, however, the exploration of hunters' ways of knowing and experiencing can teach us much about the broader human condition of being in and coming to know the world around us.

Acknowledgments This work could not have been carried out without the financial support provided through a National Science Foundation Dissertation Improvement Grant OPP-0715158. Equally Kawerak's Eskimo Heritage program provided financial support at an important moment in the field. Igor Krupnik of the Smithsonian Institution invited me to participate in SIKU and suggested that I use my field opportunity and experiences to formally document sea ice terminology in Shishmaref. This proved to be a recommendation that radically enriched my project. His editorial comments were likewise invaluable toward enriching this chapter. The Shishmaref tribal government and the Shishmaref Elders Council have been continuously supportive of my research efforts. The Shishmaref tribal council's efforts to ensure elders was compensated for all of their participation in monthly gatherings was significant toward our success. Finally to Clifford and the Weyiouanna family, none of this would have been possible without your continuing support. You know most of all the value that my experiences Shishmaref have come to have in my life. The digital image of the Eskimo Mermaid was provided by Molly Lee and is used here with permission.

Notes

1. Shishmaref residents self-designate themselves and are referred to by other Bering Strait residents as *Kigiqtaamiut* (people of the island) or simply as "Shishmaref people."
2. During my doctoral fieldwork I lived with Clifford and his family, hunting with him and his sons for three consecutive spring hunting seasons. I had previously stayed with Clifford and his family, traveling with them as I conducted iterative fieldwork over two field seasons leading up to my formal doctoral fieldwork. In the most informal sense his family "adopted me." His primary mode of instruction was not unlike that he used with his sons and nephews, which was to simply expect me to pay attention and figure things out on my own, and to reprimand me when I did things wrong. Following hunting trips we would discuss variety of topics connected to our shared experiences.

3. The Inupiaq spelling of bearded seals used here refers specifically to the Kigiqtaamiut dialectical variation. Across northwestern Alaska, ugzruk is more commonly pronounced and spelled as ugruk.
4. The *qasgig* or men's houses were in both Yup'ik and Inupiaq societies. Buildings in which men lived communally and where ritual activities, as well as those of daily life, such as manufacturing hunting equipment took place.
5. *Anmiak* is considered a "Shishmaref word" and it is not generally used by other nearby Inupiaq communities to describe a young ugzruk. Kigiqtaamiut hunters have many other terms that speak to different age classifications of ugzruk, and *anmiak* refers to a specific life stage. The use of this specific term speaks to the degree of sensitivity of hunters' experience with bearded seals. *Anmiat* is a plural form of *anmiak.*
6. There has been little-to-no systematic analysis of bearded seal populations or health in Bering and Chukchi seas since early 1980s. Changes in the Chukchi and Bering Sea ice coverage and benthos may affect the future of bearded seals. In light of their economic importance throughout Western Alaska, understanding changing bearded seal ecology remains an important research area.
7. The first walrus caught in the new hunting season is also distributed. Some younger hunters, however, do not actively participate in distribution of "firsts." Older hunters are often glad to learn when a more mature hunter caught the first ugzruk or walrus of the season in order that it will be distributed. Equally important is when a younger hunter distributes without prompting from his elders. A friend told me that after recently letting his boys get their first caribou, some of them were interested in selling them back in the village rather than giving them away, though he insisted that they be distributed.
8. An *umiaq* is an open skin-covered boat. In Bering Strait they were often built to more than 30 ft in length, as families would use them to transport their supplies for the ice-free months, while living in camps. *Umiat* is the plural from of *umiaq.*
9. For a more in-depth examination of recent historic descriptions of ice hunting in northwestern Alaska and of Inupiaq sea ice terminology, see Nelson (1969).
10. The word "play" is commonly used in today's Shishmaref to describe many different activities, i.e., "I play shot that seal"; "don't play try to help"; "guess I'll go play see what they are doing." As used in hunting stories. it often serves to distant the hunter from his success. By saying "play," hunters often imply they had success not through their own prowess, but because of factors beyond their control.
11. Yup'ik and Inupiaq people often use the word Eskimo when self-referencing. It is used here in the same context as it is used in Shishmaref to describe local (non-white) conceptualizations and understandings and interactions between human actions and intentions and natural phenomena.
12. Luck is a subtle and central component to Kigiqtaamiut hunting practices.
13. This does not refer to Clifford Weyiouanna with whom I worked, but refers to an earlier deceased person, whose name Clifford Weyiouanna inherited.
14. Fienup-Riordan (1994:88) similarly notes that in Yup'ik experience that improper behavior toward animals could result in *ellam yua* the person of the universe "waking them up." Assumptive behavior toward outcomes of a hunt is often seen as indicative of forthcoming technical failure such as outboards or snow machines breaking down, which can result in life-threatening situations.
15. Additionally I have identified and documented 16 terms directly related to qualities of snow in both terrestrial and marine environment.
16. This narrative is presented here in a highly condensed form, and outside the specific social conversational setting of its telling. In doing so much of the richness and assumed implicit understandings are not attended to here due to spatial requirement.
17. Qulliaġzruit refers to Native people living on the western side of Bering Strait. Historically this may have directly referred to Chukchi, in contemporary usage it is applied more broadly.

18. The numerical closeness of these two lists of ice terms should not necessarily be extrapolated as representative of all Inuit ice terminology. Other communities have their own terminologies that describe their localized experiences with sea ice phenomena.

References

Armstrong, T. and Roberts, B. 1956. Illustrated ice glossary. *Polar Record* 8(52): 4–12.

Briggs, J.L. 1991. Expecting the unexpected: Canadian inuit training for an experimental lifestyle. *Ethos* 19(3): 259–287.

Burch, E.S., Jr. 1998. *The Iñupiaq Nations of Northwest Alaska.* Fairbanks: University of Alaska Press.

Burch, E.S., Jr. 2006. *Social Life in Northwest Alaska: The Structure of Iñupiaq Eskimo Nations.* Fairbanks: University of Alaska Press.

Burns, J.J. 1967. The Pacific Bearded Seal. *Alaska Dept of Fish & Game, Pittman-Robertson Project.* Rep. W-6-r and W-14-R. 66p.

Casey, E. 1996. How to get from space to place. In *Senses of Place.* S. Fled and K. Basso (eds.), Santa Fe: School of American Research Press.

Fienup-Riordan, A. 1983. *The Nelson Island Eskimo: Social Structure and Ritual Distribution.* Anchorage: Alaska Pacific University Press.

Fienup-Riordan, A. 1994. *Boundaries and Passages: Rule and Ritual in Yup'ik Eskimo Oral Tradition.* Norman: University of Oklahoma Press.

Fienup-Riordan, A. 2007. *Yuungnaqpiallerput the Way We Genuinely Live: Masterworks of Yup'ik Science and Survival.* Seattle: University of Washington Press.

Hallowell, I.A. 1976 (1960). Ojibwa ontology, behavior and world view. In *Contributions to Anthropology: Selected Papers of A. Irving Hallowell.* Chicago: University of Chicago Press, pp. 357–390.

Ingold, T. 2000. *The Perception of the Environment: Essays in Livelihood, Dwelling and Skill.* London: Routledge Press.

Nelson, R. 1969. *Hunters of the Northern Ice.* Chicago: University of Chicago Press.

Rosaldo, R. 1986 Ilongot hunting as story and experience. In *The Anthropology of Experience.* V. Turner and E. Bruner (eds.), Urbana: University of Illinois Press.

Sartre, J.-P. 1965. *The Philosophy of John Paul Sartre.* In R. Cumming (ed.), New York: Vintage Books.

Wisniewski, J. 2005 "We're always going back and forth": Kigiqtaamiut Subsistence Land Use and Occupancy for the Community of Shishmaref. Report for the US Army Corps of Engineers, Alaska District. Anchorage.

Chapter 13
The Ice Is Always Changing: Yup'ik Understandings of Sea Ice, Past and Present

Ann Fienup-Riordan and Alice Rearden

Abstract This chapter presents a descriptive summary of Yup'ik elders' observations of sea ice formation and change along the Bering Sea coast of Southwest Alaska. In doing so, hunters modestly describe their efforts to negotiate a dangerous and ever-changing ice environment. While it is sometimes assumed that Bering Strait hunters on Diomede, St. Lawrence Island, and King Island hunt in the most diverse and demanding ice conditions in Alaska, conditions on the lower Bering Sea coast are equally if not more challenging due to the complex interplay between tides, currents, and wind. Moreover, as sea ice conditions change, Yup'ik elders' experiences at the southern limit of shorefast ice take on special significance.

Keywords Sea ice · Bering Sea · Yup'ik elders · Indigenous knowledge · Climate change

Introduction

The following discussion is based on gatherings with Yup'ik elders between 2000 and 2008 sponsored by the Calista Elders Council (CEC, the primary heritage organization of Southwest Alaska, representing the 1,300 elders 65 years and older) and funded by the National Science Foundation's (NSF) Office of Polar Programs (Grant No. 9909945) (Fienup-Riordan 2007). It also draws on ongoing work on Nelson Island beginning in 2006, carried out under NSF's BEST (Bering Ecosystem Study) Program (Grant No. 0611978). A much more detailed discussion of Yup'ik understandings of ice, including a dictionary of more than 70 sea ice terms, is included in *Ellavut/Our World and Weather* (Fienup-Riordan and Reardan in press). *Ellavut* documents the *qanruyutet* (instructions and admonishments) that continue to guide Yup'ik interactions with the land and sea, including snow, ice, and environmental change. Elders' concern has always been that young people understand

---

A. Fienup-Riordan (✉)
Calista Elders Council, Anchorage, AK 99507, USA
e-mail: riordan@alaska.net

I. Krupnik et al. (eds.), *SIKU: Knowing Our Ice*,
DOI 10.1007/978-90-481-8587-0_13, © Springer Science+Business Media B.V. 2010

how to treat one another. Yet elders also suffer over the fact that contemporary youth lack knowledge of *ella* – translated variously as weather, world, or universe – which many continue to view as responsive to interpersonal interaction.

Joint authorship reflects the close working relationship between cultural anthropologist Ann Fienup-Riordan and Yup'ik translator Alice Rearden. Ann and Alice attended gatherings together. Alice then carefully transcribed and translated information shared in the Yup'ik language, upon which Ann based her narrative.

Yup'ik elders actively support the documentation and sharing of traditional knowledge, which all view as possessing continued value in the world today (Fig. 13.1). In gatherings they speak their past selectively, and what is not said is often as significant as what is said. Long and careful listening to these conversations provides unique perspectives on Yup'ik knowledge. In these forums elders teach more than facts, they teach listeners how to learn. They share not only what they know, but how they know it and why they believe it is important to remember.

This mandate to document and share *qanruyutet* pertaining to proper social relations was in part realized in CEC's first two major publications (Fienup-Riordan 2005; Rearden and Fienup-Riordan 2005). Yet, as elders emphasized from the beginning, everything has rules: No one book could hold all a person needed to

Fig. 13.1 Elders and youth discuss place names during a CEC gathering in Chefornak community hall, March 2007. From *left* to *right*, Mark John, David Jimmie Sr., David Chanar, and John Eric. (Photo: Ann Fienup-Riordan)

know. Just as *qanruyutet* embodied time-tested rules for interacting with one's fellow humans, *qanruyutet* likewise guided one's interaction with one's environment. Paul Tunuchuk of Chefornak emphasized the importance of teaching young people about the world around them – not merely its physical features, but the ways in which one's actions elicit reactions in a sentient and responsive world. The reason for educating young people on *qanruyutet* is far from academic. Such knowledge, many believe, can change their lives.

Yup'ik sea ice knowledge may also hold value for other Arctic communities. Glaciologist Hajo Eicken (2009, personal communication) points out that as temperatures warm and sea ice conditions change, the experiences of Bering Sea hunters traveling in ice near its southernmost limit may help those living farther north adapt to changing conditions. Certain qualities of the sea ice cover and its seasonality exhibit strong gradients with latitude, and there is some indication that these different zones are now shifting north as ice conditions get milder in the Arctic itself. Thus the ice features that Bering Sea hunters know well may have special relevance for northern hunters in the future.

Imarpik Cimirturalartuq/The Ocean Is Always Changing

Elders repeatedly shared the observation that unlike the land, the ocean is always changing. In this they touch upon a defining characteristic of the Bering Sea coast relative to both Bering Strait and the Arctic coast of north Alaska (Fig. 13.2). There, tidal variation is modest and the landfast ice typically much thicker during most of the season, making it less likely to break and deform. On the Bering Sea coast, however, tides can vary as much as 3 ft. Moreover, the coastline's low elevation can translate into extensive mudflats during low tides, and fall storm surges can push water and ice inland up to 30 miles. Finally, Nelson Island and lower Kuskokwim coastal hunters can rely on neither multi-year ice nor a well-defined separation between freeze-up, winter, spring, and breakup regimes that, through the 1990s, helped impart more predictable conditions in most high Arctic regions (Eicken 2009, personal communication).

An added element of complexity is the important role played by deformation in the thinner (less than 3 ft.), weaker, and hence more dynamic ice of the Bering Sea coast. The result, as Yup'ik hunters know well, is a remarkably rough ice environment, where there is really no safe place, as all types of ice have some propensity to break up or deform or behave in potentially dangerous ways. Elders are indeed correct when they say the Bering Sea is always changing. In Barrow, the rule of thumb is that older, first-year ice, and multi-year ice thicker than 3 ft. is usually safe from deformation (Eicken 2009, personal communication). Along the Bering Sea coast, however, the ice rarely reaches that thickness through calm growth and hence even level ice is likely to deform in some fashion. As a result, hunting in and around the ice requires a wealth of knowledge regarding its formation, physical characteristics

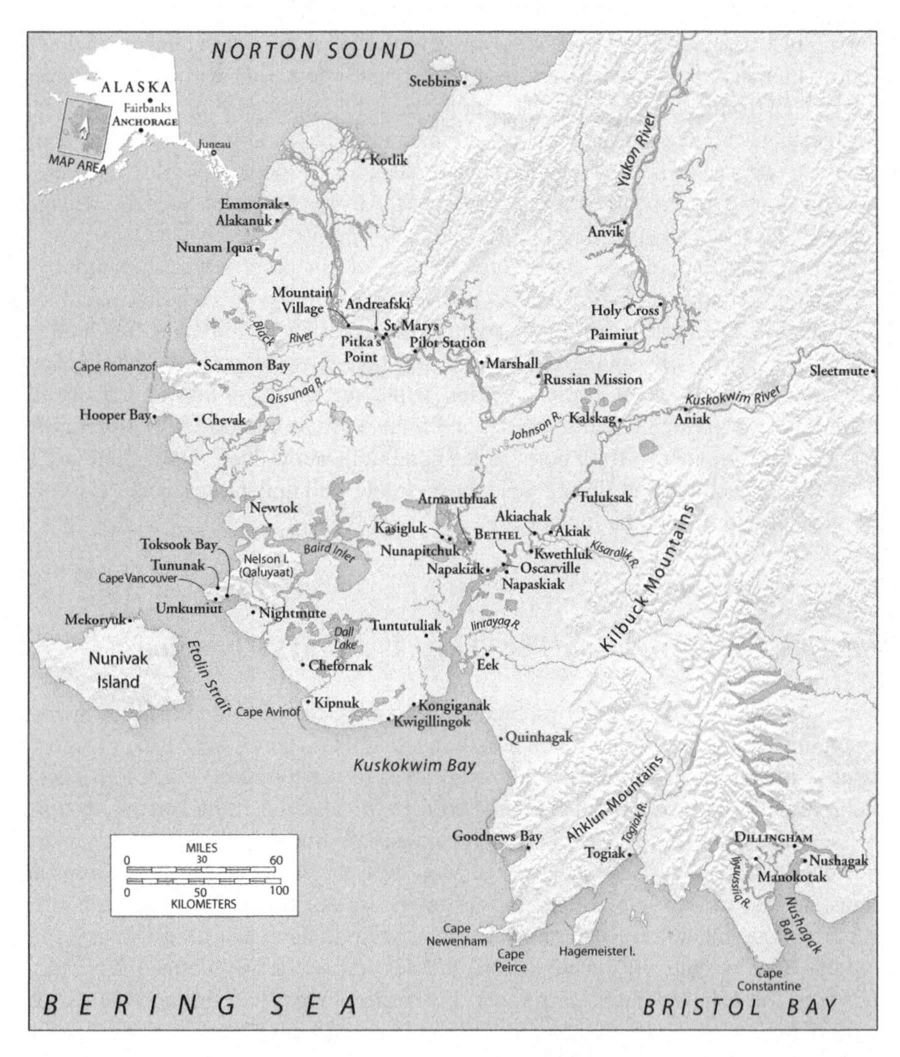

Fig. 13.2 The Yukon-Kuskokwim delta, 2009. Patrick Jankanish and Matt O'Leary

and behavior, and its dangers. John Eric (March 2008:293) of Chefornak noted that although a person could learn the normal seasonal cycle, no year was ever the same: "When an entire year has passed, the [ice formations] cannot be exactly the same as the year before. A person cannot learn it, but we can talk about it based on our observations of what it looked like in the past."

Before ever venturing out on the ocean, young men in the past were taught what to look for and what to do there. Simeon Agnus (December 2007:91) of Nightmute remembered, "Inside that very large *qasgi* [communal men's house] during spring, they would listen without making any noise at all when elderly men spoke. I think those people knew everything there was to know, since they paid close attention to things, since they listened intently to their instructors." Such instruction is no longer universal. John Eric (December 2007:83) remarked,

> When someone encountered danger before GPS devices were around, I asked him, "What type of ice is there where you are located? Are there *kaulinret* in your area, are there *et'galqitat* that are large?" He told me, "I don't know what those are because I've never heard about them." Indeed, although we ask these people who have no one to instruct them about what the wilderness is like, they won't know. It would be good to tell these young men the names of things on the ocean.

Imarpiim Cikulallra Uksuarmi/Fall Formation of Sea Ice

Along the Bering Sea coast sea ice forms anew every year, usually in November or early December. Paul John (December 2007:66) of Toksook Bay described how fall rains contribute to rapid ocean freeze-up: "Toward fall, when it starts to rain a lot, fresh water accumulates on top of the salt water. They say that leads to the ice freezing at a faster rate during fall." Fresh water accumulating along the ocean leads to the formation of *cikullaq* (newly frozen ice, frozen floodwater on the ocean, lit., "thing of *ciku* (ice)"), also known as *nutaqerrun* (new ice) or frazil ice or grease ice in English (Fig. 13.3).

Paul John (December 2007:66) recalled the dangers associated with traveling in fall when *cikullaq* began to accumulate along tidelands: "They said when the tide comes up [during high tide], the water that covers the mud also freezes. Back

Fig. 13.3 Newly frozen ice, known as *cikullaq* or frazil ice, January 2008. (Photo: Nick Therchik Jr. 001(2))

when they used kayaks, they warned not to travel anywhere when it started to get cold and *cikullaq* started to form because it would tear the kayak [skin coverings]." Paul Jenkins (March 2007:254) of Nunapitchuk confirmed that travel in *cikullaq* was dangerous: "A kayak could not travel anywhere when *cikullaq* was around. The weather might be calm and windless, but a person couldn't [hunt]. But when the sun got warm and the *cikullaq* started to soften, we'd travel in a group down on the ocean."

Hunters could also find themselves stuck when rapidly freezing *cikullaq* prevented them from going to shore. Paul John (December 2007:67) described how hunters could line up to travel through *cikullaq*: "If there were a number of people, they would connect their kayaks together with ropes. They lined up behind one another. One of the men would place his wooden plank seating [along the front of his kayak] to block [the ice] and would paddle through that ice. That was the instruction they gave back when *cikullaq* formed quickly during fall."

Paul John (December 2007:67) noted other essential equipment during early freeze-up: "Also, they told them to start bringing their snowshoes with them when *cikullaq* was forming. If *cikullaq* happened to obstruct his path to shore, those snowshoes would prevent him from falling through when he put them on and towed his kayak on top of the *cikullaq*."

Following the onset of cold weather, *cikullaq* steadily built up along the shore. Peter Dull (March 2007:574) of Nightmute explained, "When it is cold and the tide comes in, the ice comes up on the mudflats and stacks up. They get thick by stacking on top of each other. Even though they are thin, they multiply. When it is freezing, one must not pretend to be fearless. Hunting for seals toward winter is more daunting than hunting in spring." *Cikullaq* does not pose the same risks to today's hunters using aluminum skiffs. Paul Jenkins (March 2007:254) concluded, "These days, there are no obstacles for those younger than us. They are starting to travel through *cikullaq*."

Paul John (December 2007:68) explained how *cikullaq* led to the formation of *tuaq* (shorefast ice) in fall: "*Cikullaq* freezes when it's cold out, and the ice gets thick, and starting along the shore, the *tuaq* gradually forms and extends out toward the ocean. When [the shore ice] reached the area where it usually ended, it stopped gradually extending. It seemed that [shore ice] couldn't extend past the area where the ocean gets deep, what they call the *iginiq* [edge of deep water]" (Fig. 13.4).

Mark Tom (March 2007:1208) of Newtok observed, "When *cikullaq* continually layers and piles, that ice gets thick right away." John Eric (December 2007:77) called this layering process *qasmegulluni*. *Cikullaq* also thickens when it breaks apart and refreezes. John Eric (March 2008:348) described *cikullallret* (lit., "old *cikullaq*"): "These [pieces of ice] that the wind broke to pieces are *cikullallret*. And if the *cikullallret* freeze again, they will freeze again." John continued, "Since there is a lot of *cikullaq* around, some [pieces] continually pile on top of one another in layers when the tide comes up. Then another piece of ice [goes in another direction]."

Fig. 13.4 The edge of the shorefast ice as seen from Ulurruk, on the north shore of Toksook Bay, February 2008. Note the formation of *elliqaun* at the ice edge. (Photo: Nick Therchik Jr. 003(2))

Elliqaun/Newly Frozen Ice Sheets Along Shorefast Ice

As the shorefast ice continued to extend, *elliqaun* (thin, newly frozen sheets of ice, also *cikunerraq* or nilas in Western sea ice typologies) began to form along its edges. John Eric (March 2007:300) explained,

> *Elliqaun* is smooth, new ice that freezes at night. It is attached to the shore ice and gradually becomes thin as it extends out to the water. But the *tuaq* that is behind it is covered with snow....After it got warm, when it got cold again, they called the ice that formed *elliqaun*. When *elliqaun* froze, the surface was always moist. And when snow covered it, the ice underneath didn't freeze, but only became solid when it was extremely cold out.

Paul Tunuchuk (March 2008:361) commented on the viscous quality of freezing water: "They say when the [ice] starts to solidify, it freezes fast. The *elliqaun* is like that. Because it's like Jell-O^{TM} [and solidifies fast], although it's thin, a person can walk on top of it." Paul John (December 2007:70) described the formation of *elliqaun*: "Back when it used to be cold, along shallow areas were *et'galqitat*, small ice sheets beached in shallow water. And when it gets cold, the newly frozen ice gets stuck along those *et'galqitat*; that [ice that forms] is called *elliqaun*. The [ice] that the *et'galqitat* prevented from moving is *elliqaun*." Unlike shorefast ice, *elliqaun* is unstable and therefore dangerous. Paul continued,

> It isn't solid and stable, it's thin. *Elliqaun* is ice that formed during the night or over several days and sticks to the shore ice. *Elliqaun* is less solid and stable than real ice. And it can float away when the weather is warm and there are wet conditions caused by melting.

Although not as safe as shore ice, *elliqaun* is denser than freshwater ice. Paul John (December 2007:130) explained,

> Our elders mentioned that since river ice is fresh water, it breaks easily. But they said that since the ocean is salt water, its ice is much denser. Even though it shifts when a person walks on it, some areas [along the ice] don't collapse right away.
>
> Once I became terrified. Thinking that it had been cold enough, I reached *elliqaun* and traveled on, back when we started to use Skidoos [snowmobiles]. Then I noticed that the ice was moving as I was traveling, so I increased my speed and reached shore. I was amazed that I hadn't fallen through, having traveled with the ice moving beneath me. If it had been river ice, I would have fallen through and drowned. Since it was saltwater [ice] I didn't fall through.

Manialkuut Manigat-llu/Rough Ice and Smooth Ice

The surface of the shorefast ice could be more or less rough or smooth. Jagged pieces of ice pushed on shore by the high tide formed *manialkuut* (rough ice) (Fig. 13.5). Paul John (March 2008:550) noted that the location of jagged ice depends on conditions during freeze-up: "It's different all the time. When the wind freezes [ice] after it's windy, the ice isn't smooth. The broken pieces of ice gather together [and freeze]. Waves made by the wind cause that to happen." Although rough ice has always been present it is becoming more prevalent as fall storms have increased.

Fig. 13.5 *Manialkuut* (rough ice), with Toksook Bay in the distance, February 2008. (Photo: Nick Therchik Jr. 004)

John Alirkar (March 2008:550) of Toksook Bay noted the importance of smooth ice for traveling: "When traveling, they follow the areas of ice that are smooth although they are constantly turning, traveling all over the place." When the trail was rough, men worked to smooth it. Paul John explained, "We make a trail, smoothing areas, chopping the worst areas with ice picks. Sometimes when they hunted by kayak, when they went down they'd say that the ice was too jagged after it had been cold. Those who had gone for seals would return without hunting. It was possible for them to go down only after fixing a trail." John Eric (March 2008:330) recalled, "Sometimes the water is completely calm, but when the ice is rough, although we want to go down to the water, when we can't travel through the trail, we don't go."

Nepucuqiq/Rough Edge of the Shorefast Ice

John Eric (December 2007:73) described the gradual expansion of the shorefast ice:

> When ice forms and becomes thick, open water gradually moves down [farther from shore]. When the tide comes up in cold weather, the ice packs and sticks [to existing ice]. [The ice] forms rough and smooth, but the ice is mainly rough inside the channel since the tide comes in for six hours. It packs up like ice that accumulates in one place, and ice is placed haphazardly and then freezes together inside the channel. Then they stay there when it gets cold although the tide goes out.
>
> And it continually freezes, and when the tide comes in again, the newly frozen ice packs up again, and then it sticks and freezes in place. It gradually becomes thick away from shore.

Wind and tides, however, could break up the edge of the shorefast ice, creating a rough, impassable ice edge known as *nepucuqiq*. John Eric (December 2007:74) explained,

> Sometimes when the wind is blowing from the south directly against the shore, when there is a lot of ice, the large ice floes head toward shore and pack up; I used to hear them called *nepucuqit*. When [ice blown against shore] becomes thick, although [ice floes] are about three to four feet thick, the south wind breaks them to pieces and packs them.
>
> They were extremely rough and jagged, and a person moved on top of them like a dog [on all fours] and couldn't walk on them. When the tide comes in and it's windy at the same time, it packs up a lot of ice, extending a mile in any direction. And the shore ice that is usually smooth, although it is thick, [the incoming tide and wind] can break it to pieces.
>
> One time the shore ice was around three feet thick. Just as the tide started coming in, the wind direction changed and it started to blow. As the water level went up, the ice that had seemed impossible to move started folding toward the south and north, breaking to pieces. And here we thought that nothing could [break the shore ice to pieces].

John Eric (December 2007:75) noted that *nepucuqiq* could be so steep it could not be broken by ocean swells: "Back when Aassanaaq was alive, he said *nepucuqiq* formed that was very steep. And he said before it melted, they came up [to the Kuskokwim River area] in June. Steep ice had packed up [against the shore] at that time. And he said that ocean swells headed toward shore from the ocean, but it couldn't break up that *nepucuqiq*."

Travel is impossible over rough ice until snow fills in the low areas. John Eric (December 2007:74) noted, "Eventually, in December and January, the snow fills the rough areas. Boats couldn't go down when [the ice] was too rough, so they'd wait for snowfall. Only the snow smooths [the rough ice]." Conversely, Paul Tunuchuk (March 2008:331) warned that water-soaked snow can make a smooth trail over shorefast ice impassable: "Sometimes really smooth ice cannot be traveled on. The snow is soaked with water. They call that *mecqiitaq*. When snow gets soaked with water, that sometimes ruins the trail." John Eric (March 2008:332) agreed, adding that *mecqiitaq* is also hazardous when covering *elliqaun*:

> The surface of *elliqaun* is usually very smooth. And since that *elliqaun* isn't like shore ice that froze during winter, when it is covered with snow, water fills [the *elliqaun*] from underneath for a great distance. Although the surface is good, the bottom is bad.
>
> When it's below zero, [the *elliqaun*] is good and solid, but when it's warm, it immediately melts [the ice] underneath the snow. It's not good to walk on. One's shoes get heavy as the snow tends to stick to them like mud. That's how I've always viewed *mecqiitaq*.

Tuam Cimillra/Changes in Shorefast Ice

Freeze-up today occurs later than in the past, and the shorefast ice does not form as far out. Nelson Island elders provided details on what other coastal residents observe. Phillip Moses (January 2007:494) of Toksook Bay noted, "Right now it takes so long to freeze up, and [the *tuaq*] stops right there by Qikertaugaq [small rock outcrop at the head of Toksook Bay]. In the past, it always froze nice and smooth up to Ulurruk. The ice extended when it kept freezing, and when it got to Ulurruk, it stayed there for quite a while." Paul John (December 2007:252) agreed, "When I came to observe it, [the edge of the shore ice] was never too far form Ulurruk. The ice reached that area and was solid like lake ice." One year, when men were still hunting with kayaks, shore ice extended all the way to Up'nerkillermiut, and all of Toksook Bay was frozen. Camilius Tulik (March 2007:540) of Nightmute recalled, "That was when the ice really extended out. Now it has come way into the bay."

Michael John (March 2007:1038) of Newtok observed the same loss of shore ice north of Nelson Island: "Sometimes the shore ice isn't very extensive. And it hardly forms extensively below the village of Tununak because there are more sandbars nowadays." Many associate lack of shore ice with warming temperature. According to Paul John (December 2007:72), "Since it no longer gets extremely cold today, the shore ice that forms is no longer extensive; it has changed. Back when it used to be [extremely] cold, there was ice [on the ocean] and *evunret* [piled ice] formed in places." Moreover, shore ice does not stay as long. Simeon Agnus (December 2007:72) recalled,

> Our bay isn't like it was when we first became aware of life. It used to form real ice in the past. And while people were still able to travel by sled and dog team, all the spring birds would arrive. Its shore ice stayed for a long time. And although [*elliqaun*] stuck [to the shore ice] it would immediately drift away. The current would take it out [to the ocean].

And the [channel] beyond it that we call Kangirpak is like that as well. These days it no longer forms real shore ice. It has started to become dangerous following the warming of the weather.

Paul John (December 2007:68) noted that shore ice was thicker in the past: "Back when it was cold, the shore ice was solid and stable; it formed thick ice. And the ice sheets that floated [along the ocean] were thick." Peter John (March 2007:1158) of Newtok added,

In those days [in the 1950s], shore ice froze about three to four feet thick. When we were on top of [the shore ice], when the tide was coming in and the current caused a large piece of ice to bump into [the shore ice], the [piece of ice] would break to pieces although it was thick. When it hit, the sheet of ice would move under the shore ice although it was large and thick.

These days, since these areas have become shallow, now that the current isn't as strong, the pieces of ice no longer constantly go underneath the shore ice. The shore ice has gotten thinner, and sometimes it's about one foot thick. This year it's thin, and the water is close to shore since the weather was constantly warm last fall.

Shore ice also no longer stays as long in spring. Edward Hooper (March 2007:1306) of Tununak remembered, "There used to be shore ice down there along the edge of the channel, almost three miles [long], when we went seal hunting [in spring]. These days, one can't go and spend a night [on the ice] thinking that the ice may detach and float away while they're on it. We used to spend nights down below the shore, right behind the water." Edward's sister, Susie Angaiak (March 2007:1306), shared her experience:

If you happen to spend a night, you will end up calling a helicopter. [*laughter*] When I became aware of life, the shore ice extended far out to the ocean. It stayed a long time and never went away. They'd tow [their kayaks] for a long time to go seal hunting [to reach the ice edge]. And how pitiful these days, the shore ice will be extensive and then the next day one will check and see it's completely dark [from open water]. Where did the ice go? These days, the shore ice can no longer stay.

Qanisqineq/Snow in Water

Both fall and spring hunters encountered *qanisqineq* (from *qanuk*, "snow"), snow that gathers on the ocean, also referred to as slushy snow or slush ice. John Eric (March 2008:302) explained, "Sometimes, before the shore ice goes away, a large amount of snow accumulates down on the ocean, when it continually snows. Although it appears thin, the snow is thick as the current [gathers] and piles it up. When the tide goes out, it takes [the snow] away but doesn't let it scatter. It accumulates deep in the water, and most of the area is covered by *qanisqineq*."

Qanisqineq can obstruct a hunter's path and is potentially dangerous. According to Simeon Agnus (December 2007:123), "The following was an admonishment down on the ocean: They say when *qanisqineq* suddenly surfaces on top of sand-bars during spring, some is thick. When *qanisqineq* accumulates, a small outboard motor cannot travel through it as it won't be able to take in water. [*Qanisqineq*] is dangerous when it is thick and white, and it's better if a person doesn't try to travel through it."

Pugtalriit Pugteqrutet-llu/Floating Ice and Those That Suddenly Surface

Other sea ice forms important to recognize were *pugtalriit* (floating ice). John Eric (December 2007:79) explained, "Some are about the size of this table. Although a boat goes on top of it, it won't break. And seals climb on top and sleep as much as they want. We call those *pugtalriit*."

Pugteqrutet were pieces of ice stuck to sandbars that suddenly surface, also known as *tumarneret* or *tumarngalriit* (lit., "ones that are assembled"). John Eric (December 2007:73) noted, "Since there is a lot of mud around, and it formed smooth ice, that *pugteqrun* that formed during fall was dangerous. They say when the tide goes out, it can float a boat out to the ocean and away from shore."

Evunret/Piled Ice

Huge piles of ice began to develop in fall in shallow areas along the coast, including channel edges and sandbars (Fig. 13.6). John Eric (March 2008:171) noted, "The ocean forms *evunret* [lit., "those that are piled"] by piling pieces of ice on top of one another and breaking them to pieces, and then the cold weather welds them and

Fig. 13.6 A hunter atop *evunret* (piled ice), using it as a lookout. Note the *negcik* (gaff) he carries for safe traveling. (Photo: Leuman W. Waugh. 1935, National Museum of the American Indian, Smithsonian Institution, L2236)

Fig. 13.7 *Marayilugneret* (piled ice mixed with mud) on the north shore of Toksook Bay, May 2009. (Photo: Mark John 1990)

freezes them in place. The water forms [*evunret*] during winter since it's the ocean. That's why there is an oral teaching that no one should underestimate the ocean at all."

Paul John (December 2007:64) described dark-colored *evunret* mixed with sand: "Those *evunret* that have sand mixed in were called *asvailnguut* [lit., "ones that are solid and immovable"]." Piled ice mixed with sand was also known as *marayilugneret* (from *marayaq*, "mud") and *tungussiqatiit* (lit.,"ones that are dark") (Fig. 13.7). Paul John (March 2008:537) noted that sandy *evunret* are sturdier than those formed in deeper water:

> Since the area around those is shallow, when the wind blew the waves and sand, the sand that splashed on there caused it to become sandy and dark and they become *marayilugneret*. They considered the ones with sand in them stable and strong, and the sand prevented them from breaking to pieces right away. [Waves] splashing on them in cold weather cause them to stay solid.
>
> They say that since the white-colored [*evunret*] are merely piled ice, they are more dangerous and break more easily than those dark ones.

John Eric (December 2007:82) noted that *evunret* could be huge: "They say when *cikullaq* comes upon *evunret*, it sticks and eventually [the *evunret*] get large. Once we came upon very large *evunret* during June. We climbed up, and when we looked down on our boat, it was small. I think we climbed up about fifty feet. They were [*evunret*] that the wind created. Since the ocean is large, since the current is strong, it creates [ice formations] that cannot be broken to pieces. It can create very large *evunret* in minutes."

Large *evunret* grew in predictable locations year after year. John Alirkar (March 2008:535) affirmed, "[*Evunret*] always formed along the *iginiq* [edge of deep water], the area where the ocean bottom suddenly gets deep. They always formed in their usual places." Things have changed (see below) for a variety of reasons, including warming weather as well as sandbars forming in shallow areas at the mouths of bays and rivers.

Evunret growing in predicable locations every year were given names, often in the plural, as they were not considered singular monoliths but groups of ice pieces. Visible from far away, *evunret* were navigation aids. Paul Tunuchuk (March 2007:304) noted, "In the past, *evunret* served as markers for us. Although they traveled past them, they knew where they were located through these *evunret*." *Evunret* were also used as lookout points and places where hunters sought safety. Paul John (March 2008:535) explained,

> *Evunret* were used for safety when ice floes suddenly prevented them from traveling. When there was a lot of ice around, they'd go along their sheltered side to wait until they had a way to go. And when it suddenly became windy and there was no ice around, they would go along their leeward side to wait for the wind to calm down, when it was impossible to paddle because of the wind.

Cenami Up'nerkam Nalliini/Spring on the Coast

> They also told us that after winter, when spring comes, the ocean doesn't stay in its usual state. For that reason, although someone says that they have learned the ocean, they will not learn to predict its conditions. (Paul Tunuchuk, March 2007)

Peter John (March 2007:1162) soberly recalled that young men were carefully instructed as spring approached: "They would have started to talk about the ocean at this time [March], giving instructions about what to do when they began to travel during spring. Teachings for fall and spring weren't the same." As sunlight returned and days lengthened to more than 12 hours by mid-March, snow melted and conditions on the sea ice rapidly changed. John Eric (March 2007:265) continued, "During the time when these bearded-seal and spotted-seal pups are born, the names [of ice formations] change."

With returning sunlight, warm weather begins to melt once-solid shore ice (Fig. 13.8). John Eric (March 2008:181) explained how in April the places where the channels are located will form *kangiqutat* (bays): "When the sun gets closer, it heats the shore ice, and it rots and becomes thin. Then [river mouths] become [concave]. The current also melts the ice. That's how the area down below my village becomes during spring." Paul Tunuchuk (March 2008:182) warned against entering river mouths and *kangiqiugneret* (coves in the ice) during spring:

> If there is a bay in the ice ahead and you unknowingly continue to travel along the ice edge and enter far into that bay, when you look back you'll see that your way out has already become obstructed, and the ice has already enclosed you. You will stay in the middle of ice, and you won't be able to go anywhere. People are told to pay attention inside bays. They say they quickly close in.

Fig. 13.8 Broken ice along the north shore of Toksook Bay, May 2008. (Photo: Nick Therchik Jr. 007)

Paul John (March 2008:552) also recalled the admonishment to pay attention to the *qapuut* (foam) that forms along ice floes in spring: "*Qapuut* start to appear through small holes. When it melts and holes appear, *qapuut* become visible." John Eric (December 2007:110) noted that *qapuut* are a sign of melting ice: "When *qapuut* start to be seen, although [the ice] is white, it becomes thin and dangerous as the current melts the ice from underneath. When foam starts to form, the shore ice rots quickly." Indeed, along the Bering Sea coast where ocean currents can bring in warm water, rapid ice melt often occurs from the bottom rather than from the top, aided by snow cover acting as a thermal insulator (Eicken 2009, personal communication). Paul Tunuchuk (March 2008:334) reiterated John's warning: "If you see foam, that area is dangerous to walk on. Starting from foam, holes form. If you see foam and happen to travel through it, you will fall through."

John Eric (December 2007:113) noted that snow might hide dangerous ice: "Dangerous areas [along ice] are hard to distinguish because it's white. Although it's [thin], it's white. But when some areas are a little dark, you can tell [that it's dangerous]."

Paul Tunuchuk (March 2007:281) described how rapid melting could take place behind the thick, rough edge of the shorefast ice, the *nepucuqiq*.

> Some time ago, when the shore formed *nepucuqiq*, when I was bringing [a seal] to shore in the evening, I started to see pools of water that had melted just as I was about to reach the boats. The next morning when I went outside, I saw that along the area that I had walked through there was an open water current. What I had thought were pools of melted water were apparently holes. When there's *nepucuqiq* along the shore, the area behind evidently melts rapidly first.

John Eric (March 2008:298) also described the effects of *nepucuqiq* on the melting process: "They said long ago, that huge *nepucuqiq* was steep, the broken [ice] that the wind and water had piled up. And they said the rivers around this area broke up and were free of ice; although the ice was gone [in the rivers], that [*nepucuqiq*] was still there. That jagged [ice] was extremely steep, and they said they eventually went to the Kuskokwim to harvest salmon at summer fish camp [while it was still there]." Alternately, some years the rivers stay frozen and the ocean melts first. John continued, "They say that sometimes the ocean [ice] melts before the river ice. [The ocean ice] will drift away as one large ice floe by detaching along the shore."

Qairvaat/Ocean Swells

The occurrence of *qairvaat* (ocean swells, lit., "big *qairet* (waves)") marks the transition from relatively stable winter hunting conditions along shorefast ice to the myriad sea ice forms of spring. These long, high-amplitude waves originating in deep water far from shore can break the ice pack hundreds of miles away from the open ocean. Swells travel relatively fast and present a direct link to processes far away (Eicken 2009, personal communication). John Eric (December 2007:113) explained,

> When *qairvaat* first arrive, conditions are bad over there, as they are powerful [when they hit]. When that starts to occur, the sun starts to heat the shore ice toward the shore. Once in a great while the *qairvaat* break the ice and take it out to sea. [The shore ice] gradually recedes, and eventually it gets closer to shore and reaches the area where it becomes land. That's how the ocean down below my village is during spring.

John (March 2008:298) observed that current and wind are as important as heat in ice breakup,

> The ocean breaks the ice to pieces only during the time *qairvaat* are around. Then after breaking them to pieces, the sun melts the areas around the rivers, it rots the ice in those areas and forms holes there, and eventually the ice starts to recede toward shore. They become bays.
>
> The current also melts the ice from underneath. Although the ice is white, it will be extremely thin. That's why they tell us that a trail that we took in the morning can melt by the afternoon and be impossible to travel on. The sun heats the area around the river, and the current melts it.
>
> But the areas on top of sandbars aren't like that. Although it isn't like that, the *qairvaat* break it to pieces when the tide is about to go out, and when the tide goes out, they take the ice] out [to the ocean].

Ocean swells are a powerful force to be reckoned with. John Eric (March 2008:296) emphasized,

> *Qairvaat* are large, and a sheet of ice that is a distance away disappears from sight when [the swell rises]. They say they come up toward land from the ocean and can rise about three feet or more. Because they are large and wide, no matter how thick an ice sheet is, [*qairvaat*] can break it.

Paul John (December 2007:106) noted that ocean swells arrive when the ice that held them back begins to recede: "When ice that is floating on the ocean starts to melt, waves that are present far from the [shore] ice no longer have ice to hold them back, and they start to reach the shore. Those waves located far from shore are formed without ice to block them. Then when the ice begins to melt, when [the ice] diminishes, [the waves] start to reach shore down there."

Ocean swells are, in fact, a greater hazard in the Canineq (lower Kuskokwim coastal) area than around Nelson Island, which is protected by nearby Nunivak. Paul John (December 2007:106) explained, "Although the weather is calm sometimes, [the waves] are deep in shallow areas. Akuluraq [Etolin Strait] has small [waves] since Nunivak Island blocks the *qairvaat* that head toward shore from the ocean. That's why there aren't large *qairvaat* around Nelson Island."

Angenqaat/Large, Drifting Ice Floes

As ocean swells break the ice they create free-floating ice forms of various sizes. Some were particularly dangerous, and all were important to recognize for safe traveling. *Angenqaat* (lit., "biggest ones of a group") are large, moving ice floes that break away from the shore ice and drift in the ocean after ocean swells. Paul John (March 2008:539) recalled, "During spring there are *angenqaat* that the south wind or north wind brings to this area." John Eric (December 2007:81) noted their size: "Before *qairvaat* broke them apart, these *angenqaat* could be a half-mile in diameter and about three feet thick." According to John Eric (December 2007:76), "They say those *angenqaat* that are about half-a-mile long are powerful and don't stop moving right away." Paul John (December 2007:71) advised against staying on *angenqaat* when drifting:

> If their trail to shore was obstructed, *atreskaki* [if they were drifting] as they say, they were supposed to stay on a sheet of ice that was thick but not too large. They say if they go on a large sheet of ice, since large ice floes tend to travel farther, when it starts to move, it parts the smaller ice around it and enters closely packed ice. They also told them not to stay on top of an *angenqaq*.
>
> They told them only to stay on ice floes that are thick but not large because those smaller ice floes actually stop when they collide against other ice, and they don't drift inside closely packed ice. That's what they said when they told about dangers.

Angenqaat might also detach from the shore ice and drift away with people on it. Paul John (December 2007:117) recalled, "They call it *angenqiurluni* when a large piece of ice detaches, sometimes along a crack over a mile or two from shore, and drifts out to the ocean." Paul Tunuchuk (March 2008:315) noted that this was more common in the Canineq area: "It takes the ice out [to the ocean]. A large ice floe detaches [from the shore ice] when the tide is very high."

Paul John (December 2007:118) noted the admonishment to head to shore when ice detaches: "When the visibility is good, they search the area behind them and see water toward shore, and a person can seek safety by heading to the other side of the ice. When the ice drifts away, when he sees the place where the ice detached, he tries to get out of that situation by heading toward shore." Paul (March 2008:542) recalled many accounts of people floating away on detached ice: "Some people would float away when the ice detached when it was too windy down on the ocean. They would drift [on the ice] when conditions weren't good for paddling. But when conditions were good for paddling when the ice detached and floated away, they would paddle and go up on top of ice that detached and head along the ice toward shore."

Akangluaryuut/Ones That Roll

Among the most important ice formations to recognize were *akangluaryuut* (lit., "ones that roll"), rounded sheets of floating ice that could tip over, referred to as pancake ice in Western sea ice terminology. Some were small, but others were quite large. Paul Tunuchuk (March 2008:311) explained, "Back when they used kayaks, they were admonished about ice called *akangluaryuut*. If they happened to [be paddling] close to the edge [of the shore] when [*akangluaryuut*] suddenly rolled, it would immediately capsize that kayak. Boaters and not just kayakers are admonished not to travel along the edge of *akangluaryuut*."

Akangluaryuut occur in spring, as ocean water melts the submerged part of the ice and makes it lighter. Paul John (March 2008:544) explained, "When [the ice] was frozen, the bottom was heavier. And then the water began to melt the bottom [of the ice] in spring. Then the top [of the ice] becomes heavier. Since the bottom down there becomes lighter, it suddenly rolls."

Marayilugneret/Piled Ice Mixed with Mud

Elders also shared valuable observations about sediment-laden ice – a common characteristic of the shallow, muddy coastal environment where clear ice is the exception rather than the rule. Eicken (2009, personal communication) notes that whereas coastal erosion is often attributed to lack of sea ice allowing fall storms to eat away the shoreline, in fact sea ice is the most effective mover of sediments in Arctic and sub-Arctic waters with seasonal ice cover.

Simeon Agnus (December 2007:145) explained how *marayilugneret* (piled ice mixed with *marayaq* (mud)) mixed with *marayaq* (mud) form along sandbars in fall and become dangerous in spring when they suddenly break loose.

> They caution people about those *marayilugneret* that surface from underwater and suddenly wobble, rising tilted to one side. A person must watch out for those; when the shore ice has gone, there are many [*marayilugneret*] along sandbars.
>
> If a *marayilugneq* happens to surface and hit a boat, it can capsize. And when it suddenly surfaces, some are extremely steep and high. They don't [surface] flat but vertically. They also caution about those *marayilugneret* that froze in place underwater along sandbars.
>
> When the ice starts to melt, they suddenly surface once in a while. If that type of ice happens to [surface], it could injure the people inside a boat. And if it hit the center of a kayak, it could break it.

Paul Tunuchuk (March 2008:321) noted that *marayilugneret* are hard to see and thus dangerous:

> They say when those dark ice sheets poke out of the water a little, they are dangerous. They are muddy and hard to see. They anchored along the mud and could suddenly surface as they detach.

Kaulinret wall' Kaimlinret/Small, Rounded Ice Pieces

Another form of floating ice in spring was small, rounded pieces of broken ice known as *kaulinret* or *kaimlinret* (from *kaimlleret*, "crumbs"), what ice scientists call brash ice or ice pebbles (Fig. 13.9). John Eric (March 2008:344) described their formation:

> Toward summer, when they begin to melt but haven't really melted, some *kaulinret* are small and round. Although some areas are a little scattered, those drift together as a pack. Some are nice pieces of small ice, and some are small pieces of ice mixed with sand. They say those are leftover from ice that melted....And when these *kaulinret* suddenly appear from underneath ice that is situated in one place, they accumulate in the surrounding area.

In the past *kaulinret* could be difficult to travel through. Paul Tunuchuk (March 2008:309) explained, "One cannot go through *kaulinret* with a kayak but only with an outboard motor. If you travel through an area that is covered with small pieces of ice, you will see *kaulinret*."

Fig. 13.9 *Kaulinret* or small, rounded pieces of ice said to be the playthings of young bearded seals, May 2009. (Photo: Mark John)

Icinret/Overhanging Ice Edges

As warming continued in spring, ice sheets began to develop *icinret*, thin, melting, overhanging edges. Paul Tunuchuk (March 2008:312) notes the danger they pose: "The water melts the ice underneath [the overhang]. When a wave slaps [against the ice], it goes underneath [the ice], and then eventually [the ice edge] becomes an [overhang]. If you step on top of that, it will break and you will slide down the ice [into the water]. One has to continually use a *negcik* [gaff] to check for those *icinret*." Paul John (March 2008:553) agreed, "We always checked [the ice edge] when we were about to get out of our kayaks. These days even a boater does the same, checking to see if the *icineq* would collapse."

Cikum Akuliikun Ayagalleq/Traveling Through Ice

As noted above, significant tidal variation along a shallow coastline can translate into miles of mud flats or "ice flats" at low tide. In addition, Eicken (2009, personal communication) notes that there are more steady currents driven by freshwater inflow into the ocean and other factors that can be strong and seasonally variable. As elders' comments make clear, the Bering Sea has strong and variable winds. These different currents, tides, and winds can drive the ice at different speeds or in different directions. Generally, the rougher, more deformed the ice, the more effective is the transfer of momentum from the air or water to the ice. Moreover, the mass of the ice and the size of individual floes determine how ice responds to changes in forces such as tidal oscillations that drive ice motion (Eicken 2009, personal communication). Elders reveal a deep and subtle understanding of these complex forces in what they say about the ice and how to travel through it.

Traveling through sea ice was fraught with peril. Men paid close attention to instructions regarding specific conditions. Paul John (December 2007:69) recalled admonishments for traveling through ice: "They were instructed to watch for the possibility of the wind and current drifting the ice [out to open water] and causing it to collide and pile [onto other ice]. If [the ice he was on] happened to hit another piece of ice and break to pieces, he was told to try to avoid breaking and piling ice from reaching him."

When traveling in and out of open water, hunters were always on the lookout for the onset of dangerous onshore winds. Hunters were warned against entering bays along the ice, as changing currents could quickly pack the ice and close their exit. Simeon Agnus (March 2007:554) commented on the fluid character of the ocean's surface: "When there is a lot of ice, it changes every single day; there is no way that anyone can learn it. When it is absolutely calm, one has no reason to fear anything about the ocean. But when there is a lot of ice, it is dangerous to be on the outer side of the ice in case the wind picks up."

Hunters paid close attention to places where currents came together, as these tended to fill with ice. Paul John (December 2007:125) explained, "The place where [the currents] meet is called *ilacarneq*. In those places, the openings of coves tend to fill with ice and close in." An outgoing tide could be particularly dangerous. Simeon Agnus (March 2007:557) said, "When the tide is going out of Kangirrluar [Toksook Bay], if the ice is trying to split, it gets really tightly jammed. And the north-bound current helps it. And the ice in Kangirrluar pushes it out toward the ocean. It is extremely dangerous."

Stanley Anthony (December 2007:93) described the need for vigilance in areas where currents meet during an outgoing tide, packing ice together and obstructing one's path: "They said not to flee toward that area where [ice] tends to meet but to go toward the ocean. They said when the tide comes in again, the area to the north forms a trail; the ice starts to scatter and separate." Camilius Tulik (March 2007:535) was also admonished to head toward open water if trapped in the ice: "They told us not to panic if the weather is fine but to flee out to the ocean. One does not experience hardship in Akuluraq." Stanley Anthony warned, "Also, farther down, when there is an outgoing tide, one has to be alert because the outgoing tide is strong and heads toward Nunivak Island. Although a boat is situated on top of floating ice, it will move at a rapid speed toward the western part of Nunivak."

Simeon Agnus (December 2007:98) recalled the instruction not to struggle if caught in ice in Kangirpak but to follow the current north: "When a person's trail is obstructed down below Kangirpak [because of the accumulation of ice] when the weather is calm, when the tide is coming in and not going out, if the ice suddenly closes in on him there, a person shouldn't try to leave. When [the current] drifts him north, in front of Kangirrluar, when the ice starts to scatter, he can get out [of the ice] without struggling. The north current is fast down there at Akuluraq."

Paul Jenkins (March 2007:255) reminded his listeners that packed ice could prevent hunting, even when animals were present: "Sometimes when [the ice] drifted [sea mammals] toward the north in the pack ice, we couldn't do anything but watch them pass." Simeon Agnus (July 2007:276) noted that even when hunting in the ice was possible, men could not take a straight path: "The ice along Etolin Strait that goes back and forth [from the ocean toward the shore] is extremely packed and dangerous during spring. Although a person knows the location of his destination, if ice obstructs his path, he will be unable to go there even if he wants to."

Nanviuqerrneret/Ice-Free Areas Within the Ice Pack

Elders also described *nanviuqerrneret*, lake-sized areas of open water within floating ice. Currents combine in some places, opening some areas and closing others. Like bays, *nanviuqerrneret* can trap hunters when ice closes in. Stanley Anthony (December 2007:93) of Nightmute noted, "Sometimes there is a large *nanviuqerrneq* when there is a lot of ice. A person must not go inside without paying attention, since the area where he entered can immediately close. He should go to that *nanviuqerrneq* with constant caution. That was an instruction they gave, that if I started to travel with a boat, I should watch the place where I entered the *nanviuqerrneq* although that place was serene, that it closes when there is a lot of ice [when the current changes]." Simeon Agnus (December 2007:97) noted that *nanviuqerrneret* in some areas are particularly dangerous: "When the ice is situated far down in the ocean, when coming upon a *nanviuqerrneq*, no one should try to go inside although it looks calm. That [type of *nanviuqerrneq*] can close up; the *nanviuqerrneret* that are situated toward Nunivak Island are dangerous. The northern current is strong there. If he tries to exit, he will have a difficult time when [the ice] closes in on him and obstructs his path."

Aarnarqellria Ciku/Dangerous Ice

Elders shared abundant information on sea ice, in large part motivated by the danger it poses to uninstructed youth today. Camilius Tulik (March 2007:558) declared, "We should certainly talk about these things at this time, since these boys have come to be daredevils. [The ocean] is dangerous." Both men and women openly discussed their experiences – including errors in judgment – in hazardous situations. Some types of ice are inherently dangerous, like *elliqaun* (thin, new ice sheets) and *kaimlinret* (small pieces of broken ice). Simeon Agnus (July 2007:277) noted the warning not to venture far from shore when *cikullaq* (frozen floodwater) formed. Paul John (December 2007:69) agreed, "When there is an accumulation of ice, when *cikullaq* tends to form toward spring, a person has to be careful." Paul Tunuchuk (March 2007:260) added, "A kayak could only go through *cikullaq* when it thawed and got soft when the sun got warm. Also, they warned us not to travel when ice was piling. When ice piles, it thickens, and they would be unable to help that person. They said when the ocean is in the process of freezing, it prevents even a paddle from moving. These days, people travel wherever they want."

Qanisqineq (snow that gathers on the ocean) could be dangerous. Peter John (March 2007:1180) recalled, "When *qanisqineq* formed and became thick during spring, it was extremely abrasive to paddle through, and it kept one from traveling fast. We were cautioned not to travel through thick [*qanisqineq*] and to watch the current when carrying our catch. And they told us not to be confident in traveling through the lighter-colored [*qanisqineq*]. Even though [*qanisqineq*] is thin, it is extremely rough to travel through. And when traveling with a boat and going through thick [*qanisqineq*] the motor can't take in water [and will overheat]."

John Eric (March 2008:318) noted the dangers of dark-colored ice: "Some pieces of ice are hard to see. That's why we tell people that one must pay close attention while down on the ocean. If a person is driving the boat fast and comes upon that kind [of ice], his motor will suddenly split or fly off. They say they are smooth pieces of ice floating a little in the water and covered by sand. We tell boaters that those are dangerous as even a small piece of ice can cause a motor to break down."

Tommy Hooper (March 2007:1303) of Tununak said that among the most important instructions was not to give up. Simeon Agnus (March 2007:562) added the importance of not showing fear: "It is also a prohibition that when one has a younger companion, even though he is afraid, he must not talk in such a way to cause his young companion to panic." As important as not showing or giving in to fear, one should not be fearless. Peter Dull (March 2007:561) said, "They are told not to be over-confident and climb up on [pieces of ice] right away. One must be cautious because some have holes and cracks that go all the way down." Young people, however, take risks. Camilius Tuluk (March 2007:562) stated, "Nowadays, thinking that they have good gear to hunt with, we know that they are fearless since they have not experienced a dangerous situation, and they probably are curious about doing some things in the ocean." Simeon Agnus recalled what his generation knows well: "The ocean is daunting. When one says that he has learned the ocean, he is gravely mistaken. Every single day, [the ocean's] condition changes when it has a lot of ice."

Ella Iqlungariuq/The Weather Is Becoming a Liar

An undercurrent of concern ran throughout elders' discussions. Everything – rivers, lakes, wind, waves, snow, ice, plants, animals, and weather in every season – seems to be changing. Many commented on the increasingly unpredictable character of today's weather. According to Paul John (January 2007:11), "They said that wild rhubarb became as tall as people back when cold weather formed frost without any erratic and stormy weather. Then they began to say that since weather conditions are now unpredictable, the wild rhubarb no longer forms frost. Back during times when the weather wasn't erratic and stormy, I think our land always stayed cold. That's why they called it *ellarrliyuitellra* [a time when there wasn't any erratic and stormy weather]."

Today many say the weather is a liar. According to Camilius Tulik (March 2007:502): "The weather, like they say, has become a liar. When it is supposed to be calm, big winds come. When it seems like it will not get calm, it does. [*chuckles*]" Paul John (March 2008:595) added: "I believe what the deceased shaman of the people of Nightmute said. He said that the weather is becoming an incessant liar. He said that although it seemed the weather was going to turn out a particular way, these days it no longer materializes. And although it appears as though the weather is going to be bad, it improves. I really do think that it is becoming an incessant liar." Paul confirms what many believe – that their own elders predicted this unpredictability.

Imarpiim Cikuan Cimillra/Changes in Sea Ice

As temperatures warm, Yup'ik elders have observed corresponding changes in sea ice and, as a result, access to the sea mammals that call the ice home. John Phillip (October 2005:116) of Kongiganak noted that recently shorefast ice is thinner and less extensive:

> The *tuaq* [shorefast ice] used to be very thick, and it froze a long distance from shore. Nowadays our ocean doesn't freeze far from shore, and our *tuaq* and rivers become unsuitable for hunting because they are too thin and dangerous. And last year, we really couldn't go out seal hunting [in Kongiganak and Kwigillingok] because the shorefast ice was too thin.

John Eric (December 2007:334) compared past and present: "The changes in the weather and the ocean have occurred in my presence. Back when the weather was cold, the ocean down below our village froze a great distance away from shore. And the shore ice would be dry; there were no wet spots on the snow. There were also many birds and seals. And after a large amount of ice had been pushing against [the shore ice], when a head could fit through, a ringed seal would immediately appear. That no longer occurs today as the weather has changed." Paul John (January 2007:9) added, "When shore ice formed a good distance out toward the ocean, they caught many sea mammals at that time and knew that they wouldn't be scarce. Nowadays, the shore ice no longer extends far out because the weather isn't as cold as it was in the past."

Paul Kiunya (January 2007:4) of Kipnuk noted that fewer *evunret* (ice piles) form. As some bays and river mouths become increasingly shallow, ice is also piling in new places. Simeon Agnus (January 2007:127) observed, "*Evunret* are starting to grow in places where they never formed before. Although no *evunret* grew in front of Kangirrluar before, *evunret* are starting to grow there nowadays. When the current moves large ice floes and they pile into a shallow area, they quickly rise and grow." Stanley Anthony (January 2007:128) noted the same is true of *et'galqitat* (ice beached in shallow areas): "These days, Kangirrluar and the areas below Umkumiut and Qurrlurta that once had no *et'galqitat* now have *et'galqitat* because the ocean is getting shallower. Since the ocean is changing, that's what's happening; some areas that were deep are now shallow, and sandbars that were never visible are starting to appear."

Warming temperatures also translate into later freeze-up and earlier breakup in coastal communities. Peter John (March 2007:1196) explained the trend he observed:

> In the past when it started to get cold, the ice froze and wouldn't stop freezing. It got cold in October and November. These days sometimes it doesn't get cold for a long time. And the cold weather isn't as severe as it used to be. Although it does get cold, it doesn't reach the cold temperatures that it reached when we traveled by dog team [in the 1960s].

Peter Elachik (January 2007:22) of Kotlik recalled comparable changes in the timing of freeze-up at the mouth of the Yukon River:

> Sixty years ago, I remember the cold around November 10, 1945, because that's when the fur trapping season began. And [at that time] a lot of traveling took place for hunting and visiting with little or no caution [because the ice was safe]. Then they started using caution in the mid-1960s. We started noticing a slight warming, and then by the 1990s, it was dangerous. And last year, November 10, 2006, no travel was being done, and the first travel began a few days before Christmas. So there's a lot of difference between 1945 and 2005. Big change.

Many observe that shorefast ice does not stay as long as in years past. Stanley Anthony (January 2007:129) said, "Kangirrluar no longer has genuine shorefast ice on it. In the past birds would arrive while ice was still there, and the ice was safe for a long time." Mark John (March 2007:1197) commented on earlier breakup in spring: "When we came home from St. Marys [boarding school] in 1968, we landed in Tununak and they brought us [to Toksook Bay] with a boat. I think it was May 28. Kangirrluar still had shore ice at the time. These days, the shore ice sometimes melts completely by early May." Even when ice forms during winter, a south wind in spring can push the pack ice close to shore, covering open water and driving seals far from shore.

Hunters lack safe ice for butchering seals. John Eric (March 2007:262) noted that seals also lack ice: "Those adult bearded seals and even walrus could lay on top of *angenqaat* when they were far from shore. Even though there were ocean swells, the water would only reach their top edges. Those [*angenqaat*] were good because they were thick and not dangerous. But ice floes are no longer like that today. They have gotten thinner, and some break to pieces when we go on top of them. They are mostly snow."

Cikullaq (newly frozen ice along open water) is also less extensive. According to David Jimmie (January 2007:133), "When *cikullaq* forms, it thickens rapidly as thin sheets of ice stack and layer over one another. When one sheet of ice piles over another, a boat is unable to travel right away. That's what happened in the past. Since *cikullaq* no longer forms extensively, that no longer occurs."

As ice conditions change, so does the presence of seals that make it their home. Stanley Anthony (January 2007:131) observed, "Sometimes they catch many [seals] when they happen to hunt at just the right time. Seals seem to pass during winter. That's what they're starting to do nowadays. They pass. They seem to arrive earlier these days." Seals were also more plentiful in the past. David Jimmie (January 2007:118) noted, "I first started seal hunting using a kayak. We just ignored these spotted seals but only tried to hunt ringed seals and young bearded seals. Spotted seals would swim alongside us, but we never hunted them. These days, they are starting to get frightened when [people] make noise from afar." Stanley Anthony (January 2007:136) added, "Back when I was young, when we lived down at Umkumiut, there were a great many seals, different kinds, emerging from the water down below the shore. These days, seals are only seen after many trips. They're becoming scarce, and their numbers are declining."

Lizzie Chimigak (January 2007:428) of Toksook Bay noted that even when seals are available, lack of ice shortens the hunting season: "Some time ago, men who went seal hunting in spring didn't catch as many young bearded seals as usual because of the ice. These days, it seems as though sometimes their seal-hunting season is cut short, as though the ice prevents them from [hunting during the entire season]. They lack ice to hunt on. And eventually all the ice melted down on the ocean. I know these things since I depend on [seal hunters] for food."

Finally, elders do not dissociate themselves from observed changes in their homeland but accept personal responsibility. They relate the negative impacts of change they observe today to their failure to instruct their younger generation in proper behavior. Now, they say, is the time to reverse this trend. Observing uninstructed young men and women, John Eric (January 2007:26) remarked, "We must talk to them, to delay them from becoming like dogs." Elders warmly embraced our work together, which they view as much more than passive documentation of change but as part of an active solution.

Acknowledgments First and foremost, we are indebted to the many men and women throughout southwest Alaska who so generously shared their knowledge. We have cited their contributions by name, place of residence, CEC gathering date, and transcript page number. We are also grateful to the Calista Elders Council, especially Mark John and to the CEC's board of elders, who guided us in this work. We are deeply grateful to the National Science Foundation, both to Polar Programs and to the Bering Ecosystem Study Program, for funding our work. Our heartfelt thanks to Igor Krupnik for inviting us to be part of the larger SIKU project and his, as well as Gita Laidler's, constant encouragement while preparing this chapter. Moreover, Igor's work with St. Lawrence Island elders (Oozeva et al. 2004) was an inspiration. And special gratitude to Hajo Eicken of the University of Alaska's Geophysical Institute for providing invaluable comments on what Yup'ik elders shared. More than anyone else, Hajo opened our eyes to the role Yup'ik knowledge may play in understanding changes in Arctic sea ice.

References

Fienup-Riordan, A. 2005. *Wise Words of the Yup'ik People: We Talk to You Because We Love You*. Lincoln: University of Nebraska Press.

Fienup-Riordan, A. 2007. *Yuungnaqpiallerput/The Way We Genuinely Live: Masterworks of Yup'ik Science and Survival*. Seattle: University of Washington Press.

Fienup-Riordan, A. and Rearden, A. in press. *Ellavut/Our World and Weather*. Seattle: University of Washington Press.

Oozeva, C., Noongwook, C., Noongwook, G., Alowa, C., and Krupnik, I. 2004. *Watching the Ice Our Way/Sikumengllu Eslamengllu Esghapalleghput*. Washington, DC: Arctic Studies Center, Smithsonian Institution.

Reardan, A. and Fienup-Riordan, A. 2005. *Yupiit Qanruyutait/Yup'ik Words of Wisdom*. Lincoln: University of Nebraska Press.

Chapter 14
Qanuq Ilitaavut: "How We Learned What We Know" (Wales Inupiaq Sea Ice Dictionary)

Igor Krupnik and Winton (Utuktaaq) Weyapuk Jr.

Contributing authors: Herbert Anungazuk and Lawrence D. Kaplan.

Abstract The chapter discusses a collaborative effort to document more than 120 local Inupiaq terms for sea ice and associated vocabulary in the community of Wales, Alaska, in 2007–2008. The value of recording indigenous words for sea ice as a key to understanding indigenous knowledge of sea ice was first tested during an earlier project on St. Lawrence Island, Alaska (2000–2002). Under the SIKU initiative, more than 20 of such local ice vocabularies were collected in indigenous communities in Alaska, Canada, Greenland, and Chukotka, Russia. In Wales, Winton Weaypuk, a boat captain and a speaker of the Kingikmiut dialect, led the effort to collect local ice terms, documented elders' knowledge about ice, and took more than 100 photos of various ice-related activities in the Wales area. Traditional words for ice, illustrations of local ice forms, and the Inupiaq explanations and English translations collected for the project would be of help to young hunters, so that the knowledge is preserved for future generations.

Keywords Sea ice · Wales · Alaska · Inupiaq · Indigenous terminologies

In the community of *Kingigin*, also known as Wales, Alaska, about 75 indigenous terms for sea ice were recorded in 2007–2008 in the local *Kingikmiut* dialect of the Inupiaq language. In addition, more than 30 terms were collected for various biological and cultural realities associated with the sea ice and ice hunting. The *Wales Inupiaq Sea Ice Dictionary* (Weyapuk and Krupnik 2008) thus became one of the first of 25 indigenous sea ice vocabularies collected in 2006–2009 for the SIKU project and associated initiatives. The chapter tells how the *Kingikmiut* ice "dictionary" (Fig. 14.1) was prepared, how indigenous sea ice nomenclatures can be analyzed, and what we learned from compiling indigenous terms for ice in Wales, Alaska, and beyond.

I. Krupnik (✉)
Department of Anthropology, National Museum of Natural History, Smithsonian Institution, Washington, DC 20013-7012, USA
e-mail: krupniki@si.edu

I. Krupnik et al. (eds.), *SIKU: Knowing Our Ice*,
DOI 10.1007/978-90-481-8587-0_14,

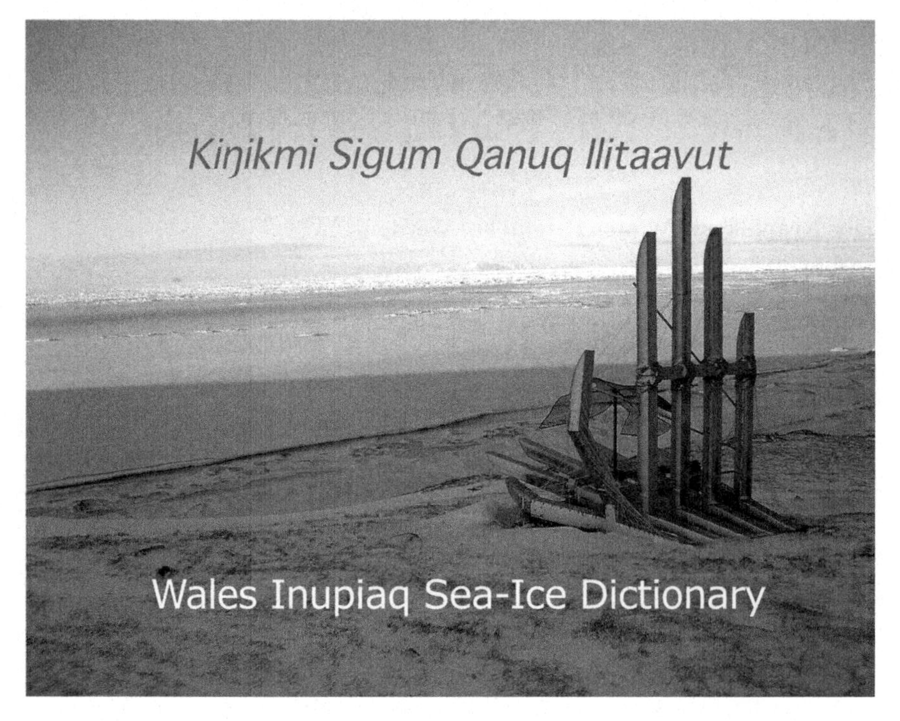

Fig. 14.1 Cover page of the Wales Sea Ice Dictionary, 2008

How the Wales *Sea Ice Dictionary* Originated

The idea to compile indigenous terms for sea ice in Wales emerged as a follow-up to an earlier knowledge documentation effort on St. Lawrence Island, Alaska, in 2000–2002. For that earlier project, Yupik hunters and elders from the communities of Gambell (*Sivuqaq*) and Savoonga (*Sivungaq*) partnered with social scientists to record local Yupik knowledge and use of sea ice (Oozeva et al. 2004).[1] A bilingual Yupik-English sea ice vocabulary of 100 terms, illustrated by pencil drawings and text photographs, was produced (Oozeva et al. 2004:26–53) based upon an earlier Yupik-only list compiled in the 1980s for the local school program (Walunga 1988). At the same time or shortly after, similar efforts to collect indigenous sea ice terminologies were launched in several other arctic communities (Aporta 2003; Gearheard et al. 2006; Laidler 2007). Also, indigenous terms for sea ice started to be used in scholarly papers and reports (Fox Gearheard 2003; George et al. 2004; Huntington et al. 2001; Metcalf and Krupnik 2003; Nichols et al. 2004; Noongwook 2000; Norton 2002). As the interest in indigenous ice terminologies was growing, the collection of Inuit ice nomenclatures and the preparation of community-focused ice "dictionaries" in local languages were placed among the key goals of the SIKU project in 2005. In the following years, the SIKU team and its partners collected more than 25 indigenous vocabularies of local terms for sea ice and associated biological and cultural phenomena in Alaska, Canada, Greenland, and Chukotka (Table 14.1; see also Chapter 1, Introduction; Chapter 18 by Tersis and Taverniers; Chapter 11 by Huntington et al.; Chapter 17 by Johns; Chapter 16 by Krupnik and Müller-Wille; and Chapter 12 by Wisniewski this volume).

The work on the draft list of ice terms in the *Kingikmiut* dialect in Wales (*Kingigin*) began in spring 2007, when Weyapuk started the collection of local ice terminology and consultations with elders about ice conditions in the Wales area.[2] An experienced hunter and speaker of the *Kingikmiut* dialect, Weyapuk, 57, led

Table 14.1 List of indigenous sea ice terminologies collected and analyzed for the SIKU project and associated initiatives

Language/ community/region	Authors/source	Year	Process	Number of terms
Alaska				
Yupik – St. Lawrence Island	Oozeva et al. (2004)	2000–2002	Compiled by local knowledge experts	100+
Yup'ik – Nelson Island	Fienup-Riordan and local elders; Chapter 13 by Fienup-Riordan and Rearden (this volume)	2008–2009	Anthropologist working with local experts	80+
Inupiaq – Wales	Krupnik and Weyapuk (this volume)	2007–2008	Compiled by local knowledge experts	120+
Inupiaq – Shishmaref	Wisniewski and local elders; Chapter 12 by Wisniewski (this volume)	2007–2008	Team meetings of several elderly experts	65+
Inupiaq – Shaktoolik	Krupnik (2008)	2008	Anthropologist working with local experts	35+
Inupiaq – Barrow	Ronald Brower and local experts	2005–2008	Native speaker working with local experts	105+

Table 14.1 (continued)

Language/ community/region	Authors/source	Year	Process	Number of terms
Inupiaq – Wainwright	Nelson (1969); Ronald Brower and local experts	1965/2009	Compiled by anthropologist through fieldwork; retransliterated and edited by native speaker and elderly experts	90+
Chukotka				
Chaplinski Yupik – Ungaziq	Vakhtin and Emelyanova (1988)	1980s	Two linguists working separately with local language experts	75+
Chaplinski Yupik – Sireniki	Aron Nutawyi and Natalya Rodionova	2007–2008	Elderly expert working with native language speaker	50+
Naukanski Yupik – Lavrentiya	Elizaveta Dobrieva and Boris Allpergen	2007–2009	Compiled by two elderly language experts	60+
Chukchi – Uelen	Roman Armaergen and Victoria Gobtseva; Bogoslovskaya et al. (2008)	2007–2009	Elderly expert working with native language speaker	120+
Chukchi – Yanrakinnot	Arthur Apalu, local elders, Natalya Kalyuzhina	2008–2009	Native language speaker working with local elders	50+
Canada				
Inuktitut – Igloolik	Claudio Aporta and local experts; Aporta (2003)	2003	Anthropologist working with local experts	100+
Inuktitut – Igloolik	Gita Laidler and local experts; Laidler (2007) and Laidler and Ikummaq (2008)	2005–2008	Human geographer working with local experts	80+

Table 14.1 (continued)

Language/ community/region	Authors/source	Year	Process	Number of terms
Inuktitut – Cape Dorset	Gita Laidler and local experts; Laidler (2007) and Laidler and Elee (2008)	2005–2008	Human geographer working with local experts	80+
Inuktitut – Pangnirtung	Gita Laidler and local experts; Laidler (2007) and Laidler et al. (2008)	2005–2008	Human geographer working with local experts	80+
Inuktitut – Clyde River	Shari Gearheard and local experts	2006–2009	Compiled by local sea ice working group	70+
Inuktitut – Sanikiluaq	McDonald et al. (2007), with additions from Schneider (1985)	1993–1995	Compiled via community meetings	100+
Inuktitut – Nunavik	Martin Tremblay, Chris Furgal, and local experts; see Appendix A by Furgal et al. (this volume)	2006–2009	Compiled by local experts via community meetings	70+
Inuktitut – Kangiqsualujjuaq	Scott Heyes and local experts; Heyes (2007)	2005–2008	Anthropologist working with local experts	60+
Inuttun – Labrador	Paul Pigott and local experts	2008–2009	Linguist working with experts from several communities	120+
Inuttut – Labrador – historical	Erdmann (1864/1866) and Peck (1925)	2008–2009	List of terms compiled from dictionary	50+
Inuktitut – Baffin Island, historical	Boas 1894; Chapter 16 by Krupnik and Müller-Wille (this volume)	1883–1884/2009	List of terms compiled from dictionary	30+
Copper Inuit/ Kangiryuarmiut – Holman	Lowe (1983)	1983/2009	Basic list of terms compiled from abridged thematic lexicon	30+

Table 14.1 (continued)

Language/ community/region	Authors/source	Year	Process	Number of terms
Siglitun – Inuvik	Lowe (1984/2001)	1984/2001/ 2009	Basic list of terms compiled from abridged thematic lexicon and dictionary	30+
Utkuhiksalingmiutitut/ Natsilingmiutut – Uqsuqtuuq (Gjoa Haven) and Qamani'tuaq (Baker Lake)	Briggs and Johns, in progress; Chapter 17 by Johns (this volume)	1970s/2009	List of terms compiled from larger lexicon/ dictionary	25+
Greenland				
West Greenlandic – Qeqertaq	Pierre Taverniers and local experts; Chapter 18 by Tersis and Taverniers (this volume)	2008–2009	Compiled by local knowledge experts	100+
West Greenlandic/ Kalaallisut	Michael Fortescue (1984)	1980s	Thematic lexicon prepared by linguist	40+
East Greenlandic/ Tunimiisut	Nicole Tersis and local collaborators; Tersis (2008); Chapter 18 by Tersis and Taverniers (this volume)	2000–2009	Thematic lexicon prepared by linguist	60+
Thule Dialect – Qaanaaq	Toku Oshima, Lene Kielsen Holm, Shari Gearheard and local collaborators; also Fortescue (1991)	2006 –2009	Compiled by local sea ice working group	70+

the SIKU documentation effort in his native community (Fig. 14.2). By summer 2007, a draft list of some 60 terms was prepared; it was reviewed and expanded in consultation with other experts in the *Kingikmiut* traditional knowledge, primarily with Herbert Anungazuk (*Aġiyaġaq*, born 1945), Pete Sereadlook (*Sauqkinna*, born 1930), and Faye Ongtowasruk (*Iqkanna*, born 1928 – see

Fig. 14.2 Winton Weyapuk, Jr., on ice, February 2007 (Photo: Matthew Druckenmiller)

Acknowledgments). The experts helped expand the list to its present 120-word format; they also supplied comments to many terms. Weyapuk also wrote Inupiaq explanations to about 20 ice forms, so that the *Kingikmiut* understanding of the main types of ice is also preserved in the native dialect.

A crucial task was to find proper illustrations, so that the *Kingikmiut* terms for ice can be identified in a life context. For that purpose, Weyapuk took more than 180 color photographs in winter–spring 2007 and in fall 2007. The pictures were taken from the beach, the mountain above the village, on the land-fast ice, and while hunting on the drifting floes and at sea. These life settings are well familiar to Wales residents and other *Kingikmiut* who now reside outside their community. On each of 80-some photographs selected as illustrations to the dictionary, Weyapuk penciled the Inupiaq terms for the types of ice that can be recognized by local viewer (Fig. 14.3). That technique made the photographs a powerful tool in introducing the *Kingikmiut* perspective on ice compared to the use of charts, satellite images, or photos taken by outside photographers.

Weyapuk and Krupnik met in Wales in September 2007 and, again, in February 2008 to combine various written and visual records into a bilingual-illustrated ice "dictionary." Several versions of the dictionary have been produced and mailed to Wales for reviewing by Weyapuk, Anungazuk, and other local experts.[3] The 112-page preprint with some 80 color and black-and-white illustrations and several text entries by Weyapuk, Anungazuk, Krupnik, Eicken, and Druckenmiller was completed in August 2008. It was reviewed by the dictionary team, also

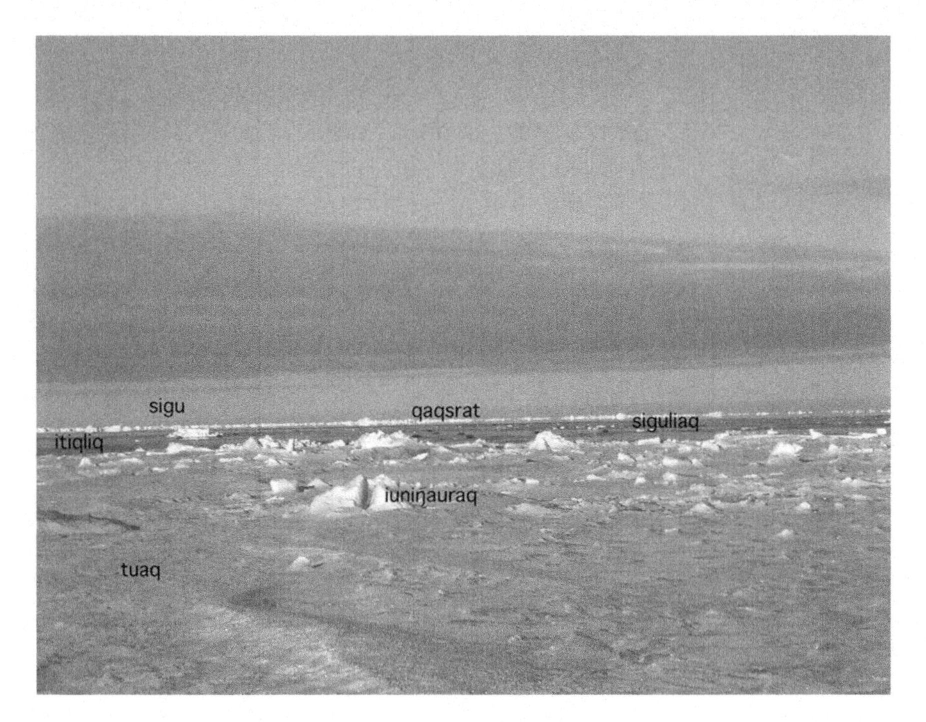

Fig. 14.3 Sample photo from the dictionary with embedded Inupiaq terms

by linguist Lawrence Kaplan and cultural anthropologist Carol Zane Jolles (see *Acknowledgments*).

In its present form, the Wales "dictionary" contains an alphabetical list of the *Kingikmiut* ice terms, with English explanations (Appendix); the same terms organized by major types of ice (young ice, pack ice, shorefast ice, ice ridges, cracks and polynyas, ice floes, etc.); and more than 60 illustrations of major ice conditions, with local terms embedded in the photographs. It also includes a set of 15 historical photographs of local scenes and of the *Kingikmiut* hunters on ice in April–June 1922 taken by visiting biologist, Alfred M. Bailey (Bailey 1933, 1943, 1971 – see below). Bailey's images were compared with photos from spring 2007 taken by Weyapuk in a similar setting. The "dictionary" will be eventually printed in several dozen color copies for the community of Wales and also produced as a PDF web-based file for interested users.

Living and Hunting in *Kingigin*

Wales (*Kingigin*) is a rural Alaskan community of some 160 residents (2000 US Census) located 180 km (110 mi) northwest of Nome, the main hub for the Seward Peninsula and the Bering Strait-Norton Sound region (Fig. 14.4). The town stretches

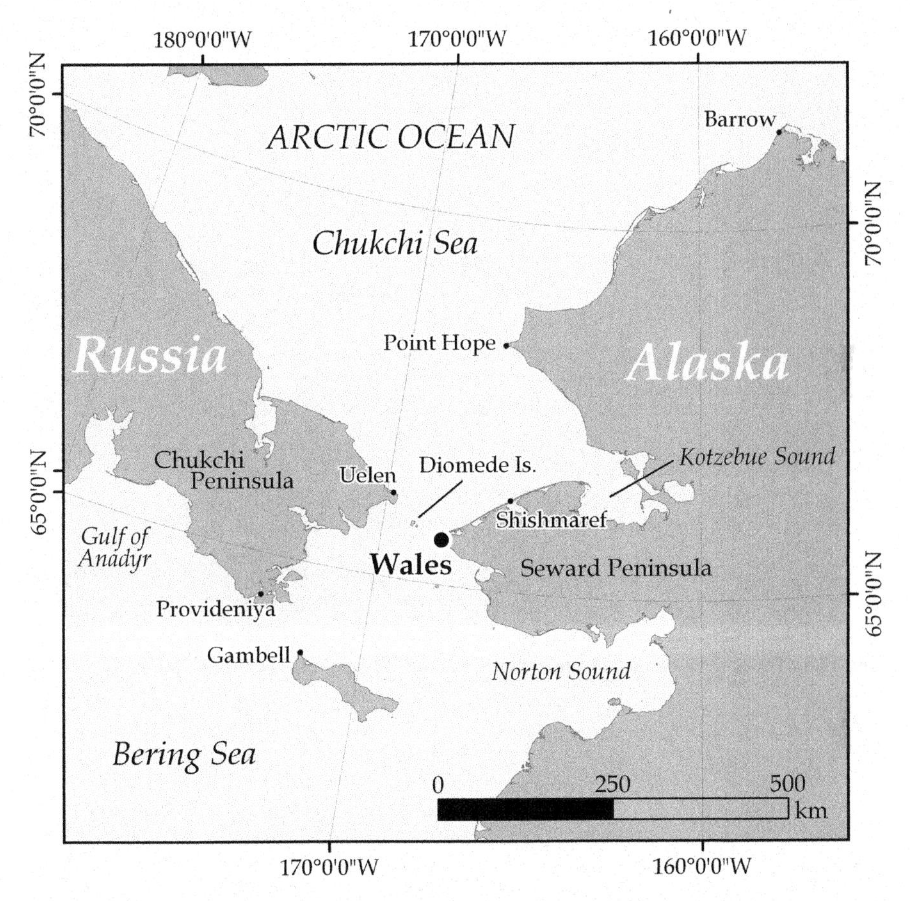

Fig. 14.4 Map of the Wales/Bering Strait area (prepared by Matthew Druckenmiller)

along a low pebble spit that forms the westernmost point of the North American continent, the Cape Prince of Wales (65° 36′44″N, 168° 5′21″W), less than 100 km below the Arctic Circle (Fig. 14.5). On a clear day, one can see two Diomede Islands in the middle of Bering Strait and a sliver of the Siberian mountains, some 80 km away on the Russian side, across the International Dateline (Fig. 14.6). The town sits literally at the junction of two oceans, the Pacific and the Arctic, and two continents, Eurasia and North America, and is thus affected by the elements from all of these directions. It is also one of the earliest major settlements of coastal sea-mammal hunters in Northwestern Alaska, as attested by its rich archaeological and historical record going back to 900 AD, if not earlier.[4] From the many ancient tools and animal remains found at Wales it is clear that hunting for marine mammals, such as walrus, whales, and seals, as well as fishing, bird and land-mammal hunting were the pillars of the village economy since the first settlers built their houses at the cape more than 1,000 years ago (Dumond 2000:136; Harritt 2003, 2004).

Fig. 14.5 The native village of Wales (Photo: Winton Weyapuk, Jr., February 2007)

Fig. 14.6 Frozen Bering Strait with two Diomede Islands and a section of the Siberian shore on the horizon (Photo: Winton Weyapuk, Jr., February 5, 2007)

Fig. 14.7 Groups of walruses (*nunavait*) on drifting ice floes (*puktaat*) off Wales on May 22, 2007. The pack ice (*sigu*) is separated from the shore by a wide open lead (*uinivak* or *imaġruk*). Small pieces of floating ice (*saŋałait*) can be seen on calm water surface (*quuniq*) (Photo: Winton Weyapuk, Jr.)

Sea ice is the environmental factor of the utmost importance to the *Kingikmiut*. Historically, various forms of sea ice were present in the area for 8–9 months of the year, usually since late October and until early July. Recently, those dates have changed, with the freeze-up time being delayed until late November or even early December and spring break-up happening in late May or early June, as reported by local hunters (Metcalf and Krupnik 2003), historical records, and Weyapuk's ice observations in 2006–2009 (see below). As the ice covers the sea, it calms the ocean storms, improves weather condition, and creates a platform for people to hunt, to fish, and to travel along the shore. Twice a year, during the fall advance and spring break-up and retreat, the ice brings the migrating stocks of marine mammals – whales, walruses, and seals – that are crucial for people's sustenance and community prosperity (Fig. 14.7).

Hunters in Wales and other nearby communities report that the timing of the ice arrival; the formation, departure, and the thickness of ice; and the duration of seasonal migration of marine mammals and of the associated spring hunting season have shifted in recent years because of climate warming (Metcalf and Krupnik 2003; Chapter 4 by Krupnik et al. this volume; Chapter 12 by Wisniewski this volume). Nonetheless, the people of Wales continue to hunt on ice and use other ice-associated resources, so that the basics of the annual subsistence cycle remain the same. Spring bowhead whaling, once major traditional activity in Wales, was resumed in 1970 and remains the apex of the community economic and social life (Anungazuk 1995; Jolles 2003), followed by the spring walrus and seal hunting, fishing, and other seasonal pursuits. In addition, people derive more income today from paid jobs, carving, pensions, and other regular transfers, various community projects, as well as from short-time jobs elsewhere off Wales.

Within the past 40 years, there have been dramatic changes in the lives of the *Kingikmiut*. Since the 1970s, new modern houses were gradually built to replace poorly insulated framed housings that some 50 years prior replaced traditional sod-covered dwellings. An electrical power plant was built in 1975, and home appliances, and, later, fax machines and computers were installed. A new airport strip is in operation and a full-size K-12 school now houses 40-some local students.

Along came many changes in the way the *Kingikmiut* live their lives. Since the late 1960s, snowmobiles have replaced dog teams; ATV's (all-terrain vehicles, four-wheelers) replaced walking to subsistence areas; and aluminum skiffs replaced traditional skin boats. Short-wave radios provide instant communication among crews and between hunters and their homes. Global Positioning System (GPS) units supplement or replace compasses and other means of navigation on land and at sea. Some changes were quite rapid (i.e., the introduction of snowmobiles in the late 1960s), whereas other progressed more gradually, like the replacement of skin boats by wooden and aluminum boats over most of the 1970s. In any case, today's hunters depend heavily upon many types of equipment bought from the outside (Fig. 14.8).

All these changes brought with them new English words that describe the new objects and their function. The changes also triggered a shift in the use of language. During the 1970s and 1980s, English gradually encroached upon and began replacing Inupiaq as the *Kingikmiut* prime means of communication. The words, stories, the instruction, and descriptions of the local environment in the *Kingikmiut* dialect began fading away. They are little used today and in many cases are already replaced by the English words. The Inupiaq language in Wales is clearly endangered, though people intersperse some Inupiaq words into their everyday conversations. Exclamations, endearments, and teasing in Inupiaq can be heard, and second Inupiaq names and nicknames are common, even among young adults. There is currently no active Inupiaq language program or native cultural curriculum at the local high school.[5]

Since most of the *Kingikmiut* are now more fluent in English than in their native Inupiaq, and since younger generation speaks and reads English only, we assumed that traditional *Kingikmiut* terms for ice could be best introduced in a bilingual context and by the use of illustrations.[6] The community was very excited about this approach and the Village Council endorsed our project at its meeting in 2007. These days, only people 50 years of age and older regularly speak in Inupiaq, and most of the *Kingikmiut* hunting crews communicate in English, even when at sea or on the ice. Nonetheless, it is still possible to document the *Kingikmiut* terminology for sea ice by working with a few dedicated elders and senior hunters, who were raised and came of age speaking Inupiaq and who preserve the knowledge of their forefathers.

Fig. 14.8 Luther Komonaseak's whaling crew getting ready to hunt with its today equipment (Photo: Winton Weyapuk, Jr., April 2007)

How Many *Kingikmiut* Words for Ice: Wales Dictionary in Comparative Perspective

By 2007, when we started work on the *Kingikmiut* ice dictionary, only a few lists of indigenous sea ice terms were available from the Western Arctic, namely, in the St. Lawrence Island and Chaplinski Yupik language (Oozeva et al. 2004; Vakhtin and Emelyanova 1988) and in the Inupiaq dialect spoken in the North Alaskan community of Wainwright (Nelson 1969). Thanks to the SIKU project efforts, many additional ice lists have been recorded, including four in the North Alaskan Inupiaq dialects from Barrow (Brower 2008), Shishmaref (see Chapter 12 by Wisniewski this volume), Wainwright (Brower 2009), and Shaktoolik (Krupnik 2008). In Chukotka, Russia, across the Bering Strait from Wales four more dictionaries have been collected: in the Naukanski Yupik language, formerly at Cape Dezhnev (Allpergen and Dobrieva 2009), in the Chaplinski Yupik in Sireniki (Nutawyi and Rodionova 2008), and in the Chukchi language from the communities of Uelen (Armaergen and Golbtseva 2009; Bogoslovskaya et al. 2008) and Yanrakinnot (Apalu and Kalyuzhina 2009). Scores of local ice vocabularies were also collected from North Alaska, Canada, and Greenland (Table 14.1). Among many local ice lists, those from Barrow, Wainwright, and Shishmaref Inupiaq; Uelen Chukchi; and St. Lawrence and Naukanski Yupik are geographically the closest to the *Kingikmiut* list. Thanks to these new materials, the *Kingikmiut* ice terminology and use of sea ice can now be assessed in a broader comparative perspective.

Most of the SIKU ice lexicons, including the one prepared in Wales, were originally formatted as bilingual alphabetical lists. They commonly include a broad spectrum of terms referring to ice types and associated processes (freeze-up, break-up, ice rafting, etc.), but also words for cultural phenomena related to sea ice (like actions, realities, and tools developed for hunting, fishing, and moving on ice), animals on ice, weather and ocean features, and even terms for freshwater ice. This "associated vocabulary" makes up to 20–30% of the words recorded in Wales, though the rate varies among many ice lists collected for the SIKU project (see below).

The ice lexicons compiled in different dialects and orthographies require certain processing steps, in order to be made compatible or even comparable to each other.[7] Terms for ice types proper and for physical features and realities associated with people and animals on ice have to be treated separately (Appendix), as they refer to different groups of phenomena. Words for sea ice and ice conditions are to be organized along certain typological groups to facilitate comparison across dialectal and language lines. In the *Kingikmiut* dictionary we introduced 11 typological groupings (or clusters): terms for fall freeze-up and young ice (8); shorefast ice (6); pack ice (16); multi-year ice (2); ice surface features: pressure ridges, rafted, rough/smooth ice (7); cracks, leads, polynyas (13); ice floes, floating ice (14); dangerous spots on ice (10); spring break-up and melt (4); other ice-related phenomena (6); animals on ice (13); moving and hunting on ice (22). These groupings may not be universally applicable to other local lists (cf. Bogoslovskaya et al. 2008; Laidler and Elee 2008; Laidler and Ikummaq 2008; Laidler et al. 2008). Nonetheless, they offer a template to assess the richness of local sea ice terminologies and compare the distribution of ice terms across the Inuit/Eskimo language area, among the neighboring language groups, and against the scientific or historical ice terminologies (see Chapter 16 by Krupnik and Müller-Wille this volume).

Overall, the *Kingikmiut* dictionary of some 120 terms, including almost 75 terms for various types of ice and ice conditions, is very much *a par* to other indigenous ice lists from the Western Arctic area, including those from Barrow (105, including about 90 terms for ice forms and processes), Wainwright (90+), Uelen (120+, including many descriptive and derivative forms), and St. Lawrence Island (over 100 – see Table 14.2). Two geographically closest lists from Shishmaref and the Naukanski Yupik, barely 100–120 km away, feature about 50 terms each for various ice types and ice formations. The difference among individual dialectal lists most certainly reflects the diversity of local ice condition but also the degree of language and knowledge preservation by today's speakers. For example, the list of recorded Yupik ice terms from Gambell, St. Lawrence Island, is almost twice larger than that from Sireniki, Chukotka, although the language spoken in both towns is all but identical. Rather, it exposes a much weaker status of the Yupik language and knowledge preservation in Sireniki, where only a few elders still speak the language. The situation is similar with the Naukanski Yupik that also has but a handful of elderly speakers and hardly any among active hunters.

Table 14.2 Major groupings of sea ice terms in the *Kingikmiut* and other Bering Strait-Chukchi sea ice lists

Major groups of ice formations	Wales Inupiaq	Barrow Inupiaq	Wainwright Inupiaq	St. Lawrence Island Yupik	Uelen Chukchi	WMO sea ice nomenclature
New ice forms, ice development	8	15	16	18	18	13
Ice cakes, broken, and floating ice	14	4	8	27	6	34
Ice surface features and processes (ridged, rafted, layered ice)	28	28	31	20	20	19
Shorefast ice features	7	7	6	12	11	10
Pack ice	15	3	5	3	24	9
Old and multi-year ice	2	2	2	–	14	2
Break-up; melting ice	4	8	3	7	–	5
Water/cracks/openings in the ice	13	11	24	4	13	16
Dangerous spots	9	2	2	5	8	–
Total	90	80	97	96	114	
General terms	6	13	9	3	6	
Terms related to animals on ice	12	5	–	–	7	–
Terms related to people's activities on ice 22	22	7	–	–		–
Other terms (total)	40	25	9+	3+	13+	
Ice of land origin	0	0	0	0	0	16
Terms related to surface shipping	0	0	0	0	0	7
Terms related to submarine navigation	0	0	0	0	0	6
Other (ice/snow forms)	0			2		2

It is clear that today's Wales list of 120 terms reproduces merely a fraction of what was once known to the *Kingikmiut* experts. Many a time elders referred to the types of ice, for which they could not recall its Inupiaq name. Some of the "old" terms for ice might have been lost following the influenza epidemic of 1918 that killed more than half of the *Kingikmiut* population, including most elders and experienced hunters in the village. The loss of knowledge could have been particularly pronounced with regard to certain groups of terms associated with hunting on the moving ice and the multi-year ice formations that were once more prevalent in the Bering Strait area. Today, only a fraction of the old words for multi-year ice are remembered in Wales. Also, traditional *Kingikmiut* ice terminology almost certainly included words relevant to the ice condition along the Siberian shore that was once familiar to the *Kingikmiut*, but not anymore.

One can get an indirect hint that the *Kingikmiut* once might have used more terms for multi-year ice, besides the two recorded in 2007.[8] The Uelen Chukchi ice terminology on the opposing side of the Bering Strait (Bogoslovskaya et al. 2008:40)[9] features ten terms for multi-year ice, including a general word for "old ice" (*petygel*) and special terms for "very heavy and high old ice that arrives like a wall" (*itchygilil*), "large and heavy old ice that crushes the shorefast ice" (*etchygelchyn'yn*), broken old ice (*talyagel*), rafted and ridged old ice (*petygelyt-van*), huge piece of floating old ice (*yanragel*) brought by the northern current, and the like. Faye Ongtowasruk, 80, one of the last expert speakers of the *Kingikmiut* dialect, referred to the pieces of old ice that her late father used to collect on the beach in mid-winter as a source of drinking water. But she could not recall its name and she has not seen such ice since the 1940s. We may only guess how many old words were similarly lost and not transmitted to today's hunters.

Many *Kingikmiut* terms are quite similar to those used in other North Alaskan Inupiaq communities, like Shishmaref, Wainwright, and Barrow, particularly the words for young ice and fall freeze-up events (Table 14.3). Nonetheless, each dialectal list displays obvious "strength" in a certain type of ice or ice-related condition that is of special importance to local users. For example, the St. Lawrence Island Yupik ice vocabulary has more than 30 terms for various types of drifting ice floes, whereas the Wainwright list is most detailed in the terminology for rafted and piled ice (20, as recorded by Nelson in 1965). The Barrow ice list is the richest for terms for the shorefast ice features (20). The *Kingikmiut* ice dictionary is very rich in cultural terms (22) and words for animals associated with ice (13). It also features 15 terms related to the pack ice and its movement (Table 14.4), which indicates the value of the knowledge of the moving ice to local hunters used to travel in the often-dangerous moving pack ice in the Bering Strait.

Lastly, when compared to the nomenclatures used in sea ice research, shipping, and ice observations (Armstrong 1956; LaBelle et al. 1983; World Meteorological Organization 1970, 2007), indigenous terminologies reveal a very distinctive "framing" of the sea ice environment. Mariners' and scientific nomenclatures are designed to be universally applicable across the Arctic (even in Antarctica), whereas indigenous terms define first and foremost, ice conditions often specific to a particular site or area.[10] As a result, indigenous terms have much higher "resolution," so that they help distinguish many more specific ice types, often within a very limited space (Fig. 14.9). To the hunters, sea ice is first and foremost a dynamic environment. Therefore, hunters' terms refer primarily to many ice processes they observe and deal with, like cracking, rafting, or ridging, rather than to certain dimensional parameters of the ice forms that are central to the scholarly or shipping ice nomenclatures.

Besides, to indigenous users, their local terms always have special meaning of safety, as they are passed from elders and senior hunters to the youth (see also Chapter 12 by Wisniewski this volume). Many *Kingikmiut* terms for ice recorded by Weyapuk (or experts in other communities – see Chapter 13 by Fienup-Riordan

Table 14.3 Inupiaq terms for young ice and freeze-up events

WMO ice nomenclature	Wales (Weyapuk and Krupnik 2008)	Shishmaref (Wisniewski et al. 2008)	Barrow (Brower 2008)	Wainwright (Nelson 1969/Brower 2009)
	Siqpaaniq – ice frozen on rocks from ocean water or slush splashing on them			
Frazil ice (fine spicules or plates of ice suspended in water)	*Mitivit* – ice crystals floating on top of an ice fishing hole			
Grease ice (later stage of freezing, when the crystals have coagulated to form a soupy layer on the surface)	*Siguliaqsraq* – "that which shall become young ice," also referred to as "grease ice" *Qinuliaq* – "becoming slush", light slush in ocean water that is just beginning to form			*Ugsruġiisaq* – grease ice; the earliest stage of freezing *Muġaliq* – slush ice or ice rind; heavy development of grease ice, almost to the point of being nilas
Slush (snow which is saturated and mixed with water as a viscous floating mass in water)	*Qinu* – slush on open water *Qaimuġuq* – berm of frozen slush ice or ocean splash built along the shore	*Qinu* – is a form of snow on water that consolidates and freezes *Qaimut* – slush ice formed by water waves that forms snow berms	*Qinu* – slush ice *Mayuqtitaq* – slush ice pushed onto the shore with wraps frozen into waves *Muġrak* – slush ice *Misałhak* – slushy top of young salt water ice *Muġaliq; muġałłiq* – slush ice on sea *Muġġuti* – to be unable to move in slush ice (*of boat*)	*Qinu* – slush ice *Agiaġniq* – slush or mush ice formed by grinding along the edges of ice pans, floes, or cracks
Shuga (accumulations of spongy white lumps, a few cm across, formed from grease ice or slush)		*Siġimaq* – small pieces of broken up ice which are too small for seals to rest upon		

Table 14.3 (continued)

WMO ice nomenclature	Wales (Weyapuk and Krupnik 2008)	Shishmaref (Wisniewski et al. 2008)	Barrow (Brower 2008)	Wainwright (Nelson 1969/Brower 2009)
Dark nilas (thin elastic crust of ice nilas which is under 5 cm in thickness and is very dark in color)	*Siguliaq* – solidified young ice (usually a few inches thick)	*Siguliaq* – this term is most often used to relate dynamics and observations connected to young or newly formed ice	*Sikuliaq* – young ice formed around edge of old solid ice on open lead	*Sikuliaq* – nilas, or black young ice; a thin flexible sheet of newly formed ice, which will not support a man,
Light nilas (nilas which is more than 5 cm in thickness and rather lighter in color)				*Sikuliaq / sikuliraq* – young ice; general term including all ice which is newly formed, from the time it becomes a cohesive mass until it has been modified by piling or rafting
Ice rind (brittle shiny crust of ice formed on a quiet surface by direct freezing or from grease ice)				*Sikuliaruq* – slush ice or ice rind; *Sikuliuraq* – similar in meaning to sikuliaq (but thinner in composition; ice rind ???)
Pancake ice (predominately circular pieces of ice from 30 cm to 3 m in diameter and up to about 10 cm in thickness)	*Nutiġaguugvik* – pancake ice; very thick slush ice	*Sauzaŋuat* – ice that originates as *siguaq* or young thin ice which eventually breaks up into multiple pieces that continuously bump into each other round and round forming circular pieces shaped like dance drums	*Qaivaġniq* – flat round cakes of ice frozen together (*kaivaġniq?*)	*Puktaaġruat* – pancake ice *Sagsraq* – pancake; circular pieces of young ice, 1–6 ft in diameter, with raise rims; the shape and appearance result from rotation and collision
		Siguaq – the very first phase of sea ice formation in either the lagoon systems or on the ocean	*Sikuaq* – thin ice, dangerous to walk on	
Gray young ice (young ice 10–15 cm thick. Less elastic than *nilas* and breaks on swell. Usually rafts under pressure			*Sikuliaġruaq* – thick ice approximately 2.5 ft thick and thicker	*Sikuliaq mapturuaq* – gray young ice; young ice which rides high enough in the water to be grayish in color and has become thick enough to support a man
		Qalliġiititaaga – the active process of *siguliaq* (young ice) becoming thicker as it		*Sikuliaġruaq* – heavy or thick young ice; according to the Eskimo informant this is

Table 14.3 (continued)

WMO ice nomenclature	Wales (Weyapuk and Krupnik 2008)	Shishmaref (Wisniewski et al. 2008)	Barrow (Brower 2008)	Wainwright (Nelson 1969/Brower 2009)
Gray-white young ice (young ice 15–30 cm thick; under pressure more likely to ridge than raft)	*Qaligit* – new ice, or young ice that has rafted on top each other to double or triple the thickness	broken by waves, winds, and tidal currents. These broken pieces move around and become thicker through building layers on top of other pieces of ice	*Kaniqtaq* – slightly refrozen ice pieces but fragile; this ice will quickly spread out when it is stepped on (ice formed by frost?)	ice about 1 ft thick
		Qalligiiktauniq – thicker *siguliaq* formed of multiple layers *siguliaq*	*Imuniq* – crushed young ice caused by moving ice	*Agiġniq* – sometimes referred to as "file ice," because it is formed by the ice "filing" itself
			Arguqtaġniq – newly formed thin ice collecting on the downwind side of a polynya or lead (arguqtinniq)	
First-year ice[a] (sea ice of not more than one winter's growth, developing from young ice; thickness 30 cm–2 m)		*Sigugiktuaq* – "good ice," clean white ice. This is a form of ice ugzruk prefer. It is not	*Iġnigluq* – thin young ice broken up or crushed and refrozen as found in cracks	
TOTAL 11	9	9	14	13

[a]In WMO nomenclature "first-year ice" is further divided into "thin first-year ice/white ice" (30–70 cm thick); "thin first-year ice/white ice first stage" (30–50 cm thick); "thin first-year ice/white ice second stage" (50–70 cm thick); "medium first-year ice" (70–120 cm thick); and "thick first-year ice" (over 120 cm thick).

Table 14.4 *Kingikmiut* terms for pack ice and associated vocabulary

Inupiaq term	English explanation
Sigu	Ocean pack ice, also a generic term that refers to all sea ice, including shorefast ice
Qamiyanaqtuaq	Pack ice that is packed tightly together
Qamaiya	Pack ice that has been packed tightly together by a large eddy current; bow waves of a seal, walrus, or whale swimming underwater visible on top the water
Qamiyaq-	To pull a boat over a section of pack ice to open water or shorefast ice
Quqsruafniq	Reflection of pack ice on low clouds on the horizon
Sibimizimaaqtuaq	Pack ice that has been grounded into small pieces by moving against shore fast ice (description, not a specific ice term)
Sigum kafibaa	The end of a floating mass of pack ice, the open ocean edge of a floating strip of pack ice
Sigu taatuaq	Pack ice that is "coming in," pack ice approaching land or shore fast ice; also known as sigu agituaq
Sigu uituaq	Pack ice that is moving out and forming a lead
Siguturuaq	Area that has a lot of pack ice
Sigutuvaktuaq	Area that has a lot of heavy pack ice
Siguiq-	To become free of sea ice
Siguitpaktuaq	Area that has very little pack ice
Tamalaaniqtuaq	Pack ice that is scattered enough to boat through
Tasrraatuaq	Pack ice that is moving tightly against the edge of shorefast ice

and Rearden this volume) are often accompanied with comments, like "this ice is very unsafe," "that ice breaks up easily," "our Elders do not recommend to go on that ice". This layer of embedded safety information makes indigenous ice classifications an invaluable source of knowledge, as the Arctic ice continues to thin and becomes increasingly more dangerous under the impact of climate warming.

Alfred Bailey at Wales: Ice Change from 1922 to 2009

On April 6, 1922, Alfred Marshall Bailey, a young biologist from the Colorado Museum of Natural History in Denver, CO, arrived in Wales. Bailey (1894–1978) and his Iñupiat companions reached Wales at the end of their 800-mile (1,300 km) dogsled journey from Wainwright on the Chukchi Sea shore (Bailey 1943:12–37). Wales was the last stop on Bailey's year-long field trip to Alaska that brought him to Bering Strait, Siberia, Barrow, Wainwright, and, finally, to Wales (Bailey 1933, 1943, 1971). As curator of birds and mammals at the Colorado Museum, Bailey aspired to collect Arctic bird and mammal specimens, and specifically, to secure skins and skulls of the largest Alaskan mammals, like walruses, polar bears, caribou, and seals, so that they may be eventually transformed into stuffed animal life groups or "dioramas" at what is now called the Denver Museum of Nature and Science (DMNS) in Denver, CO.

Fig. 14.9 The "high resolution" of indigenous terms for sea ice often allows distinguish numerous types of ice and related phenomena within a small area. On this photograph, a female walrus and a calf (*isavgalik*) are resting on the ice (*nunavait*) in the midst of scattered pack ice (*tamalaaniqtuaq*) interspersed with patches of calm flat water (*quuniq*). The mass of floating ice (*sigu*) consists of various ice formations such as *puktaat* (large floes), *puikaanit* (vertical blocks of ice), *kaŋiqłuit* (floes with overhanging shelves) *taaglut* (pieces of darker or dirt ice), and *saŋalait* (small floating pieces of ice) (Photo: Winton Weyapuk, Jr., May 21, 2007)

For that task, Bailey and his companion, Russell W. Hendee wintered in Wainwright and, as the days grew longer in early March 1922, Bailey made a dogsled trip to Wales. Over the next 2 months, from May till late June 1922, he went hunting with the *Kingikmiut* crews, until the spring hunting was over. Besides the needed museum specimens, Bailey also took some 120 black-and-white photographs of the *Kingikmiut* hunters on ice and in boats, and of birds and the scenery around Wales.[11] During that trip, Bailey also kept a daily journal (preserved at DMNS and partly published in Bailey 1943:35–77) that contains references to ice and weather condition, hunting activities, marine mammal and bird migration. Bailey's journal entries helped identify many of his pictures, even by day. Overall, Bailey's 4-month records from Wales are invaluable resource to modern studies of indigenous knowledge, arctic ice, and climate change.

In August 2005, prints of some 30 Bailey's photographs of 1922 were sent to Wales, where Herbert Anungazuk and Winton Weyapuk, Jr., shared those photos with local elders. Matthew Druckenmiller and Hajo Eicken continued that work in

May 2007. Altogether, six Wales experts – Herbert Anungazuk, Winton Weyapuk, Jr., Pete and Lena Sereadlook, Raymond Seetook, Sr., and Faye Ongtowasruk – made comments to Bailey's photographs. For *Wales Sea Ice Dictionary*, we selected 11 images that feature ice conditions and hunting activities in spring 1922. Many dates on Bailey's photographs came as a big surprise to Wales residents, as many of today's activities usually take place a few weeks, if not a month earlier.

To articulate the scope of change, we positioned Bailey's photographs in the "dictionary" next to contemporary pictures taken by Weyapuk in 2007, often in similar settings. A dramatic shift in ice and weather conditions from Bailey's time is quite obvious. For example, in Bailey's picture of the village taken on June 1, 1922, all of the houses are covered in deep snow, with an unbroken expanse of shorefast ice and no open water seen on the horizon (Bailey 1943:34). According to Bailey's diary, the snow started to melt on June 5, 1922 (Bailey 1943:58); today it usually melts up rapidly in the last week of May. In 1922, the shorefast ice was still intact by June 16; these days, there is always open water in front of the village by mid-May and sometimes even in late April. In the last 3 years of Weyapuk's observations, the shorefast ice broke off on May 29, June 7, and June 13, in 2007, 2008, and 2009, respectively, and the sea was clear off drifting ice by June 8, June 9, and June 17, in these 3 years. Therefore, Bailey's photograph featuring *Kingikmiut* hunters butchering walrus on drifting ice floes in late June 1922 (Fig. 14.10) left many Wales residents dazzled. In recent years, by mid-June there is usually no ice to speak of. In 1922, Bailey reported the first walrus of the season killed on May 23, and the first major walrus hunt taking place on June 2–3. Today, the peak of walrus migration comes around May 15, and the hunting is normally over by the first week of June.

Though Bailey did not report air temperature in spring 1922, he made numerous references in his diaries to wind direction, status of the sea ice, ocean currents, overall weather condition, as well as to the arrival of many migratory birds and mammals. Thanks to this data, Bailey's record offers a rare glimpse on the seasonal transition from winter to spring to summer some 85 years ago and to the overall scope of modern climate change at Wales.

Two additional sources of historical "proxy" data support Bailey's one-spring observation record. For almost a full decade, 1892–1902, Ellen Kittredge Lopp, early schoolteacher and missionary's wife, lived in Wales. She recorded local environmental conditions in her diaries, letters, and handwritten newsletters (Lopp 2001), so that substantial information on ice and subsistence hunting in Wales can be extracted from her writings. Lopp's records confirm that Bailey's dates for *Kingikmiut* walrus hunting in drifting ice in late June 1922 were in line with the condition 20 years prior to his visit (of course, accounting for annual fluctuations – Table 14.5).

Another set of historical "proxies" is the tabulated daily temperature record from Wales spanning July 1943 to February 1952, on file at the Alaska Climate Research Center (ACRC), at the Geophysical Institute, University of Alaska Fairbanks.[12] During springtime, the sea ice – both the drifting pack ice (*sigu*) and the shorefast ice (*tuaq*) – has substantial cooling effect on local air temperature. Thus, the break of

Fig. 14.10 Skinning walrus on ice. Late June 1922. Photograph by Alfred Bailey, DMNS IV BA-21-591. "This is late June?! Wow! That is pretty late still to have pack ice compared to now. By late June in recent years there has been no ice at all to speak of. That looks like the way it was yesterday and the day before" (Winton Weyapuk, Jr., May 23, 2007)

Table 14.5 Spring walrus hunting season in Wales, 1893–1922 (Lopp 2001; Bailey 1943)

Year	References to walrus hunting and ice condition	"Peak" hunting time
1893	Preparations in May (p. 66). June 15: Thornton is hunting for walruses with Wales crew; slept on the floating ice. Got a female walrus with a baby (p. 61)	Around June 15 (?)
1895	May 10: crews hunting walruses for a few days	After May 10 (?)
1897	May 25: 7 walruses and "good many" large seals are killed so far	May 25–?
1898	June 12: people hunting walruses but could not get any because the strait is full of ice (p. 186). June 17: many walruses killed	June 17 and after
1899	May 28: the whaling season is about over and they are beginning to get walrus (p. 218). One umiak came today with 10 seals	June (?)
1901	June 1: first walruses killed this season (p. 309). June 5: good hunting, with already some 40 walruses killed (p. 310). June 18: several boats returned from a successful hunting near Diomede (p. 311)	June 5–20
1922	May 23: first walrus of the season killed; still hardly any have been seen. June 1–3: major walrus hunt off Diomede. June 18: one boat was almost blown away north with the moving pack ice; has to abandon three walruses killed. June 24–26: last major walrus hunting of the season, going to Fairway Rock and toward Siberia; plenty of walruses on ice in the middle of the strait	June 1–25

the shorefast ice and the retreat of the pack ice are usually followed by a noticeable spike in the daily temperature, as it occurred after May 30, 2007, June 11, 2008, and, May 31, 2009, according to Weyapuk's observation. The 8-year record for the month of June in 1944–1951 reveals similar 5–8°F (3–4°C) temperature "spikes" after June 15, 1944; June 23, 1945; June 12, 1946; June 17, 1947; June 12, 1948; June 30, 1949; June 8, 1950; and June 24, 1951. These dates are remarkably consistent. Though the 1943–1952 record has no reference to sea ice, it may be a proxy indicator of the dates of the supposed pack ice retreat and, subsequently, of the end of the active ice-hunting season, in line with Bailey's and Lopp's data.

Conclusions: "How We Learned What We Know"

The *Kingikmiut* way of knowing, much like that of other Arctic people, was traditionally based upon careful listening to elders' stories (see Chapter 12 by Wisniewski this volume). Teens and young hunters also watched more experienced people and followed their steps since early age. As Weyapuk recalled in his introduction to the "Wales Inupiaq Sea Ice Dictionary":

> When I began hunting with my father, my uncles, other adults, and Elders in my youth, our communication was totally in Inupiaq, in our *Kingikmiu* or Wales dialect. Those words and sentences in Inupiaq concisely relayed information necessary to be successful as a hunter and just as importantly, to be safe on the ice and at sea. The behavior of the animals we hunted, their biology and the environment they lived in could be quickly described and understood in our language. Understanding led to success and safety of individuals and hunting crews. I was eleven years old when I started going with my father's crew and because Inupiaq was our family's primary language, I was fluent and could fully understand the hunters' conversations and their instructions on ice and in the boat. I was able to grasp many of our Inupiaq sea ice terms simply by hearing them spoken around all the time and by using them myself (Weyapuk and Krupnik 2008:8).

Many of the *Kingikmiut* and other indigenous ice terms explain conditions that pose hazards or are extremely dangerous to hunters. Years ago, the commonality of the Inupiaq usage meant that all hunters, young and old, knew of, or would readily learn about, these dangerous situations. Young people listened as adults and elders talked every day about hunting and the conditions they encountered, and thus they learned how to watch for those conditions and to use proper words to describe them. This was the "Inupiaq" way to pass the knowledge. It also ensured that younger people followed the safety rules learned by experienced hunters and elders and that young hunters were not exposed to unnecessary risks while walking and hunting on ice. Today, because English is the prime means of communication, it is more difficult for elders and older hunters to pass such information in full. Indeed, elders say that when they speak in Inupiaq to young people they receive blank, uncomprehending stares. Younger people today tell stories about their hunt and what they have seen on hunting trips exclusively in English. That is a very different way of learning about safety and ice condition than one practiced in the "old days."

Of course, there are many words and phrases in English to describe sea ice, and modern hunters have mastered and applied certain English terms for crucial ice- and weather phenomena (see Chapter 12 by Wisniewski this volume). People also listen to the daily weather forecast over the radio and on TV; many hunters are using the Internet to download satellite images and to assess weather and ice conditions. However, the information on the radio, TV, or over the Internet is never related specifically to the local ice patterns in Wales. Also, with some rare exceptions, the English terms for ice and weather do not transmit any information regarding the safety or stability of the ice for travel, camping, or watching for marine mammals.

Hunters also regularly communicate among themselves and with families and friends in the community, over short-wave radio. In this way, more experienced people may be consulted and the message of danger or increased risk can be passed to the hunters in boats or on ice. One can argue, though, that the number of terms and definitions of ice used by hunters speaking in English is significantly smaller than those known by elders in the Inupiaq language (see also Chapter 12 by Wisniewski this volume). Also, many stories associated with particular ice and weather conditions, specific places, or safety measures in emergency situations cannot be retold in full to younger hunters because of the language barrier. This is why knowing traditional words for sea ice and listening to elders' talking about ice is still the best way to become a successful hunter and to minimize the risk for one's partners and crew.

Therefore, it is our hope that the *Kingikmiut* words for sea ice, illustrations of local ice forms, and the Inupiaq explanations and English translations collected in the dictionary can help young hunters supplement what they have learnt in English about ice in their native area. There have been no adult Inupiaq language classes in Wales (until winter 2008/2009) and no youth-elders language camps, as practiced in many communities in Alaska and Canada. We may only hope that the young hunters will learn and then begin to use some of the old Inupiaq words for sea ice as a way to teach those younger than themselves, so that the *Kingikmiut* ice terms will be preserved for future generations. By no means is this sea ice "dictionary" an attempt to revive the Inupiaq language in Wales, as only constant usage by adults and youth can achieve that.

As Weyapuk wrote in his Preface to the *Wales Inupiaq Sea Ice Dictionary*, "[This book] can be viewed as a link between the way our Elders communicated in the past and today's way of communicating. This book, in its way, is a heartfelt tribute to our Elders who taught us so much. Without their dedication and instruction our life would be dramatically different today. This book can also be seen as praise for the youth who continue our way of life and whom we love deeply. As they continue to hunt on the sea ice we hope that they do it safely. Language, any language, is beautiful in its own way. Inupiaq, because of its construction and its concise description of the natural environment is no less beautiful. This dictionary may help preserve parts of it for generations to come." The *Kingikmiut* sea ice dictionary – as well as other indigenous sea ice lexicons collected for the SIKU project – thus become a direct contribution of IPY 2007–2008 to the preservation of Arctic people's languages and knowledge. It is our shared hope that by the time of the next International Polar Year the *Kingikmiut* words for sea ice will be spoken in boats and on the ice in the Bering Strait.

Acknowledgments The work on the *Wales Inupiaq Sea Ice Dictionary* was sponsored by the grant from the "Shared Beringia Heritage Program," National Park Service, Alaska Office, with additional support from the *SIZONet* project at the Geophysical Institute, University of Alaska Fairbanks (NSF OPP 0632398) and the National Museum of Natural History, Smithsonian Institution. We are grateful to our team members, Herbert Anungazuk, Hajo Eicken, Matthew Druckenmiller, Lawrence D. Kaplan for their constant support and for many helpful edits to an earlier draft of this chapter. Our prime Inupiaq consultants, Faye Ongtowasruk and Pete Sereadlook, generously shared their knowledge with us. We appreciate the support of the Native Community of Wales and of the Wales Village IRA Council that endorsed our work in 2007. Many people made valuable contribution to the production of the dictionary, including Elizabeth Clancy, Rene Payne, and Lisa Crunk at the Denver Museum of Nature and Science; Martha Shulski at the Alaska Climate Research Center at the University of Alaska Fairbanks, and Carol Zane Jolles at the University of Washington. Ronald H. Brower, Sr., and Josh Wisniewski compiled sea ice lists from Barrow, Wainwright, and Shishmaref we used to compare people's knowledge of ice in various Alaskan communities. Hajo Eicken, Shari Gearheard, and Gita Laidler kindly read the first draft of this chapter; Carol Jolles, Amber Lincoln, and Matthew Druckenmiller shared their photographs for the dictionary. Matt also produced a map that is used as illustration. We thank you all, *Quyana!*

Notes

1. The project was sponsored by the grant from Marine Mammal Commission (MMC), as an outcome of a workshop, *Impact of Changes in Sea Ice and Other Environmental Parameters in the Arctic* (Huntington 2000) organized by MMC in 2000.
2. By that time, Weyapuk was already participating in another NSF-funded project focused on the seasonal ice dynamics in the Alaska coastal zone (*SIZONet*) (see Chapter 15 by Eicken this volume).
3. Throughout this process, sea ice scientists Hajo Eicken and Matthew Druckenmiller were active partners in our effort.
4. On the rich cultural history of the ancient and historical village of Wales, see Harrit 2003, 2004; Koutsky 1981; Ray 1975.
5. In winter 2009, Winton Weyapuk taught the first adult Inupiaq language class for his fellow villagers, out of the Wales high school.
6. We used this approach in the St. Lawrence Island Yupik ice dictionary (Oozeva et al. 2004), even though on the island the Yupik language is better preserved and is actively used by hunters.
7. For example, Richard Nelson's ice list from Wainwright, Alaska (1969) required a complete re-transliteration, since no standard orthography for Alaskan Inupiaq existed at that time. Nelson's list was re-transliterated in 2009 by Ronald H. Brower, Sr., fluent Inupiaq speaker on staff at the Alaska Native Language Center, University of Alaska Fairbanks, with the help from Wainwright elder Rossman Peetook and other experts from Wainwright and Barrow.
8. *Inipkaq* (??) and *utuqaq* (old multi-year ice that used to arrive in the Bering Strait in the last week of November). The latter term was recorded by Herbert Anungazuk and is not shared by other senior experts. According to Anungazuk, this ice is moving in the north-to-south direction (*uŋavriq*) and carries sea mammals, like walrus and polar bear; it may be also spelled as *utukaq*.
9. Uelen Chukchi list of sea ice terms was originally prepared in spring 2008 by local elder and experienced hunter Roman Armaergen, 73 (see Bogoslovskaya et al. 2008). The list was checked and expanded by other Uelen elders and hunters in 2009.

10. For example, the term kaivsraaqtuaq refers to the ice floe that is rotating in the current, usually in the lead between the pack and shorefast ice (literally, to go round and round). There is an area about 7–8 miles north of Wales where in the springtime the current moves in a circle and ice gets tightly packed up. About 20 years ago one hunting crew got caught in the middle and they had to pull their boat over the floes to get out. It took them most of the day to get out. It does not happen too often and it depends on how much drifting ice there is. Older hunters and elders always warned about that area (Winton Weyapuk to Hajo Eicken, June 8, 2007).
11. The photographs are currently preserved in the Alfred Bailey Archives at the Denver Museum of Nature and Science (DMNS) in Denver, CO. Bailey's photographs have been digitized and can be accessed online on the DMNS public web site at http://www.bcr.org/cdp/search.html. Over the years, Bailey used his Alaskan photos as illustrations to his publications, including his book on Wales (Bailey 1943) that featured major portions of his Wales diaries from 1922. Prior to our project, his photos were unknown in Wales. We are grateful to the then DMNS photo archivist Elizabeth Clancy for offering copies of Bailey's photographs to the community.
12. We are grateful to Dr. Martha Shulski, Geophysical Institute, UAF, for kindly sharing Wales historical temperature records with us.

Appendix: Alphabetical List of *Kingikmiut* Sea Ice Terms and Related Vocabulary, 2008

(Prepared by Winton Weyapuk, Jr., with edits by Herbert Anungazuk and Lawrence D. Kaplan. Plural forms are given in parenthesis; verb forms are followed by a hyphen)

Ice Forms and Conditions

Analuaq (analuat) piece of floe ice that has walrus droppings on it

Auksaaniq (auksaanit) hole melted into or through shorefast ice or floe ice

Auniq (aunġit) "rotten" ice, very unsafe, shorefast ice or pack ice that is thin and has many melted holes in it

Ikalitiq (ikalitit)/ikalitaq ice floe grounded in shallow water (usually refers to smaller floes)

Iluqnauq (iluqnaut) large ice floe, up to one-half square mile in size or larger

Iluqnauqpak (iluqnauqpait) very large ice floe, up to one mile long or more

Imaġruk very large and wide lead; also *uinivak*; a large pond of open water within pack ice

Imauraq small open pond of water within pack ice or a large ice floe

Inipkaq (inipkat)?? multi-year ice, or old ice (HA); check Shishmaref *kiniqtaq*

Itiġliq (itiġlit) large opening in shorefast ice, such as a bay or cove; also refers to the end of a lead with shorefast ice on one side and pack ice on the other (literally "a place you can enter")

Iuɫuk- the sound of ice piling up in ridges [a verb stem, meaning "make the sound of... " The correct form is *iuɫuktuq*, "it (ice) is making noise as it piles up"]

Iuniq (iunġit) pressure ridge formed on shorefast ice

Iunilauraq (iunilaurat) small pressure ridge formed within shorefast ice

Iunivak (iunivait) very large pressure ridge on shorefast ice

Kaivsraaqtuaq ice floe that is rotating in the current; large ice floe that is rotating as one end nudges other ice floes or shorefast ice (literally, to go round and round)

Kaŋiqɫuk ice with overhanging shelf, dangerous to step on (in the north slope dialect this term means "bay along edge of the ice")

Mitivik ice crystals floating in the ocean or a fishing hole

Mituglak surface of shorefast ice or pack ice that has been changed by rain from a smooth surface to a rough surface with ice crystals

Nazirvik ice floe or floe berg with a pressure ridge that can be climbed to look around; means "look out place" and can be also used on land; see also *puktaaq* (a look out place)

Nuti'aġuugvik very thick slush; pancake ice

Puikaaniq (puikaanit) piece of vertically lifted ice; standing chunk of ice

Puilauq (puilaut) piece of ice that has broken off underwater from an ice floe or shorefast ice and surfaced, can be dangerous if it hits a boat or outboard motor

Puktaaġruaq (puktaaġruat) small ice floe, small floe berg

Puktaaq (puktaat) ice floe, floe berg

Puktaaqpak (puktaaqpait) very large ice floe, very large floe berg

Puktaaqpak ikalititaaga iceberg that has become "stuck" by grounding in shallow water

Qaimuġuq (qaimuġut) berm along shore formed when slush and brash ice and the water from waves freezes on the beach; snow drifts on the shorefast ice and tundra 1–2 ft high that resemble waves

Qamiyanaqtuaq pack ice that is packed tightly together

Qamaiyaq pack ice that has been packed tightly together by a large eddy current; bow waves of a seal, walrus, or whale swimming underwater visible on top the water

Qaŋataruaq ice floe floating free after high tide; overhanging ice floe

Qaaptiniq overflow on shorefast ice or ice on a lagoon; water splashed on top of ice around a seal breathing hole

Qaupik/qauɫuk ice shove; pack ice or shorefast ice that is pushed onto land by wind and current; cf. *Ivu* – the term used in the north slope dialect; can be very dangerous if it threatens people who are unaware it is happening

Qinu slush, slush ice on open water; see also *nutiġaġuugvik*

Qinuliaq "becoming slush"; light slush in ocean water just beginning to form

Quppaq (quppait) crack in the shorefast ice or ice floe

Qupniq (qupnit) crack on shorefast ice or ice floe that has re-frozen

Saalguraq thin fresh water ice on still waters. Also applies in spring conditions when melt water on sea ice freezes (HA)

Saŋałak (saŋałait) "dirt," small broken pieces of brash ice

Siġimizimaaqtuaq pack ice that has been grounded into small pieces by moving against shorefast ice (description, not a specific ice term)

Sigu ocean pack ice, also a generic term that refers to all sea ice, including shore fast ice

Siguliaksraq "that which will become young ice," grease ice

Siguliaq young ice, gray or gray-white in color

Sigum izua the end of a floating mass of pack ice, the open ocean edge of a floating strip of pack ice (HA);

Sigum kaŋiġaa the extent of the pack ice (FO, PS, and WW); *sigum kaŋia* (HA)

Sigu qayua qaatuaq ice that has sand on top

Sigu taatuaq pack ice that is "coming in," pack ice approaching land or shorefast ice; also known as *sigu agituaq*

Sigu uituaq pack ice that is moving out and forming a lead

Siguturuaq area that has a lot of pack ice

Sigutuvaktuaq area that has a lot of heavy pack ice

Siguiq- to become free of sea ice

Siguitpaktuaq area that has very little pack ice

Siuġaq point along the edge of shorefast ice that sticks out from the rest of the ice; also *nuvuk* in the north slope dialect (HA)

Suŋatlaq scatter consisting of slush and ice that is weakly connected to land-fast ice or large ice floes; a dangerous place (HA); see *sanałak* above

Tamalaaniqtuaq pack ice that is scattered enough to boat through

Tasrraatuaq pack ice that is moving tightly against the edge of shorefast ice

Tuaġituq smooth, flat section shorefast ice

Tuaq shorefast ice, landfast ice

Tuwaiq- to break up (of shorefast ice); for shorefast ice to break off and drift away from the shore

Tuwaiġniq (tuwaiġnit) piece of shorefast ice that has broken off from the shore and is floating free

Tunŋuruaq grounded pressure ridge or ice floe

Uiniq open lead that has formed between pack ice and the shorefast ice

Uiniġum izua end of the open lead, the corner of the open lead with pack ice on one side and shorefast ice on the other side; also known as *uiniġum kaŋiġaa*

Uinġum taġġaa "the open lead's shadow"– a dark reflection of open water in a lead on low clouds; see also qissuk

Uinavak very large and wide lead; also *ima'ruk* – heavy water in lagoon (HA)

Utuqaq old winter ice or also known as multi-year ice, which usually arrives in the Bering Strait in the last week of November (HA)

Uuyuaq scattered ice and slush cemented onto land-fast ice and subject to removal by current or wind action; a very dangerous type of ice to walk on (HA); "extension ice," literally an extension of something

Ice-Associated Terms

Allu (allut) seal breathing hole in the shore ice

Alluaq (alluat) fishing hole chipped out on the shorefast ice or in cracks in new ice (*siguliaq*)

Anisaaq visible breath of whales or walrus; the breath of a whale visible from a long ways away without the whale being visible; the sound of a whale breathing within pack ice

Attaaq- for a sea mammal, such as a bowhead whale or beluga whale, to swim underneath shorefast ice (*attaaqtuq-* "it swims underneath")

Atiqtaq (atiqtat) person or persons who drift away on the moving ice (HA)

Aułhaaniq (aułhaanit) fresh melt water on top of sea ice

Aŋuun paddle or oar; can be used to listen for sea mammal vocalizations underwater

Ayaupiaq slender pole with an ice testing tip on one end and a hook at the other end; it is dangerous to walk on questionable ice without an *ayaupiaq*

Auyuuq male bearded seal making its mating calls underwater; to "sing" of an ugruk

Auyuuġluk- to listen for seal, walrus, and whale calls underwater using a paddle with one end placed against an ear and the other end underwater

Iluaq seal's maternal den; a den near a pressure ridge where a newborn seal resides until it is old enough to leave

Imaq ice-free ocean or sea

Inipkaq for there to be a mirage; ice, land, or water loom above the horizon; when pack ice appears as a white line along the horizon

Iqłak-, igłaktaaq- to catch a fish by jigging through a fishing hole or crack on shorefast ice; to jig for fish

Iziq "smoke"; frost smoke over open water

Issuaq- to look through a fishing hole made for spear fishing to the bottom of shallow water; to look underwater

Kagiaq- to go spear fishing on shorefast ice (*kagiaqtuq-* he caught something spear fishing)

Mauŕhigutaaq-/mauraaq to cross open water by jumping from one ice floe to another

Mapsa overhanging snow cornice on the edge of shorefast ice; very dangerous

Mauqsruq- to watch for seals or other sea mammals from the edge of shore-fast ice or pack ice

Mauqsravik place on shorefast ice or pack ice from which to watch for seals or other sea mammals

Mitiŋiun/mitiġmiutaq screened ice scoop for removing ice crystals from a fishing hole

Naġituaq low place on shorefast ice or on ice floe, where a boat can be pulled out

Nakkaq- for a sea mammal to dive into water from shorefast ice or pack ice

Nalunaitkutaq marker on trail; marker at the boat launch site

Nannum ilua polar bear maternal den on shorefast ice

Nunavak (nunavait) walrus on top of an ice floe

Pituqi (pituqit) ramp cut into the edge of shorefast ice for launching boats

Pituqiuġvik marker at the boat launch site

Pituqiliuq to make a boat ramp on the edge of shorefast ice

Puizri (puizrit) sea mammal, or seal that surfaces in a lead or open water

Puiyaq- for a sea mammal or piece of ice to surface in a lead or open water

Qagi- for a sea mammal to climb onto shorefast ice or pack ice

Qaksraq (qaksrat) seal sleeping on shorefast ice or pack ice (WW); any sea mammal on ice until positively identified (HA)

Qamiyaq- to pull a boat over a section of pack ice to open water or shorefast ice

Qissuk "water sky"; reflection of open water on low clouds or frost smoke

Quqsruaŋniq reflection of pack ice on low clouds on the horizon; also *qauksraaġniq* (HA)

Saġvaq current; ocean current, also verb *saġvaq-* to flow (of current)

Taglu (tagluk) snowshoe made for walking on new ice (*siguliaq*); dl. – pair of snowshoes

Tuwaiyauti- for a person, animal, or thing to drift away on a broken piece of shorefast ice

Tutqiksrivik place with a layer of snow on shorefast ice, near a pressure ridge, where a boat and gear can be stored up side down, and the sides of the boat banked with snow to protect gear under the boat. The bow of the boat is anchored to a piece of ice.

Tuuq, tuuq- long-handled ice chipping tool; to chip an ice fishing hole on shorefast ice

Tumi (tumit)/tuvi trail over shorefast ice; cf. *tumisaaq* – a trail on land

Tuvli- to cut or make a trail (over shorefast ice)

Uqqutaq windbreak wall built around a fishing hole; a windbreak wall built on shorefast ice or pack ice

Uqsruġaq oil slick in an open lead or open water, often left by a wounded animal

References

Allpergen, B. and Dobrieva, E. 2009. List of the Naukanski Yupik Sea Ice terms. Unpublished manuscript prepared for the SIKU project (in Russian and Naukanski Yupik).

Anungazuk, H.O. 1995. Whaling: A ritual of life. In *Hunting the Largest Animals: Native Whaling in the Western Artic and Subarctic*. A.P. McCartney (ed.), Studies in Whaling, 3, Edmonton: Canadian Circumpolar Institute, University of Alberta, pp. 339–345.

Apalu, A. and Kalyuzhina, N. 2009. List of the Yanrakinnot Chukchi Sea Ice Terms. Unpublished manuscript prepared for the SIKU project (in Chukchi, with Russian parallels).

Aporta, C. 2003. Old Routes, New Trails: Contemporary Inuit Travel and Orienting in Igloolik, Nunavut. Unpublished Ph.D. thesis, Department of Anthropology, University of Alberta, Edmonton.

Armaergen, R. and Golbtseva, V. 2009. List of the Uelen Chukchi Sea Ice Terms. Unpublished manuscript prepared for the SIKU project (in Chukchi, with Russian parallels).

Armstrong, T. 1956. Illustrated ice glossary. *Polar Record* 8(52): 4–12.

Bailey, A.M. 1933. A cruise of the "Bear." *Natural History* 3(5): 497–510.

Bailey, A.M. 1943. The birds of Cape Prince of Wales, Alaska. *Proceedings of the Colorado Museum of Natural History* 18(1). Denver.

Baily, A.M. 1971. Field work of a museum naturalist. *Museum Pictorial* 22. Denver.: Denver Museum of Natural History.

Bogoslovskaya, L.S., Vdovin, B.I., and Golbtseva, V.V. 2008. Izmeneniia klimata v regione Beringova proliva. Integratsiia nauchnykh i traditsionnykh znanii (Climate change in the Bering Strait region: Integration of scientific and traditional knowledge). *Ekologicheskoe planirovanie i upravlienie* 8(3): 36–48. Moscow (in Russian).

Brower, R.H., Sr. 2008. Barrow Inupiaq Sea Ice Terminology. Unpublished manuscript prepared for the SIKU project.

Brower, R.H., Sr. 2009. Inupiaq Sea Ice Terms from Wainwright, Alaska. Retransliterated from Nelson 1969, with the assistance of Rossman Peetook, Leo Panick, Jonathan Aiken, and Lawrence D. Kaplan. Unpublished manuscript prepared for the SIKU project.

Dumond, D.E. 2000. Henry B. Collins at Wales, Alaska, 1936. *University of Oregon Anthropological Papers* 56. Eugene: University of Oregon.

Erdmann, F. (ed.) 1864, 1866. Eskimoisches Wörterbuch gesammelt von Missionaren in Labrador, revidirt und herausgegeben von Friedrich Erdmann. Erster Theil: Eskimoisch-Deutsch (1864, 350 pp.), Zweiter Theil: Deutsch-Eskimoisch (1866, 242 pp.). Budissin [Bautzen]: Ernst Moritz Monse.

Fortescue, M. 1984. *West Greenlandic*. London, etc.: Groom Helm.

Fortescue, M. 1991. I'nuktun, An Introduction to the Language of Qaanaaq, Thule. *Institut for Eskimologi publication series* 15. Copenhagen.

Fox (Gearheard), S. 2003. *When the Weather is Uggianaqtuq: Inuit Observations of Environmental Change*. Boulder, Colorado USA: University of Colorado Geography Department Cartography Lab. Distributed by National Snow and Ice Data Center. CD-ROM.

Gearheard, S., Matumeak, W., Angutikjuaq, I., Maslanik, J., Huntington, H.P., Leavitt, J., Matumeak Kagak, D., Tigullaraq, G., and Barry, R.G. 2006. "It's not that simple": A collaborative comparison of sea ice environments, their uses, observed changes, and adaptations in Barrow, and Clyde River, Nunavut, Canada. *Ambio: A Journal of the Human Environment* 35(4): 203–211.

George, J.C., Huntington, H.P., Brewster, K., Eicken, H., Norton, D.W., and Glenn, R. 2004. Observations on shore-fast ice dynamics in Arctic Alaska and the responses of the Inupiat hunting community. *Arctic* 57(4): 363–374.

Harritt, R.K. 2003. Re-examining Wales' role in Bering Strait prehistory: Some preliminary results of recent work. In *Indigenous Ways to the Present: Native Whaling in the Western Arctic* (Studies in Whaling No. 6, Occasional Publication No. 54, The Anthropology of Pacific North America Series). A.P. McCartney (ed.), Edmonton: Canadian Circumpolar Institute (CCI) Press and University of Utah Press, pp. 25 –67.

Harritt, R.K. 2004. A preliminary re-evaluation of the Punuk-Thule interface at Wales, Alaska. *Arctic Anthropology* 41(2): 163–176.

Heyes, S.A. 2007. *Inuit Knowledge and Perceptions of the Land-water Interface* Ph.D. Dissertation (McGill University, Department of Geography: Montreal, Canada).

Huntington, H.P. 2000. Impact of Changes in Sea Ice and Other Environmental Parameters in the Arctic. Report of the Marine Mammal Commission Workshop. Girdwood, 15–17 February 2000. Marine Mammal Commission, Bethesda.

Huntington, H.P., Harry, B., Jr., and Norton, D.W. 2001. The Barrow Symposium on sea ice, 2000: Evaluation of one means of exchanging information between subsistence whalers and scientists. *Arctic* 54(2): 201–2006.

Jolles, C.Z. 2003. When whaling folks celebrate: A comparison of tradition and experience in two Bering Sea whaling communities. In *Indigenous Ways to the Present: Native Whaling in the Western Arctic* (Studies in Whaling No. 6, Occasional Publication No. 54, The Anthropology of Pacific North America Series). A.P. McCartney (ed.). Edmonton: Canadian Circumpolar Institute (CCI) Press and University of Utah Press, pp. 307–340.

Koutsky, K. 1981. Early Days on Norton Sound and Bering Strait. vol 2, The Wales Area. *Anthropology and Historic Preservation. Comparative Park Studies Unit. Occasional Paper* 29. University of Alaska, Fairbanks.

Krupnik, I. 2008. Draft list of the Inupiaq sea ice terms from Shaktoolik, Alaska. Unpublished manuscript prepared for the SIKU project.

LaBelle, J.C., Wise, J.L., Voelker, R.P., Schulze, R.H., and Wohl, M.M. 1983. Alaska Marine Ice Atlas. Anchorage: Arctic Environmental Information and Data Center, University of Alaska, 1983.

Laidler, G.J. 2007. Ice, Through Inuit Eyes: Characterizing the importance of sea ice processes, use and change around three Nunavut Commuties. Unpublished Ph.D. Thesis, University of Toronto, Department of Georgraphy.

Laidler, G.J. and Elee, P. 2008. Human geographies of sea ice: Freeze/thaw processes around Cape Dorset, Nunavut, Canada. *Polar Record* 44: 51–76.

Laidler, G.J. and Ikummaq, T. 2008. Human geographies of sea ice: freeze/thaw processes around Igloolik, Nunavut, Canada. *Polar Record* 44: 127–153.

Laidler, G.J., Dialla, A., and Joamie, E. 2008. Human geographies of sea ice: Freeze/thaw processes around Pangnirtung, Nunavut, Canada. *Polar Record* 44(231): 335–361.

Lopp, E.L. 2001. *Ice Window: Letters from a Bering Strait Village*. K. Lopp-Smith and V. Smith, (eds.), Fairbanks: University of Alaska Press.

Lowe, R. 1983. *Kangiryuarmiut Uqauhingita Numiktittitdjutingit*, Basic Kangiryarmiut Eskimo Dictionary. Inuvik, Committee for Original Peoples Entitlement.

Lowe, R. 1984. *Siglit Inuvialuit uqausiita kipuktiritait. Basic Siglit Inuvialuit Eskimo Dictionary*. Inuvik, Committee for Original Peoples Entitlement.

Lowe, R. 2001. *Siglit Inuvialuit Uqautchiita Nutaat Kipuktirutait Aglipkaqtat. Siglit Inuvialuit Eskimo Dictionary*. Québes, Nota Bene.

McDonald, M.A., Arragutainaq, L., and Novalinga, Z. 2007. Voices from the Bay. Traditional Ecological Knowledge of Inuit and Cree in the Hudson Bay Bioregion. Ottawa, Canadian Arctic Resources Committee.

Metcalf, V. and I. Krupnik, (eds.), 2003. Pacific walrus. Conserving our culture through traditional management. Report produced by Eskimo Walrus Commission, Kawerak, Inc. under the grant from the U.S. Fish and Wildlife Service, Section 119, Cooperative Agreement # 701813J506.

Nelson, R.K. 1969. Hunters of the Northern Ice. Chicago and London: University of Chicago Press.

Nichols, T., Berkes, F., Jolly, D., and Snow, N.B., and the Community of Sachs Harbour. 2004. Climate change and sea ice: Local observations from the Canadian Western Arctic. *Arctic* 57(1): 68–79.

Noongwook, G. 2000. Native observations of local climate changes around St. Lawrence Island. In *Impacts of Changes in Sea Ice and Other Environmental Parameters in the Arctic*. H. Huntington, (ed.), Bethesda: Marine Mammal Commission, pp. 21–24.

Norton, D.W. 2002. Coastal sea ice watch: Private confessions of a convert to indigenous knowledge. In *The Earth Is Faster Now. Indigenous Observations of Arctic Environmental Change*. I. Krupnik and D. Jolly (eds.), Fairbanks: ARCUS, pp. 127–155.

Nutawyi, A. and Rodionova, N.. 2008. List of the Sireniki Yupik Sea Ice Terms. Unpublished manuscript prepared for the SIKU project (in Yupik, with Russian parallels).

Oozeva, C., Noongwook, C., Noongwook, G., Alowa, C., and Krupnik, I. 2004. *Watching Ice and Weather Our Way. (Sikumengllu Eslamengllu Esghapalleghput)*. Washington: Arctic Studies Center, Smithsonian Institution.

Peck, E.J. Rev. E. 1925. Eskimo-English Dictionary. (Compiled from Erdman's Eskimo-German Edition 1864 A.D.). Published by the Church of the Ascension Thank-Offering Mission Fund, Hamilton. Toronto.

Ray, D.J. 1975. *The Eskimos of Bering Strait, 1650–1898*. Seattle and London: University of Washington Press.

Schneider, L. 1985. *Ulirnaisigutiit. An Inuktitut-English Dictionary of Northern Quebec, Labrador and eastern Arctic Dialects*. Québec, Les presses de l'Université Laval.

Tersis, N. 2008. Forme et sens des mots du Tunumiisut, Lexique Inuit du Groenland Oriental: Lexique-tunumiisut-anglais-danois. Societe D'etudes Linguistiques et Anthropologiques de France 9. Louvain, etc., Peeters Leuven.

Vakhtin, N.B. and Emelyanova, N.M. 1988. *Praktikum po leksike eskimosskogo iazyka* (Practical Aid to the Eskimo Lexicon). 1988, Leningrad: Prosveshchenie Publishers.

Walunga, W., comp. 1988. *St. Lawrence Island Curriculum Resource Manual*. Gambell.

Weyapuk, W., Jr., and I. Krupnik (eds.), 2008. *Kiŋikmi Sigum Qanuq Ilitaavut/* Wales Inupiaq Sea Ice Dictionary (preprint). Washington, DC: Arctic Studies Center, Smithsonian Institution, 112 pp.

Wisniewski, J. and local collaborators. 2008. List of the Inupiaq sea ice and snow terms from Shishmaref, Alaska. Unpublished manuscript prepared for the SIKU project.

World Meteorological Organization. 1970. WMO Sea-Ice Nomenclature. Terminology, codes and illustrated glossary. *WMO/OMM/BMO* 259, *Technical Paper* 145. Secretariat of the World Meteorological Organization, Geneva, 147 pp.

World Meteorological Organization. 2007. WMO Sea-Ice Nomenclature. WMO/OMM/BMO 259. Suppl. No. 5 – Linguistic Equivalents. http://www.jcomm-services.org/documents.htm?parent=136 [Accessed June 24, 2009].

Part IV
SIKU and *Siku*: Opening New Perspectives

Chapter 15
Indigenous Knowledge and Sea Ice Science: What Can We Learn from Indigenous Ice Users?

Hajo Eicken

Abstract Drawing on examples mostly from Inupiaq and Yupik sea ice expertise in coastal Alaska, this contribution examines how local and indigenous knowledge (LIK) can inform and guide geophysical and biological sea ice research. Part of the relevance of LIK derives from its linkage to sea ice use and the services coastal communities derive from the ice cover. As a result, indigenous experts keep track of a broad range of sea ice variables at a particular location. These observations are embedded into a broader worldview that speaks to both long-term variability or change and the system of values associated with ice use. The contribution examines eight different contexts in which transmission of LIK is particularly relevant. These include the role of LIK in study site selection and assessment of a sampling campaign in the context of inter-annual variability, the identification of rare or inconspicuous phenomena or events, the contribution by indigenous experts to hazard assessment and emergency response, the record of past and present climate embedded in LIK, and the value of holistic sea ice knowledge in detecting subtle, intertwined patterns of environmental change. The relevance of local, indigenous sea ice expertise in helping advance adaptation and responses to climate change as well as its potential role in guiding research questions and hypotheses are also examined. The challenges that may have to be overcome in creating an interface for exchange between indigenous experts and sea ice researchers are considered. Promising approaches to overcome these challenges include cross-cultural, interdisciplinary education, and the fostering of Communities of Practice.

Keywords Sea ice geophysics · Sea ice use · Local indigenous knowledge · Sea ice system services · Arctic observing network

---

H. Eicken (✉)
Geophysical Institute, University of Alaska Fairbanks, Fairbanks, AK 99775-7320, USA
e-mail: hajo.eicken@gi.alaska.edu

I. Krupnik et al. (eds.), *SIKU: Knowing Our Ice*,
DOI 10.1007/978-90-481-8587-0_15,

Introduction

Over the past few years, Arctic sea ice has received increasing attention by the public, mostly in the context of climate change. Media coverage typically discusses the shrinking and thinning of Arctic sea ice by referring to scientific studies based on satellite data or computer models of the climate system. While such information is generally scientifically precise at the large scale, it provides only a limited view of the characteristics of the ice cover itself and the processes that shape its seasonal evolution and role in Arctic ecosystems. Consider the expanse of coastal sea ice and ocean shown in the photograph taken during fall freeze-up at the Alaskan community of Wales in Bering Strait (Fig. 15.1a). The different ice types, open water, and the stretch of coast visible in the photo cover only a small fraction of the area that makes up a single data point (pixel) in the satellite imagery typically used to determine ice concentration and extent for studies of sea ice climatology. Several of these pixels, 25 by 25 km in extent, are shown in Fig. 15.1b, for a satellite scene acquired on the same day, November 9, 2007, that Winton Weyapuk, Jr., a SIKU project participant from Wales took the photograph (see Chapter 14 by Krupnik and Weyapuk, this volume). In fact, the satellite imagery does not indicate any presence of ice near Wales for this same date, mostly as a result of the small width of the narrow belt of coastal ice but also due to other factors that make new ice difficult to detect in nearshore environments.

The coarse observational scale of the satellite data is sufficient for broad studies of how Arctic sea ice helps regulate Earth's climate. However, on their own such data are of lesser value if the aim is, for example, to learn more about the processes that control seasonal ice growth and decay, its role as a platform for marine mammals or its importance in the context of coastal erosion. In a changing north that experiences not only substantial sea ice retreat but also increasing

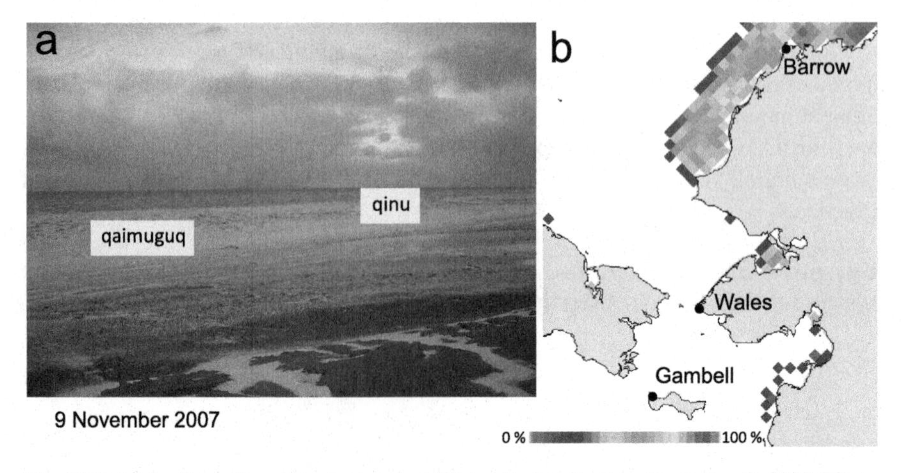

Fig. 15.1 (**a**) Ice formation along the beach at Wales, Bering Strait on November 9, 2007 (Photo: W. Weyapuk, Jr.). Note the formation of a slush ice berm (*qaimuguq*) that offers some protection to the coast from waves. (**b**) Ice concentrations for the same day obtained from passive microwave satellite data (Special Sensor Microwave/Imager, SSM/I) show no detectable ice near Wales

ship traffic and industrial activities, demand is great for more detailed information about the characteristics of the ice cover and its seasonal waxing and waning. Here, a more comprehensive, multifaceted perspective on sea ice as both a material and a process – a freezing water, or a melting, moving or deforming ice – is of value. Indigenous ice experts may provide such a perspective both through long years of detailed observation and through transmission and evaluation of knowledge from elders and peers.

This contribution touches on the question of what sea ice scientists, in particular geophysicists, oceanographers, and meteorologists (and to some extent biologists), can learn from indigenous sea ice experts and ice users. This question has been examined in detail in the broader context of local or indigenous knowledge and has been addressed in anthropological, geographic, or social science studies (Agrawal 1995; Berkes 1999; Krupnik and Jolly 2002). Here, I take less of a scholar's and more of a practitioner's approach and discuss how learning from indigenous ice experts may enhance, deepen, or broaden sea ice geophysical or biological research. By drawing on examples from field research or the literature this contribution aims to

- provide a perspective on the insights and understanding collaboration with indigenous ice experts may generate – mostly for those engaged in sea ice geophysical, climatological, or biological research but less familiar with local, indigenous knowledge;
- identify promising areas for further work where local, indigenous knowledge can contribute substantially to guiding observations and furthering understanding;
- develop a rough outline of what an interface between indigenous and geophysical–biological knowledge of sea ice may look like and what may be required to foster transmission and exchange across this interface.

For those interested in a comprehensive picture of sea ice knowledge in a coastal Arctic community, Richard Nelson's classic study documenting sea ice use in Wainwright, Alaska, is still highly relevant (Nelson 1969). Norton (2002) summarized insights gained from a symposium held in Barrow in 2000 that provided a good perspective on a more diversified approach to documenting and discussing indigenous and geophysical sea ice knowledge. Finally, a summary by Henry Huntington and several colleagues active in community-based observing programs is an excellent, up-to-date resource (Huntington et al. 2009).

Use of Sea Ice and Local, Indigenous Knowledge: Key Concepts and Terminology

As implied by the title of this contribution, a fundamental aspect of learning from indigenous or local ice experts is the recognition that their knowledge in large part derives from the use of sea ice. The term "use of sea ice" refers to more than simply using the ice, e.g., as a platform for travel or hunting as described by Druckenmiller and others and Gearheard and others in this book. Rather, it describes the suite of services that communities or individuals derive from the sea ice zone, including the ecosystems associated with it. Here, sea ice services include tangible and intangible benefits such as protection from waves and coastal erosion or the important place sea ice occupies in the lives of Arctic coastal residents. Moreover, the concept of sea ice system services, extending the theory of ecosystem services to the Arctic ice cover (Eicken et al. 2009), also includes the hazards and threats emanating from the ice. Within this framework, indigenous people typically observe and keep track of a range of different phenomena, processes, and animals as they relate to the specific services derived from sea ice.

This is not only apparent in the typically more than one hundred terms indigenous languages reserve for sea ice features and ice-associated phenomena (see Chapter 4, Chapter 2 by Taverniers; Chapter 14 by Krupnik and Weyapuk, this volume), from which a "map" or schematic of key ice processes and interactions can be constructed. In working with sea ice experts in the communities of Gambell, Wales, and Barrow, who are making observations of coastal ice as relevant to their communities' activities (see Chapter 4, Krupnik et al. this volume), we find references to a large number of animals observed in conjunction with the ice cover or specific ice processes. This includes not only obvious mention of ice-associated seals or walrus but also observations of fish and bird species that display preference for specific ice types and exhibit a seasonality coupled to the ice, which in turn determines how they may be harvested from the ice platform.

Intimate and long-standing familiarity with a specific place, often based on the use of resources at that location, is generally referred to as local knowledge. The power of such local knowledge was demonstrated on an icebreaker cruise with a German vessel into Siberian waters that the author participated in some years ago. Near the Franz-Josef-Land archipelago the vessel was making no progress in very heavy ice, despite non-stop ramming and breaking. A Russian icebreaker captain – onboard as an observer and familiar with the tidal currents in the region and their often barely perceptible impact on the periodic opening and closing of cracks and leads – was finally able to pick a path through cracks and narrow passageways.

Indigenous knowledge embeds local knowledge and other insights, beliefs, and values into a worldview that extends into the human and spiritual realm and is shared by a larger community. It builds on a tradition of environmental observations at a given place, providing a backdrop of greater temporal depth and topical breadth. Through ties to specific applications and uses, local and indigenous knowledge is subject to repeated critical review and reaffirmation, both in the field and by the elders and recognized experts in a community. Such knowledge is commonly also referred to as traditional (ecological) knowledge (see Agrawal 1995; Berkes 1999; Huntington et al. 2005). Both local and indigenous knowledge are relevant in this chapter and will be summarily abbreviated as LIK.

Kawagley (1995) describes how the Yupiaq worldview in western Alaska is supported by a balance between the human, natural, and spiritual realms. Since this type of knowledge or understanding cannot be compartmentalized or categorized in the manner that western (i.e., Euro-American) science ingests and evaluates information, it may be challenging or possibly discomforting for scientists to even begin to learn from sea ice users whose expertise is firmly embedded in such a holistic worldview. Hence, it is not uncommon for scientists to dismiss indigenous expertise. It often happens because the scientific method (which of course is also firmly embedded in a worldview of its own) does not come with the tools that allow one to process quantitative information about the nature of the physical environment while at the same time accepting the idea, say, that the division between the realms of people and animals is permeable and indistinct. Even though it is less rich and deep, local knowledge – such as that of the icebreaker captain referred to previously – is often easier to accept in such an LIK-skeptical context because it is divorced from any specific worldview.

The threshold that has to be crossed in order for a substantive discourse to occur is typically higher for the physical than the biological sciences. The study of ecosystems or animal physiology and behavior lends itself more readily for exchange among LIK and western science experts because of similar methodologies employed, e.g., in capturing animals, than the study of the physical environment, in particular with an increasing specialization and reliance on remote-sensing methods and model simulations. This situation is unfortunate because the latter stands as much to gain from LIK as the former.

A Survey of Indigenous Expertise and Knowledge Relevant to Sea Ice Research

One approach to help those in the physical sciences engage with indigenous sea ice experts is to identify or delineate subject areas or ways in which LIK can inform, guide, or enhance research. Eight such ways of engagement are sketched out in the brief survey below, organized to progress from specific, obvious applications to broader, potentially more complicated and less well-explored categories. It needs to be recognized, however, that by its very nature indigenous knowledge does not lend itself to compartmentalization or integration into the framework of western science (Kawagley 1995; Nadasdy 1999). Thus, the categories explored below are meant to guide the gaze as one attempts to catch a glimpse of different facets of indigenous knowledge; by no means are they intended to classify the knowledge itself. The discussion below reflects instruction by Inupiaq and Yupik sea ice experts, insights gained from field work in coastal Alaska, and discourse with colleagues working in the region. Nevertheless, similar categories have been arrived at by others, such as Berkes (2002) who recognized five potential areas of convergence between climate research and traditional knowledge that are in many ways equivalent to those delineated below.

Is This a Good Spot? LIK and Study Site Selection

A key aspect of field measurements on sea ice is the selection of an appropriate study site, which is often challenging due to the heterogeneity of sea ice as a material and the patchiness of physical and biogeochemical properties (e.g., Granskog et al. 2005). Hence, some measurements, such as the determination of ice algal biomass, may require large numbers of samples to allow statistically significant conclusions (McMinn et al. 2009). Knowledge of the ice growth history and targeting of sites with homogeneous snow cover and absence of – often hidden – deformation features can help minimize the number of samples that need to be taken and enhance insights gained from their analysis (Eicken 2009). Here, local, indigenous ice experts are in a position to offer valuable guidance. Many of the processes and ice features that complicate analysis of field data and potentially invalidate conclusions, such as ice deformation features, sediment inclusions, patchy snow cover, and anomalous growth history impacted by freshwater runoff or flooding, are familiar to local ice users in coastal communities.

As discussed by Aporta (2002), Norton (2002), Laidler and others (2009) and Druckenmiller and others (Chapter 9, this volume), use of trails over sea ice requires monitoring of ice evolution throughout the ice season to anticipate potential hazards as the season progresses. Travel and hunting on the sea ice foster close observation and tracking of snow cover and ice deformation features, potentially of great value to researchers planning a field campaign with specific requirements for the ice types that are to be sampled. A considerable challenge, however, exists in having researchers communicate their interests and then translating relevant aspects of local, indigenous knowledge that would help in the planning and execution of a field campaign. While it is common practice to hire local experts as guides, it is much less common for field parties to engage in an appropriate form of communication (see Huntington et al. 2009) with local experts during the early stages of project design and field trip planning in order to evaluate whether suitable ice types are present and if so where and how best to access them.

These difficulties are less of a challenge if the features or processes to be studied partly overlap with the interests of indigenous ice use. Let us consider two examples to illustrate this point. Multiyear sea ice (piqaluyak being an Inupiaq equivalent) helps ensure the stability of landfast ice (see Chapter 9 by Druckenmiller et al. this volume) and provides a preferred source of freshwater to hunting crews on the ice and elders in the village (Nelson 1969; George et al. 2004). Over the past two decades, access to multiyear ice in some Arctic locations, such as coastal Alaska, has become more difficult due to changes in Arctic ice circulation and dramatic loss of multiyear ice (Maslanik et al. 2007). Even small multiyear floes few tens of meters across can be of great value in a range of studies of sea ice (Eicken 2009) but are difficult to detect in satellite imagery or even on the ground. At Barrow and other Alaska coastal locations, however, many hunters and ice experts can accurately point to such old ice fragments embedded in the landfast ice over a stretch of tens of kilometers in the vicinity of town. An example of such a multiyear floe fragment is shown in Fig. 15.2. This site had been scouted from the air after Billy Adams, a seal hunter and ice expert from Barrow, had pointed it out to us. Seal hunters active in winter or whaling crews scouting potential sites for trails make note of such occurrences of piqaluyak. Typically, knowledge of even such temporary landmarks

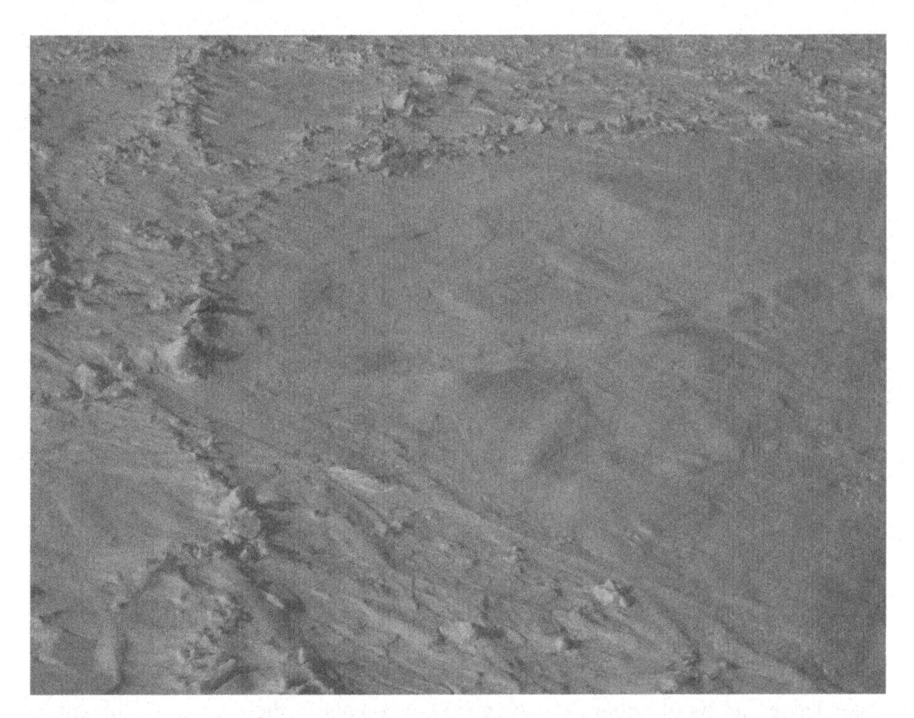

Fig. 15.2 Multiyear ice fragment (roughly 100 m wide, occupying the right-hand three-quarters of the image) in landfast sea ice north of Barrow, Alaska, in an aerial photograph taken on April 14, 2009. Note the rolling topography of hummocks in contrast with the rough, ridged, and rubbled surrounding ice. Navigating across the ice at ground level, only the most experienced of experts would be able to identify such smaller fragments of multiyear ice

is so accurate that they can be visited and talked about without needing to refer to a Global Positioning System (GPS) device.

A second example of overlap in Inupiaq ice use and scientific research builds on our early work in Barrow, when we wanted to install instruments in ice that had formed locally early in the year and remained stable throughout winter. Stability and early or late access to sea ice as a platform is naturally of great interest to local hunters. Hence, Kenneth Toovak, a Barrow ice expert who through years of working at the Naval Research Laboratory had also honed his skills as a mediator between Inupiaq and geophysical ice science, was able to advise us of a location that was protected by a shoal (not evident in charts of the area) and tended to form ice early in the year with little risk of later ice break-out. This site was chosen also for its proximity to the laboratory, even though another location further away had been pointed out as being more suitable because the ice there was more stable, less heterogeneous and formed in situ in the vast majority of years. Our first deployment of sensors at the closer site was into roughly 40 cm thick sea ice on November 11, 1999. In the following seasons, it became increasingly difficult to deploy before the end of the year because of lack of (stable) ice. Now, we typically deploy instruments in late January at the second, more distant location further up the coast that Mr. Toovak had recommended as the more appropriate site. He was right – we should have deployed our instruments at that location in the first year and would have had a longer, internally consistent time series for it.

Is This Normal? Inter-annual Variability and Guidance from LIK

Many field projects may only visit for a brief period during a single ice season. Under such circumstances, conclusions of more general validity about a specific process or ice characteristic can be affected by the occurrence of anomalous events or the prevalence of atypical conditions. To be sure, process studies ultimately are linked to a specific process and not a particular site or year. However, large, expensive programs such as the Study of the Heat Budget of the Arctic Ocean (SHEBA, Perovich et al. 1999), typically only run for a single field season. Representations (so-called parameterizations) of important processes in numerical models are then based on the suite of interrelated processes and phenomena sampled at that particularly site in the given year. Here, both direct personal experience in the form of local knowledge and more so the longer-term history embedded in indigenous knowledge can be of value in placing measurements in the context of the annual cycle and inter-annual variability.

The example of landfast ice protected by a shoal quoted above is also relevant here, since Mr. Toovak was able to recommend the sampling site based on his insight into recurring features. The absence of protective ridges and grounding ice in a particular year would hence have been identified as anomalous and not representative of long-term conditions. Another, intriguing and less clear-cut case relates to the question of potentially anomalous sediment entrainment into coastal sea ice. Sediment inclusions in sea ice can have a tremendous impact on optical or biogeochemical properties and sea ice microbial communities (Light et al. 1998; Gradinger et al. 2009). However, until late in the melt season, and even then, they are not always easy to detect or are ignored in studies focusing on other aspects of the ice cover that are nevertheless potentially impacted by their presence. Fienup-Riordan and Rearden (Chapter 13, this volume) highlight how Yupik elders in the Nelson Island region pay close attention to sediments in sea ice. Similarly, two hunters in the Norton Sound and Bering Strait region commented to the author unprompted on how they had noticed an increase in the amount of sediment-laden ice, an observation that matches indications from a study in the Chukchi and Beaufort Sea (Eicken et al. 2005). Such knowledge can be key in the selection of suitable study sites that need to either preclude or include the presence of sediments in the ice. It can also provide guidance on whether a particular sampling may have been impacted by the presence or absence of sediment-laden ice, a phenomenon difficult to observe through remote sensing and notoriously patchy in time and space, and hence ideally suited for sharing of insight by indigenous, local experts.

The Hidden Whales and the One-Hundred Year Ridge: Rare or Inconspicuous Phenomena and Events

One of the more powerful ways in which LIK can help scientists or engineers gain a new level of understanding is also quite accessible to those unfamiliar with LIK. Here, two classic examples will be discussed. Both relate to significant events that are difficult to observe either because they are uncommon or because they are concealed from the eyes of casual observers or scientists relying on inappropriate methodology.

The Bowhead whale has been hunted by Iñupiat and Yupik Eskimo for centuries. When the US National Marine Fisheries Service (NMFS) threatened to close the hunt because of low whale population estimates (less than 2,000 animals), indigenous experts disputed both these numbers and the methodology used to arrive at them (Albert 2004). A long-term research program supported and guided by the Iñupiat of the North Slope of Alaska demonstrated that indigenous knowledge had been correct, asserting that whales do not shy away from ice, which they can break at 30 cm thickness or more. Hence visual counts in open leads had missed many animals. Currently the stock is well above 10,000. The application of acoustic under-ice tracking techniques prompted by Inupiaq experts has now confirmed the presence of whales other than bowheads in the winter ice pack, attributed to changing ice and ocean conditions (Stafford et al. 2007).

The second example dates back to the first wave of oil and gas exploration along the North Slope of Alaska. The risk of so-called ice ride-up events presented a significant potential threat for coastal installations and infrastructure placed in shallow water on artificial islands. To allow engineers and regulators assess the hazards associated with such events and develop appropriate structural designs in order to minimize the risk to the installation, specific data on the frequency and severity of such ice ride-ups were required. With such rare but severe events as the equivalent of a storm of the century, and considering the near-complete lack of environmental engineering studies in the coastal environment at that time, this situation presented a substantial challenge. In a classic study, Lew Shapiro and Ron Metzner of the University of Alaska Fairbanks (UAF) and Kenneth Toovak of Barrow set out to interview local experts along the North Slope to record their knowledge of a range of important aspects of such events, most importantly their frequency of occurrence and severity (Shapiro and Metzner 1979). Due to the threat these events represented, they were able to compile an impressive and useful record that extended back well into the era prior to World War II. While the interviews, transcripts, and translations, residing at UAF, have proven useful in the context of the engineering design studies and hazard assessments, they represent a wealth of knowledge that remains largely untapped. In interpreting such interviews it is important to note that often the day of the year on which specific events occurred is recalled much more accurately – because it relates to the seasonal cycle of ice use – than the year itself, in particular if the event is several decades in the past (see also George et al. 2004).

Is It Safe to Go Out? Hazard Assessment and Emergency Response

The example discussed above has hinted at the potential value of LIK for engineering applications. The question of how LIK may relate to ice engineering and industrial activities in the north is one of the more difficult and important aspects of the topic at hand. Both LIK and engineering represent applied, use-driven knowledge systems. Hence it is reasonable to expect that transmission across the LIK-engineering interface would be more straightforward than exchange between climatologists and Iñupiat sea ice experts. Thus, industry commonly relies on Iñupiat guides to ensure safety of field parties and may include advice from recognized indigenous experts to arrive at major decisions concerning deployments in potentially unsafe areas. At the same time, concern by coastal communities over potential hazards associated with oil and gas development greatly complicates exchange between the different expert groups. This problem is exacerbated by the challenges encountered by indigenous experts in seeing their expertise represented in the decisions made by regulatory agencies and industry, in particular with respect to coastal and offshore development. The latter problem stems not necessarily from willful exclusion or dismissal of evidence based on LIK, but is often a result of the inability of the regulatory or scientific apparatus to come to terms with the nature of indigenous knowledge (e.g., Nadasdy 1999; Usher 2000).

Fig. 15.3 Aerial photograph from April 12, 2008, of a whaling trail (Jacob Adams crew, see also Figs. 9.11 and 9.12 in Chapter 9 by Druckenmiller et al. this volume) winding its way through pressure ridges in coastal landfast ice near Barrow, Alaska. The trail originates in the lower left corner of the image and can be seen in the center of the image as it traverses a stretch of level ice in a near-straight line. Note the band of clouds visible over the lead toward the top of the photograph

This complicated set of issues transcends the scope of this chapter and is examined in detail elsewhere (Eicken et al. 2010). However, the example of Iñupiat ice trails discussed in depth by Druckenmiller and others (Chapter 9, this volume) highlights the similarities between indigenous ice use and engineering applications. Thus, escape, evacuation, and rescue (EER) across landfast ice for coastal and offshore installations in Canada and the U.S. is a key aspect of their design and operation (Barker et al. 2006). EER also plays into safety requirements of Arctic tourism or shipping, both of which are on the rise. The expertise that enters into the operation of hunting camps and over-ice travel by the Inupiat is highly relevant in such a safety context (Eicken et al. 2010). Moreover, as pointed out by Huntington and others (2005), expertise on events that represent hazards or have threatened people in the past, like catastrophic ice break-outs (George et al. 2004; Chapter 9 by Druckenmiller et al. this volume), is particularly relevant to ice users and holders of LIK. Thus, trails placed on the ice to provide access to open water (Fig. 15.3) are specifically designed to also serve as efficient evacuation routes, hence embodying knowledge and skills relevant for EER applications as well.

Extending the Record

The discussion of local, indigenous ice experts' perspective on inter-annual variability and anomalous events leads us to the broader theme of LIK and environmental change on climatological time scales, covering several decades or more. This topic is distinct from the other two, since it requires more than the recollection of memorable events (such as the threat to life and property represented by an ice ride-up event) or the evaluation of weather and ice conditions in the current season relative to those a few years back. Moreover, objectivity and accuracy need to be considered carefully in collecting and interpreting information on constancy or change of climate variables transmitted through LIK. Cruikshank's (2005) study of oral history in the upper Yukon demonstrates just how difficult of a question this can be, partly since consideration of memories accumulated over such longtime intervals requires introspection and places greater emphasis on the human and spiritual components of indigenous knowledge (see also Huntington et al. 2009).

In the course of his analysis of historical records of weather and climate during the first International Polar Year 1882–1883, Kevin Wood has pointed to the value of indigenous expertise that allowed researchers to place their brief observation interval into the reference frame of the local indigenous community (Wood and Overland 2006). This is relevant as nineteenth century Arctic climate was emerging from the Little Ice Age. The latter period was characterized by lower temperatures and more severe ice conditions between the seventeenth and the nineteenth centuries, driven in part by insolation anomalies due to volcanism and sunspot activity (Overpeck et al. 1997). In his journal of 2 years (1852–1854) spent near the present town of Barrow, Rochfort Maguire refers to an Inupiaq ice expert's knowledge of more severe conditions, both with respect to weather and with respect to mammal harvests: "Erk-sin-ra our great authority, thinks they had worse seasons before the ship came" (Bockstoce 1988:365).

How accurate are such observations when examined from the perspective of geophysical or climatological research? As discussed by Nelson (1969), observations by individuals who are recognized in their community as sea ice experts are generally quite accurate, even when describing environmental phenomena that seem to contradict prevailing scientific thought, such as bowhead whales breaking 30 cm of ice. The continuous cycle of review and reaffirmation or revision of indigenous sea ice knowledge by the community, in particular elders and recognized experts, helps calibrate and provides accurate baselines for assessments of long-term change in ice conditions or local climate. Observations of gradual change are furthermore calibrated by keeping track of, e.g., changes in the treeline, permafrost, multiyear ice presence, and other longer-term integrators of subtle, but significant climate change that underscore and relate to sea ice change. While such observations of co-varying or concerted change are important in their own right (see below), they can provide a reference framework that allows indigenous experts to be cognizant of change that in the absence of instrumental records would otherwise go undetected (see Huntington 2000).

LIK and Detection of Subtle, Intertwined Patterns of Sea Ice and Environmental Change

The interaction between ice, ocean, and atmosphere results in sea ice distribution patterns that are frequently complicated and typically subject to great inter-annual variability. For example, the (then) record minimum in Arctic summer sea ice extent in 1995 (since the start of systematic satellite observations in 1979) was followed by a record maximum summer ice extent in 1996. In northern Alaska, the (then) record minimum Arctic summer ice extent of 2005 was followed by a summer in 2006 where lingering multiyear ice slowed seasonal ice retreat, fostering marine mammal harvests, and hampering summer shipping for the first time in well over a decade. Superimposed on this inter-annual variability is a trend toward decreasing summer ice extent. Local, indigenous ice experts are keenly aware of the predominance of inter-annual variability in ice conditions. The potential presence of cooling or warming trends on top of such pronounced variability may hence be examined in great detail before any LIK experts reach conclusions about observations of climate change. This is reflected in Shari Fox Gearheard's study of climate change in the Canadian Arctic in the late 1990s and early 2000s where different communities debated such questions of variability and change (Fox 2002). Similarly, on St. Lawrence Island comparison of present-day observations with records from the past indicates a more complicated picture than simple steady change (Chapter 4, Krupnik et al. this volume; Chapter 5 by Kapsch et al. this volume).

At the same time, indigenous knowledge and local expertise are in a unique position to assess even subtle changes and to separate patterns of variability from signals associated with longer-term change. This is a result of the holistic, comprehensive nature of observations that go along with a subsistence lifestyle in coastal Arctic communities. Thus, indigenous sea ice experts – and in contrast with, say, sea ice geophysicists – are not merely experts in one discipline engaging in a small set of observations. They have typically acquired expertise that spans a range of disciplines and phenomena through hunting and participating in other activities that are part of Arctic village life. In discussing climate change in the Canadian Arctic, Fox's (2002) collaborators considered a complex of 11 factors that included not just ice, but also observations of rain, insects, and birds. While LIK does not necessarily employ such observations to establish causal relationships, they allow indigenous experts to detect subtle, but coordinated change that spans a broader physical or biological system.

For St. Lawrence Island, Krupnik and co-workers (Chapter 4, this volume) refer to the utility of a range of "benchmarks" that can serve as important indicators in tracking changes in the seasonal cycle. Observations of how such indicators are linked to other physical or biological phenomena provide indigenous sea ice experts with a holistic perspective of change and variability relevant to the services they derive from their environment. Leonard Apangalook of Gambell identified roughly 30 such indicators in his log. His observations of open water in the ice pack during spring, a phenomenon referred to as *kelliighineq* (or *gelleghenak*), illustrate this point:

> April 27, 2007: West wind at 3 mph, 23F
>
> *Gelleghenak* about 3 miles offshore with pods of walrus hauled out on outer edge of ice pack. Yesterday (4/26) eight boats went out and sighted few [bowhead] whales southwest of Gambell moving north along with many beluga whales. Some loose ice packed in against shore preventing boats from getting out. Numerous walrus hauled out on ice. Mostly female walrus with juveniles hauled out. (L. Apangalook unpubl. observations)

Here, a recurring ice pattern that is closely monitored during each annual cycle and that depends on the interaction of atmosphere, sea ice, and ocean is linked to the distribution of three different species of marine mammals. Walrus and other animals like to congregate in such areas, and Mr. Apangalook provides further detail on the gender and age of walrus encountered as well as specifics on the ice conditions relevant to walrus and hunters from Gambell. Such richness of information and the identification of key processes or linkages that co-vary with this pattern cannot be extracted readily from remote sensing of ice conditions or oceanographic measurements. Hence, detection of changes in the seasonality of these patterns by indigenous communities can be a powerful indicator of systemic change that eludes disciplinary scientific approaches (see also Huntington 2000, and Krupnik et al. Chapter 4, this volume).

Adaptation to Climate Variability and Change

Indigenous sea ice knowledge derives from centuries of sea ice use at a particular location, and thus embeds in some form expertise in adaptation to a variable or potentially changing environment (Nelson 1969; Krupnik and Jolly 2002; Laidler et al. 2009). Given the magnitude of recent changes observed in Arctic sea ice and coastal environments (Rachold et al. 2000; Serreze et al. 2007), scientific research on and development of strategies for effective adaptation to a changing physical, socio-economic, and geopolitical environment is of increasing importance (e.g., Berkes 2002; Adger et al. 2007). Here, local, indigenous sea ice use and the associated body of expertise have much to offer, in particular at a time where appropriate response strategies are very much under discussion and a theoretical framework is only slowly evolving.

Indigenous communities, though hard hit by some of the negative impacts of climate change, in particular by those limiting access to sea ice, are quick to adapt. They are responding by relying on different technologies or through modifications of their lifestyle. The rapidity of such adaptation, evident, e.g., in some of the shifts in hunting patterns or hunting methodology (George et al. 2004; Chapter 4 by Krupnik et al. this volume; Chapter 5 by Kapsch et al. this volume), may be due in part to the ability of indigenous environmental experts to keep a finger on the pulse of variability and change. Moreover, LIK is very much both a fundamental and an applied knowledge, thus placing its holders in a uniquely qualified position to respond to change (see Eicken et al. 2009). This opens the door for rich exchange of information and synthesis of indigenous and western scientific approaches to increase community resilience in the face of change and improve people's ability to respond to a broader range of challenges associated with a rapid transition.

Building on the work by Druckenmiller and others (Chapter 9, this volume), the photograph showing an Iñupiat whaling trail on landfast sea ice (Fig. 15.3) illustrates how the combination of LIK and geophysics can benefit both scientists and a range of sea ice users. Trail routing is based on a comprehensive assessment of a range of factors that determine the ease of travel, safety, stability, and persistence of the trail throughout the season. These factors, as pointed out by Druckenmiller and others, may be related to key geophysical variables such as the thickness distribution, morphology, and physical properties of the ice. Repeat annual surveys of the trail systems thus represent a form of highly integrated information that may serve as an indicator of both the nature of environmental change and the indigenous community's response to such change. Building on the concept of benchmarks or indicators discussed by Krupnik and others (Chapter 4, this volume), I postulate that such use-based indicators represent a higher level of integration that may help sea ice science in grappling with the difficult question of how to move from observations of climate change to an understanding of the underlying processes and the development of response strategies.

Indigenous ice expertise has something else to offer that is often lacking in large-scale studies of climate variability and change: The detailed understanding of how large-scale processes work in concert to produce impacts at the local scale. Returning to Fig. 15.1 and Winton Weyapuk's observations of ice formation in the coastal environment, such records of fall freeze-up can help validate and improve remote-sensing approaches, which may not necessarily capture the relevant processes or phenomena. Moreover, such local observations can also help with downscaling from observations or projections of climate variability and change, typically registered at a much coarser scale (Fig. 15.1).

LIK and the Development of Research Questions and Hypotheses

Those willing to engage with local, indigenous ice experts from the very outset of a scientific study can gain much from having these experts and the knowledge shared by them and their community define or inform the development of research questions and hypotheses. As different scientific programs and nations focus on the development of an Arctic Observing Network, much emphasis is placed on not merely satisfying the interests of the researchers relying on such a system to answer fundamental science questions, but to consider the information needs of those impacted by or in some way linked to Arctic change (e.g., Committee on Designing an Arctic Observing Network 2006; Eicken et al. 2009). These goals can be achieved through the joint development of research questions and hypotheses by sea ice users and sea ice scientists. While such approaches are only in their infancy, past experience suggests that they hold much promise. Here, the classic case of building on Inupiaq traditional knowledge of the bowhead whale's use of the ice environment to develop and test novel (at least to western science) hypotheses that led to a revision of western scientific thought is a good example (Albert 2004). Along similar lines, George et al. (2004) explored different Inupiaq and western science postulates explaining the causes of large landfast ice break-out events. Carmack and Macdonald (2008) adapted their hydrographic measurement program to explore local environmental knowledge in the Mackenzie Delta region from a geophysical perspective. However, as pointed out by indigenous experts (Gearheard et al. 2006), often "it's not that simple." Rigorous and thorough exchange across the interface between LIK and the biological–geophysical sciences is as challenging, if not more, than complex, highly involved interdisciplinary research and may not necessarily yield nuggets of insight that can be directly translated into research questions or hypotheses. Here, perseverance and innovative approaches are needed. This book highlights such perseverance that bears the fruits of substantial exchange between different knowledge systems.

Conclusions

In concluding this survey of ways sea ice science can learn from local, indigenous sea ice users, I hope at least some benefits have become clear. This contribution has glossed over some of the challenges that may await those who are willing to explore the interface between LIK and western sea ice science.

While written about anthropological work with indigenous historians, I consider the following summary by Ernest S. Burch, Jr., as highly relevant to the issues discussed above:

> Thirty years ago, my view was that all narrative history which challenged my notions of common sense should be regarded as false until confirmed as true. Unfortunately, what

> originally passed for common sense proved to be little more than nonsense. In 1991, I would restate my position as follows: information that is provided by people whom the Iñupiat consider competent historians should be regarded as true until proven false, no matter how extraordinary what they say may first appear. [. . .]
>
> I include the caveat "whom the Natives consider competent historians" because there are incompetents and charlatans, as well as genuine experts, among Alaska Native peoples, just as there are among all peoples. The Natives know who is which, although, out of politeness, they generally listen to every elder who expounds on legendary or historical matters, regardless of the truth value of that person's remarks. In any event, oral accounts, like written accounts, must be subjected to rigorous historiographic evaluation and critical analysis. (Burch 1991:13)

Replace "historian" with "ice expert" in this quote and there are arguably few other summaries of the issue that make the pertinent points as succinctly and eloquently. An important issue implicit in Burch's remarks and highly relevant for audiences in the physical and biological sciences is that work at the interface between LIK and western science is most effective and rewarding in a truly interdisciplinary setting that pairs the natural sciences with the social sciences in order to open up an unobstructed channel for communication and transmission between knowledge systems. Cruikshank's (2005) work – building on the highly relevant study of glacier history in the Yukon – illustrates just how complex and sophisticated of an endeavor is required to achieve such goals, but also suggests that similar work in the realm of oral histories of sea ice holds much promise.

This contribution has been concerned mostly with *what* sea ice science can learn from sea ice users, less so with *how* such knowledge can be transmitted. Huntington and others (2009) have recently made the methodology of community-based ice observations more accessible to a broader audience. Here, I wish to conclude with a few remarks on how to promote two-way transmission of local, indigenous knowledge and physical–biological expertise. An effective way to progress is to bring together recognized LIK and sea ice geophysics or biology experts in a teaching and learning environment to share their expertise with students. Field courses, typically an integral part of many polar geophysics and biology curricula, can play a major role by including indigenous ice experts among the instructors and allowing them to share their knowledge in a culturally appropriate setting. Figure 15.4 highlights the role of experts, such as Richard Glenn from Barrow, Alaska, who are versed in the language, ways, and methodology of both indigenous and western sea ice science. The openness of students toward new approaches and their ability to side-step many of the traps that have plagued transmission of relevant knowledge across cultural divides in the past should not be underestimated.

The example of Richard Glenn also points to the important role of mediators or experts versed in both knowledge systems for transmission across a cultural interface. One of the most effective means of enhancing the role of such mediators is to entrain indigenous students into academia, allowing them to develop skills that are relevant to both the LIK and western science spheres. A promising arena to entrain local talent and promote exchange and learning is the nascent Arctic Observing Network, with its component projects of community-based observations. Several chapters in this book highlight the benefits gained from observations made within

Fig. 15.4 Richard Glenn (*center*), Inupiaq sea ice expert and academically trained geoscientist, teaching students in an international sea ice field course held at Barrow in May 2004

the community by recognized experts as well as the younger generation (see also Huntington et al. 2009). While the community-based components of these studies are increasingly robust and effective, further work is needed to improve the integration of such projects into the overall network in a meaningful way.

A concept and practical approach that holds much promise in weaving together these different activities and perspectives is to foster the development of the so-called Communities of Practice. Wenger and others (2002) have highlighted the importance of such informal groups of experts sharing a common interest or passion in advancing both theoretical understanding and practical progress with respect to pressing, difficult problems. A similar approach has been identified as highly promising in fostering exchange between engineers, regulators, and indigenous environmental experts in the context of offshore and coastal oil and gas development (Eicken et al. 2010). The challenges facing the establishment of such Communities of Practice are mostly geographic and cultural separation (both indigenous vs. western and academic vs. non-academic cultures). However, the field courses referred to above, community-based ice observations (Huntington et al. 2009) and efforts such as the Barrow Sea Ice Symposium that brought together a diverse group of sea ice experts (Norton 2002) have demonstrated their promise and potential value.

A highly promising topic area that arguably has already benefited greatly from the emergence of Communities of Practice is the problem of coastal dynamics and coastal retreat. As highlighted above, a range of geological, geophysical, biologic, and human processes conjoin along the Arctic coastline. With enhanced rates of coastal retreat furthered by declining sea ice and thawing permafrost, indigenous expertise, with a holistic perspective on coastal dynamics that still eludes academic and engineering approaches, has much to offer. Efforts such as the relocation of the Alaska coastal community of Newtok where native elders, young community leaders, scientists, and engineers from academia and state and federal agencies and countless others are taking a pragmatic, hands-on approach to the problem hold significant promise for a new era in Arctic fundamental and applied research.

Acknowledgments This chapter draws from numerous conversations, instructions, and support by a range of Inupiaq and Yupik sea ice experts. I am particularly grateful to Winton Weyapuk, Jr., Joe Leavitt, Leonard Apangalook, Sr., the late Kenneth Toovak, the late Arnold Brower, Sr., and Richard Glenn for sharing their insights and providing guidance: Quyanaqpak and Quyanaghhalek! This contribution would not have been possible without the encouragement by Igor Krupnik, who introduced me to the finer details of working with indigenous experts: Thank you! Claudio Aporta, Igor Krupnik, and Matthew Druckenmiller provided helpful comments on the manuscript. The work reported on in this contribution has been supported by the University of Alaska and the National Science Foundation (grants OPP-0632398 and 0805703). The views reflected in this contribution are the author's and do not necessarily reflect those of the aforementioned people and organizations.

References

Adger, W.N., Agrawala, S., and Mirza, M.M.Q., et al. 2007. Assessment of adaptation practiees, options, constraints and capacity. In *Climate change 2007: Impacts, adaptation, and vulnerability. Contribution of Working Group II to the Fourth Assessment Report of the Intergovernmental Panel on Climate Change*. Parry, M.L., Canziani, O.F., Palutikof, J.P., van der Linden, P.J., and Hanson, C.E. (eds.), Cambridge: Cambridge University Press, pp. 717–743.

Agrawal, A. 1995. Dismantling the divide between indigenous and scientific knowledge. *Development and Change* 26: 413–439.

Albert, T.F. 2004: Long-term research program verifies specific aspects of Eskimo Traditional Knowledge regarding estimating bowhead whale population size. *55. AAAS Arctic Science Conference*.

Aporta, C. 2002. Life on the ice: Understanding the codes of a changing environment. *Polar Record* 38(207): 341–354.

Barker, A., Timco, G., and Wright, B. 2006. Traversing grounded rubble fields by foot – Implications for evacuation. *Cold Regions Science and Technology* 46: 79–99.

Berkes, F. 1999. *Sacred ecology: Traditional Ecological Knowledge and Resource Management*. Philadelphia: Taylor and Francis, 209 pp.

Berkes, F. 2002. Epilogue: Making sense of Arctic environmental change? In *The Earth is faster now: Indigenous observations of Arctic environmental change*. I. Krupnik and D. Jolly (eds.), Fairbanks: Arctic Research Consortium of the United States, pp. 335–349.

Bockstoce, J. (ed.), 1988. *The Journal of Rochfort Maguire 1852–1854: Two years at Point Barrow, Alaska, Aboard H.M.S. Plover in the Search for Sir John Franklin*. London: The Hakluyt Society, 2 vols.

Burch, E.S., Jr. 1991. From skeptic to believer: The making of an oral historian. *Alaska History* 6: 1–16.

Carmack, E. and Macdonald, R. 2008. Water and ice-related phenomena in the coastal region of the Beaufort Sea: Some parallels between native experience and western science. *Arctic* 61: 265–280.

Committee on Designing an Arctic Observing Network, N. R. C. 2006. *Toward an Integrated Arctic Observing Network*. Washington: National Academies Press, pp. 1–182.

Cruikshank, J. 2005. *Do Glaciers Listen? Local Knowledge, Colonial Encounters, & Social Imagination*. Vancouver: UBC Press.

Eicken, H., Gradinger, R., Graves, A., Mahoney, A., Rigor, I., and Melling, H. 2005. Sediment transport by sea ice in the Chukchi and Beaufort Seas: Increasing importance due to changing ice conditions? *Deep-Sea Research II* 52: 3281–3302.

Eicken, H. 2009. Ice sampling and basic sea ice core analysis. In *Field Techniques for Sea Ice Research*. Eicken, H., Gradinger, R., Salganek, M., Shirasawa, K., Perovich, D.K., and Leppäranta, M. (eds.), Fairbanks: University of Alaska Press, pp. 117–140.

Eicken, H., Lovecraft, A.L., and Druckenmiller, M. 2009. Sea ice system services: A framework to help identify and meet information needs relevant for Arctic observing networks. *Arctic* 62: 119–136.

Eicken, H., Ritchie, L.A., and Barlau, A. 2010. The role of local, indigenous knowledge in Arctic offshore oil and gas development, environmental hazard mitigation, and emergency response. In *North by 2020: Perspectives on a Changing Arctic*. A.L. Lovecraft and H. Eicken (eds.) (in press).

Fox, S. 2002. These are things that are really happening: Inuit perspectives on the evidence and impacts of climate change in Nunavut. In *The Earth is Faster Now: Indigenous Observations of Arctic Environmental Change*. I. Krupnik and D. Jolly (eds.), Fairbanks: Arctic Research Consortium of the United States, pp. 13–53.

Gearheard, S., Matumeak, W., Angutikjuaq, I., Maslanik, J., Matumeak Kagak, D., Tigullaraq, G., and Barry, R.G. 2006. "It's not that simple": A collaborative comparison of sea ice environments, their uses, observed changes, and adaptations in Barrow, Alaska, USA, and Clyde River, Nunavut, Canada. *Ambio* 35: 204–212.

George, J.C., Huntington, H.P., Brewster, K., Eicken, H., Norton, D.W., and Glenn, R. 2004. Observations on shorefast ice dynamics in Arctic Alaska and the responses of the Iñupiat hunting community. *Arctic* 57: 363–374.

Gradinger, R.R., Kaufman, M.R., and Bluhm, B.A. 2009. The pivotal role of sea ice sediments for the seasonal development of near-shore Arctic fast ice biota. *Marine Ecology Progress Series* 394: 49–63.

Granskog, M., Kaartokallio, H., Kuosa, H., Thomas, D., Ehn, J., and Sonninen, E. 2005. Scales of horizontal patchiness in chlorophyll a, chemical and physical properties of landfast sea ice in the Gulf of Finland (Baltic Sea). *Polar Biology* 28: 276–283.

Huntington, H., Fox, S., Krupnik, I., and Berkes, F. 2005. The changing Arctic: Indigenous perspectives. In Arctic Climate Impact Assessment (ed.), *Arctic Climate Impact Assessment*. Cambridge: Cambridge University Press, pp. 61–98.

Huntington, H.P., 2000: *Impacts of changes in sea ice and other environmental parameters in the Arctic. Report of the Marine Mammal Commission Workshop, 15–17 February 2000, Girdwood, Alaska*. Bethesda: Marine Mammal Commission, iv + 98 pp.

Huntington, H.P., Gearheard, S., Druckenmiller, M., and Mahoney, A. 2009. Community-based observation programs and indigenous and local sea ice knowledge. In *Sea Ice Field Research Techniques*. H. Eicken, R. Gradinger, K. Shirasawa, M. Salganek, D. Perovich, and M. Leppäranta (eds.), Fairbanks, AK: University of Alaska Press, pp. 345–364.

Kawagley, A.O. 1995. *A Yupiaq Worldview: A Pathway to Ecology and Spirit*. Prospect Heights: Waveland Press, 166 pp.

Krupnik, I. and Jolly, D. 2002. *The Earth is Faster Now: Indigenous Observations of Arctic Environmental Change*. Fairbanks: Arctic Research Consortium of the United States, 384 pp.

Laidler, G.J., Ford, J.D., Gough, W.A., Ikummaq, T., Gagnon, A.S., Kowal, S., Qrunnut, K., and Irngaut, C. 2009. Travelling and hunting in a changing Arctic: Assessing Inuit vulnerability to sea ice change in Igloolik, Nunavut. *Climatic Change,* 94: 363–397.

Light, B., Eicken, H., Maykut, G.A., and Grenfell, T.C. 1998. The effect of included particulates on the optical properties of sea ice. *Journal of Geophysical Research* 103: 27739–27752.

Maslanik, J.A., Fowler, C., Stroeve, J., Drobot, S., Zwally, J., Yi, D., and Emery, W. 2007. A younger, thinner Arctic ice cover: Increased potential for rapid, extensive sea ice loss. *Geophysical Research Letters* 34(L24501): doi:24510.21029/22007GL032043.

McMinn, A., Gradinger, R., and Nomura, D. 2009. Biogeochemical properties of sea ice. In *Sea Ice Field Research Techniques*. H. Eicken, R. Gradinger, K. Shirasawa, M. Salganek, D. Perovich, and M. Leppäranta (eds.), Fairbanks: University of Alaska Press, pp. 259–282.

Nadasdy, P. 1999. The politics of TEK: Power and the "integration" of knowledge. *Arctic Anthropology* 36: 1–18.

Nelson, R.K. 1969. *Hunters of the Northern Ice*. Chicago: University of Chicago Press.

Norton, D.W. 2002. Coastal sea ice watch: Private confessions of a convert to indigenous knowledge. In *The Earth is Faster now: Indigenous Observations of Arctic Environmental Change*. I. Krupnik and D. Jolly (eds.), Fairbanks: Arctic Research Consortium of the United States, pp. 126–155.

Overpeck, J., Hughen, K., Hardy, D., Bradley, R., Case, R., Douglas, M., Finney, B., Gajewski, K., Jacoby, G., Jennings, A., Lamoureux, S., Lasca, A., MacDonald, G., Moore, J., Retelle, M., Smith, S., Wolfe, A., and Zielinski, G. 1997. Arctic environmental change of the last four centuries. *Science* 278: 1251–1256.

Perovich, D.K., Andreas, E.L., Curry, J.A., Eicken, H., Fairall, C.W., Grenfell, T.C., Guest, P.S., Intrieri, J., Kadko, D., Lindsay, R.W., McPhee, M.G., Morison, J., Moritz, R.E., Paulson, C.A., Pegau, W.S., Persson, P.O.G., Pinkel, R., Richter-Menge, J.A., Stanton, T., Stern, H., Sturm, M., Tucker, W.B., III, and Uttal, T. 1999. Year on ice gives climate insights. *Eos, Transactions, American Geophysical Union* 80: 481, 485–486.

Rachold, V., Grigoriev, M.N., Are, F.E., Solomon, S., Reimnitz, E., Kassens, H., and Antonow, M. 2000. Coastal erosion vs. riverine sediment discharge in the Arctic shelf seas. *International Journal of Earth Sciences* 89: 450–460.

Serreze, M.C., Holland, M.M., and Stroeve, J. 2007. Perspectives on the Arctic's shrinking sea ice cover. *Science* 315: 1533–1536.

Shapiro, L.H. and Metzner, R.C., 1979. Historical references to ice conditions along the Beaufort Sea coast of Alaska. *University of Alaska, Geophysical Institute, Scientific Report*.

Stafford, K.M., Moore, S.E., Spillane, M., and Wiggins, S. 2007. Gray whale calls recorded near Barrow, Alaska, throughout the winter of 2003–2004. *Arctic* 60: 167–172.

Usher, P.J. 2000. Traditional ecological knowledge in environmental assessment and management. *Arctic* 53: 183–193.

Wenger, E., McDermott, R., and Snyder, W.M. 2002. *Cultivating Communities of Practice: A Guide to Managing Knowledge*. Boston: Harvard Business School Press.

Wood, K.R. and Overland, J.E. 2006. Climate lessons from the first international polar year. *Bulletin of the American Meteorological Society* 87: 1685–1697.

Chapter 16
Franz Boas and Inuktitut Terminology for Ice and Snow: From the Emergence of the Field to the "Great Eskimo Vocabulary Hoax"

Igor Krupnik and Ludger Müller-Wille

Abstract Franz Boas, the "founding father" of North American anthropology, has long been credited with many pioneer contributions to the field of Arctic anthropology, as a result of his first and only fieldwork among the Inuit on Baffin Island, following the First International Polar Year 1882–1883. In this new "polar year" the SIKU project has initiated several studies of the Inuit terminology for sea ice and snow, including in the areas of Baffin Island once surveyed by Boas, as well as in the nearby regions of Nunavut, Nunavik, Labrador, and Greenland. Also, in the past decade the story of Boas' fieldwork on Baffin Island has become known in full, in diaries, personal letters, and field notes. This chapter capitalizes on these new sources: it examines Boas' knowledge of the Inuit terminology for sea ice and snow and its value to current discussion about language, indigenous knowledge, the Inuit, and beyond. It also addresses the so-called Great Eskimo Vocabulary Hoax debate of the past decades that misconstrues Boas' use of the Inuit terms and the analysis of the contemporary Inuit ice and snow vocabulary.

Keywords Franz Boas · Inuktitut · Baffin Island · Ice and snow terminology

Ever since the special issue of the journal *Études/Inuit/Studies* dedicated to the centennial of Franz Boas' work on Baffin Island (1883–1884) was published 25 years ago (Freeman 1984), Arctic anthropologists have claimed Boas as the "founding father" of their discipline. Not only was Boas praised for his pioneer contribution to the wide range of topics (Cole and Müller-Wille 1984:51–53), his name was also firmly linked to the first International Polar Year (IPY) 1882–1883, of which he was rather a post factum beneficiary than a scientific contributor.[1] A hundred-and-twenty-five years later, Boas' link to the first IPY served as a source of inspiration and a justification for social scientists to argue for their role in the Fourth International Polar Year 2007–2008 (Krupnik 2003; Krupnik et al. 2005).

I. Krupnik (✉)
Department of Anthropology, National Museum of Natural History, Smithsonian Institution, Washington, DC 20013-7012, USA
e-mail: krupniki@si.edu

I. Krupnik et al. (eds.), *SIKU: Knowing Our Ice*,
DOI 10.1007/978-90-481-8587-0_16,

In this current IPY, several projects follow in Boas' footsteps by exploring polar peoples' knowledge and use of the sea ice and/or snow habitats.[2] SIKU project, in particular, has initiated several studies of the Inuit terminology for sea ice and snow, including in the areas of Baffin Island once surveyed by Boas (Laidler et al. 2008), as well as in the nearby regions of Nunavut, Nunavik, Labrador, Greenland, and elsewhere across the Arctic (see Chapter 1, Introduction; Chapter 17 by Johns this volume; Chapter 14 by Krupnik and Weyapuk this volume). In addition, voluminous new details related to Boas' fieldwork on Baffin Island in 1883–1884 have become available over the past 10 years (Harper 2008; Müller-Wille 1998, 2008, 2009; Müller-Wille and Weber Müller-Wille 2006; Müller-Wille and Gieseking 2008). This chapter capitalizes on these new sources; it also examines for the first time Boas' knowledge of the Inuit terminology for sea ice and snow and its value to today's discussion about language, indigenous knowledge, the Inuit, and related issues.

In late 2008, one of us (I.K.), while working with Friedrich Erdmann's early Eskimo–German and German–Eskimo dictionaries from Labrador (Erdmann 1864–1866), tried to match Erdmann's Labrador Inuit terms for sea ice with the Inuktitut words listed in various Boas' publications, diaries, and short papers (i.e., Boas 1888, 1894, 1911; Müller-Wille 1998:273–276). Of those, Boas' Inuit "dictionary" (lexicon) from Cumberland Sound, Baffin Island (Boas 1894), offered by far the largest corpus consisting of more than 2,000 Inuit words, and personal and place names recorded during his work with the local Inuit during 1883–1884. From these sources, the initial sample of about 20 terms for types of sea ice and associated phenomena was compiled and translated from the German original. Then the second co-author (L.M-W) joined the study leading to a new SIKU research focused on the early historical lists of Inuit ice and snow terms. Boas' list of the Baffin Island Inuktitut terms emerged as a valuable source to compare to both historical and modern Inuit ice lexicons from the adjacent regions of Canada and Greenland. This chapter assesses Boas' material from 1883 to 1884/1894, his relations with the Inuit, and his grasp of their language and of their snow and sea ice terminology, specifically. We also address the so-called Great Eskimo Vocabulary Hoax debate of the past decades (that misconstrued Boas' use of the four Inuit terms for snow from his Baffin Island fieldwork) and approach it from the analysis of the contemporary Inuit ice and snow vocabularies collected during the SIKU project. We argue that the latter "debate" is completely misdirected and that the Inuit (Eskimo) have many more terms associated with the sea ice than with snow.

Franz Boas and Inuit Languages

It is one of the many legacies of Franz Boas that despite having collected and published extensive materials on Inuit anthropology, ethnography, and geography, his degree of linguistic competence and fluency in Inuktitut, the language spoken by the Inuit he worked with on southern Baffin Island, has been questioned (Harper 2008). Learning more about his linguistic competence would help evaluate the validity of the materials he documented and interpreted. In order to understand Boas' level of immersion into Inuktitut, the record of his linguistic engagement can be traced in archival documents, such as in the Boas Papers in the American Philosophical Society (APS), in Philadelphia, U.S. and the Hinrich Rink Papers in The Royal Library, Copenhagen, Denmark, as well as in his publications (Boas 1885, 1888, 1894, 1911).

Although his original research interests were in natural sciences, in particular in physical geography and cartography, Boas understood fully the necessity of learning the languages of the people he intended to study and to live with during his full-year sojourn in the Canadian Arctic, i.e., Inuktitut (in German *Eskimoisch* at that time) with the Inuit and English with the American and Scottish whalers. He had a solid, but passive grounding in classical Greek and Latin, as well as in French. Due to family connections in New York and private tutoring, Boas obtained functional knowledge of English early on; it was still limited but improving while he was on Baffin Island. Boas also mentioned that he started to learn Russian during his first term at university, but that did not seem go far.

More important, in 1881–1883, while preparing for his research in the Arctic, he began to learn Danish and Inuktitut. He knew very well through his literary studies that practically all materials concerning Inuit languages, such as grammars and lexicons, were published in these languages, mainly on Greenland, except for works that were written in German and collated by Moravian missionaries who had been stationed in Greenland and Labrador. Based on Boas' comments in his letters and diaries, he made some limited progress in learning Inuktitut. In a letter to his parents in November 1882 sent while he was in Berlin, he stated proudly: "In English, Danish and Eskimo I am rather diligent."

Although Boas referred to only a few sources directly, it can be safely assumed that he had access to the literature in this area in German libraries, mainly at the University of Göttingen. In the summer of 1881 he went on a trip to Copenhagen with the purpose of visiting the National Museum and university library to learn about Inuit culture and language. He left no record of what he actually achieved.

There were three authors whose works Boas most likely consulted and studied for the purpose of becoming familiar with the Inuit and, in particular, their language. Friedrich Erdmann (1810–1873), a Moravian missionary who worked in Labrador, compiled extensive "Eskimo–German"[3] (1864) and "German–Eskimo" (1866) dictionaries, which was later translated into English and revised by Edmund James Peck (1850–1924) in (1925). Samuel Kleinschmidt (1814–1886), also a Moravian missionary who was born and stayed all his life in Greenland, published an exhaustive and seminal grammar (1851) and a Greenlandic–Danish dictionary (1871) expanding on two earlier dictionaries published in the late 1700s. Kleinschmidt's two books put Greenlandic on a solid footing as a written language. Lastly, Hinrich Johannes Rink (1819–1893), geologist, geographer, ethnologist, linguist, and administrator in Greenland, wrote in detail about the Inuit and Greenland in Danish and English (1866, 1871, 1875 and, after Boas' research, 1887). Boas was to establish direct contact with Rink in 1885, sending him Inuit texts and word lists for his review and corrections (Rink 1887:39), visiting him in Copenhagen, and maintaining correspondence with him until Rink's death in 1893.

During his 1883–1884 stay on Baffin Island, Boas expanded his efforts to learn Inuktitut by associating with James S. Mutch, the Scottish manager of the Kekerten whaling station at Kingua Fjord, Cumberland Sound. Mutch was fluent in Inuktitut and functioned as a translator and facilitator between the Inuit and Boas for a long time thereafter (Harper 2008). From the start, Boas kept word lists in alphabetical order, systematically mapped and collected place names, and recorded stories and legends in Inuktitut. Some of this original linguistic material is in the archives of the Royal Library (Rink Papers) in Copenhagen and the APS (Boas Papers), but it has not been studied extensively.

In his own estimation, Boas did not feel that he had acquired satisfactory expertise in Inuktitut and sought all the help he could get from knowledgeable people. Here Hinrich Rink became the person who helped him the most in his effort to overcome his linguistic shortcomings offering friendly advice with regard to translation, interpretation, and orthography. Recognizing his limitations, Boas wrote to Rink in April 1885:

> You overestimate my knowledge of the Eskimo language, because my understanding of the songs is immensely deficient, some of them I almost do not understand at all; those which I master according to their content, I know due to the thorough narrations by the natives.... Some unintelligible words might have originated from the erroneous recognition of the sounds on my part, which happens easily when knowledge of a language is incomplete.... Indeed, I feel that I am in no way up to this task.[4]

Franz Boas' Publications of Original Inuit Linguistic Material

Today, the name of Franz Boas is strongly associated with the emergence and development of the linguistic study of aboriginal languages in North America. Therefore, in retrospect, it is curious that he, despite some major personal efforts, published few of the original Inuit texts or even word lists. The publications were restricted to listing of place names, personal names, or vocabularies/lexicons (Boas 1885, 1888, 1894) with translation into German, but without any assessment or interpretation. The materials Boas included in "The Central Eskimo" (1888), particularly on religion, traditions, and arts, are solely presented in English except for a glossary of Inuit terms with derivations (1888:659–666). Still neither yet edited nor published, the Rink Papers in Copenhagen contain Inuit texts of tales (myths) recorded and transcribed by Boas in his own handwriting, which he sent to Rink for examination in 1885.

Clearly, Boas was reluctant to publish the original Inuit texts. However, he included a detailed list of Inuit place names (toponyms) he had collected in his cartographic survey and from published maps (1885), which remains a lasting and important source and heritage of Inuit geographical knowledge to this day. Then in 1894, 10 years after his return from Baffin Island, he published a list of Inuit words, with some conjugations and derivations, place names, and personal names (ca. 2,200 entries), with German translation. This list most likely comprised all that he had collected during his stay with the Inuit on Baffin Island. He cross-checked his list with the sources from Greenland and Labrador mentioned earlier and demonstrated a closer linguistic relationship between Labrador and Baffin Island Inuktitut than with Greenland.

In order to understand Boas' reason for publishing this list in 1894, it is useful to refer to his short introduction in which he provided some explanation regarding the "dialect" of the Inuit on southern Baffin Island. Boas' linguistic recording was the first ever that was carried out to such a broad extent in that particular region of the Inuit homeland. In that way the word list is of special historic interest both linguistically and culturally, particularly with respect to human–environmental relations and interactions as experienced by the Inuit during the 1880s. Although Boas stated explicitly his intention to go deeper into a linguistic discussion of the Inuit languages at a later stage, this intention never materialized. His mentor in Inuit languages, Hinrich Rink, who had seen the original list had died in 1893, just before it was published.

As Boas wrote in his introduction (1894:97, English translation by Ludger Müller-Wille):

> The following material was collected during a journey to Baffin Land during the years of 1883 and 1884. The dialect, to which this material refers, is spoken in Cumberland Sound and in the parts of the west coast of Baffin Bay lying somewhat farther north. The dialect is closer related to the one in Labrador than to the one in Greenland, which should not surprise us since the customs and traditions of these tribes are also more similar to the ones of Labrador than of Greenland.
>
> I have hesitated long with the publication of the material collected by me because it is in many respects deficient and imperfect. These are the results of an initial journey and the collector's insufficient experience shows the material's lacunae and imperfection. During my whole journey I was not fully aware of the importance of linguistic studies since I believed, that the studies by missionaries in Greenland and Labrador provided a sufficient image of the Eskimo language, and therefore concentrated on geographic and ethnological problems. Only after my return it became apparent, when I tried to get the obtained texts translated, what the ancient texts, the peculiar secret language of the angakut [shamans] and the dialectical variations offered for an interesting study.
>
> After careful consideration it seems to me that the collected material still offers enough new insight to justify its publication.
>
> The present article includes collected vocabulary that is compared with the lexicon from Labrador and Greenland. A subsequent article is to contain the texts, phonetics, grammar and a discussion of the vocabulary.
>
> I have used the orthography developed by missionaries in Greenland, however, the long vowels are presented only by their lengths. The accents within the words are provided often. The velar k is expressed by q and the German ch like in Bach is rendered by χ. All other letters are pronounced like in German.

List of Words and Place Names: Ice, Snow, and Related Phenomena

The following list of the Inuktitut words and names related to ice and snow presented in alphabetical order has been taken directly from Boas' publications along with his original German translation (1885:90–95; 1894). The English translation from German marked by slashes is by Müller-Wille partially based on a preliminary translation by Krupnik in 2008. Words beginning with a capital letter are place names. The abbreviations for the provenance of words used by Boas (1894) are as follows: G. – Greenland (based on Kleinschmidt 1851, 1871; Rink 1866, 1871, 1875, 1887), L. – Labrador (based on Erdmann 1864, 1866). In the few cases where Boas uses the Greek letter χ it is replaced here with the standard q.

Boas must have collected as much as he could possibly obtain and grasp. The resulting collection was clearly influenced by his own experience of living and traveling extensively with the Inuit for a period of 12 months, which were most of the time defined by ice and water. Furthermore, it should be noted that the list includes several hundred toponyms and personal names that are normally not part of a lexicon.

Ice

a'jorang Spalt im Eise (Sprungspalt, nicht die Spalten am Ebbestrande)/crack in the ice (extensive crack, not the cracks in the low tidal flats)/L. *ajorak.*

A'jorang die Spalte/the crack/

Aqti'nirn wo das Eis vor der Flussmündung schmilzt/where the ice melts in front of the mouth of a river/

Angmaritung das Offene, nicht Ueberfrorene/the open one, not frozen over/L. *angmarok.*

Angmartung das Offene (nicht überfrorene)/the open one (not frozen over)/

igjijug dick, dickes Eis/thick, thick ice/L. *ivjo'vok*, G. *ivssuvoq.*

imakti'nirn Eis auf schmelzendem Schnee, stark genug Schlitten zu tragen/ice over melting snow, strong enough to carry a sled/

ivu'dnirn Grundeis am Strande/ground ice on the beach/,L. *ivuvok* – Eis treibt am Strande übereinander/ice piling upon the beach/

kaqvaq Packeis /pack ice/L. *kackvak.*

manituā'dlu hügeliges Land, rauhes Eis/hillocky land, rough ice/

mase'lirang dünnes Eis, das sich im Frühjahre auf dem Schnee bildet, nasses Moos/thin ice that is formed on the snow in spring, wet moss/L. *masalerak.*

miso'majung in's Meer reichender Gletscher/glacier reaching into the sea/G. *misugpa* – er taucht es sein/he dips it into [something]/

nilang Süsswassereis/freshwater ice/L. *nillak*, G. *nilak.*

penartua'dlu Wasser steht auf dem Eis/water on ice/

piqalu'jang Eisberg/iceberg/L. *Pekkalujak.*

Piqaluirtung reich an Eisbergen/abundant with icebergs/

qaqbang mehrjähriges Eis, Packeis (siehe *kaqvaq*)/multi-year ice, pack ice (see *kaqvaq*).

qati'dinrn das Eis am Strande im Herbste, das bei Fluth schwimmt, bei Ebbe strandet/the ice on the beach in the fall that floats at high tide and is stranded at low tide/L. *kattinek.*

qavirpi'jung Grundeis/ground ice/G. *qaungoq.*

qu'gnirn Spalte im Grundeis/crack in the ground ice/L. *Kongnek.*

quta'rong Eisfuss an einer Steilküste/ice foot at a steep coast/

sī'ko Eis/ice/L., G. same.

Sikosuilaq das Eislose/[location is] free of ice/

sikoqa'ngenut über das Eis/across the ice/

siko'qoang 1-1 1/2 Fuss dickes Eis im Herbste/1 to 1 ½ ft of thick ice in the fall/

sī'koaq dünnes Eis beim ersten Gefrieren/thin ice formed by the first frost/

siku'kulu kleine Stücke Treibeis/small pieces of drift ice/

sikū'liaq dünnes Eis an der Eiskante im Winter/thin ice at the ice edge during the winter/

Sikosū'ilaq das Eislose/[place] without ice/

sinā' seine Kante, Eiskante/his edge, ice edge/

Sirmilling mit einem Gletscher versehen/place where there is a glacier/

si'rming dünnes Eis, Firn, Gletscher/thin ice, firn, glacier/L. *sermek*, G. *sermeq.*

tu'vang dickes Wintereis/thick winter ice/L. *tuvak.*

tuvarea'qtung dickes Eis im Spätherbste /thick ice at the end of the fall season/

Snow

aqilokoq weich gefallener Schnee/softly fallen snow/(G. *aqipoq* – es ist weich/it is soft/), L. *akkiilokàk.*

apun liegender Schnee/snow on the ground/L. *aput*, G. *aput; aput* (Boas 1911:25).

Auqardnelling mit im Frühling schmelzenden Stellen/with spots that thaw in the spring/

ikijuq, ikivū' Wind höhlt Schnee aus /wind hollows out snow/

mauja weicher Schnee/soft snow/L. *maujak* (derives from G. *mauvoq* – er sinkt mit den Füßen ein/he sinks in with his feet/).

Maujatung Reich an weichem Schnee/abundant with soft snow/

piegnartoq der Schnee ist gut zum Schlittentreiben/the snow is good for driving sled/L. *piarngnartoq*.

pi'rtsirpoq der Schnee treibt/drifting snow/L. *perkserpok*, G. *Persoq; piqsirpoq* (Boas 1911:26).

qaneq fallender Schnee/falling snow/L. *kannek*, G. *qanik*. *qa'nerpok* – es schneit /it is snowing/, *qana* (Boas 1911:26).

Qimi'sung Schneewehe (?)/snowdrift (?)/, Hundsfell (?)/dogskin(?)/(see *qimu'qdjung*).

Qimissung die Schneewehe/the snowdrift/

qimu'qdjung Schneewehe/snowdrift/L. *kimuksuk*; *qimuqsuq* (Boas 1911:26).

savujua'rtuang Schneeblock/snow block/(from *savik* – Messer/knife/).

siorpā'lirpoq Athem gefriert zu Eis/breath freezes to ice/

Related Phenomena

agdlu Seehundsloch/seal's breathing hole through the ice/L. *aglo*, G. *agdlo*.

Audnerbing Wo man nach Seehunden kriecht/where one crawls for seals [on the ice]/

aunerpoq er kriecht auf dem Eis nach Seehunden/he crawls over ice for seals/L. *Aungniarpok*.

igdlu Schneehaus/snow house/L. *igloo*, G. *igdlo*.

Igdluviaujang das einem Schneehaus Ähnliche/the one similar to a snow house/

Sarbartijung reich an Stromschnellen/abundant in currents/[referring to open places in sea ice, i.e., polynya], from *sarbaq* – starke Strömung im Meer/strong sea current/

taglun Schneeschuhe/snowshoes/L. *taglut*.

tilu'qtun Schneeklopfer/snow beater/L. *tilluktũt*, G. *tilugtût*.

ukiuq Winter/winter/L. *okkiuk*, G. *ukioq*.

ū'toq Seehund auf dem Eise bei seinem Loche liegend/seal on the ice basking at its [breathing] hole/L. *otok*, G. *utoq*.

The terminology on ice, snow, and related phenomena that Boas recorded in the 1880s showed that the Inuit knowledge of their physical environment was detailed, precise, and intricate. Boas understood the importance of this socioecological setting. Therefore, his linguistic record is a testimony of these complex human–environmental interactions at a particular period in Inuit history. Furthermore, by looking at terms from Labrador and Greenland, Boas detected variations and similarities to demonstrate connections among various Inuit groups.

Of particular interest are the toponyms that included references to ice and snow, and which are in their content consistent from region to region in the Arctic (cf. Müller-Wille 1985; Müller-Wille and Weber 1983). From the material available it cannot be determined if Boas put a particular emphasis on collecting ice and snow terminology; however, this specific vocabulary was an important part of the general Inuit lexicon and could not be overlooked. As historic material, this word list, which in Boas' eyes was a humble effort, is an important record for the development of Inuit linguistics and for our contemporary understanding of the Inuit (and other Arctic people's) knowledge of ice, snow, and related phenomena.

Boas' Contribution to Inuit Terminology of Ice and Snow: Present-Day Perspective

Boas' interest in the Inuit terminology for sea ice and snow was by no means accidental if one recalls his background in physics and geography and the topic of his doctoral dissertation, "Contributions to Understanding the Color of Water" (1881). While going over various Inuit dictionaries in preparation for his Baffin Island trip, Boas certainly became aware that the Inuit had many more terms for ice and snow than were available in his native German (or other European languages). Erdmann's German Labrador Eskimo dictionary with its 17 terms for various types of "ice" (*Eis*), plus a dozen more terms for ice-associated phenomena, and 8 terms for snow (Erdmann 1866:65, 163) could have been an obvious source of inspiration. Other prospective source, besides several Danish dictionaries of the West Greenlandic/Kalaallisut (see earlier), might have been Émile Petitot's (1876) French–Eskimo dictionary from the Mackenzie Delta, with its 12 terms for ice and 13 terms for snow. It comes as no surprise that Boas, a meticulous fieldworker, produced an Inuktitut lexicon from Cumberland Sound that included more than 30 terms for ice and related phenomena and more than a dozen terms for snow.

Boas' Inuktitut terms for ice and snow can be assessed following the same procedures we applied to contemporary sea ice and snow vocabularies collected for the SIKU project (see Chapter 14 by Krupnik and Weyapuk this volume). First, ice terms should be studied separately from the snow terms, since they use different stems and refer to different phenomena. Second, the analysis is more productive when the terms are organized and compared within certain typological groups, such as "young ice," "shore-fast ice, "pressure ridges," and "layered ice". Third, cultural and human-/animal-associated terms should be treated separately, since they are often built as "derivatives," due to the patterns of word formation in the Eskimo/Inuit languages.

During his Baffin Island trip, Boas had ample opportunity to hear many Inuktitut words for ice and snow, thanks to his prolonged dog sled and boat trips with Inuit

Table 16.1 Terms for young ice and freeze-up formations in some historical and contemporary Inuit/Inuktitut/Inupiaq Ice vocabularies

Cumberland Sound: (Boas 1894)	Pangnirtung: (Laidler 2007)	Labrador: (Erdmann 1864, also Peck 1925)	Igloolik: (Aporta 2003)	Nunavik: (McDonald et al. 1997)	West Greenland (Taverniers 2009)	Barrow (Brower 2008)
	atuqsaruqtuq when the ice is thick enough to walk on			*agutitaat* thin layers of new ice		*arguqtaġniq* newly formed thin ice collecting on the downwind side of a polynya or lead
	iluvaliajuq the first film of ice that starts forming		*illuvaliajuq* the sides of the river when it starts to freeze	*akgutitak, akgutinik* slushy mixture of ice and snow frozen into flexible ice		*atiġniġaq* new ice forming a smooth apron around pre-existing ice
	nutaaminiq newly formed ice in the fall, about a week old			*minguirniq* slush ice *milutsinik* snow-soaked water that freezes at the floe edge	*nutarneq* recently formed sea ice.	*iġnigluq* thin young ice broken up or crushed and refrozen as found in cracks
qati'dinrn ice on the beach in the fall that floats at high tide and is stranded at low tide	*qainngu* ice ledge formed by water overflowing and refreezing	*qaingok* the ice on the banks, shores	*qainnguq* when the shorelines start to freeze	*qainguniq, qainguq* slush washed ashore that freezes on the beach	*qaanngoq* icefoot, a narrow fringe of ice attached to the coast	*imuniq* crushed young ice caused by moving ice
	qillirusijuq an early stage of ice formation, when ice begins forming along the edge of the shoreline		*qinnuaq* primary ice formation on the shores when the sea starts to freeze	*qinuq* slush ice formed along shorelines and at the floe edge		*kaniqtaq* slightly refrozen ice pieces but fragile; this ice will quickly spread out when it is stepped on
			qinuaq soft or slushy snow in the water	*qiqngurusirtuk* new ice formed from slush in narrow water bodies, like inlets		*mayuqtitaq* slush ice pushed onto the shore with wraps frozen into waves
	qinnuaq slushy suspension in the water, the beginning of ice formation	*kinuak/kKinnuak* thin ice when the water begins to freeze *kinnuarpok* the water becomes ice	*qinulimajuviniit* the ice that is caused by the snowfall when it begins to freeze	*qaiquit* new ice *putatak* new ice, *puttaq*, autumn floating ice, the first ice floating in	*qinuuvoq* there is slush	*misaĺhak* slushy top of young saltwater ice
	quppirquaq very thin sheet of ice that looks like an oil slick on top of the water	*kinnumarllerpok* it is making strong ice; it is freezing hard (*lit.* it is making full-grown ice)	*quasalimajuq* crystal-free fall freeze up, so it becomes slippery			*muġaliq; muġaĺliq* slush ice on sea *muġrak* slush ice
siko'qoang 1 to 1 ½ ft thick ice in the fall	*sikuallaajuq* big plates of new ice; traditional term for pancake ice	*sermek (E)* new ice on the boat, etc.	*quvviqua* strip of ice on the water, when the sea ice is about to freeze		*sikorlaaq* recently formed sea ice. *sikuarllaaq* dark nilas (a thin elastic crust of ice) few millimeters thick.	*qaivaġniq* flat round cakes of ice frozen together *qinu* slushy ice

Table 16.1 (continued)

Cumberland Sound: (Boas 1894)	Pangnirtung: (Laidler 2007)	Labrador: (Erdmann 1864, also Peck 1925)	Igloolik: (Aporta 2003)	Nunavik: (McDonald et al. 1997)	West Greenland (Taverniers 2009)	Barrow (Brower 2008)
sī'koaq thin ice formed by the first frost	*sikuaq* new, thin brittle sheet of ice; the first continuous layer of ice	*sikkoak* quite thin ice	*sikuaq* new ice forms on the bays; first stage of freeze up	*sikuak* very thin layers of new ice that are formed on a calm day	*sikuaq* dark nilas (a thin elastic crust of ice) 1–5 cm thick.	*sikuaq* thin ice, dangerous to walk on
		sikkoarpok there is quite thin ice				
	sikujuq the water has frozen over; there is no more open water	*sikkulliak* fresh (new) ice on the shore (or anywhere), young		*sikulirutit* new ice, freshly frozen from part saltwater and part freshwater along shorelines and within inlets.	*sikujartuaarpoq* sea ice forms and becomes thicker (on several days)	*sikullaġruaq* new thick ice approximately 2.5 ft thick and thicker
	sikuluqtuq bad, thin, dangerous ice	*sikkovok* it gets ice			*sikujumaataarpoq* sea ice delays in forming	
	sikurataaq sea ice that has just recently formed	*sikkolauvok* it is frozen over		*sikutak* new ice that forms from *sikuak*, once inlets are frozen		
	sikutaq first ice to form in bays, inlets, and heads of fjords		*singniqsaq* similar to *qinu* but much more solid, so that one can travel through	*sikutait* solid ice in small inlets or bays formed before the land-fast ice	*sikunippoq* floating pieces of sea ice arrive (moved by wind or current) when the sea was ice-free	
sikū'liaq thin ice at the ice edge during the winter	*sikuvaliajuq* early process of the ice formation; first layers of ice starting to form		*sikuliaq* new ice on which it is possible to walk	*sikuliak* newly formed ice with no snow on top; safe to walk or travel on.	*sikuliaq* recently formed sea ice	*sikuliaq* young ice formed around edge of old solid ice on open lead
	sikuvaalluuti the first ice that forms and stays until the following year		*tuqujaktinniq* the area of the newly formed ice that is darker than the rest	*ukiurjait* new ice that forms from *sikuak*;	*sikulaaq* recently formed sea ice	
	sivaujanguaq pancake ice	*siva* fresh ice; (grease ice?, also cod oil)				
	uiguaq new ice that forms along the lead edge, "an extension"		*uiguaq* new ice formed at the edge of the floe edge *uiguaviniq* when the new ice *uiguaq* becomes older	*uiguaq* smooth, solid extension ice		*iiguaq* ice extension

companions. His travels also exposed him to the richness of the Inuit environmental knowledge, such as the many names for winds, directions, constellations, and other natural phenomena (Boas 1888/1964:235–236). Boas' interest in Inuit place names and personal names, many of which were built by using various base terms for ice and snow, could likewise have offered insight into the Inuit perspectives on ice and snow formation. What is most striking in Boas' list of Inuktitut ice and snow terms from Baffin Island is that it is not very extensive, particularly the snow list.

More than 30 Cumberland Sound Inuktitut ice terms recorded by Boas form a rather skeletal, albeit a solid, vocabulary particularly when arranged by major typological groupings (Appendix). Boas did document many basic terms, but several key words are missing. Overall, the remarkable richness of the Inuit ice terminology is barely visible in his material. Boas' list is particularly thin in the "young ice," "ice floes/drifting ice," and "spring melting, breakup" categories, when compared with Erdmann's dictionary of 1864 and several modern ice vocabularies collected for the SIKU project (Table 16.1). Boas' list lacks many ice terms that are familiar to today's residents of the Cumberland Sound area (see Laidler 2007; Laidler et al. 2008). The small number of derivative terms and words for human activities associated with the sea ice in Boas' lexicons from 1885 and 1894 is also noteworthy. In many contemporary lists collected for the SIKU project such terms contribute up to 25–30% of the total.

The limited nature of Inuktitut ice terminology collected by Boas supports our assessment that he did not achieve sufficient fluency in Inuktitut during his fieldwork and that he was recording the terms in a context of his traveling and learning rather than via a systemic survey. Also, he might have lacked good interpreter(s) to reach out to knowledgeable local collaborators or elderly experts, in the way the collection of indigenous terminologies is usually done today.

The same is also true with regard to Boas' much shorter list (13 entries) of the Cumberland Sound Inuktitut snow and snow-associated terms. Several contemporary Inuit/Eskimo snow lexicons (see Table 16.2) range from 20 to 35 terms, i.e., West Alaskan Yup'ik (Woodbury 1991, based upon Jacobson 1984), Barrow Inupiaq (Brower 2008), Inuvialuit/Siglutun (Lowe 1984), Copper Inuit (Lowe 1983); Nunavik Inuktituk (Dorais 1996: 145), Thule Inuit (Fortescue 1991), West Greenlandic (Fortescue 1984). More extensive lists feature up to 60–80 terms, like Siberian Yupik (Vakhtin and Emelyanova 1988:24–27), St. Lawrence Island Yupik (Walunga 1988; Womkon Badten et al. 1987), Barrow Inupiaq (Sturm 2009a, b), historical Labrador Inuttut (Erdmann 1864/Peck 1925). The largest known list, compiled from the Nunavik Inuktitut dictionary (Schneider 1985), has more than 100 terms, including numerous derivative forms.

When those contemporary lists are organized in major typological groups, like "types of snow on the ground" (Table 16.2), "falling and drifting snow," "snow forms," "snow-associated phenomena," "human activities associated with snow," it becomes obvious that Boas missed several critical terms, particularly related to the snowstorm and snowfall patterns (3 words versus 18 words in Erdmann/Peck's Labrador dictionary), terms to describe forms built on the snow surface by wind blowing and accumulation (3 terms only), melting and freezing, and the like. Many

Table 16.2 Number of terms for major types of snow and associated phenomena in historical and contemporary Eskimo/Inuit vocabularies

	Snow on the ground	Falling snow, snowfall	Snow on ice and water	Snow forms	Others[a] (including derivatives)	Total
Inuktitut, Cumberland Sound, 1883 (Boas 1894)	3	3	–	3	2	11
Inuttut, Labrador 1850s (Erdmann 1864/Peck 1925)	12	18	–	3	23	56
Siglitun, Mackenzie Delta, 1860s (Petitot 1876)	10	3	–	–	1	14
Kalaallisut, West Greenland, 1970s (Fortescue 1984)	6	11	3	2	9	31
Yupik, St. Lawrence Island, 1980s (Walunga 1988)	15	3	13	4	21	56
Central Alaska Yup'ik (Jacobson 1984/Woodbury 1991)	7	3	7	3	6	26
Siberian Yupik, 1970s (Vakhtin and Emelyanova 1989)	12	24	–	8	15	59
Inuktitut, Nunavik, 1970s (Schneider 1985)	26	23	–	4	53	106
Inupiaq, Barrow, 1990s (Brower 2008)	16	3	5	5	7	36
Inupiaq, Barrow, 2000s (Sturm 2009b)	17	8	1(?)	15	18	59+

[a]Including terms for hail, frost, rime, snow crystals; human activities associated with snow.

terms that are familiar to today's residents of the Cumberland Sound area are not in Boas' list, like *apputtattuq*, snow that accumulates on the newly formed ice and causes its thinning; *kiviniq*, wet snow sinking into the sea ice; *qissuqaqtuq*, snow that has frozen/hardened at night and is good for travel (Laidler et al. 2008). The most logical explanation, again, is that Boas did not have adequate access to local experts who could have helped expand his snow list.

It is no wonder that Boas was humbled by the quality of his Cumberland Sound Inuktitut lexicon. He was in no rush to publish it, and he did so only 10 years after his initial fieldwork. Evidently, only the death of Hinrich Rink in 1893 and/or a temporary break in correspondence with James Mutch (1847–1931), Boas' only solid contact in Baffin Island (Harper 2008), finally convinced Boas that his material could not be improved any further. The appearance of the Cumberland Sound lexicon in the Vienna-based *Mittheilungen der Anthropologischen Gesellschaft* (*Proceedings of the Anthropological Society*), with an abridged introduction and no analysis, was Boas' acknowledgment that it was the most he could produce from that field trip.

Boas clearly learned his lesson. Soon after, he developed a long-term collaboration with two outstanding language and knowledge experts from the Northwest Coast area, George Hunt (1854–1933) and James Teit (1864–1922). Long-term partnerships with Hunt, Teit, and other local collaborators (Henry Tate, Louis Shortridge, James Mutch) became the basis of Boas' work in the documentation of indigenous cultures, texts, languages, and spiritual systems in the later stages of his professional career (Berman 2001; Rohner 1969).

The Cumberland Sound Inuktitut lexicon of 1894 also testifies to Boas' prudence as a researcher. Though he inserted, whenever possible, parallels with other Inuit dialects (Labrador Inuttut, West Greenlandic), he made no effort to expand his record by adding forms from other sources. Surprisingly, his fieldwork "prudence" was challenged some 90 years, after the publication of his Inuktitut lexicon from Cumberland Sound.

"The Great Eskimo Vocabulary Hoax": A Rejoinder from the SIKU Project

In 1986, Laura Martin, linguistic anthropologist and specialist in Mayan languages, produced a short essay titled "Eskimo Words for Snow" (Martin 1986). Martin traced the origins of what she called "the Genesis and Decay of an anthropological example," that is, of the popular belief that the Eskimo have "dozens, if not hundreds words for snow" to Franz Boas. According to Martin (1986:418), Boas' quite casual and not very solid reference to *four* lexically unrelated words for snow in the Eskimo (language) (Boas 1911:25–26) was later picked up by Benjamin Whorf (1940) and was subsequently exploited in dozens of textbooks and popular writings. Through this repeated and often thoughtless recycling, Boas' original four terms eventually "snow-balled" into up to "two hundred" terms for snow that reportedly were known by the Eskimo.

Martin's arguments were soon amplified by linguist Geoffrey Pullum, who used a catchy title, "The Great Eskimo Vocabulary Hoax" for his rejoinder to Martin's critics (e.g., Murray 1987). Pullum recycled that title in several subsequent reprints and online postings, including his book of linguistic essays on other unrelated topics (Pullum 1989,1990,1991a, 1991b, 1994,1996, 2003). Pullum's publications stirred a passionate debate, primarily among linguists and popular writers, that continues to this day.[5] Meantime, cohorts of students who have taken classes in *Introductory Linguistics* since 1991 have been trained to believe that Pullum and Martin had put to rest the myth originating from Boas (and Whorf) that the Eskimos had "many hundred terms for snow."[6] Even more, Pullum and his followers argue that the Eskimo snow vocabulary has roughly the same number of words as English. In fact, most of the contributors to the "Great Eskimo Vocabulary Hoax" debate hardly ever ventured into an Eskimo dictionary and none cited Boas' Baffin Island lexicon of 1894, though a few Eskimo linguists took part in the discussion (i.e., Woodbury 1991; de Reuse 1994; Kaplan 2003, 2005).

Our analysis of the Inuit words for ice and snow recorded by Boas may clarify some misunderstandings common to the "Great Eskimo Vocabulary Hoax" debate. First, Boas certainly knew *more* than the four Eskimo terms for snow that he cited as an example of the "differences in how the groups of ideas are expressed by specific phonetic groups in different languages" (Boas 1911:25). We do not know why he selected these particular words (*aput*, *qana*, *piqsirpoq*, and *qimuqsuq*).[7] He could have easily picked several more words for snow from his Baffin Island lexicon, like *axilokoq/aqilokoq* – "softly falling snow," *mauja* – "soft snow on the ground," *piegnartoq* – "the snow (which is) good for driving sled," or from Erdmann's Labrador dictionary that had more terms, such as *pukak* – "crystaline snow," *sakketok* – "fresh fallen snow," *machakit* (*masak/masayak*) – "wet, mushy snow." Hence, Martin's (1986:418) criticism that "Boas makes little distinction among "roots," "words," and "independent terms" is a gross misinterpretation, as Boas was very careful not to use any derivative snow terms to illustrate his point.

Second, all Eskimo/Inuit languages have many more words for snow and snow-related phenomena than the four terms cited by Boas. Although none of the known Eskimo snow lists expands into "many hundreds," local snow lexicons commonly feature several dozen terms for types of snow and snow-related phenomena, including specific patterns of snowfall and/or snowmelt, forms created by wind and other agents over the snow surface. For example, there are eight to 12 independent words for various types of snow on the ground (see Table 16.3).[8] True, the majority of the Eskimo/Inuit snow terms are built by adding suffixes to a certain number of basic stems. Nonetheless, each represents a meaningful and clearly distinguishable phenomenon to indigenous speakers, very much like the terms "new snow" and "old snow" or "small" and "medium ice floe" have special meaning to meteorologists and ice/snow specialists in the scientific nomenclatures (Anonymous n.d.; Armstrong 1958; WMO 1970). This relevance to the language speakers has been completely lost in the "Great Eskimo Vocabulary Hoax" debate, whose participants had little experience with the Eskimo/Inuit knowledge of snow (or ice).

Third, all Eskimo languages possess many more words for various types of sea ice and associated phenomena than for snow, as first argued by de Reuse (1994) and Michael Krauss (several personal communications during the 1990s) and amply illustrated by the SIKU project materials. Most of the 20-some indigenous sea ice vocabularies collected for the SIKU project feature 60–80 (some of them up to 110–120) terms, including dozens of terms for various types of ice and ice-associated phenomena (see Chapter 18 by Tersis and Taverniers this volume; Chapter 14 by Krupnik and Weyapuk this volume; Chapter 17 by Johns this volume; Laidler 2007). Boas was certainly well aware of that richness and he, perhaps, would have put himself on a more solid footing had he picked the Inuit terms for ice, rather than for snow, as an illustration. His own lexicon from Cumberland Sound features numerous independent terms, like *Siku* (ice), *igjijiuq* (thick ice), *kakvaq* (pack ice), *piqalu'jang* (floating iceberg), *tu'vang* (thick winter ice, shore-fast ice), *nillang* (freshwater ice), *si'rming* (glacier ice), *ivu'dnirn* (ground ice on the beach), and many more that are common across the Inuit/Inupiaq language area.

Lastly, contrary to Pullum's much-reiterated claim based on the comment by Woodbury (1991), the English vocabulary for snow and related phenomena is clearly inferior to those recorded in several Eskimo/Inuit languages and dialects. The terms for snow and associated features are almost universally produced in English by using the stem "snow" plus additional stem or a separate word with another meaning, i.e., "snowstorm," "snowdrift," "snowflake," "snowball," "snow-bank," "snowcap," and "snowfall".[9] The examples Pullum cited (following Woodbury) to illustrate many

Table 16.3 Boas' Inuktitut terms for snow on the ground compared to historical and modern Inuit vocabularies

Baffin Island (Boas 1894)	Labrador (Erdmann 1864)	Nunavik (Schneider 1985)	West Greenland (Fortescue 1984)	Barrow (Sturm 2009b)	Inuvik (Lowe 1984/2003)
		aniu snow for making water		*aniu* snow, soft snow	*aniu* packed snow to make water
				aniuvak hard packed snow in a gully	
		apijaq snow covered by bad weather	*apirlaat* new-fallen snow	*apivaaluqqaaq* first snow of the year	*apilaraun* first snow layer in fall
apun snow on the ground	*aput* snow (generally)	*aput, aputi* snow on the ground	*aput* snow (on ground)	*apun* snow cover	*apun* snow lying on a surface
			aput masannartuq slush/wet snow on ground		
aqilokoq softly fallen snow	*akkilokak* soft snow, not hard	*aqilluqaaq* mixed snow and water that is thawing		*auktaaq* melting snow	*aqiluraq* light soft snow
		atsaakatsaaq snow good for making a giant snowball		*imaktinniq* very wet snow, slush	
		isiriartaq snow that falls yellow or reddened		*isrriqutit* diamond dust or ice crystals in the air	
		katakartanaq snow with a hard crust that gives way under footsteps			
		kavisirlaq snow rendered rough by rain and freezing			
		kinirtaq anything, including dump snow that is compact because soaked			
		mannguq melting snow			
		mannguumaaq snow softened by warm weather	*mangiqqak* hard snow (also *manngikaajaaq*)		
		masak snow soaked in water, wet snow falling from the sky		*masayyak* damp snow	*maqayak* watery mud or snow

Table 16.3 (continued)

Baffin Island (Boas 1894)	Labrador (Erdmann 1864)	Nunavik (Schneider 1985)	West Greenland (Fortescue 1984)	Barrow (Sturm 2009b)	Inuvik (Lowe 1984/2003)
	machakit wet snow (also porcelain, china, earthenware)	*matsaaq* snow soaked in water, half melted on the ground			*masak* mushy, waterlogged snow
	mangokpok the snow is soft, watery	*matsaaruti* wet snow prepared for pouring on a sled runner so as to ice them thickly			
mauja soft snow	*mauja* soft deep snow	*maujaq* any ground that gives under one' steps: mud, marsh, soft snow		*nutabaq* fresh powder snow	*mauya* deep soft snow
	maujak soft snow		*nittaalaaqqat* hard grains of snow (pl.)	*nutaaq* soft new snow	
				nuturuk good snow for making snow house; firm, yet not too hard	*misak* wet snow
piegnartoq snow is good for driving sled		*pukaangajuq* snow that is sufficiently crystallized, good enough to make a snow house			
	pukak snow looking like salt, not cleaving together	*pukak* crystalline snow that breaks down, separates, and looks like rough salt	pukak snow crust	*pukak* bottom layer of the snow cover, made up of large loosely bonded crystals, easily broken up	*pukak* sugar snow
		piirturiniq thin coat of light, soft snow deposited by a snow flurry on an object, ice	*putsinniq/puvvinniq* wet snow on top of ice		
	kersok crust on the ice in the spring	*qannitaq* snow recently fallen to the ground			
	kerksokak frozen crust of the snow	*qiasuqaq* snow that had thawed and is refrozen with an iced surface		*qiqsruqqaq* glazed snow	
	kersokpok frozen snow, in which there are tracks	*qiqumaaq* snow whose surface is frozen	qinuq rotten snow/slush on sea		
	sakketok fresh fallen snow, not blown or drifted	*qirsuqartuq* snow that thawed freezes stiff and hard		*sixxigruk* old icy snow, extra-hard	

independent English stems for snow are mostly faulty.[10] The only independent terms for snow in today's English from Pullum–Woodbury's list are "blizzard" (severe snowstorm), "avalanche" (a borrowing from French), and "hardpack."[11]

If we are to count independent stems only, the diversity of the English snow nomenclature is indeed quite limited, compared not only to the Inuit/Eskimo but also to several other languages, including Indo-European ones, spoken by people having greater exposure to snow and severe winter conditions. In Russian, for example, most snow-related terms are built, like in English, by using the stem *sneg/snezh* ("snow") with various added suffixes or independent stems, i.e., *snegopad* (snowfall), *snezhinka* (snowflake), *snezhok* (snowball), *snezhnik* (leftover snow in summer on the mountain slope), *snezhura* (wet mushy snow, close to the Inuit *qinu*), *snezhnitsa* (melted water from snow). In addition, Russian has several more independent terms for snow, such as *porosha* (first light snow on the ground), *sugrob* (heap of snow), *nast* (snow crust), *naduv* (pile of blown or drifted snow), *zastruga* (snow wave, linear-shape snowdrift), *pozemok* (low-level snowdrift), *protalina* (open ground where the snow has melted).[12] Russian speakers use *four* independent terms for various types of snowstorm or blizzard, including *metel'* (the most common word), *viyuga* (strong snowstorm, usually with a connotation of noisy blowing wind), *buran* (loan from Turkic, violent snowstorm, usually in the open space, like the steppe), and *purga* (loan from Finnish *purka*, prolonged and violent snowstorm commonly seen in northern parts of Russia and Siberia). This illustrates that some languages with more (or longer) exposure to snow and/or sea ice than English naturally develop detailed and meaningful terminologies for those phenomena that are of practical value to its speakers, even if some linguists claim otherwise.

Conclusions

It is obvious that during and after his fieldwork on Baffin Island, Boas was aware of his limitations in communicating with the Inuit and was reserved in assessing the value of his field material. Like many scientists on a first trip to a new area, he tried to overcome these limitations by traveling extensively across the terrain with local companions and by focusing on Inuit settlement patterns and mapping of local tribal groupings, place names, mobility, and land use. In his later research, Boas used a very different approach in forging long-term collaborations with locally based experts who were fluent in indigenous languages and had access to knowledgeable elders. This is also the approach used by many SIKU teams for this project.

It is not surprising that the number of local Inuktitut terms for ice and snow that Boas recorded on his first and only field work among the Inuit was, perhaps, less than one-third of what is known to today's speakers. The message from the first to the fourth IPY 125 years later is clear. It is not how much or how long we travel over the Arctic terrain that defines the quality of our material, but rather the depth of our collaboration with local experts and the amount of information they are willing to share with us. Six generations after Boas's fieldwork, and despite many transitions in the Inuit lifestyles, there is a tremendous pool of environmental expertise in many local communities, including that on ice and snow habitats. In many areas people still have a full command of rich indigenous terminologies that are often more extensive and nuanced than those developed by ice and snow scientists for their research.

Unlike Boas, we may rely these days not only on observations and recordings made by local partners and indigenous cultural specialists but also on the data collected at community meetings or stored in local schools and heritage curricula; and on the use of electronic maps and other modern digital technologies mastered by many of today's northern residents. These and other new resources have been actively employed to document indigenous knowledge and use of sea ice in modern polar communities during the SIKU project.

Lastly, Boas' recording of the Inuit terms for ice and snow had little relevance to what 100 years later became known as "The Great Eskimo Vocabulary Hoax." This entire story is a misnomer and it is based on the misunderstanding of Boas' legacy. Also, this debate has little to contribute to the study of Inuit knowledge of snow and ice, except that it subjects this knowledge to misinterpretation and condescending boasting. So, if some linguists and journalists are interested in counting someone's "words for snow" we have a message for them. Please switch to another language! The Norwegian Sámi, who tend to their reindeer herds over the northernmost realm of Europe, reportedly have 100 words for snow (Magga 2006).

Acknowledgments We are grateful to our colleagues, Claudio Aporta, Ernest S. Burch, Jr., Louis-Jacques Dorais, Ives Goddard, and Michael Krauss for many valuable comments to the first draft of this chapter. Mark Halpern was a source of inspiration on many issues related to the "Great Eskimo Vocabulary Hoax" debate. Matthew Sturm shared with us his lexicon of the Inuit/Inupiaq terms for snow collected among knowledgeable elders in Barrow, Alaska; and Noel Broadbent shared and translated the Swedish list of ice terms from Edlund's dissertation (2000). Our colleagues in the SIKU project – Claudio Aporta, Ron Brower, Gita Laidler, and Pierre Taverniers – kindly offered Inuit ice vocabularies they collected in 2003–2008 in Igloolik, Barrow, Pangnirtung, and Qeqertaq, respectively, for our comparative analysis with Franz Boas' 1894 lexicon. All shortcomings in interpreting the Inuit knowledge or ice and snow are of our own.

Notes

1. Boas went to his first fieldwork in Baffin Island in June 1883 on board the supply ship *Germania* sent to bring home the scientific personnel of the German IPY station at Kingua Fjord. His work was supported by the German Polar Commission that was in charge of the German research activities during IPY 1882–1883. Boas also benefited from some supplies, equipment, and logistics left behind by the German IPY expedition (Cole and Müller-Wille 1985:41–45).
2. SIKU (#166); ELOKA (Exchange of Local Knowledge in the Arctic, #187); EALAT (#399); Inuit, Narwhal and Tusks (#164), and others.
3. Throughout this chapter, the term "Eskimo" is used when referring to historical sources from the 1800s and 1900s and also when applied to the entire "Eskimo" language area that includes both the Inuit/Inuktitut/Inupiaq and Yupik/Yup'ik languages. In all other cases, indigenous residents of the Canadian Arctic are called Inuit and their language, Inuktitut.
4. Franz Boas/Hinrich Rink, Minden/Copenhagen, April 28, 1885; also A. F. Elsner (Moravian missionary to Labrador)/Franz Boas, Bremen/Minden, June 26, 1885; both in the Rink Papers, Archive of The Royal Library, Copenhagen. Translation from the German by L. Müller-Wille.

5. See De Reuse (1995); Derose (1999/2005); Fayhee (2009); Halpern (2008); Kaplan (2005); Liberman (2003, 2006); Muldrew (1997/2000); Pullum (2003, 2004); Woodbury (1991, 1994). For the most thorough and balanced review of the various terminologies for snow, including the origins of the "Great Eskimo Vocabulary Hoax" – see Mergen (1997:159–182).
6. See, lecture, "Language and Thought," by Colin Phillips, University of Delaware, 11.04.1999 http://www.ling.udel.edu/colin/courses/ling101_f99/ lecture18.html (accessed April 28, 2009).
7. *Aput* is the most general Inuit term for snow on the ground; *qana* (?) evidently is the wrong form for *qaneq*, "falling snow" (Boas 1894:104); *piqsirpoq* (*pi'rtsirpoq* in Boas 1894:110) refers to the process of snow blowing or drifting under the force of wind (Germ. *der Schnee treibt* – Boas 1894:110); *qimuqsuq* means "snowdrift," or rather, a wavy surface built by the snowdrift.
8. Most of those terms, like *aniu* (soft snow good for drinking water), *apun* (the most general term for snow), *aqilluqqaq/aqilluraq* (light soft snow), *masak* (mushy, water-logged snow), *mauya/mauja* (deep soft snow), *pukak* (crystalline snow), *qiqsruqqaq* (glazed snow) are common across the Inuit/Inupiaq/Inuktitut area.
9. As Mark Halpern rightly points out (personal communication, August 4, 2009), many English words involving "snow" do not describe the type of snow but refer to something else, like "snowball" is not a kinds of snow, but a kind of ball, just as "meatloaf" is not a kind of meat but a kind of loaf. Same applies to snow goose, snowmobile, snowbird, snowbell, snowplow, etc.
10. English words like "powder," "crust," or "dusting" cannot be associated with snow without special context. "Sleet" is more a term for "freezing rain" or ice pellets, rather than snow; and "slush" is used for all kinds of mushy substances, besides partly melted snow and ice, including soft mud, and paper pulp (http://www.merriam-webster.com/dictionary/slush).
11. The latter, besides being a definition for compacted snow (http://www.merriam-webster.com/dictionary/hardpack), is commonly used for any firm ground (hardpack soil) or even for special brands of bag-packs and bicycles.
12. All explanations of the Russian snow terms are from Murzaev (1984).

Appendix: Franz Boas' List of Cumberland Sound Ice Terms (1894) by Major Groupings

Ice

sī'ko ice/ L., G. same.

Young Ice

qati'dinrn the ice on the beach in the fall that floats at high tide and is stranded at low tide/L. *kattinek.*
siko'qoang 1 to 1 ½ ft of thick ice in the fall
sī'koaq thin ice formed by the first frost
sikū'liaq thin ice at the ice edge during the winter

Pack Ice, Old Ice

kaqvaq pack ice/L. *kackvak.*
qaqbang multi-year ice, pack ice (see *kaqvaq*).
tu'vang thick winter ice/L. *tuvak.*
tuvarea'qtung thick ice at the end of the fall season

Land-Fast Ice

qavirpi'jung ground ice/G. *qaungoq.*
quta'rong ice foot at a steep coast

Pressure Ridges, Rafted and Layered Ice

igjijug thick, thick ice/L. *ivjo'vok*, G. *ivssuvoq.*
ivu'dnirn ground ice on the beach/, L. *ivuvok* /ice piling upon the beach/

Various Ice Forms

manituā'dlu – hillocky land, rough ice/

Cracks, Leads, Polynyas

a'jorang crack in the ice (extensive crack, not the cracks in the low tidal flats)/L. *ajorak.*
A'jorang Place name: the crack
Aqti'nirn Place name: where the ice melts in front of the mouth of a river/
Angmaritung Place name: the open one, not frozen over/L. *angmarok.*
Angmartung Place name: the open one (not frozen over)/
qu'gnirn crack in the ground ice/L. *Kongnek.*
sinā' his edge, ice edge

Ice Floes, Floating/Drifting Ice

piqalu'jang iceberg/L. *Pekkalujak.*
Piqaluirtung Place name: abundant with icebergs
siku'kulu small pieces of drift ice

Spring Ice Melt

imakti'nirn ice over melting snow, strong enough to carry a sled/
mase'lirang thin ice that is formed on the snow in spring, wet moss/L. *masalerak.*

Other Types of Ice and Ice-Related Phenomena

miso'majung glacier reaching into the sea/G. *misugpa* – he dips it into [something]/
nilang freshwater ice/L. *nillak*, G. *nilak.*
penartua'dlu water on ice
Sirmilling Place name: place where there is a glacier/
si'rming thin ice, firn, glacier/L. *sermek*, G. *sermeq.*

Derivatives

Sikosuilaq Place name: [location is] free of ice/
sikoqa'ngenut across the ice/
Sikosū'ilaq Place name: [place] without ice/

References

Anonymous. N.d. Glossary (of) Snow and Avalanches. Working group (of the) European Avalanche Forecasting Services. WSL Institute for Snow and Avalanche Research, Davos, Switzerland. http://waarchiv.slf.ch/index.php?id=278#5416 (Accessed May 9, 2009).

Armstrong, T. 1958. Illustrated ice glossary. Pt. 2. *Polar Record* 9(59): 90–96.

Aporta, C. 2003. Old Routes, New Trails: Contemporary Inuit Travel and Orienting in Igloolik, Nunavut. Unpublished Ph.D. thesis. Department of Anthropology, University of Alberta, Edmonton.

Berman, J. 2001. Unpublished Materials of Franz Boas and George Hunt: A Record of 45 Years of Collaboration. *Gateways: Exploring the Legacy of the Jesup North Pacific Expedition, 1897–1902. Contributions to Circumpolar Anthropology* 1. I. Krupnik and W.W. Fitzhugh (eds.), Washington: Arctic Studies Center, pp.181–213.

Boas, F. 1881. *Beiträge zur Erkenntnis der Farbe des Wassers.* Doctoral dissertation, Faculty of Philosophy, University of Kiel, Germany. Kiel: Schmidt & Klaunig.

Boas, F. 1885. Baffin-Land. Geographische Ergebnisse einer in den Jahren 1883 und 1884ausgeführten Forschungsreise. *Petermanns Mitteilungen*, 80: 1–100. Ergänzungsheft.Gotha: Justus Perthes.

Boas, F. 1888. The Central Eskimo. *Sixth Annual Report of the Bureau of Ethnology* 1884–1885:399–675. Washington: Bureau of Ethnology (Reprint 1964. Toronto: Coles).

Boas, F. 1894. Der Eskimo-Dialekt des Cumberland-Sundes. *Mittheilungen der Anthropologischen Gesellschaft in Wien*, Band XXIV (Neue Folge, Vol. XIV): 97–114. Alfred Hölder, Wien.

Boas, F. 1911. Introduction. In *Handbook of American Indian Languages*. Part 1. Smithsonian Institution, *Bureau of American Ethnology, Bulletin* 40: 1–84. B. Franz (ed.), Washington: Government Printing Office.

Brower, R., Jr. 2008. (Barrow) Inupiaq snow terminology. In *Light Snow, Deep for Walking.* M. Sanchez, *Weekend America*, January 12. http://weekendamerica.publicradio.org/ display/web/2008/01/11/snowwords/ (Accessed April 27, 2009).

Cole, D. and Müller-Wille., L. 1984. Franz Boas' expedition to Baffin Island, 1883–1884. *Etudes/Inuit/Studies* 8(1): 37–63.

De Reuse, W. 1994. Eskimo Words for 'Snow,' 'Ice,' etc. *Linguist List* 5.1293 (Accessed April 26, 2009 at http://www.linguistlist.org/issues/5/5-1293.html)

Derose, S.J. 1999/2005. Eskimo Words for Snow. http://www.derose.net/steve/guides/snowwords (Accessed June 2, 2009)

Dorais, Louis-Jacques 1996. *La parole inuit. Langue, culture et société dans l'Arctique nord-américain.* Louvain-Paris: Peeters Press.

Edlund, A.-C. 2000. *Sälen och jägaren. De bottniska jägarnas begreppssystem för säl ur ett kognitivt perspektiv*. Norrlands Universitetsförlag, Umeå.

Erdmann, F. (ed.), 1864, 1866. Eskimoisches Wörterbuch gesammelt von Missionaren in Labrador, revidirt und herausgegeben von Friedrich Erdmann. Erster Theil: Eskimoisch-Deutsch (1864, 350 pp.), Deutsch-Eskimoisch, Zweiter Theil (1866, 242 pp.). Ernst Moritz Monse, Budissin [Bautzen].

Fayhee, J.M. 2009. Snow by Any Other Name. *Mountain Gazette* 151. http://www.mountain gazette.com/exclusive/features/snow_by_any_other_name.

Fortescue, M. 1984. *West Greenlandic*. London, etc: Groom Helm.

Fortescue, M. 1991. I'nuktun, An introduction to the language of Qaanaaq, Thule. *Institut for Eskimologi publication series* 15. Copenhagen.

Freeman, M.M.R. 1984. Franz Boas on baffin island: A centennial observed. (In Boas footsteps: 100 years of Inuit anthropopology). *Études/Inuit/Studies* 8(1): 11–12.

Halpern, M. 2008. *Language and Human Nature*. Transaction Publications.

Harper, K. 2008. The collaboration of James Mutch and Franz Boas, 1883–1922. *Études/Inuit/Studies* 32(2): 53–71.

Jacobson, S.A. 1984. *Yup'ik Eskimo Dictionary*. Fairbanks: Native Language Center.

Kaplan, L. 2003. Inuit snow terms: How many and what does it mean? In *Building Capacity in Arctic Societies: Dynamics and shifting perspectives. Proceedings from the 2nd IPSSAS Seminar. Iqaluit, Nunavut, Canada: May 26–June 6, 2003*. F. Trudel (ed.), Montreal: CIERRA. Facultés sciences sociales Université Laval.

Kaplan, L. 2005/2008. Inuit Snow Terms: How Many and What does it Mean? http://www.uaf.edu/anlc/snow.html#citation (Accessed April 26, 2009).

Kleinschmidt, S. 1851. *Grammatik der grönländischen Sprache mit teilweisem Einschluss des Labradordialektes*. Berlin: Walter de Gruyter & Co, (Reprint: Hildesheim: Georg Olms Verlagsbuchhandlung 1968).

Kleinschmidt, S. 1871. *Den grønlandske Ordbog*. København: Louis Kleins Bogtrykkeri.

Krupnik, I. 2003. Fourth International Polar Year. *Northern Notes* Fall, 6–7.

Krupnik, I., Bravo, M., Csonka, Y., Hovelsrud-Broda, G., Müller-Wille, L., Poppel, B., Schweitzer, P., and Sörlin., S. 2005. Social sciences and humanities in the International Polar Year 2007–2008: An integrating mission. *Arctic* 58(1): 91–97.

Laidler, G.J. 2007. Ice, Through Inuit Eyes: Characterizing the Importance of Sea Ice Processes, Use and Change Around Three Nunavut Commuties. Unpublished Ph.D. Thesis, Department of Georgraphy, University of Toronto (Accessed July 14, 2009 at http://eratos.erin.utoronto.ca/robinson/dissertation/).

Laidler, G.J., Dialla, A., and Joamie., E. 2008. Human geographies of sea ice: Freeze/thaw processes around Pangnirtung, Nunavut, Canada. *Polar Record* 44(231): 335–361.

Liberman, M., 2003. 88 English Words from Snow. *Language Log* December 7, 2003 (Accessed April 26, 2009 – http://itre.cis.upenn.edu/~myl/languagelog/archives/000200.html).

Liberman, M. 2006. The Proper Treatment of Snowclones in Ordinary English. *Language Log*, February 4, 2006 – Accessed April 26, 2009 http://itre.cis.upenn.edu/~myl/languagelog/archives/002806.html.

Lowe, R. 1983. *Kangiryuarmiut Uqauhingita Numiktittitdjutingit*. Basic Kangiryarmiut Eskimo Dictionary. Published by the Committee for Original Peoples Entitlement. Inuvik, Canada.

Lowe, R. 1984. *Siglit Inuvialuit Uqausiita Kipuktirutait*. Basic Siglit Inuvialuit Eskimo Dictionary. Published by the Committee for Original Peoples Entitlement. Inuvik, Canada.

Magga, O.H. 2006. Diversity in Saami terminology for reindeer, snow, and ice. *International Social Science Journal* 58(187): 25–34.

Martin, L. 1986. Eskimo words for snow: A case study in the genesis and decay of an anthropological example. *American Anthropologist* 88(2): 418–423.

Mergen, B. 1997. *Snow in America*. Washington and London: Smithsonian Institution Press.

Muldrew, K. 1997/2000. A Lexicon of Snow. http://www.ucalgary.ca/~kmuldrew/cryo_course/snow_words.html (Accessed May 10, 2009).

Müller-Wille, L. 1985. Snow and Ice in Inuit Place Names in the Eastern Canadian Arctic. In *43rd Eastern Snow Conference Proceedings 1984*, pp. 55–57. Department of Geography, Montréal: McGill University.

Müller-Wille, L. 2008. Franz Boas and the Inuit. *Études/Inuit/Studies* 32(2): 9–12.

Müller-Wille, L. 2009. Franz Boas' Beitrag zur Ethnologie der Inuit: Methodik und Ansätze zwischen Geographie und Völkerkunde. *In Franz Boas (1858–1942). Wissenschaft, Politik, Mobilität*. H.-W. Schmuhl (ed.), Bielefeld: Transcript Verlag, pp. 303–310.

Müller-Wille, L., (ed.) 1998. *Franz Boas among the Inuit of Baffin Island, 1883–1884: Journals and Letters*. Toronto: University of Toronto Press.

Müller-Wille, L. and Gieseking, B. (eds.), 2008. *Bei Inuit und Walfängern auf Baffin-Land (1883/1884). Das arktische Tagebuch des Wilhelm Weike*. (Mindener Beiträge, Band 30). Minden: Mindener Geschichtsverein.

Müller-Wille, L. and Weber, L. 1983. Inuit Place Name Inventory of Northeastern Québec-Labrador. Marburger Geographische Schriften 89/McGill Subarctic Research Papers 37: 151–222.

Müller-Wille, L. and Weber Müller-Wille, L. 2006. Inuit geographical knowledge one hundred years apart: Place names in Tinijjuarvik (Cumberland Sound), Nunavut. In *Inuit Studies in an Era of Globalization*. P. Stern and L. Stevenson (eds.), Lincoln: University of Nebraska Press, pp. 217–229.

Murray, S.O. 1987. Snowing canonical texts. *American Anthropologists* 89(2): 443–444.
Murzaev, E.M. 1984. *Slovar' narodnykh geograficheskikh terminov* (Glossary of folk geographic terms). Moscow: Mysl Publishers.
Peck, E.J. Rev. 1925. *Eskimo-English Dictionary*. (Compiled from Erdman's Eskimo-German Edition 1864 A.D.). Published by the Church of the Ascension Thank-Offering Mission Fund, Hamilton. Toronto.
Petitot, É. 1876. *Vocabulaire francaise-esquimau*. Paris: Ernest Leroux.
Pullum, G. 1989. "The Great Eskimo Vocabulary Hoax." *Natural Language and Linguistic Theory* 7(2): 275–281.
Pullum, G. 1990. "The Great Eskimo Vocabulary Hoax." *Lingua Franca* 1: 28–29.
Pullum, G. 1991a. *The Great Eskimo Vocabulary Hoax and Other Irreverent essays on the Study of Language*. Chicago: University of Chicago Press.
Pullum, G. 1991b. The Great Eskimo Vocabulary Hoax. In *The Shape of Reason: Argumentative Writing in College*. New York: Macmillan Publishing, pp. 33–38.
Pullum, G. 1994. The Great Eskimo Vocabulary Hoax. *Probable Cause: A Literary Review* Winter: 3–6.
Pullum, G. 1996. 100 Wörter für Schnee – der grosse Eskimo-Bluff. *Weltwoche Supplement*. March, 22–25.
Pullum, G. 2003. Bleached conditionals. *Language* Log (October 21) http://itre.cis.upenn.edu/~myl/languagelog/archives/000049.html (Accessed April 26, 2009).
Pullum, G. 2004. Sasha Aikhenvald on Inuit Snow Words: A Clarification. *Language Log*, January 30, http://158.130.17.5/%7Emyl/languagelog/archives/000405.html (Accessed April 26, 2009)
Rink, H.J. 1866. Eskimoiske Eventyr og Sagn: oversatte efter de indfødte Fortælleres Opskrifter og Meddelelser af H. Rink. København: C. A. Reitzel Boghandel.
Rink, H.J. 1871. Eskimoiske Eventyr og Sagn: Supplement indeholdene et tillage om eskimoerne, deres kulturtrin og ovrige eiendommeligheder samt formodede herkomst. København: C.A. Reitzels Boghandel.
Rink, H. 1875. *Eskimo with a Sketch of Their Habits, Religion, Language and Other Peculiarities*. Translated from the Danish by the author. R. Brown (ed.), Edinburgh and London: William Blackwood and Sons, 472 pp.
Rink, H. 1887. *The Eskimo tribes their Distribution and Characteristics, Especially in Regard to Language with a Comparative Vocabulary*. Copenhagen: C.A. Reitzel, and London: Longmans and Green.
Rohner, R.P. (ed.), 1969. *The Ethnography of Franz Boas*. Translated by H. Parker. Chicago: University of Chicago Press.
Schneider, L. 1985. *Ulirnaisigutiit: An Inuktitut-English Dictionary of Northern Quebec, Labrador, and Eastern Arctic Dialects*. D. Collis (ed.), Quebec: Les Presses de l'Université Laval.
Sturm, M. 2009a. *Apun: The Arctic Snow.* Fairbanks: University of Alaska Press.
Sturm, M. 2009b. Composite List of Inupiaq Snow words (March 2009). Unpublished manuscript cited with author's permission.
Vakhtin, N.B. and Emelyanova., N.M. 1988. *Praktikum po leksike eskimosskogo iazyka* (Practical Aid to the Eskimo Lexicon). Leningrad: Prosveshchenie Publishers.
Walunga, W. comp. 1988. *St. Lawrence Island Curriculum Resource Manual*. Gambell.
Whorf, B.L. 1940. Science and linguistics. *Technological Review* 42(6): 229–231.
Womkon Badten, L., Kaneshiro, V.O., and Oovi., M. 1987. *A Dictionary of the St. Lawrence Island/Siberian Yupik Eskimo Language*. Fairbanks: Native Language Center.
Woodbury, A. 1991. Eskimo Words for 'Snow'. *Linguist List* 5-1239 (Accessed April 26, 2009) http://www.ecst.csuchico.edu/~atman/Misc/eskimo-snow-words.html.
Woodbury, A. 1994. [Tony Woodbury's] 'Snow' lexemes in Yup'ik [Accessed June 15, 2009] http://www.prairienet.org/prairienations/woodbury.htm.
World Meteorological Organization. 1970. WMO sea ice nomenclature. Terminology, codes and illustrated glossary. Geneva, Secretariat of the World Meteorological Organization, *WMO/OMM/BMO*, 259, TP. 145.

Chapter 17
Inuit Sea Ice Terminology in Nunavut and Nunatsiavut

Alana Johns

Abstract This chapter provides a linguistic perspective on recent research by anthropologists and human geographers about indigenous sea ice terms in Nunavut and Nunatsiavut (Labrador), providing a basic introduction to pertinent linguistic properties of Inuktitut and arguing that they shed further light on Inuit sea ice knowledge. A number of sea ice terms from the largely unknown Utkuhiksalingmiut dialect are provided.

Keywords Inuktitut · Nunavut · Nunatsiavut · Polysynthesis · Terminology · Lexicalization

Introduction

Inuktitut[1] dialects contain a myriad of sea ice terms, but this is not surprising, given that sea ice plays a pivotal role in the Inuit economy, transportation, and sustenance (see Laidler and Elee 2008; Laidler and Ikummaq 2008; Laidler et al. 2008; Chapter 7 by Aporta, this volume; Chapter 3 by Laidler et al. this volume; Chapter 11 by Huntington et al. this volume). Once the newly formed fall ice has frozen enough, transportation becomes much easier, since the same vehicle can traverse both land and ice. At the same time, underneath and on the ice are Arctic char, various seals, walrus, and other food resources. Knowledge of ice conditions and how to navigate the ice safely is critical to human survival in the Arctic environment. Thus the knowledge that lies behind the words used to refer to and describe sea ice is very important to both Inuit and non-Inuit. Collecting, preserving, and analyzing these terms are worthwhile endeavors.

Inuit qauijimajaqtuqangit, or Inuit traditional knowledge, is becoming more and more prominent, both in the north and in the south. The knowledge is there. The questions are how to transmit it, to whom, and in what form? Traditional methods

---

A. Johns (✉)
Department of Linguistics, University of Toronto, Toronto, ON M5S 3G3, Canada
e-mail: ajohns@chass.utoronto.ca

I. Krupnik et al. (eds.), *SIKU: Knowing Our Ice*,
DOI 10.1007/978-90-481-8587-0_17,

of teaching younger generations are still very strong, but Inuktitut is waning as a spoken language in some regions and most members of the younger generation report that they know fewer specialized or technical words than their elders. New methods of knowledge transmission are emerging. At the same time, non-Inuit scientists are hoping to augment their own science of ice through learning about the ice knowledge of Inuit (see Chapter 15 by Eicken, this volume). Collecting and recording Inuit ice terminology is a crucial step in this research. From an anthropological perspective, collecting these terms offers insight into Inuit cultural knowledge, tradition, and ways of learning (Chapter 12 by Wisniewski, this volume).

From a linguistic perspective, there is, perhaps, less scientific purpose in collecting all the available words for ice. Most linguists are weary of futile, albeit passionate, discussions about the number of words for snow in Inuktitut (see Chapter 16 by Krupnik and Müller-Wille, this volume). The exact number of words for "snow" (or ice) in Inuktitut is beyond anyone's count, as in any language where snow is a significant cultural factor. The inability to be precise regarding the number of terms results from dialect differences combined with the limitations of the researcher's ability to collect all the words in all dialects. In addition, in the Inuktitut language, the grammatical complexity of words, where the same root might be followed by what seems like an infinite combination of additional elements (see below), further thwarts any principled goal of counting. For a representative discussion of a linguist's perspective on counting words for snow, see Woodbury (1991) and Kaplan (2003). As a result, we will wisely put aside here any goal of determining the number of words for sea ice. Instead we can examine these sets of words for more important reasons, that is, to explore their nature and what they tell us about the ice conditions they describe.

Questions regarding differences between dialects in the use of ice terms are also daunting.[2] For each list of words discussed below, we cannot be sure if other words could have also been included, or sometimes even whether or not a particular word is indeed an ice term. For example, the Utkuhiksalik[3] word *aniu*[4] generally refers to snow that is to be melted for water; however, it can sometimes be used for ice which was originally snow (e.g., in a ravine). Is this then an ice or a snow term? Questions such as this and others abound in the collection of any sets of cultural terminology, particularly when organized by scientists along certain topical lines (i.e., "ice" versus "snow").

Aspects of Inuktitut Linguistics Pertaining to Collecting Ice Terminology

As analyses of Inuit sea ice terminologies have not involved linguists before, it is worthwhile to consider a few linguistics concepts that may be relevant to those whose goal is to compile or examine such lists. This discussion is necessarily incomplete, as a full treatment is beyond the scope of this chapter or even a single linguist.

Inuktitut words range from simple to highly complex. An Inuktitut word can be equivalent to a full and complex English sentence, as in the South Baffin Inuktitut word *qaujigumatuinnarattaqtunga,* "I just wanted to know." This property in a language is known as polysynthesis. Thus, the Inuktitut word is often not a monolithic object but is composed of sub-units, each bearing a piece of meaning, often abstract, which linguists call morphemes. In almost all words, the first element is the root – a very important morpheme, which carries a central piece of meaning. The simplest words will consist of only a root, as in the noun *siku* "ice." If the Inuktitut word is a verb or adjective, it must also have, in addition to the root, an ending (see below). The root may be immediately followed by any number of modifiers and grammatical morphemes to add extra meaning. These intermediate elements sometimes change the category of the word from noun to verb or vice versa. It has long been observed that these word-internal morphemes appear in the reverse order to that found in an English sentence. Consider again, the example from above: *qauji-guma-tuinna-rattaq-tunga,* which can be roughly glossed as *know-want-merely-habitual-1singular.statement*. We can read the pieces of meaning in the gloss, starting from right to left, and begin to understand how these pieces add up to the colloquial English translation "I just wanted to know."

As a result, it will increase our understanding of Inuktitut sea ice terms if we examine what the meaning of the root of each word is, and also the meaning of any other morphemes that are attached. To ignore this additional information is to treat a complex object as a simplex one.[5]

Inuktitut words can be entities (nouns), descriptions (adjectives or verbs), or actions (verbs). Often we see words in ice lists that have simple intransitive verb endings in the third-person singular (*-juq* after vowels or *-tuq* after consonants) added on to an entity to make the word into an action. For example, in Cape Dorset *sikuaq* is "the first thin layer of frozen ice"; therefore, we are not surprised to see the action word *sikuaqtuq,* "the process of sikuaq forming" (Laidler and Elee 2008:55). We need to determine whether verbs such as these are indeed independent ice terms, since the addition of these verbs endings is a predictable and transparent process.

Another linguistic issue that can arise is the differing phonology across dialects. In many Baffin dialects the final /q/ in a word is not pronounced, so we are not surprised to see *qamittu* (as in Laidler and Elee 2008), in the Cape Dorset word meaning "ice with a little bit of water on it." This word likely should be *qamittuq* with the intransitive verbal ending *-tuq* (see above). The root of the Cape Dorset word is interesting because *qamittuq* usually means "put out a light" or "a light goes out" (Spalding 1998). We find also a similar Utkuhiksalingmiut word *qamittuq* "it's extinguished (a motor, light, flame)." The same root can sometimes also mean "close an eye," as in the Utkuhiksalingmiut word *qamititsiřuq* "He shuts one eye." Thus we gain a deeper understanding of the Cape Dorset ice word as alluding to the covering effect of the water on the ice.

Dedicated and Contextual Terms

One aspect to consider in understanding the ice terms in these and other available local lists is the issue of whether or not the Inuktitut word for ice is indeed "dedicated" to the concept of ice. By "dedicated" I mean that the use of this word necessarily denotes some actual ice no matter what the context of the utterance. For example, the most general Inuit term *siku* "ice" is a dedicated ice word that almost always refers to ice and, most often, specifically to sea ice (but see the Utkuhiksalingmiut word *hiku/siku* in the Appendix). On the other hand, the word *aulajuq* "moving ice" (Laidler and Ikummaq 2008:139) is not a dedicated word. It is based on the verb root *aula-* "move" (Spalding 1998; Andersen et al. 2007). Out of context, this word can be used to describe anything from wood floating in water to a piece of paper in the wind on the land; but in the ice context, this word refers to the process by which the ice as a whole moves c.f., *aulaniq* "moving ice field" (Spalding 1998), which contains the nominalizer *-niq*). So are these non-dedicated ice terms really ice terms?

Many non-dedicated Inuktitut ice terms are like terms we use in certain special contexts in English. Within these special contexts, there is no ambiguity whatsoever. We often see English terms of this nature in game settings, where all the players knows the special terms or learn them through playing with more experienced players. To take an example, terms used in the game of curling (which is coincidentally played on ice) are probably unfamiliar to most English speakers (other than some Canadians or Scots). Within the game of curling, the terms *button*, *shot rock*, *pebble*, *weight*, *house*, and *curl* each have a unique meaning within the game environment. Within curling, the term *house* refers to the ring of circles at the far end of the ice, including a central spot within the circles. The whole area constitutes the target range for the play.[6] If you regularly participate in the game of curling, the noun *the house* is completely unambiguous while you are within the curling context, and another speaker would have to provide extra detail and information if they intend to refer to the more common definition of "house" (building). In the same way, once out hunting, fishing, or traveling on the ice, the set of Inuit ice terms have little competition from the general meanings that co-exist in the language alongside these terms. From this perspective, both dedicated and some non-dedicated ice terms are equally deserving of an independent status, although in the case of a non-dedicated term, it is enlightening to keep track of its more general sense.

Spalding (1998) clearly recognizes the above distinction between general terms and semi-dedicated terms. Laidler and Ikummaq (2008:139) give the Igloolik ice term *puktaaq*, "free-floating ice" (which is also a common term for ice floes in many other dialects from Alaska to Greenland – see Chapter 2 by Taverniers, this volume; Chapter 14 by Krupnik and Weyapuk, this volume). The root in this word is *pukta-* meaning simply "float (in water)," so that in principle, words containing it could refer to anything floating in water, e.g., a dead seal. Spalding (1998:102) gives this general sense through the words *puktajuq* "it floats or buoys itself up in the water" and also *puktaakkut* "float; buoy; fish-net float." Importantly, Spalding also gives the word *puktaaq*, marked explicitly as a meteorological word, indicating that it has a specialized meaning, in this case "flat drift ice." We find this same distinction in Andersen et al. (2007:177), where *puttavuk* means "it is floating" but *puttâk* means "small or large ice pans floating freely in the fall or spring." In Utkuhiksalingmiutitut, *puktaaq* is "a piece of flat drifting ice used as a raft," even though the root clearly refers only to floating, as in the Utkuhiksalingmiutitut word *puktalaaqtut* "they are floating (e.g., tea leaves in kettle)."

So it may be that the word *puktaaq*, even though it is built on the general root "float," is actually a dedicated ice term, exclusive to the world of ice. This could be confirmed with speakers by asking in a non-ice context, e.g., inside a house, what a *puktaaq* is. If they respond by describing only ice, then *puktaaq* has become a lexicalized word, i.e., it is a word that has acquired more meaning that just its formal components. This is reminiscent of semantic narrowing, such as we see in language change generally, e.g., *hound* used to refer to any dog in earlier versions of English but now it usually refers to certain breeds of dog.

Another example of lexicalization may be the word *sikujuq,* which is composed of *siku* "ice" + *juq*, the intransitive verbal ending (third-person singular). While we might imagine this combination could have a wide range of potential meanings, e.g., "it's icy," etc., in some dialects it in fact means that the ice has formed into a thick enough object as to allow travel, i.e., "ice that is travelable," as in Cape Dorset (Laidler and Elee 2008:55). In the Pangnirtung list (Laidler et al. 2008:339), we are only told that the "water has frozen over," with no mention of whether or not one can travel. This more cautious meaning is also found in Spalding (1998:132), Jeddore (1976:129), and Andersen et al. (2007:185). An Utkuhiksalingmiut version of this word has not been found, although there is a root for "freeze/harden," which is used for objects *qiqiřuq* "it is hardened or frozen."

Some ice terms seem to be found in all dialects of Inuktitut. The word for "ice" *siku* (or *hiku* in western dialects, where single [s] is found as [h]) is such a word. We might wonder if cross-dialectal terms are only dedicated terms. There are also some gaps. For example, *sinaaq* "ice edge" (where the ice meets open water) is very common in eastern dialects and is recorded in three Nunavut lists, in Igloolik, Pangnirtung, and Cape Dorset (Laidler et al. 2008:339; Laidler and Ikummaq 2008:131; Laidler and Elee 2008:55), and also in west Greenlandic (Chapter 2 by Taverniers, this volume). But its counterpart is not found in the Utkuhiksalik dialect. Utkuhiksalingmiutitut has only *hinnaa* "its edge," which can be applied to a wide variety of edge contexts, including both shoreline and floe edge, i.e., it has no special relation to ice. Spalding (1998) has *sinaa* as a meteorological word for "floe edge," and also *sini* "edge" and *sinaarut* "beaded edging of a garment or *atigi*." We understand *sinaa* to possibly mean "its edge" with the third-person possessive *-a* morpheme attached.[7] Perhaps the final /q/ is a late addition.[8] Labrador (Andersen et al. 2007) also has the same word as in the Baffin dialects, *sinâk* "edge of the shore ice" (where â = aa) and /k/ is the final consonant of all singular non-possessed nouns.[9] Jeddore's Labrador Inuttun list (1976) gives it as *sinâ*, again suggesting that it contains a possessive ending, since there is no final **k**.

Some Remarks on Ice Terms in Various Dialects

As non-Inuktitut speakers, we gain extra insight into the nuances of Inuktitut ice terms through more careful linguistic analysis. For example, the word *nigajutaq* is "an area of sea ice that freezes later than others" (Laidler and Elee 2008:55; Laidler and Ikummaq 2008:131; Laidler et al. 2008:339). The root of this word is *nigaq* "snare" (Spalding 1998:68) and *-taq* is the passive participle form, which indicates that it is a noun that has undergone the particular action. This means that an unfrozen area of water is described as being ensnared by the surrounding ice (literally: the one ensnared).[10] The exact same description is found in the Utkuhiksalingmiut dialect. The word *nigaqtuq* refers to an action where something has been trapped by placing a rope around its burrow opening, i.e., snare. At the same time, *nigajutaq* means "a place (places) that haven't yet iced over (in autumn, when ice is first forming)." The fact that dialect regions so distant from one another make identical distinctions means that this is a long-standing word for this ice phenomenon.[11] In Labrador Inuttun, however, while we find the same root used for snaring in Labrador *nigan-niajuk* "he has put out snares" (Andersen et al. 2007:147), we do not see any variant listed in local ice terminologies.

The Igloolik word *uukkaqtuq* "the ice breaking off from the *sinaaq*" in Laidler and Ikummaq (2008:139) appears to be a dedicated ice term, in the sense defined above. It is a verb or description exclusively referring to the action of ice breaking off. We can well imagine that this action would be very important to survival, potentially leaving a person helplessly stranded on a piece of ice drifting out to sea. This seems to be a dedicated ice word in other dialects as well; Spalding (1998:193) gives *uukkaqtuq* as "the solid sea ice breaks off" with neither meteorological marking nor competing definitions. Laidler and Ikummaq (2008:139) also give the related term *ukkaruti* "the ice that has broken off due to *uukkaqtuq*, and is now free floating." The related word contains the instrumental morpheme *-uti*, so the word literally means "the instrument causing the ice breaking off." They also give another related word *uukkaqtaqtuq* "ice continuously breaking from the *sinaaq*." This last word contains within it the morpheme of repetition *-taq-*. These two words, transparently related to *uukkaqtuq*, are not really independent ice words, but simply predictable variations of *uukkaqtuq*. We can call these "satellite" terms.

The word *millutsiniq* "a slushy patch on the ice caused by snowfall on thin ice" is found in Cape Dorset (Laidler and Elee 2008:55), but not found in other dialect lists. On investigation, this word nonetheless turns out to conjure up a potentially perilous situation, for it is composed of *milluk-tsi-niq*, or [suck in/under -actor-event], i.e., an event where something sucks in another entity.[12] Indeed, Laidler and Elee (2008:59) discuss this term in the context of dangerous melting conditions, where snow serves as insulation to either warm the ice or prevent it from freezing further. The English translation of "slushy patch" does not seem as dangerous as the term in Inuktitut.[13]

The dedicated ice term meaning "iceberg" is *piqalujaq* in Igloolik (Laidler and Ikummaq 2008:139) and also in several other local lists,[14] but is the slightly different *piqulajaq* in Utkuhiksalingmiut. We see the word *piqalujak* in Clyde River (Gearherd et al. in progress). In Labrador we are not surprised to see the form *piKalujak* (Jeddore 1976:98; Andersen et al. 2007:168), where K = q. The final [k] is expected in Labrador because of the neutralization of final **k/q** (mentioned in note 9).

Another term seemingly unique to Cape Dorset in Laidler and Elee (2008:55) is *sallivaliajuq* "ice thinning due to rain, wind or snowfall." This term appears to contain *-vallia-* "more and more"[15] and the root is likely *salik-* "wipe away, clean off, scrape" (see Spalding 1998:126), so the Cape Dorset word may actually be *salivalliajuq*, literally meaning "it has been worn away." If this is so, then whether or not it is really a genuine ice term in the sense defined here or is simply an on-the-spot description, remains to be clarified.

The word *sinaaviniq* "a former *sinaaq*" is found in Laidler and Ikummaq (2008:131). This word is transparently composed of *sinaaq* "floe edge" + *viniq* "former." Because this word is produced through a regular word formation process, there seems little need to make an independent entry for it, any more than there is for the many English phrases with *former* added to them in English (e.g., *former house*). Note that *-viniq* is much more commonly found in Inuktitut than *former* is in English (and has a wider range of meaning). We can apply the exact same reasoning to *nigajutaviniq* "a former *nigajutaq* that has frozen over" (Laidler and Ikummaq 2008:131). Both *sinaaviniq* and *nigajutaviniq* appear to be satellite words, as discussed above, rather than independent ice words.

Another Igloolik ice term is *aggurtippalliajuq*[16] "the process of ice freezing in an upwind direction" (Laidler and Ikummaq 2008:131). This word is literally *aggur-tit-pallia-juq* or in [the face of the wind-make-more and more-verb ending], suggesting it means literally "it is becoming progressively facing into the wind." It would be nice to have evidence that this is more like a conventional term rather than merely a description. By "merely a description," I mean that another person from the same community could potentially refer to the same process with a different turn of phrase. This same issue is found in the Utkuhiksalingmiutitut word *aaqluqtittuq* "a piece of ice that has been stuck to the shore, comes unstuck from the land under the water, separates from the land-bottom and rises up at one end more than at the other." It is based on the root *aaqluq* "to raise one's face up."

The word *qanguti* is found in Igloolik (Laidler and Ikummaq 2008:131) and Cape Dorset (Laidler and Elee 2008:55) and as *qanngut* in Pangnirtung (Laidler et al. 2008:357). These words refer to ice with crystal-like snow formations on top. We might expect them to have the root *kani(q)* since the root meaning "frost" is *kaniq* in Spalding (1998:39) and also in Utkuhiksalingmiutitut. The Igloolik ice word also seems to contain the instrumental morpheme *-ut(i)*. In Labrador, *kani* is "hoar frost, frost crystals that form due to moisture in the air" (Andersen et al. 2007:84). Labrador also has *KaKunnak* "hoarfrost" in Jeddore (1976:110), which starts with q (recall K = q). So it looks as if variation between k/q in the initial consonant of the root for "frost" is common.[17] There is a chilling definition in Spalding (1998:39), where the plural form of frost *kanit* is defined as "great caverns in the sea ice formed by the collection of soft slush ice in the fall that, when frozen in winter, separates, forming deep caverns; these in turn fill up with hoar frost (*kanit*) and the surface is covered with snow, giving the appearance of solid ice – very dangerous for sled drivers and sea ice hunters."

The word *sikuqaq* "new sea ice, a few days old" in Laidler and Ikummaq (2008:131) and Laidler and Elee (2008:55) would not seem to be a term but a shortened version of the phrase *sikuqaqtuq* "There is (or it has) ice." It is not found in either Spalding (1998), Andersen et al. (2007), Jeddore (1976), or Utkuhiksalingmiutitut. It is possible that this is a neologism, but this needs to be investigated more. It might also be some form of *sikuaq* (see above). Interestingly, Aporta (2003) has *sikutuqaq* as "multi-year ice" in Igloolik, and Gearheard et al. (in progress) also have the word *sikutuqaq* in Clyde River. Utkuhiksalingmiutitut does not have a variant of this word, but does have the morpheme *-tuqaq* in the word *iituqaq* "yes, a long time ago," so *-tuqaq* seems to refer to something that is long standing or that took place a long time ago. This is verified in Labrador, as Smith (1978:109) gives the morpheme *-tuKak* "a long time, old" as in *nunalituKak* "one who has had land a long time."

Spalding (1998) has some terms not seen in other Central or Eastern Inuktitut lists. There is *qaaptiniviniq* (p. 105) or "ice formed from seeping or bubbling water over cracks of old sea ice; white chalk-colored ice formed from water seeping up over old ice." Spalding gives *qaaptiniq* as the process "water welling through cracks in the sea ice." A related word is found in Utkuhiksalingmiutitut: *qaaptinniq* "white ice that results when water bubbles up through ice (either through a crack or when a fishing hole is dug), and floods snow on top of ice, which (snow) then freezes." In this instance, the simpler word is the process, not the result. It is possible that these words relate to *qaa-* "top, surface" and contain a third-person possessive in the locative.

Some Considerations on Methodologies and Disciplines

The majority of ice terms have until recently been collected either by anthropologists, geographers or by Inuit people themselves. Linguists are less likely to focus on specific sets of terminology. More Inuit are co-authoring with other researchers to make these collections, as in the work by Gaidler and her partners (Laidler and Elee 2008; Laidler and Ikummaq 2008; Laidler et al. 2008; Chapter 14 by Krupnik and Weyapuk, this volume), Gearheard (Chapter 11 by Huntington et al. this volume), and McDonald et al. (1997). Most of the Inuit side of the co-authored research in published forms seems to be the words themselves and their definitions, often in alphabetical order (i.e., Oozeva et al. 2004). We do not see much in the text in any of these contributions that was clearly written from an Inuk perspective, other than definitions or quotes. I am curious about how Inuit view these lists, whether they have noticed some of the issues discussed above, and whether they would further refine the discussion of the words.[18] Recall that words are often sentences in Inuktitut so the status of word between the two languages is quite different. It is unlikely that Inuit would collect lists of sentences in English.

We get a sense of the differences between anthropology and geography as fields as we examine the data, thus confirming the suspicion that a researcher's background can influence the collection of words. Geographers seem to start with specific geophysical distinctions in mind (such as ice types and processes), while anthropologists collect terms in the context of hunting, fishing, and travel. Inuit lexicographers Jeddore (1976) and Andersen et al. (2007) do not seem to have paid any special attention to ice terminology, but that might be changing as the specialized work on ice gains more attention.

An increase in collaboration across disciplines, perhaps producing co-authored work, would be welcomed. Evans (2009:111) in particular states clearly that in language documentation, specialists from different areas should be involved in close cooperation (as is the goal of this book). He considers interdisciplinary work "essential." [19] He cites a story told in Diamond (1991), who describes an instance in Papua New Guinea, where the experienced anthropologist Ralph Bulmer was initially told there was only one word for rocks in the Kalam language. The anthropologist reports that the following year, a geologist was given a long list of names of different rocks only because the geologist's questions showed that he knew about rocks. This kind of experience can happen even within a single discipline.[20]

The collection of ice knowledge is a relatively new area. Collaboration and communication are essential. New tools will likely be developed. I would be interested in hearing first-hand ice accounts from Inuit, where contexts and actions are described in detail. Video may be a useful way of preserving and sharing such knowledge. It is an exciting area in which we can all learn from each other.

Acknowledgments Many thanks to those who shared their Inuktitut sea ice lists with me. This includes Gita J. Laidler, Shari Gearheard, Claudio Aporta, and Paul Pigott. I especially thank my colleagues Jean Briggs and Conor Cook. This chapter would not exist without the work of all these researchers. I am also grateful to our editor Igor Krupnik for so much help, advice, and encouragement.

Notes

1. In this chapter I use the term "Inuktitut" to refer to the language and all its dialects spoken across the Canadian Arctic, knowing that some regions reject this cover term in favor of their own dialect name.
2. Even within a single dialect, it is well known that lexical differences exist, so a linguist might consider some sort of sociolinguistic study, such as those that are done to track lexical differences in English, e.g., sneakers/running shoes; soft drink/pop, etc.
3. The Utkuhiksalingmiut people originally lived in the area of the Back River in Nunavut. They moved into existing communities and now live mostly in Uqsuqtuuq (Gjoa Haven) and Qamani'tuaq (Baker Lake). All Utkuhiksalingmiut dialect data in this chapter come from joint work with Jean Briggs on the Utkuhiksalingmiutitut Dictionary Project (Briggs and Johns in progress). All the data in this project were collected by Jean Briggs through fieldwork in Utkuhik (Chantrey Inlet), Uqsuqtuuq, and Qamani'tuaq. I would like to thank Jean Briggs for allowing me to use this data here and Conor Cook for initially extracting a set of ice data from the database, from which I have selected a subset (see Appendix).
4. In the following, all Inuktitut examples are given in the orthography of the dictionary or paper in which they are written, with the exception of Jeddore (1976) and Smith (1978), which I transliterate into a modern Labrador orthography.
5. As in all languages, speakers of a language are not naturally able to provide principled analyses of morphemes without training, special skills, or experience. Many Inuktitut speakers do not perceive the individual pieces within the word, even as they use them with great expertise.
6. For the definitions of other curling terms, see http://www.curling.ca/content/GoCurling/glossary.asp
7. Possessives are much more frequent in Inuktitut than English. All the expressions of positional location with respect to an object or person are constructed with possessives, e.g., "on top of" is *qaanga* (*qaa + nga*) "its top."

8. The western possessive ending *-a* has become *-nga* in eastern dialects so that it is possible that eastern dialect speakers do not recognize the *-a* in a word as a possessive marking.
9. There is no syllable final K (q) in Labrador; we find k instead.
10. English had no word for this concept so borrowed the Russian word *polynya*, which is based on the meaning "hollow." So while Russian focuses on the hole, Inuktitut focuses on the circling or delineation.
11. We see the word exists in Clyde River in the plural *nigajutat* (Gearherd et al. in progress) and also confirmed in the plural for Igloolik *nigajutait* (Aporta 2003).
12. Spalding (1998:56) gives both *milluartuq* "it is sucked in or ingested with a series of sucks" and *miluktuq* "it is sucked in; ingested (as blood, milk, fumes, juices, etc.)."
13. Perhaps I am going too far here. We need to consult Inuktitut speakers further.
14. This term is verified by the Aporta's Igloolik list, which contains *piqalujaviniit* "ice floe that has broken off from an iceberg," where the removal of the plural morpheme *-vinniit* "former (pl.)" gives us *piqalujaq* (presumably ending in **q**).
15. We see this same morpheme in the Igloolik word *sikuvalliajuq* "the process of the ocean freezing over (freeze-up)" in Laidler and Ikummaq (2008:131). Based on the classification outlined in this chapter, we now see that this word as a satellite term, which does not require an independent entry; it is the predictably augmented form of *sikujuq* (see above).
16. I have inserted an extra **p** and **l** in the middle of this word because *-vallia-* "more and more" only becomes *-pallia-* following a consonant, which would then assimilate, giving **pp**.
17. Jean Briggs (p.c.) confirms this to be true. Another example she gives is "snow tongues" (wind-formed ridges), which is *qahuq&ait* in Utkuhiksalingmiutitut and *kahuqsait* in Qamani'tuarmiutitut.
18. Jay Arnakak is working on a dictionary where reduction of word entries is given considerable thought.
19. In fact, Briggs and Johns (in progress) involve a fruitful collaboration between an anthropologist, fluent in Inuktitut, and an Inuktitut linguistics specialist.
20. I had been unsuccessful in eliciting third-person agent ergative constructions in South Baffin for a number of years, when a colleague, Peter Hallman, started asking for them as the second (not first) sentence in a consecutive series. All of sudden, the constructions were fine.

Appendix: A Sample Set of Utkuhiksalingmiutitut Ice Words

These words come from the database compiled by Jean Briggs. I have ordered them alphabetically, except that I have placed words beginning with **h** in the position of **s**, since they correspond to **s** words in other dialects (1) and I have placed words beginning with **ř** in the position of **j** since they correspond to **j** in other dialects (2) (From Briggs and Johns in progress).

aaqluqtittuq a piece of ice that has been stuck to the shore, comes unstuck from the land under the water, separates from the land-bottom, and rises up at one end more than at the other. cf. **aaqluqlutit** raise your head/face up

ainniq a crack in sea ice, narrow in the fall but in the spring it is open, sometimes 1 to 2 ft wide; contains water and is dangerous. They widen as the ice deepens.

akluaq a hole in the ice for fishing.

atuarut (1) crack that runs between shorefast ice and ice that is not attached to the shore, parallel to the shore; (2) cracks that run as (approx.) right angles to (1), starting from a rock that is underwater but attached to the shore. These *atuarutit* are rich in fish, good places for fishing holes. They exist all winter and are not dangerous. It is a specific instance of a *qu'niq* or general word for crack.

iřitittuq thin ice weighted down by snow or many fish piled beside fishing hole so that has sunk; and surface (if it's by a fishing hole) has become water-covered. c.f. **iřiqtut.** They are hiding.

ikiqtiniq the channel of water that is formed between shore and sea/lake/river ice after the ice has loosened and floated up from the bottom of the water in spring.

ikkalruq a place in the sea where ice rests on top of an underwater "hill" (stays resting on the "hill") when the surrounding water level has dropped. cf. **ikkattuq** it is shallow (water or sleep).

illaurat plural. *illauraq* singular. Vertical ice needles that result from (or that constitute the ice surface, on both sea and lake, when) water has drained off in spring.

ipřuaqtittuq the ice has gotten thick; said when the ice is about 2 ft thick. Exclamatory: *ipřuaqtilla&ranguřaqquq* The ice is really getting thick! c.f. *ipřutaq* thick (cloth, pile of papers, etc.)

ivulaaqtuq distant roaring sound of river ice breaking up in the spring; can also be applied to any distant roaring noise, such as an airplane motor.

kaanniq a place where ice has detached and risen up from the river or sea bottom, i.e., from the land under the ice.

kipuktitaqtut plural. Many pieces of ice have run under/over each other from opposite directions to create layered ice. cf. **kipuktut** they pass, coming from opposite direction without seeing or taking note of each other (if they do see each other).

kuaha ice with no snow on top (i.e., slippery).

kuřřiniq a concavity/depression in sea ice in vicinity of a seal hole or crack. They can be wide, long, winding, and (after a time) as deep as 3 ft. Water stands in them and drains into the hole or crack.

maniillat plural. Uneven ice, forced up and broken by pressure. cf. *maniituq* it (surface) is rough or uneven

nataaq a thin underlayer of ice between two layers of water in river; from top down: ice – water – ice [nataaq] – water – river bottom

piqulajaq iceberg.

puktaaq a piece of flat drifting ice used as a raft; people can fish from it.

qaaptinniq white ice that results when water bubbles up through ice (either through a crack or when a fishing hole is dug), and floods snow on top of ice, and the snow then freezes.

qaimnguq New-forming ice at edge of river/lake; ice that forms on top of shore rocks and on shore, as a result of tides. This ice forms in early fall and is uneven and bumpy; it forms only on seashore and not in Chantrey (which lacks tides). It remains attached to shore in spring when the rest of the ice floats out to sea.

haaviliqtuq the ice is moving away from the land (in spring) or has completed moving away from land. cf. **haavittuq** the (food – and/or other object) is put out in full view/central position.

hakliq thin autumn ice. cf. **hakliqtaq** a thin piece of board (e.g., plywood); also a thin braid; plywood.

hiku ice; glasses; watch face; lantern globe.

uiguaqtuq Long thin strips of new, very thin ice form on the surface of water that is just beginning to freeze – so thin they look like calm water (when the water surface is wind-ruffled). cf. **uiguřut** several pieces (of something) have been laid end to end to lengthen something. The Netsilik equivalent to this word is **qimiraqhiřuq**.

References

Andersen, A., Kalleo, W., and Watts B. (eds.) 2007. Labradorimi Ulinnaisigutet: An Inuktitut-English Dictionary of Northern Labrador Dialect. Nain, Nunatsiavut: Torngâsok Cultural Centre.

Aporta, C. 2003. Old Routes, New Trails: Contemporary Inuit Travel And Orienting In Igloolik, Nunavut. Ph.D. thesis, University of Alberta.

Briggs, J. and Johns, A. (in progress) *Utkuhiksalingmiutitut Dictionary: A Dictionary and a Postbase Dictionary*. http://www.chass.utoronto.ca/~inuit/UIDP/index.html

Diamond, J. 1991. Interview techniques in ethnobotany. In *Man and a Half: Essays in Pacific Anthropology and Ethnobiology in Honour of Ralph Bulmer*. A.M. Pawley (ed.), Auckland: The Polynesian Society, pp. 83–86.

Evans, N. 2009. *Dying Words: Endangered Languages and What They Have to Tell Us*. Malden Wiley and Blackwell.

Gearheard, S. et al. (in progress) *Clyde River Ice Terms*.

Jeddore, R. 1976. Labrador Inuit uqausingit. Nain: Labrador Inuit Committee on Literacy.

Kaplan, L. 2003. Inuit snow terms: How many and what does it mean? In *Building Capacity in Arctic Societies: Dynamics and shifting perspectives*. Proceedings from the 2nd IPSSAS Seminar. Iqaluit, Nunavut, Canada: May 26 – June 6, 2003. François Trudel. (ed.), Montréal: Université Laval, see also http://www.uaf.edu/anlc/snow.html.

Laidler, G.J., Dialla, A., and Joamie, E. 2008. Human geographies of sea ice: Freeze/thaw processes around Pangnirtung, Nunavut, Canada. *Polar Record* 44(231): 335–361.

Laidler, G.J. and Elee, P. 2008. Human geographies of sea ice: Freeze/thaw processes around Cape Dorset, Nunavut, Canada. *Polar Record* 44(228): 51–76.

Laidler, G.J. and Ikummaq, T. 2008. Human geographies of sea ice: Freeze/thaw processes around Igloolik, Nunavut, Canada. *Polar Record* 44(229): 127–153.

McDonald, M., Arragutainaq, L., and Novalinga, Z. comps. 1997. *Voices from the Bay. Traditional Ecological Knowledge of Inuit and Cree in the Hudson Bay Bioregion*. Ottawa: Canadian Arctic Resources Committee.

Oozeva, C., Noongwook, C., Noongwook, G., Alowa, C., and Igor, K. 2004. *Watching Ice and Weather Our Way.* (*Sikumengllu Eslamengllu Esghapalleghput*). Washington, DC: Arctic Studies Center, Smithsonian Institution.

Smith, L.R. 1978. A survey of the derivational postbases of Labrador Inuttut (Eskimo). *National Museum of Man Mercury Series*.

Spalding, A. 1998. *Inuktitut: A Multi-dialect Outline Dictionary*. Iqaluit: Nunavut Arctic College.

Woodbury, A. 1991. Counting Eskimo Words for Snow: A Citizen's Guide. *Linguist List* http://linguistlist.org/issues/5/5-1239.html.

Chapter 18
Two Greenlandic Sea Ice Lists and Some Considerations Regarding Inuit Sea Ice Terms

Nicole Tersis and Pierre Taverniers

Abstract The analysis of a set of Inuit words associated with the "ice," collected in west and east Greenland (Kalaallisut and Tunumiisut), shows how polysynthesis works in lexical morphology. Many lexical items can be easily segmented, and a single root can be used as a base for a number of lexical items with a variety of senses. These words often take the form of an explanatory and/or descriptive comment on the reality referred to. They frequently express some impression or certainty deriving from the observation of nature. They may also convey something of the functions and attributes of their referents.

Keywords Sea ice · Inuit · Polysynthesis · Greenlandic · Kalaallisut · Tunumiisut

The following two lists of the Greenlandic Inuit sea ice terms are the result of field research in Greenland, and they do not pretend in any way to be exhaustive. The first list relates to the language of west Greenland, spoken by approximately 52,000 people, and recognized since 1979 as the official language of Greenland under the name of *Kalaallisut* (Berthelsen et al. 2004, Sadock 2003). The version presented here was recorded in the community of Qeqertaq in the Disko Bay area of northwest Greenland (see Chapter 2 by Taverniers this volume) and it reflects what is called the "northwest Greenlandic" subdialect of the *Kalaallisut* language (Dorais 2003:136). The second list presents the terms of the language of east Greenland, or *tunumiisut*, spoken by approximately 3,500 people in the municipalities of Ammassalik and Ittoqqortoormiit (Gessain et al. 1986, Robbe and Dorais 1986, Victor and Robert-Lamblin 1989, 1993) (see Fig. 18.1). These two languages are part of a continuum of Inuit dialects that extends from northern Alaska to Greenland, spoken by approximately 80,000 people. This variety of dialects belongs to the linguistic family known as *Eskaleut* (Eskimo-Aleut).

All of the dialectal forms of this language continuum belong to the polysynthetic type, in the sense that a unique word base (verb or noun) is followed by a large number of morphemes (Mahieu and Tersis 2009). This results in several characteristics

N. Tersis (✉)
CELIA, National Center for Scientific Research (CNRS), 94800 Villejuif, France
e-mail: tersis@vjf.cnrs.fr

I. Krupnik et al. (eds.), *SIKU: Knowing Our Ice*,
DOI 10.1007/978-90-481-8587-0_18,

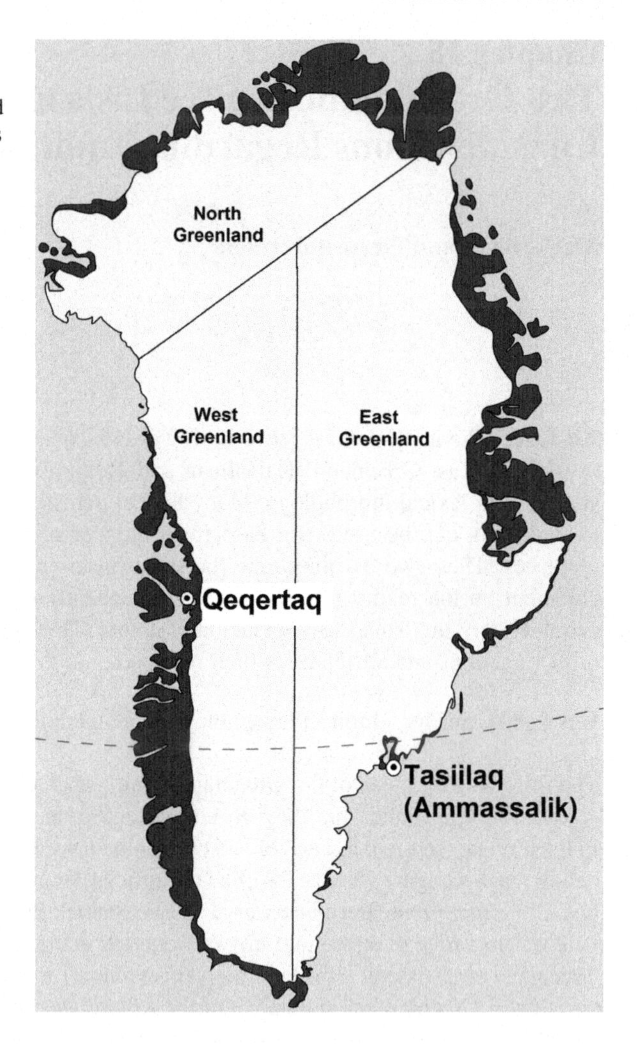

Fig. 18.1 Map of Greenland showing location of the communities of Qeqertaq and Tasiilaq, in which local terms for sea ice have been collected (Prepared by Pierre Taverniers)

that we find in the Inuit (Eskimo) words for sea ice: the roots that stand alone are limited in a synchronic analysis, while there is a large number of derived forms (see also Chapter 17 by Johns this volume). The same base word can also bring up a series of more or less fossilized forms, as the examples derived from the base word siku, ice, show. We could even build other terms, which shows that some lexical formations are really open lists: *sikkiar-poq*: [he, she] goes to a place where there is sea ice; *sikorip-poq*: sea ice is good, *sikorlup-poq*: sea ice is bad (thin and unsafe), etc.

Inuit sea ice words are explanatory and descriptive commentaries to the physical reality that they name, that is, to the many types of ice, ice processes, and associated phenomena. They often reveal a perception or knowledge that relates to people's observations of the environment, often for many generations. They can equally refer to the function or the description of the object or entity described in the term.

Whenever possible, we found it essential to provide an analysis of the components of derived terms, as well as the translation of each unit. Some interpretations of the components of terms that are more hypothetical are followed by a question mark. Differences that may exist between the final form of the word and the analyzed elements are due to morphophonological rules (elisions, vowel, or consonant assimilations). This way of presenting the material allows us to demonstrate the transparence and expressiveness of numerous Greenlandic sea ice terms.

References to the *Comparative Eskimo Dictionary* of proto-forms (Fortescue et al. 1994) have been introduced in certain cases where there are documented differences between the word in Greenlandic and other dialects. The lexicon in *Kalaallisut* adopts the official Greenlandic orthography, while the lexicon in *Tunumiisut* is rendered phonologically because the orthography of this language is not fixed. In addition, verbs are presented in their indicative form (*-poq*/*-voq*) in *Kalaallisut* (cf. *aagup-poq*), and in their radical form in *Tunumiisut*, followed by a dash (*siki*-freeze).

Abbreviations

3sg	third-person singular
3sg.3sg	third-person singular subject acting on second-person singular object
(–)	verb and noun
C	indeterminate consonant
ind	indicative mood
pl	plural
V	indeterminate vowel

Qeqertaq Sea Ice Dictionary (2008) – Kalaallisut (Western Greenlandic)

This list was compiled in 2008 in the community of Qeqertaq by Pierre Taverniers and his local collaborators, Aka Tobiassen, Thora Tobiassen, Zakkak Tobiassen (see Chapter 2 by Taverniers this volume). It has been revised and analyzed by Nicole Tersis in collaboration with Naja Frederikke Trondhjem (Trondhjem 2009), from the University of Copenhagen (see Acknowledgments).

aaguppoq: the hole in sea ice enlarges.
/aap-gup-poq melt-tend to-ind.3sg/

aakkarneq: melted sea ice (by current or weather action).
/aap-kkar-neq melt-causative- abstract participium/

aakkarpoq: [sea ice] melts (by current or weather action).
/aap-kkar-poq melt-causative-ind.3sg/

aallaaniagaq: there is an animal (seal or sea mammal) in a hole in sea ice.
/aallaa-niar-gaq shoot -will-passive participium/ *one, which should be shoot*

aanneq: fracture (any break or rupture through fast ice or a single floe resulting from deformation processes).
  /aap-neq melt-abstract participium/

aannersaq: small polynya (polynya: any non-linear shaped opening enclosed in ice).
  /aap-ner-saq melt-abstract participium-passive participium/ *which are melted*

allu: seal breathing hole in sea ice.

allualiorpoq: [he, she] builds an **alluaq** (fishing hole chipped out in sea ice).
  /alluaq-lior-poq fishing hole-make-ind.3sg/

alluaq: fishing hole chipped out in sea ice.
  /allu-aq seal breathing hole-alike/

allusiorpoq: [he, she] looks for an **allu** (seal breathing hole in sea ice).
  /allu-sior-poq breathing hole-look for-ind.3sg/

ammalataq: hole in sea ice built by seal or sea mammal.
  /amma-la-taq be open-little?-passive participium/

ammavoq: there is a passage-way through sea ice which is navigable by boat.
  /amma-voq be open (state)-ind.3sg/

appakarpoq: [a boat] moves through sea ice.
  /appakar-poq come through (often a narrow pass)-ind.3sg/

aqqartarfik: passage from land to fast ice for sled (through icefoot and tide crack).
  /aqqar-tar-fik go down- habitual-place/

aserorterpaa: [a boat or an iceberg] breaks it [sea ice].
  /aseror-ter-paa be broken-gradually-ind.3sg.3sg/

ilarpaa: [he, she] removes ice from a fishing hole chipped out in sea ice with an **ilaat** (bailer).
  /ilar-paa remove ice-ind.3sg.3sg/

ilaat: bailer used to remove ice from a fishing hole chipped out in sea ice.
  /ilar-ut remove ice-mean for doing/

iluliaq: iceberg (a massive piece of ice, more than 5 m above sea level, which has broken away from a glacier); when surrounded by solid fast ice, icebergs are climbed and used as a promontory to watch sea ice (to hunt and travel), also used to provide fresh water. At spring, when sea ice becomes less solid, iceberg moved by current break sea ice and open leads.
ilulissap eqqaa: area around the iceberg (where sea ice is often thin and unsafe, or deformed or hummocked).
  /ilulissa-p eqqaa iceberg-genitive around.3sg/

iluliusaq: bergy bit (a large piece of floating glacier ice, generally showing less than 5 m above sea level but more than 1 m and normally about 100–300 sq. m in area). The area around a bergy bit is a good place to set net under sea ice to catch seals.
  /iluliaq-usaq iceberg-look like/

imarnersaq: puddle (an accumulation on ice of melt-water, mainly due to melting snow, but in the more advanced stages also to the melting of ice). Refers to **imaq**: sea.
/imar-ner-saq sea-abstract participium-passive participium/

imarorpoq: sea becomes ice free. Refers to **imaq**: sea.
/imaq-ror-poq sea-become-ind.3sg/

inguneq: ridge (a line or wall of broken ice forced up by pressure).
/ingu-neq form pressure ridges-abstract participium/

inguvoq: there is a pressure process by which sea ice is deformed.
/ingu-voq form pressure ridges-ind.3sg/

itivippoq: [he, she] travels by land and sea ice.
/itivip-poq pass over-ind.3sg/

kassoq: translucent piece of floating glacier ice (melted water which has re-frozen in a glacier's crack).

kikkuleq: seal hole in sea ice.

manerak: flat area on sea ice.
/manip-rak be flat-area/
Cf. Proto Eskimo *qar "area or part (in direction)" (Fortescue et al. 1994:421).

maniillat: uneven ice or deformed ice (ice which has been squeezed together and in places forced upwards; ridged ice or hummocked ice).
/manip-ip-lat be flat-negative-passive participium.pl/

maniippoq: [land, sea ice] is not flat.
/manip-ip-poq be flat-negative-ind.3sg/

manippoq: [land, sea ice] is flat.
/manip-poq be flat-ind.3sg/

naggutit: small pieces of sea ice.
/naggur-tit break into pieces-participium.pl/

nallorpoq: [sea ice, fast ice] breaks up in many pieces.
/nallor-poq break (in several places)-ind.3sg/

napasoq: vertical chunk of ice installed close to a fishing hole chipped out in sea ice used to fix a fishing line and also to built a placemark.
/napa-soq be standing-participium/

nilat: small pieces of glacier ice; collected to provide fresh water. When included in sea ice they make uneven ice.

nilattarpoq: [he, she] collects **nilat** (on the shore or on the sea when there is no fast ice).
/nilat-tar-poq pieces of glacier ice-collect-ind.3sg/

nivinngarpaa: [he, she] sets a net under sea ice.
/nivinngar-paa hang up-ind.3sg.3sg/

nutarneq: recently formed sea ice.
/nutaar-neq new- abstract participium/

paarmuliaq: seal which crawls on sea ice.
/paarmur-liaq crawl- made thing/

puttaaq: small piece of floating sea ice. Usually ice cake (any relatively flat piece of sea ice less than 20 m across) or small ice cake (an ice cake less than 2 m across). Cf. **putta**- be afloat

puttaarpoq: [he, she] jumps from a **puttaaq** to an other.
/puttaar-poq jump from a piece of floating sea ice to another-ind.3sg/

puttaqut: float made with a wood piece fixed on the long-handled (wood) ice chipping tool to set net under sea ice.
/putta-qut be afloat-instrument/

puttineq: melting snow area on sea ice.
Cf. pugtípoq, puvfípoq, Water oozes up through the ice, forming a king of slush with the snow on it, Schultz-Lorentzen 1927:197.
Cf. **pui**- swell

puttippoq: sea ice surface becomes wet (melting snow).
/pui-tip?**-poq** rise to surface-cause?-ind.3sg/

qaanngorsiorneq: a way onto the **qaanngoq** (icefoot).
/qaanngor-sior-neq icefoot-look for-abstract participium/

qaanngoq: icefoot (a narrow fringe of ice attached to the coast, unmoved by tides and remaining after the fast ice has moved away).

qaannguеruppoq: the **qaanngoq** (icefoot) breaks up (usually in spring, after have been submerged by high tides).
/qaanngoq-erup-poq icefoot-have no more-ind.3sg/

qaannguniрpoq: the **qaanngoq** (icefoot) forms.
/qaanngoq-nip-poq icefoot-get-ind.3sg/

qaatersuarpoq: [a person or a dog] falls through sea ice.
/qaaser-ter-rsuar-poq? be wet-gradually-much-ind.3sg/

qaatsinneq: wet snow area or flooded ice area or re-frozen area along the tide crack (crack at the line of junction between an immovable icefoot and fast ice, the latter subject to rise and fall of the tide).
Cf. Proto-Inuit *qaaptet- "overflow (water over ice)" in connection with *qaa "top or surface of s.th" (Fortescue et al. 1994:274).

qaattarpoq: [he, she] places a net under sea ice.
/qaattar-poq hunt with a net-ind.3sg/
In some dialect (Upernavik) **qaattar**- "catch a seal by net"

qinuuvoq: there is slush (snow that is saturated and mixed with water on land or ice surfaces, or as a viscous floating mass in water after a heavy snowfall).
/qinoq-u-voq snow on the ice-be-ind.3sg/

qitulligarpoq: [sea ice surface] has an undulation (aged ridge: ridge which has undergone considerable weathering. These ridges are best described as undulations).
/qitulligar-poq bend when someone treads on it-ind.3sg/

quasaq: slippery ice (often bare ice).

quasaliarpoq: [he, she] goes to slippery ice.
/quasaq-liar-poq slippery ice-go to-ind.3sg/

quasasiorpoq: [he, she] is on slippery ice.
/quasaq-sior-poq slippery ice-experience-ind.3sg/

sarfaq: current. Current under sea ice thins the sea ice and makes it dangerous.

sassat: sea mammals trapped by ice (when sea ice forms) and who have only a small hole or polynia to breath.

seersinneq: sea ice (and river ice) formed in sheets or thin layers.
/seer-sip-neq sizzle-cause?-abstract participium/

seersippoq: sea ice (and river ice) forms in sheets or thin layers.
/seer-sip-poq sizzle-cause?-ind.3sg/
Cf. **seersoq** something that flashes; willow, Schultz-Lorentzen 1927:212.

sermiisorpoq: [he, she] collects blocks (to provide fresh water) on an iceberg (only if iceberg is surrounded by solid fast ice).
/sermeq-isor-poq ice block-fetch-ind.3sg/

sequmippoq: [sea ice] is broken (by waves) in a multitude of small pieces.
/sequmit-poq break into pieces-ind.3sg/

sequmissimavoq: fast ice is broken up.
/sequmit-sima-voq break into pieces-perfective-ind.3sg/

sikkiarpoq: [he, she] goes to a place where there is sea ice.
/siku-liar-poq ice-go to-ind.3sg/

sikorippoq: sea ice is good (more than 20 cm in thickness).
/siku-gip-poq ice-be good-ind.3sg/

sikorluppoq: sea ice is bad (thin and unsafe).
/siku-rlup-poq sea ice-have a bad-ind.3sg/

sikkuppaa: [a boat] is beset (surrounded by ice and unable to move).
/siku.up-paa ice.with it-ind.3sg3sg/

sikkussaavoq: it is trapped by ice.
/siku.up-saq.u-voq ice.with it-passive participium.be-ind.3sg/

sikkutak: thick ice cake (any relatively flat piece of sea ice less than 20 m across) or thick small floe (any relatively flat piece of sea ice 20–100 m across).

sikutaq: ice cake (any relatively flat piece of sea ice less than 20 m across).
/siku-taq ice-piece of/

sikoqasaarpoq: there are floes (or floating glacier ice) everywhere.
/siku-qa-saar-poq ice-have-a few here and there-ind.3sg/

sikulaaq: recently formed sea ice.
/siku-laaq ice-new/

sikoqannginnersaq: ice free area inside pack ice after fast ice break up.
/siku-qa-nngin-ner-saq ice-have-negative-abstract participium-passive participium/

sikorsuarsiorpoq: [a boat] navigates through ice (sea ice or floating glacier ice).
/siku-rsuar-sior-poq ice-big-move about in-ind.3sg/

sikorsuit: thick sea ice (in an isfjord where sea surface temperature is cold because of floating glacier ice and where sea ice forms early and becomes thick).
/siku-rsuit ice-big.pl/

siku: ice (sea ice, river ice, lake ice, ice-cream)

sikuarllaaq: dark nilas (a thin elastic crust of ice) few mm thick.
/sikuar-llaaq dark nilas-newly/

sikuaq: dark nilas (a thin elastic crust of ice) 1–5 cm thick.

sikuarpoq: **sikuaq** forms.
/sikuar-poq form nilas-ind.3sg/

sikuerpoq: sea ice breaks up and goes away.
/siku-er-poq ice-remove-ind.3sg/

sikueruppoq: sea is ice-free.
/siku-erup-poq ice-have no more-ind.3sg/

sikuiuippoq: there is sea ice "which doesn't melt": old ice (sea ice which has survived at least one summer's melt).
/siku-er-juip-poq ice-remove-never-ind.3sg/

sikuiuitsoq: sea ice "which doesn't melt": old ice (sea ice which has survived at least one summer's melt).
/siku-er-juit-soq ice-remove-never-participium/

sikujartuaarpoq: sea ice forms and becomes thicker (on several days).
/siku-jartuaar-poq form ice-gradually-ind.3sg/

sikujumaataarpoq: sea ice delays in forming.
/siku-jumaataar-poq form ice-a long time to get-ind.3sg/

sikujuippoq: there is no more sea ice.
/siku-juip-poq form ice-never-ind.3sg/

sikuliarpoq: [he, she] goes to the fast-ice.
/siku-liar-poq ice-go to-ind.3sg/

sikuliaq: recently formed sea ice.
/siku-liaq ice-made/

sikumukarpoq: [he, she] goes to the fast-ice.
/siku-mu-kar-poq ice-allative-go to-ind.3sg/

sikumiippoq: [he, she, it] is on the fast-ice.
/siku-mi-ip-poq ice-locative-be-ind.3sg/

sikunippoq: floating pieces of sea ice arrive (moved by wind or current) when the sea was ice free.
/siku-nip-poq ice-get-ind.3sg/

sikusiorpoq: [boat] moves through sea ice.
/siku-sior-poq ice-move about in-ind.3sg/

sikut (pl): drifting ice (pieces of sea ice or small floating glacier ice).

sikutannguеruppoq: there is no ice at all on the sea.
/siku-taq-nguaq-erup-poq ice-piece of-little-have no more-ind.3sg/

sikuvoq: fast ice is formed.
/siku-voq form ice-ind.3sg/

sinaaliarpoq: [he, she] reaches the **sinaaq** by travelling on the fast ice.
/sinaaq-liar-poq ice edge-go to-ind.3sg/

sinaalippoq: [boat] reaches the **sinaaq**.
/sinaaq-lip-poq ice edge-reach-ind.3sg/

sinaaniippoq: [he, she] is at the **sinaaq**.
/sinaaq-ni-ip-poq ice edge-locative-be-ind.3sg/

sinaaq: fast ice edge (the demarcation between fast ice and open water).

sinaasiorpoq: [he, she] is walking or travelling along the **sinaaq**.
/sinaaq-sior-poq ice edge-move about in-ind.3sg/

sinaasserpoq: the **sinaaq** forms after break up.
/sinaaq-ser-poq ice edge-move along-ind.3sg/

tiggunnerit: small pieces of ice (sea ice or floating glacier ice) compacted by wind or current.
/tiggut-ner-it split- abstract participium-pl/

toorpaa: [he, she] chips ice (sea ice or glacier ice) with a **tooq**.
/toor-paa chip ice with a tool-ind.3sg.3sg/

tooq: long-handled (wood) ice chipping tool; to chip an ice fishing hole on fast ice; to test sea ice strength when walking on an area which can be dangerous; to set a net: the **tooq** is pushed under sea ice from a fishing hole to an other (using **puttaqut**); to collect blocks from an iceberg to provide fresh water.

uisaavoq: [he, she] goes away on a floe when fast ice breaks up.
/uiar-saq-u-voq go around headland-passive participium-be-ind.3sg/

umiarluppoq: [he, she] uses a small ice cake as a boat (moves by paddling). Refers to **umiaq**: (boat); to reach a seal which was shot from the fast ice edge (using a rifle as a paddle).
/umiar-rlup-poq boat-have a bad-ind.3sg/

unerraq: seal's track on sea ice.
/uniar-gaq? drag-passive participium?/

unneraarsuppaluk: grate sound or creak from sea ice moving up and down along the icefoot (by wave or tide action).
/unneraarsuk–paluk the spirit of the beach-sound of/

uukkaappaa: fast ice edge breaks up.
/uukkar-up-paa calve-for him-ind.3sg.3sg/

uuttoq: seal which basks in the sun on fast ice.
/uut-toq bake-participium/ *one who bakes himself on the ice in the sun*

uuttorniaq: hunter which hunts **uuttoq**.
/uuttoq-niar seal-hunt/

Tunumiisut (Eastern Greenlandic)

The Tunumiisut vocabulary of terms related to "ice" was collected in the course of a wider study of the structure of the Tunumiisut lexicon (Tersis 2008). This study was conducted between 1990 and 2000 in collaboration with speakers from Tasiilaq in eastern Greenland. Among the most important contributors were Kathrine Svanholm, Isais Kuitse, Elisa Maqi, and Marie Otuaq.

aakkaqniq: melted sea ice (which becomes soft and dark, hence rotten).
/aaC-kkaq-niq melt-causative-abstract notion/

aaqniaat: harpoon for hunting seal on ice
/aaq-niaq-Vt? crawl-intention-means of/

ammatitaq: hole in sea ice built by narwhal or sea mammal
Cf. **amma**- be open

aniq: gap in sea ice or channel ice

aniqsiq: floating sheet of ice broken off from ice floe and coming out of fjord
Cf. **ani**- go out

apusiiq: inlandsis, ice cap

aputtiq: snow bridge hiding crevasses in melting ice
Cf. **aput** snow on ground

attiq: seal's breathing hole in ice

attisuut: harpoon for hunting seal at breathing hole
/attiq-qsuq-Vt hole-use-means of/

ayaappiaq: icepick
Cf. **ayaC** push

iimaataq: part of harpoon formerly used for hunting seal on ice
/iimaq-Vtaq foreshaft of harpoon- look like/

imaaq-: be ice-free (sea)
Cf. **imaq** contents, sea

imaayuk: mixture of sea water and melted ice, treacle, molasses
Cf. **imaq** contents, sea

ima-qa-nngitaq: there is no sea (only ice)
/sea-have-negative.ind.3sg/

ima-qqi-qaaq: sea is ice free, calm
/imaq-qqiC-qaa-wuq sea-do well-intensive-ind.3sg/

imaqniqsaq: unfrozen part of sea, opening in ice
Cf. **imaq** contents, sea

ingiiNiaq: three-footed stool used by hunter at breathing hole
/ingiiq-ngiaq thrust (hips forward)-place/

issinniq: pieces of overlapping sea ice on shore
/issiC-niq lean forward- abstract notion/
Cf. **iiC**- sit down

issiwaqtiit: harpoon for hunting seal through hole in ice
Cf. Proto Eskimo: *iyyuR - "poke head out for a look" (Fortescue et al. 1994:150)

itaaqniaat: wooden or bone ice scraper used to remove coat of ice from kayak
/itaaqniq-Vyaq-Vt hoarfrost-remove-means of/

itaaqniq: hoarfrost, rime
Cf. **itiC**- inside

ititiaq: iceberg
Cf. Proto Eskimo: *ilu "inside", *ilul(l)iRaR "s.th.inside?" (Fortescue et al. 1994: 128–129)

kattiitaq: floating pan of dark ice (small size)
/kattiq-Vtaq ice-look like/

kattiq: floating pan of fresh-water ice

kattiwat: small fragment of iceberg

kikkitiq: breathing hole in ice where seal comes up

kinniq: ice edge (where ice and sea meet), fast-ice edge

maniitsiq: glacier berg (irregularly shaped iceberg)
/maniC-nngit-siq be flat-negative-attributive/

maniittat: hummocky ice
  /maniC-nngit-tat be flat-negative-resultative.pl/

manniNaq: tabular berg (flat-topped iceberg)
Cf. **maniC**- be flat

nappat: small floes, flatter pieces of sea ice

nitak: 1. freshwater ice 2. ice coating

nutaqniq: new ice (layer of thin, pliable ice)
  /nutaaq-niq new- abstract notion/

pattingatiq: seal lying on ice
  /pattiq-nga-tiq lie down-resultative-attributive/

pukkuwik: breathing hole made in ice by seal
Cf. –**wik** place

qaaNuq: shore sea ice

qii- freeze

qiissiiaq: thing frozen, (meat, fish, etc.)
  /qii-ssiiaq freeze-left/

qinitiaq: slush ice (dense mass floating on water after snowfall)

qassimatiq: (ringed) seal up on ice
  /qassi-ma-tiq climb up on ice-state-attributive/

qiqqiniq: mass of freshwater ice encased in piece of seawater ice

qiqsaqniq: ice cover frozen after thaw

qiqsiqqaqtaq: ice cover hardened by wind

quasaq(-): 1. bare ice 2. be slippery

saqpaq(-): 1. current (of water) 2. flow quickly (current)

sassat: hole in ice used by narwhal

sikaqniq: hard surface (snow, ice, bread, etc.)
  /sikaq-niq be stiff- abstract notion/

siki-: freeze, ice up

sikiitaq(-): 1. thin ice on water, 2. skate (on ice)

sikiitaqattaaNisit: (ice) skate
  /sikiitaq-qattaaq-ngisit skate on ice-repetitive-means of.pl/

sikiq: sea ice

siki-qa-nngitaq: there is no ice (ice-free area)
/ice-have-ind.negative.3sg/

siki-qa-qaaq: there is a lot of ice (thick ice)
/ice-have-intensive.ind.3sg/

siki-wiq-puq: sea ice goes away
/ice-remove-ind.3sg/

sikitaq: gray ice (new ice that can be broken by heave)

sikiwiit: ice pack, ice floe
/sikiq-tiwiit ice-big.pl/

siqmiq: freshwater ice from mainland

siqmiqsuaq: ice cap
/siqmiq-suaq ice-big/

tuaq: pack ice, (thick) motionless sea ice, ice which has thickened for only one winter

tukkaqtiit: ice pick
/tukkaq-tiq-Vt thrust in to make hole-repetitive-means of/

tuuq: ice pick

Acknowledgments Nicole Tersis wants to thank particularly Kathrine Svanholm who was a close collaborator on the Tunumiisut lexicon from 1990 to 2000, and also Isais Kuitse, Elisa Maqi, and Marie Otuaq who were kind enough to lend their assistance during the fieldwork in Tasiilaq, Eastern Greenland. She is also grateful to Naja Trondhjem, Greenlandic linguist at the University of Copenhagen, for helping analyze words from Western Greenland. Fieldwork in Greenland and Denmark was funded by the National Center for Scientific Research-CNRS (LACITO Laboratory for Oral Tradition Languages and Civilizations and CELIA- Center for the Study of he Indigenous Languages of The Americas). Raymond Boyd (CNRS, LLACAN Laboratory for Languages and Cultures of Black Africa) was helpful with the translation of the Tunumiisut ice lexicon into English. Pierre Taverniers is grateful to Aka, Jakob, Thora, and Zacharias Tobiassen for kindly sharing their knowledge about the ice terminology, and also to Igor Krupnik and Shari Gearheard for their helpful comments.

References

Berthelsen, C., Jacobsensen, B., Petersen, R., Kleivan, I., and Rischel, J. 2004 (1977 first edition). *Oqaatsit, Kalaallisuumiit Qallunaatuumut/Grønlandsk Dansk Ordbog*, Nuuk, Ilinniusiorfik.

Dorais, L.-J. 2003. *Inuit Uqausiqatigiit. Inuit Languages and Dialects*. Iqaluit: Nunavut Arctic College.

Fortescue, M., Jacobson, S.A., and Kaplan, L., 1994. *Comparative Eskimo Dictionary with Aleut Cognates*, Fairbanks, University of Alaska, Alaska Native Language Center Press, Research Papers 9.

Gessain, R., Dorais, L.-J., and Enel, C. 1986. *Vocabulaire du Groenlandais de l'est,* Paris, Documents du Centre de Recherches Anthropologiques du Musée de l'Homme 5.

Mahieu, M.-A. and Tersis, N. (eds.), 2009. *Variations on Polysynthesis: The Eskaleut Languages*, Typological Studies in Language. Amsterdam/Philadelphia: John Benjamins.

Robbe, P. and Dorais, L.-J. 1986. *Tunumiit oraasiat/Tunumiut oqaasii/Det østgrønlandske sprog/The East Greenlandic Inuit Language/La langue inuit du Groenland de l'Est,* Québec, Université Laval, Nordicana 49.

Sadock, J. 2003. *A Grammar of Kalaallisut (West Greenlandic Inuttut)*. Languages of the World /Materials 162, Lincom Europa.

Schultz-Lorentzen, C.W. 1927 (reprint 1967). *Dictionary of The West Greenland Eskimo Language*. Copenhagen: Meddelelser om Grønland.

Tersis, N. 2008. *Forme et sens des mots du tunumiisut, lexique inuit du Groenland oriental*. Louvain, Paris: Peeters.

Trondhjem, N.F. 2009. The marking of past time in Kalaallisut, the Greenlandic language. In Mahieu, M.-A. and Tersis, N. (eds.), *Variations on Polysynthesis: The Eskaleut Languages*, Typological Studies in Language. Amsterdam/Philadelphia: John Benjamins, pp. 171–182.

Victor, P.-E. and Robert-Lamblin, J. 1989. *La civilisation du phoque. Jeux, gestes et techniques des Eskimo d'Ammassalik*, vol 1. Paris: Armand Colin-Raymond Chabaud.

Victor, P.-E. and Robert-Lamblin, J. 1993. *La civilisation du phoque. Légendes, rites et croyances des Eskimo d'Ammassalik*, vol 2. Paris: Raymond Chabaud.

Chapter 19
Partnerships in Policy: What Lessons Can We Learn from IPY SIKU?

Anne Henshaw

> *As states increasingly focus on the Arctic and its resources, and as climate change continues to create easier access to the Arctic, Inuit inclusion as active partners is central to all national and international deliberations on Arctic sovereignty and related questions, such as who owns the Arctic, who has the right to traverse the Arctic, who has the right to develop the Arctic, and who will be responsible for the social and environmental impacts increasingly facing the Arctic. We have unique knowledge and experience to bring to these deliberations. The inclusion of Inuit as active partners in all future deliberations on Arctic sovereignty will benefit both the Inuit community and the international community.*
>
> Circumpolar Inuit Declaration on Sovereignty in the Arctic 2009

Abstract The role and participation of indigenous peoples in international arctic policy matters represents a critical element of meeting the future governance challenges in the region. This chapter describes how the nature of partnerships between scientists and northern indigenous peoples can serve as a model for partnerships of a more political nature. Such partnerships, like those developed as part of IPY SIKU, increasingly have a commitment to sharing and reciprocity that is grounded by in-depth documentation of indigenous knowledge, intergenerational engagement, and investments in capacity. Using the Arctic Council and the status of Permanent Participants as a case study, the chapter examines the current challenges and opportunities in translating these kinds of partnerships into an international policy context. It argues that opportunities for political partnerships between indigenous peoples and nation-states do have the potential to grow if these core tenants are supported.

Keywords Inuit place names · Indigenous knowledge · Arctic Council · Permanent participants · Nunavut

---

A. Henshaw (✉)
Oak Foundation USA, Portland, ME 04101, USA
e-mail: anne.henshaw@oakfnd.org

I. Krupnik et al. (eds.), *SIKU: Knowing Our Ice*,
DOI 10.1007/978-90-481-8587-0_19, © Springer Science+Business Media B.V. 2010

While the rapid decline of multi-year polar sea ice is increasingly recognized as the global bellweather for a warming world, it is also a barometer for a hotly charged political climate characterized by competing claims of sovereignty among individual nation-states and the indigenous peoples who call the Arctic home. In November 2008, the Inuit Circumpolar Council (ICC) held a closed meeting in Kuujuaq, Canada, to address their growing concern over the increasing focus of the international community on the Inuit homeland known as *Inuit Nunaat*.[1] The gathering was mainly prompted in response to the claims of sovereignty made in the *Ilulisaat Declaration*[2] (May 2008) by the five coastal nation-states that border the Arctic Ocean including Canada, Denmark, Norway, the Russian Federation, and the United States. In large part, the *Ilulisaat Declaration* angered Inuit leaders because it failed to recognize pre-existing international agreements that promote and protect the rights of indigenous peoples including the important and unique role Inuit play as Permanent Participants in the Arctic Council. What the *Circumpolar Inuit Declaration on Sovereignty*,[3] the policy document stemming from Kuujuaq ultimately calls for, is the need for nation-states to recognize Inuit as full and active partners in the "protection and promotion of indigenous economies, cultures and traditions ... [and] that industrial development of the natural resource wealth of the Arctic can proceed only insofar as it enhances the economic and social well-being of Inuit and safeguards our environmental security" (*Circumpolar Inuit Declaration on Sovereignty in the Arctic* 2009).

While partnerships of a different nature make up the projects described in this volume, those between research scientists and local Inuit sea ice experts, the basic premise remains the same. Inuit have unique knowledge and experience that contribute not only to understanding Arctic environmental change; they also have their own vision on how the Arctic should be governed in the years to come. This chapter describes how the nature of partnerships between scientists and northern indigenous peoples can serve as a model for partnerships of a more political nature. Such partnerships, like those developed during IPY SIKU, increasingly have a commitment to sharing and reciprocity that is grounded by in-depth documentation of indigenous knowledge, intergenerational engagement, and investments in capacity. The chapter discusses the current challenges and opportunities in translating these kinds of partnerships into an international policy context using the Arctic Council as a case study. It argues that opportunities for political partnerships between indigenous peoples and nation-states do have the potential to grow if these core tenants are supported.

Partnerships in Science

Anthropology and other social sciences have a varied track record with respect to their relationships with indigenous groups in the Arctic. They are also fields increasingly defined by collaborative research that fosters stronger relationships between scientists and northern communities. These relationships can take a variety of forms but at their core all seek to engage and recognize indigenous experts as active partners in the discovery and documentation process. Some of these partnerships develop through formal academic channels and granting opportunities, and most recently in the form of co-authorship on publications (Krupnik et al. 2005; Brook et al. 2006; Tremblay et al. 2006; Laidler and Elee 2006; Ford et al. 2006a, b; Gearheard et al. 2006; MacDonald et al. 1997; Oozeva et al. 2004; Henshaw et al. 2007). These partnerships often form over many years, even decades, and represent an important paradigm shift in the ways science is conducted, documented, and shared with local communities and the academic community at large. Within the context of the International Polar Year (IPY) 2007–2008, the inclusion of the human-oriented research, including the documentation of indigenous knowledge, speaks to how this new paradigm is part of a larger political shift in the way science is being conceived and defined by non-state actors (Shadian 2009).

This is not to say collaborative research linking different knowledge structures is not without significant challenges (Nadasdy 1999; Gilchrist et al. 2005; Brook et al. 2006; Gearheard and Shirley 2007). However, as others have noted, its most productive applications come when there are overlapping interests and time-tested trust among participants (Huntington 1998, 2000; Huntington et al. 2002). Much of the work included in this volume speaks to the high level of commitment and dedication the social science research community has to these research ideals and the openness on the part of their indigenous counterparts. Furthermore, I would argue, underlying these ideals are three important elements key to fostering long-term cooperation and collaboration: in-depth documentation of indigenous knowledge, intergenerational engagement, and investments in capacity. Below I will discuss these key elements both in the context of my own research in Nunavut, Canada, and in the projects described in this volume.[4]

While much has been written about what constitutes indigenous knowledge as compared to western scientific knowledge (Agrawal 1995; Huntington and Fox 2005; Nadasdy 1999; Sillitoe 1998), what my own research and that contained within this volume demonstrate is that it is rich with cultural meaning, cumulative, experiential, and pragmatic, incorporating many forms of knowing. This research also shows that work committed to understanding and documenting the way communities experience the world around them requires an in-depth and culturally sensitive approach. For me, this community-oriented approach developed over the course of 20 years while conducting archaeology and oral history research in Qiqiktaaluk (Baffin Island), Nunavut. The relationships I developed, and the knowledge documented during the course of the research, ultimately came together in the context of an environmental knowledge study focused on the gathering of Inuit place names and their ecological "history" (Henshaw 2006a, b).

The Sikusilaarmiut Place-Name Project drew from the expertise of Inuit living in the present-day community of Kinngait (Cape Dorset), meaning mountains or high hill (Fig. 19.1). Kinngait is a medium-sized hamlet with approximately 1,200 residents. The region has a rich archaeological record dating back at least 3,000 years with Pre Dorset, Dorset, and Thule sites scattered across the historical landscape. The Hudson's Bay Company opened a post in the present-day community

Fig. 19.1 Kinngait, Nunavut, May 2005. Photograph by Anne Henshaw

in 1913, followed by a Roman Catholic mission that operated from 1938 to 1960 and an Royal Canadian Mounted Police attachment that opened in 1965. People began permanently settling in the community in the 1960s with the advent of government housing, health, educational programs, and the West Baffin Cooperative, which is the main distributor of soapstone carvings, prints, drawings, and engravings (Kemp 1976). Many of the eldest members of the community interviewed as part of this study grew up in outpost camps located along the coastline. They moved regularly across vast areas of the land and sea to hunt and trap animals such as ringed seals, bearded seals, walrus, beluga, caribou, and arctic fox, among others.

For Kinngaitmiut (inhabitants of Kinngait), place names are part of a rich cultural heritage that defines their history and knowledge of the land, sea, and ice. Until recently Inuit place names did not appear on any published maps, yet they are an integral part of how Inuit move through, monitor, and share stories about the world they inhabit. Place names recorded during the course of this project also marked the importance of topographic and ecological features as sensitive climatic indicators and as reference points for Inuit wayfinding in a highly dynamic environment (Henshaw 2006a; Fig. 19.2). These findings are similar to other regions including those documented in the Inuit Sea Ice Use and Occupancy Project (see Chapter 1, this volume). Today, these maps, now published through the Canadian Geological Survey not only recognize claims Inuit have to their homeland but also serve an immediate need (Henshaw et al. 2007). One of the main community drivers of the project was the local Hunter and Trapper Association, an organization that believed

Fig. 19.2 Photograph of Kinngait in July 2007 showing the main Sikusillarmiut place names surrounding the community. Photograph by Anne Henshaw

the maps would be important for search and rescue teams who increasingly are called on to locate stranded travelers (Henshaw 2006a).

During the course of this project I also recognized the central role of local interpreter- translators in building strong relationships with community members. While most researchers have some working knowledge of the distinct languages and local dialects in which they work, only a handful are fully fluent. In this context, close partnerships develop with interpreter-translators who do much more than simply translate words from one language to another; they serve as an important bridge between different ways of knowing and interacting with members of the community, most importantly with elders. For example, over the course of the Sikusilaarmiut Place-name Project, the interpreter-translator, Aksatungua Ashoona, served as much as a collaborator as a translator. She was a highly skilled interviewer and listener who truly appreciated the opportunity to learn from elders and to share that knowledge with myself and the community (Fig. 19.3). As Aksatungua states, "*Tukimuatitauniq* – 'Elders guide us' – They are like Inuksuit on the land that provide direction and give us knowledge that grows in us like seeds inside. They are the inspiration for what we know" (Henshaw and Ashoona 2006). In these project-specific examples, it is important to note both tangible resources that included real-life applications (maps used in search and rescue) and language preservation (Inuktitut place-name documentation directed by elders with the help of a skilled interpreter) and the intangible trust and dedication among participants in seeing the work through to completion were all paramount to building partnerships that enabled the project to ultimately succeed.

Fig. 19.3 Aksatungua Ashoona documenting Inuktitut place names with elders Iqalluq Nungusuituq and Sagiatu Sagiatu. Photograph by Anne Henshaw

SIKU and Other Related Community-Oriented Research

Similar approaches to collaborative research are also seen among many of the SIKU projects described in this volume especially as they relate to language preservation, intergenerational transference of knowledge, and capacity building. Just as language represents the core of Inuit toponymy, it also forms the core of how indigenous people around the Arctic understand, interpret, and move through the environment around them, including sea ice. The dictionaries stemming from the SIKU projects speak to the richness of this knowledge (Chapter 14 by Krupnik and Weyapuk this volume, Johns (Chapter 17 this volume); Tersis and Taverniers (Chapter 18 this volume)). Specifically, indigenous sea ice terminologies and nomenclature are much more than mere dictionaries of terms but represent a "high resolution" way to understand place-specific sea ice dynamics critical for safety and overall hunting success (Chapter 14 by Krupnik and Weyapuk this volume). As Holm (Chapter 6 this volume) also point out, for many Inuit communities invoking sea ice language also represents more than its physical properties and use, but embodies important cultural, emotional, and personal meanings to the people who need and use it most.

The other aspect of the SIKU research that stands as a core component of relationship building is that the in-depth documentation of knowledge carried with it real world applications. For example, the mapping of ice trails and camps undertaken by Mathew Druckenmiller and this team (Chapter 9, this volume) outside Barrow, Alaska, not only helped scientists understand the thickness distribution of sea ice but also showed how different ice types were used by the community. Such knowledge and related maps will be particularly important for communicating and understanding the safety and risk associated with the placement of whaling crews. As Druckenmiller et al. state (Chapter 9, this volume), "With input from the community and iterative improvements, these maps have evolved into a product that is useful for on-ice navigation, general ice-type description (flat ice versus rough ice), and as a reference for Barrow's Search and Rescue." More broadly, as Laidler et al. (Chapter 3, this volume) also points out, the process by which such local perspectives are incorporated into vulnerability assessments will be critical to their long-term adaptive capacity.

Aside from the technological applications and products of the various – of the various SIKU projects, the process of conducting research allowed for true substantive input from community participants. For example, the local sea ice expert working groups set up as part of the Siku-Inuit-Hila project (Chapter 11 by Huntington et al. this volume), meant that local Inuit, Inupiaq, and Inughuit experts could "document their own knowledge in their own way." Such freedom, away from the methodological boxes of academically trained researchers, allowed participants to discuss not only the physical changes and properties of sea ice but what the impacts of these changes mean to people. The products of such engagement in both written and oral form will make a lasting contribution not just for the researchers but for local communities themselves.

This kind of engagement represents another core component of community-based research, namely, the creation of partnerships that pay special attention to passing knowledge from one generation to the next. Intergenerational connections take a variety of forms including educational outreach in formal school settings as well as more experiential learning on the land and ice. Within formal school curriculum, researchers together with their Inuit partners are communicating and sharing knowledge through community-oriented publications as well as electronic multimedia. For example, many community-based place-name projects are coming to life through electronic web-based media that move the research beyond fixing names on maps. Projects like the Kitikmeot Atlas Project[5] and others supported by the Inuit Heritage Trust Place Name Program in Nunavut incorporate sound and text through geographically referenced hyperlinks using Google Earth™ and the Atanaattiaq Map Viewer.[6] In the western Arctic, Project Jukebox's Nunivak Island Place Names Project at the University of Alaska[7] has taken similar multimedia approaches. In other examples, northern youth are able to learn about Inuit wayfinding skills using a multimedia CD ROM that provides in-depth documentation and curriculums for teaching important concepts and beliefs in a K-12 setting (Aporta 2006); and another provides video, audio, and interactive maps that show Inuit elders discussing their observations of climate change (Fox 2003). The Cybercartographic Atlas of Sea Ice, a project stemming from SIKU, also represents a new means for sharing and representing knowledge using innovative technologies (Chapter 10 by Pulsifer et al. this volume).

These types of tools can only help if they are linked with more experiential forms of knowledge acquisition based on a culture rooted in oral tradition. Such efforts are beginning to take place with land-based programs like the Junior Rangers and within some educational contexts where land/sea/ice skills are increasingly being recognized as important ingredients to sustainability and resiliency amidst growing social and environmental uncertainty. In the *Nelson Island Cultural and Natural History Project*, Ann Fienup-Riordan and Alice Rearden (Chapter 13 this volume) together with her Yup'ik partners embarked on a month-long journey with elders and youth to document this knowledge "in situ" where stories of the landscape and history were shared in the actual places where things happened. A similar "in situ" approach was also used in the Siku-Inuit-Hila project where scientists and Inuit learned from each other, shared expertise, as well as new experiences together while traveling on the sea ice extensively in Alaska, Canada, and Greenland. Different generations were part of the project and the intercultural knowledge sharing in such intense settings led to the creation of new knowledge spanning age groups and cultural backgrounds (cf. Chapter 11 by Huntington et al. this volume).

Investments in capacity form another core part of relationship building in Inuit–science partnerships. Within a research context such investments can take a variety of forms depending on the nature of the project but typically involve sharing methodological approaches to knowledge documentation, technology, networking and educational program development. Core to each of these investments is access to equipment and training. For example, in Nunavik, Tremblay et al. (2006) trained Inuit to monitor ice conditions and to conduct their own interviews with experienced Inuit hunters on how routes are affected by changing ice conditions. Increasingly such research also involves the integration of new spatial technologies such as Global Positioning Systems (GPS), Satellite Imagery, and Geographic Information Systems (GIS) (Meier et al. 2006; Chapter 8 by Gearheard et al. this volume; Henshaw 2006a; Chapter 3 by Laidler et al. this volume). In the *Igliniit* Project, geomatic engineers and Inuit hunters worked together to design and implement a new integrated GPS system that was mounted on snow machines to log the location/trails of hunters, local weather conditions, and any observations recorded by hunters (Chapter 8 by Gearheard et al. this volume). Important as these new technologies are, community access to them is not typically widespread, and most are not adapted to Inuit use or language, as in the *Igliniit* Project. While initial investments by individual research projects help communities learn how these technologies can be used to document their own knowledge and inform their decision making, their long-term use and access remain questionable. Within Canada, for example, investments in science and monitoring are largely dictated by political interests and modeled after large-scale resource-intensive research stations where widespread impacts and benefits for communities are not always felt (Bravo 2009). Instead, approaches like that taken by the Siku-Inuit-Hila project, which take a more simple but robust approach to creating sea ice observation networks, are not given adequate attention or recognition (Chapter 11 by Huntington et al. this volume).

Given the opportunities and challenges inherent in Inuit science partnerships, how can we begin to take the best practices from these relationships and replicate them in the more heated terrain of policy-making? The answer is not simple and significant challenges need to be recognized before achieving substantive progress. Below I outline some of those challenges and conclude with some potential pathways forward.

Partnerships in Policy

While partnerships in science revolve around the pursuit of knowledge, partnerships in policy take on a fundamentally different character by largely focusing on the building of coalitions to achieve common goals. Such coalition building is not new to indigenous communities in the arctic (Shadian 2006), but the nature of these coalitions depend on what entities are involved and to what political end. For the purposes of this chapter, the political end point relates to the central right of indigenous peoples to self-determination. According to the *Circumpolar Inuit Declaration on the Sovereignty of the Arctic* (2009), Inuit define this concept as follows: "It is our right to freely determine our political status, freely pursue our economic, social, cultural and linguistic development, and freely dispose of our natural wealth and resources." Putting into practice the values inherent in self-determination get translated on many policy fronts from the rights to co-manage wildlife resources to economic development opportunities related to the extraction of non-renewable resources making it sometimes difficult for indigenous groups to speak with one voice on such complex issues. Partnerships function at multiple levels of governance including international regimes to nation-states and the sub-entities within them (provinces, territories, states, and regions). The intent here is not to describe how relationships operate at each of these levels, but rather to examine whether in-depth knowledge documentation, intergenerational engagement, and investments in capacity, all important to Inuit–science partnerships, are playing a role today in the relationships Inuit and other Arctic indigenous peoples are forming with nation-states at the international level through the Arctic Council.

The Arctic Council and the Role of the Permanent Participants

The Ottawa Declaration of 1996 formally established the Arctic Council, originally known as the Arctic Environmental Protection Strategy, "as a high level intergovernmental forum to provide a means for promoting cooperation, coordination and interaction among the Arctic States, with the involvement of the Arctic Indigenous communities and other Arctic inhabitants on common Arctic issues, in particular issues of sustainable development and environmental protection in the Arctic." Member States of the Arctic Council are Canada, Denmark (including Greenland and the Faroe Islands), Finland, Iceland, Norway, Russian Federation, Sweden, and the United States of America. In addition to the Member States, the Arctic Council has the category of Permanent Participants (PP). As indicated on the Arctic Council web site, "the category of Permanent Participation is created to provide for active participation of, and full consultation with, the Arctic Indigenous representatives within the Arctic Council. This principle applies to all meetings and activities of the Arctic Council."[8]

Currently there are six Permanent Participants (PP) to the Arctic Council including Aleut International Association (AIA), Arctic Athabaskan Council (AAC), Gwich'in Council International (GCI), Inuit Circumpolar Council (ICC), Saami Council, and Russian Arctic Indigenous Peoples of the North (RAIPON). The ICC,

RAIPON, and the Saami Council were admitted as Permanent Participants when the Arctic Council was created in 1996, the other PPs were admitted shortly there after in 1998 (AIA), 1999 (GCI), and 2000 (AAC). As an international body the Arctic Council is unique in the fact it recognizes indigenous peoples as legitimate non-state actors in the international policy context where opportunities for joint activities and policy coordination are abundant (Semenova 2005). Despite these opportunities, the PPs, ability to engage as active and full partners in the Arctic Council process are not without challenges. Below I relate these challenges to the fact that partnerships built on some of the same premises of those described in the science context have yet to gain the support they deserve and need in the policy realm.

In-Depth Indigenous Knowledge Documentation

Much of the work of the Arctic Council gets completed within six working groups that include the Arctic Contaminants Action Program (ACAP), Arctic Monitoring and Assessment Program (AMAP), the Conservation of Arctic Flora and Fauna (CAFF), the Emergency Prevention, Preparedness and Response (EPPR), Protection of the Arctic Marine Environment (PAME), and the Social Development Working Group (SDWG). Each of the working groups typically has its own Chair and Executive Secretary. The hallmark of these working groups are the scientific assessments and guidelines they publish to inform policy-making. The level of involvement of the Permanent Participants in these various reports and the working groups themselves varies on an individual basis depending on their own interests and capacity to participate. Oftentimes outreach with local communities, in terms of both contributing content to the reports and creating opportunities for feedback, are limited because of inadequate budgetary support for travel, staff, and research. For example, one of the best-known assessments commissioned by the Arctic Council, the Arctic Climate Impact Assessment (ACIA 2005), completed by AMAP and CAFF, in association with the International Arctic Science Committee (IASC), was the first to include substantive and in-depth documentation of indigenous knowledge (Huntington and Fox 2005). However, since the publication of the ACIA, two reports approved by the Arctic Council, the Arctic Marine Shipping Assessment (AMSA 2009), and the Arctic Offshore Oil and Gas Guidelines (2009), both the work of PAME, involved input from the Permanent Participants but did not include the same degree of in-depth knowledge documentation as the ACIA, largely due to lack of funding (Jimmy Stotts, personal communication).

So while the Arctic Council does provide space for indigenous voices to be heard on a variety of issues (with respect to both science and policy), some but not all of the PPs often lack the organizational capacity to conduct adequate outreach, research, and in-depth commentary on issues that play out at the local level. Without such support, the weight of the policy recommendations that stem from these reports will have limited traction on the ground and/or not receive the attention they deserve within communities. Additionally, these reports have the tendency to overgeneralize the complex issues indigenous communities face, particularly with regard to industrial-scale activities forecasted to take place in the years ahead in the Arctic. As I have previously argued, indigenous peoples represent a heterogenous group who regard issues like climate change as both a threat and an opportunity (Henshaw 2009). In this context, we need to differentiate the voice of indigenous NGOs from indigenous-led governments who can come to the issues from distinct vantage points (Nuttall 2009). So while the PPs certainly have a "place at the table," their ability to inform and direct the course of governance and decision making cannot be fully realized until they have full administrative, outreach, and research support. This sentiment echoes that of Stephen Ellis's (2005) analysis of traditional knowledge in environmental decision making which contrasts "top down" strategies (preparing environmental governance authorities to receive traditional knowledge) with "bottom up" strategies (fostering the capacity of indigenous people to bring traditional knowledge to bear on environmental decision-making). Until recently, the Arctic Council has largely been characterized by a "top down" organization, leaving the capacity issue largely ignored.

However, in recent years, the Arctic Council has shown growing interest in some new fields, such as indigenous observations, indigenous knowledge and languages, and community well-being that attempt to address issues important to communities. For example, the Arctic Social Indicators project (ASI) is a new project following up on the Arctic Human Development Report (AHDR). The ASI project seeks to devise indicators to facilitate the tracking and monitoring of human development in the Arctic and is being developed under the auspices of the Sustainable Development Working Group (SDWG) of the Arctic Council.[9] In addition, CAFF supports projects including the Bering Sea Sub Network (BSSN) – a community-based monitoring program led by Executive Director of the AIA, Victoria Gofman. Several IPY 2007–2008 projects including SIKU also helped build capacity through projects such as the *Sila-Inuk* led by Lene K. Holm, ICC Greenland Director of Environment. Both of these projects provide a great example of PP-led efforts that increase involvement and participation of communities in research that pays special attention to the ways Traditional Knowledge can influence and directly inform policy. Importantly, the ICC 2012 Climate Change Roadmap, developed aboard the CCGS *Amundsen* as part of the IPY – Circumpolar Flaw Lead System project, clearly indicates the importance of such engagement and cooperation beyond the IPY timeline. Funding support will be critical to ensure such programs have sustainability over the long term.

Intergenerational Engagement

The opportunity for younger generations of indigenous peoples to engage with the Arctic Council is undertaken largely under the direction and initiative of individual PPs. While the general level of engagement of young people in the Arctic Council is relatively low, and is indeed part of a larger capacity issue, it may also be due to the fact that politically active indigenous youth are overburdened with regional issues ranging from the implementation of land claims agreements and devolution to immediate social concerns, such as high rates of suicide and domestic violence. Such concerns reflect the fundamental challenges of negotiating and understanding the world of their ancestors and the new social, political, and economic context in which they live (Obed 2009; Alexie 2009). As Udluriok Hanson (2009:500), a senior policy liaison for Nunavut Tunngavik Inc. notes,

> Many [northerners] have gone from igloos to the internet in one generation. The challenge now is to determine how to maintain the social fabric of decision-making and community processes in a new world of land claims agreements, treaties, cultural and geographic distinctions, and cash-based economies. Our vision for the future is based on finding a balance—one that enforces old practices and effectively introduces new ones; one that uses the tools passed down from generations to serve and empower "the collective." In the past, this meant that successes were shared and challenges met on the basis of adaptability.

Given this vision, it makes sense to create a space for more indigenous youth to engage at the international level. Certainly, young Inuit like Ms. Hanson remain inspired by leaders like Mary Simon and Siila Watt-Cloutier who precede them, yet there is little direct support for them to participate in international venues like the Arctic Council. This is not to say individual PPs like the Inuit Circumpolar Council and the Saami Council do not recognize the importance of intergenerational programs through their Youth and Elder Councils, but more simply that the Arctic Council and the governments they represent need to provide more direct support for such engagement.

In this regard, the Arctic Council could potentially look to other international forums like the United Nations for potential models to include more youth in their activities. The Youth Program at the United Nations "aims to build an awareness of the global situation of young people, as well as promote their rights and aspirations. The Program also works towards greater participation of young people in decision-making as a means of achieving peace and development."[10] Through this program, the United Nations hosts a cyberschoolbus web site to help promote educational outreach and youth can apply to attend international meetings. The United Nations also publishes a guide to help youth negotiate the complexities of attending an international meeting.[11] More recently, the UN Permanent Forum on Indigenous Issues also recently launched a program specific to indigenous youth to promote their participation in the Forum as a way to highlight issues important to them.[12] Although the purpose of involving youth in many of these programs is to give voice to concerns and issues effecting the younger generations, it is less so about intergenerational knowledge transference. Nevertheless, such programs are important because they recognize that youth deserve "a place at the table" and will be increasingly important to consider if the Arctic Council grows its political weight in the years to come. The Model Arctic Council, a youth oriented training program being developed by the Northern Forum, represents a step in the right direction.

Investments in Capacity

Investments in capacity for the PPs are required on many levels. Funding for travel, administrative support, and outreach are all necessary for effective participation, yet capacity on an individual PP basis varies enormously. Some have more resources than others and the IPS, as it is currently structured, cannot address this problem. At present, the IPS is based in Copenhagen and funded through the Danish Government. It serves as a support Secretariat for the International Indigenous Peoples' Organizations of the Arctic Council but is not a legally independent entity and has no mechanism to deliver services and funding to the PP organizations.

The IPS was designed to ensure that the PPs receive documents and reports of the Arctic Council, helping Permanent Participants to present their views to the Arctic Council and its working groups, coordinating meetings for the PPs to meet with each other and communicating information about the Arctic Council to indigenous peoples in the Arctic.[13] IPS activities are carried out by the IPS staff and contractors, under the direction of the Executive Secretary who reports and is accountable to the Danish/Greenland Homerule Government. Although the IPS does have a board consisting of PP representatives including a Chair which rotates every 2 years, it has only consultative functions, but no financial control. This poses a significant challenges to running the organization from the standpoint of the PPs. For example, basic travel costs for the IPS board members to travel to IPS meetings fall on individual member organizations to fund. Last year, under the leadership of IPS Chair Patricia Cochran there was an attempt to address this problem. She asked AC SAOs to adopt a Memorandum of Understanding which would give the IPS Board powers inherent to any governing board, including financial control, and to lift the IPS board travel ban. Unfortunately no action was taken by the SAOs and it remains to be seen if the new IPS Chair, Geir Tommy Petersen from the Saami Council, will continue to raise the issue with the AC or the Danish Government (Victoria Gofman, personal communication).

Despite these structural challenges, the IPS serves as an important bridge between the PPs and it will be interesting to see how, and if, the role of the IPS board strengthens in the coming years. Considering the Arctic Council stands alone as the only international body where Arctic indigenous organizations can directly interact with governments and scientists to influence policy relevant recommendations, it will be important for the PPs to find a way to ensure this interaction is substantive (Lindroth 2006). The other important area to recognize is that the PPs, and the indigenous peoples they represent, are not homogenous and do not speak with one voice but many. Whether together through the IPS and/or as individual member organizations, it is important that the PPs build their capacity through direct support from their home nation-states and other sources to ensure each can fully participate in Arctic Council activities.

Discussion and Conclusion

During a time of unprecedented social and environmental change, Inuit knowledge and use of sea ice represents an important part of our human history. But the careful work involved in documenting this knowledge needs to go beyond "salvage ethnography." We need to take what we are learning about how Inuit understand and experience their world as well as the "lessons learned" about what it means to develop research partnerships based on mutual respect and reciprocity to help inform how parallel relationship can develop within the realm of public policy.

In the context of IPY 2007–2008, the recognition that indigenous knowledge represents a valid and important form of knowing the environment was a significant contribution compared to past international efforts (Shadian 2009; Nilsson 2009). SIKU projects, in particular, highlight the importance of bringing tangible resources – in the form of expertise, local knowledge and observers, language preservation, education and outreach, and the incorporation of modern technology – in a way that ultimately benefits both scientists and indigenous communities. In terms of relationship building, SIKU scientists provide an important lesson in crossing boundaries between cultures, academic disciplines, national research organizations, funding programs, and data management – as well as the importance of networking and sharing data with the communities in which they worked. For Inuit involved in SIKU and related community-oriented projects, knowledge sharing is integral to the research process but most important in the context of passing knowledge down to younger generations. Nunavut's new Piqqusilirivvik Cultural School, which will offer programs to help capture and pass cultural knowledge, skills, and language to younger generations, represents a great example of how this kind of sharing is becoming more integral to the educational system in northern Canada. Similarly, the Hunters School in Ummanak, Greenland, is charged with a similar mandate to pass knowledge down to future generations.

What are the prospects for translating this inclusionary approach within a policy context? The materials collected through IPY SIKU will be of tremendous importance to many individual communities but will also provide an opportunity for Inuit governing bodies to learn about the impact of sea ice and climate change, language/knowledge preservation, and people's observations of change across the entire Inuit homeland. This is particularly important given the fact that policy makers will need to be well informed by local lifeways as much as they do by climate science, economics, and politics in order to be successful at the local level (Mehdi et al. 2006).

While politicians, diplomats, academics, and NGOs debate the future of Arctic governance (Rayfuse 2008; Jabour and Weber 2008; Koivurova 2008; Berkman and Young 2009; Huebert 2009; Young 1992, 1998, 2009), some indigenous leaders are calling for cooperation that builds off a model of comanagement. As Siila Watt-Cloutier states,

> Recognizing the importance of the Arctic for the whole of the planet, and the historical stewardship of indigenous peoples of the Arctic ecosystem, consider an Arctic treaty that charges circumpolar indigenous peoples with the stewardship through comanagement, of

> the Arctic for the continued benefit of humankind. These proposed international comanagement boards, on which the indigenous peoples of the Arctic would be guaranteed majority representation, would integrate traditional and scientific knowledge to ensure sound and peaceful management of the Arctic's natural resources (Watt-Cloutier 2009:74).

Although the solution(s) to Arctic governance will likely be more complex and mulitfacted than an "Arctic Treaty" or greater investments in the capacity of the Arctic Council, the basis of cooperation Watt-Cloutier describes is important because it recognizes the importance of process and consultation, not simply the institutionalization of indigenous knowledge. And, I would argue, these are the very same processes that are all key ingredients to the new science paradigm of inclusion and recognition. In the end strengthening the participation of indigenous peoples at the international level will be key to ensuring a long-term legacy of adaptability and resilience on behalf of the people who call the Arctic home region.

Acknowledgments The research conducted as part of the Sikusiilarmiut Place-Name Project was supported through grants made by the National Science Foundation; the Nunavut Department of Culture, Language, Elders and Youth; and the Geological Survey of Canada. I would like to thank the Community of Cape Dorset for their support of the Sikusiilarmiut Place-name Project and I would like particularly acknowledge the hard work of my colleague, interpreter, and friend Aksatungua Ashoona. I would particularly like to thank Victoria Gofman (AIA) and Jimmy Stotts (ICC) for sharing their insights into the role of Permanent Participants on the Arctic Council and for providing helpful comments on this chapter. Igor Krupnik, Oran Young, and Shari Gearheard also kindly read and provided feedback on drafts of this chapter; I particularly want to thank Igor, without his encouragement this manuscript would not have been written.

Notes

1. This term refers to Inuit living in the modern nation-states of Russia, the United States, Canada, and Greenland/Denmark.
2. "Denmark Hosts the Five Nations: Arctic Ocean Conference in Greenland". Ministry of Foreign Affairs of Denmark. 2008-05-28. Retrieved on April 30, 2009. http://www.ambwashington.um.dk/en/menu/TheEmbassy/News/NewsArchive2008/DenmarkHoststheFiveNationsArcticOceanConferenceInGreenland.htm
3. Circumpolar Inuit Declaration on Sovereignty. http://www.itk.ca/circumpolar-inuit-declaration-arctic-sovereignty Inuit Tapiriit Kanatami. April 28, 2009.
4. My discussion on international arctic policy and the Arctic Council is informed by my work as a Program Officer of Arctic Marine Conservation at the Oak Foundation, a philanthropic organization dedicated to committing its resources to address issues of global social and environmental concern.
5. This work is being led by Darren Keith of the Kitikmeot Heritage Society in Nunavut. http://www.kitikmeotheritage.ca/atlas.htm
6. Inuit Heritage Trust Placename Program. http://www.ihti.ca/eng/home-new.html. Retrieved July 21, 2009.
7. Nunivak Island Place Name Project Jukebox web site (http://www.nunivak.org/jukebox/index.html).
8. Arctic Council. http://arctic-council.org/article/about. Retrieved July 21, 2009.
9. Arctic Observing Network Social Indicator Project. (http://www.iser.uaa.alaska.edu/projects/search-hd/index.htm).

10. Youth Program at the United Nations (www.un.org/esa/socdev/unyin/mandate.htm). Retrieved July 16, 2009.
11. “Navigating International Meetings: A Pocketbook Guide to Effective Youth Participation” (2002).
12. The Youth Program at the United Nations (www.un.org/esa/socdev/unpfii/en/children.html).
13. Indigenous Peoples Secretariat web site (http://www.arcticpeoples.org/about/)

References

Agrawal, A. 1995. Dismantling the divide between indigenous and scientific knowledge. *Development and Change* 26: 413–439.

Alexie, E. 2009. Arctic indigenous peoples and the reconciliation of the past and present. In *Northern Exposure: Peoples, Powers and Prospects*. F. Abele, T.J. Courchene, F.L. Seidle, and F. St. Hilaire (eds.), Montreal, QB, Canada: The Institute for Research and Public Policy, pp. 501–506.

Aporta, C. 2006. “Anijaarniq: Introducing Inuit Landskill and Wayfinding. Multimedia CD-ROM Released by the Department of Education, Nunavut Government and the Nunavut Research Institute, Iqaluit, NT, Canada.

Berkman, P. and Young, O. 2009. Governance and environmental change in the Arctic Ocean. *Science* 324, Apr 17: 339.

Bravo, M. 2009. Arctic science, nation-building and citizenship. In *Northern Exposure: Peoples, Powers and Prospects*. F. Abele, T.J. Courchene, F.L. Seidle, and F. St. Hilaire (eds.), Montreal, QB, Canada: The Institute for Research and Public Policy, pp. 141–168.

Brook, R., M’Lot, M., and McLachlan, S. 2006. Pitfalls to avoid when linking traditional and scientific knowledge. In *Climate Change: Linking Traditional and Scientific Knowledge*. R. Riewe and J. Oakes (eds.), Winnipeg: University of Manitoba, Aboriginal Issues Press, pp. 139–146.

Ellis, S. 2005. Meaningful consideration? A review of traditional knowledge in environmental decision making. *Arctic* 58(1): 66–77.

Ford, J. and the community of Arctic Bay. 2006a. Hunting on thin ice: Risks associated with the Arctic Bay narwhal hunt. In *Climate Change: Linking Traditional and Scientific Knowledge*. R. Riewe and J. Oakes (eds.), Winnipeg: University of Manitoba, Aboriginal Issues Press, pp. 139–146.

Ford, J. and the community of Igloolik. 2006b. Sensitivity of Iglulingmiut hunters to hazards associated with climate change. In *Climate Change: Linking Traditional and Scientific Knowledge*. R. Riewe and J. Oakes (eds.), Winnipeg: University of Manitoba, Aboriginal Issues Press, pp. 139–146.

Fox, S. 2003. *When the Weather is uggianaqtuq: Inuit Observations of Environmental Change*. University of Colorado Geography Department Cartography Lab. Distributed by National Snow and Ice Data Center. CD-ROM, Boulder.

Gearheard, S., Matumeak, W., Angutikjuaq, I., Maslanik, J., Huntington, H.P., Leavitt, J., Matumeak Kagak, D., Tigullaraq, G., and Barry, R.G. 2006. “It’s not that simple”: A comparison of sea ice environments, uses of sea ice, and vulnerability to change in Barrow, Alaska, USA, and Clyde River, Nunavut, Canada. *Ambio* 35(4): 203–211.

Gearheard, S. and Shirley, J. 2007. Challenges in community-research relationships: Learning from natural science in Nunavut. *Arctic* 60(1): 62–74.

Gilchrist, G., Mallory, M., and Merkel, F. 2005. Can local ecological knowledge contribute to wildlife management? Case studies of migratory birds. *Ecology and Society* 10(1): 20. http://www.ecologyandsociety.org/vol10/iss1/art20/ (Accessed April 4, 2008).

Hanson, U. 2009. The next generation. In *Northern Exposure: Peoples, Powers and Prospects*. F. Abele, T.J. Courchene, F.L. Seidle, and F. St. Hilaire (eds.), Montreal, QB, Canada: The Institute for Research and Public Policy, pp. 389–394.

Henshaw, A. 2006a. Learning landscapes: Pausing along the journey: Learning landscapes, environmental change and place names amongst the Sikusilarmiut. *Arctic Anthropology* 43(1): 52–66.

Henshaw, A. 2006b. Winds of change: Weather knowledge amongst the Sikusilarmiut. In *Climate Change: Linking Traditional and ScientifiC Knowledge*. R. Riewe and J. Oakes (eds.), Winnipeg: University of Manitoba, Aboriginal Issues Press, pp. 177–188.

Henshaw, A. 2009. Sea ice: The sociocultural dimensions of a melting environment in the Arctic. In *Anthropology and Climate Change: From Encounters to Actions*. S.A. Crate and M. Nuttall (eds.), Walnut Creek: Left Coast Press, pp. 153–165.

Henshaw, A. and Ashoona, A. 2006. Learning landscapes: place, time and place names amongst Sikusilarmiut of Nunavut. Paper presented at the 15th Inuit Studies Conference, Paris, France. October 26–28, 2006.

Henshaw, A., Ashoona, A., and Gilbert, C. 2007. Inuktitut Place Names, Sikusiilarmiut, Nunavut. Open Files 5615–5620, Scale 1:250,000. Geological Survey of Canada, Ottawa.

Huebert, R. 2009. Science, cooperation and conflict in the polar region. In *Legacies and Change in Polar Sciences: Historical, Legal and Political Reflections*. J. Shadian and M. Tennberg (eds.), Hampshire, UK: Ashgate Publishing Limited, pp. 63–72.

Huntington, H.P. 1998. Observations on the utility of the semi-directive interview for documenting traditional ecological knowledge. *Arctic* 51(3): 237–242.

Huntington, H.P. 2000. Using traditional ecological knowledge in science: Methods and applications. *Ecological Applications* 10: 1270–1274.

Huntington, H.P., Brown-Schwalenberg, P.K., Fernandez-Gimenez, M.E., Frost, K.J., Norton, D.W., and Rosenberg, D.H. 2002. Observations on the workshop as a means of improving communication between holders of traditional and scientific knowledge. *Environmental Management* 30(6): 778–792.

Huntington, H. and Fox, S. (Lead Authors). 2005. The changing Arctic: Indigenous perspectives. In *Arctic Climate Impact Assessment*. Cambridge: Cambridge University Press, pp. 62–98.

Jabour, J. and Weber, M. 2008. Is it time to cut the Gordian knot of polar sovereignty? *Review of European Community and International Environmental Law* 17(1): 27–40.

Kemp, W. 1976. Inuit land use in south and east Baffin Island. In *Inuit Land Use and Occupancy Project*. M. Freeman (ed.), Department of Indian and Northern Affairs: Ottawa.

Koivurova, T. 2008. Alternatives for an Arctic treaty – Evaluation and a new proposal. *Review of European Community and International Environmental Law* 17(1): 14–26.

Krupnik, I., Bravo, M., Csonka, Y., Hovelsrud-Broda, G., Müller-Wille, L., Poppel, B., Schweitzer, P., and Sörlin, S. 2005. Social sciences and humanities in the International Polar Year 2007–2008: An integrating mission. *Arctic* 58(1): 91–101.

Laidler, G.J. and Elee, P. 2006. Sea ice processes and change: Exposure and risk in Cape Dorset. In *Climate Change: Linking Traditional and Scientific Knowledge*. R. Riewe and J. Oakes (eds.), Winnipeg: University of Manitoba, Aboriginal Issues Press, pp. 55–176.

Lindroth, M. 2006. Arctic Indigenous Peoples' Organizations in Arctic and Global Arenas: Indigenous NGOs in Intergovernmental Organizations. Abstract from the Conference Proceedings of the International Studies Association.

MacDonald, M., Arragutainaq, L., and Novalinga, Z. 1997. *Voices from the Bay: Traditional Ecological Knowledge of Inuit and Cree*. Canadian Arctic Resource Committee and Environmental Committee of Municipality of Sanikiliuaq, Ottawa, Ontario and Sanikiluaq, N.W.T.

Mehdi, B., Mrena, C., and Douglas, A. 2006. *Adapting to Climate Change: An Introduction for Canadian Municipalities* C-CAIRN (Canadian Climate Impacts and Adaptation Network).

Meier, W.N., Stroeve, J., and Gearheard, S. 2006. Bridging perspectives from remote sensing and Inuit communities on changing sea ice cover in the Baffin Bay region. *Annals of Glaciology* 44(1): 433–438.

Nadasdy, P. 1999. The politics of TEK: Power and the "integration" of knowledge. *Arctic Anthropology* 36: 118.

Nilsson, A. 2009. A changing Arctic climate: More than just the weather. In *Legacies and Change in Polar Sciences: Historical, Legal and Political Reflections*. J. Shadian and M. Tennberg (eds.), Hampshire, UK: Ashgate Publishing Limited, pp. 9–34.

Nuttall, M. 2009. Living in a World of movement: Human resilience to environmental instability in Greenland. In *Anthropology and Climate Change: From Encounters to Actions*. S.A. Crate and M. Nuttall (eds.), Walnut Creek: Left Coast Press, pp. 292–310.

Obed, N. 2009. Inuit values and the implementation of land claims agreements. In *Northern Exposure: Peoples, Powers and Prospects*. F. Abele, T.J. Courchene, F.L. Seidle, and F. St. Hilaire (eds.), Montreal, QB, Canada: The Institute for Research and Public Policy, pp. 511–514.

Oozeva, C., Chester, N., George, N., Christina, A., and Igor, K. 2004. *Watchng Ice and Weather Our Way/ Sikumengllu Eslamengllu Esghapalleghput*. Washington and Savoonga: Arctic Studies Center, Smithsonian Institution and Savoonga Whaling Captain Association.

Rayfuse, R. 2008. Protecting marine biodiversity in polar areas beyond national jurisdictions. *Review of European Community and International Environmental Law* 17(1): 3–13.

Semenova, T. 2005. Indigenous Organizations as Participants to the Arctic Council. Abstract from Conference Proceedings International Studies Association.

Shadian, J. 2006. Remaking Arctic governance: The construction of an Arctic Inuit polity. *Polar Record* 42(222): 249–259.

Shadian, J. 2009. Revisiting politics and science in the poles: IPY and governance of science post Westphalia. In *Legacies and Change in Polar Sciences: Historical, Legal and Political Reflections*. J. Shadian and M. Tennberg (eds.), Hampshire, UK: Ashgate Publishing Limited, pp. 35–62.

Sillitoe, P. 1998. The development of indigenous knowledge: A new applied anthropology. *Current Anthropology* 39(2): 223–235.

Tremblay, M., Furgal, C., LaFortune, V., Larrivée, C., Savard, J.-P., Barrett, M., Annanack, T., Enish, N., Tookalook, P., and Etidloie, B. 2006. Communities and ice: Linking traditional and scientific knowledge. In *Climate Change: Linking Traditional and Scientific Knowledge*. R. Riewe and J. Oakes (eds.), Winnipeg: University of Manitoba, Aboriginal Issues Press, pp. 123–138.

Watt-Cloutier, S. 2009. A principled path. In *Northern Exposure: Peoples, Powers and Prospects*. F. Abele, T.J. Courchene, F.L. Seidle, and F. St. Hilaire (eds.), Montreal, QB, Canada: The Institute for Research and Public Policy, pp. 69–76.

Young, O. 1992. Arctic *Politics: Conflict and Cooperation in the Circumpolar North*. Hanover: University of New England Press.

Young, O. 1998. *Creating Regimes: Arctic Accords and International Governance*. Ithica: Cornell University Press.

Young, O. 2009. Whither the Arctic? Conflict or Cooperation in the Circumpolar North. *Polar Record* 45: 73–82.

Chapter 20
Epilogue: The Humanism of Sea Ice

Michael T. Bravo

Abstract This chapter assesses the contributions to understanding sea ice in terms of the concept of *social ontology*, which refers to the web of social relations that give objects their meaning and significance. In the Inuit world, sea ice has a clearly defined set of nomenclatures and toponymies, and is embedded in a rich system of meanings and significance. By contrast the meaning of sea ice in the sciences is different and varied, but it is not simply mechanical or lifeless. Natural historians and natural philosophers are shown to have long contested the broader significance of a sea ice and its philosophical significance in the history of the earth sciences. The predominant interpretation of sea ice as an inert obstacle to progress reflects the social and religious contexts in Europe and America where scientific progress was often closely linked to commitments to economic improvement through commerce, trade, and profitable shipping routes. The essays in this volume, taken together, also represent, a sustained a thoroughly researched contribution to a humanistic understanding of the High Arctic and knowledge of it during International Polar Year 2007–2008.

Keywords Inuit · Sea ice · Arctic · Social ontology · Indigenous knowledge

In this collection of essays, the reader is privileged to approach the subject of sea ice through the eyes of "expert" indigenous travelers, hunters, and researchers. This "expert knowledge" is the product of a centuries-old practical tradition of meticulous firsthand study by indigenous experts. For the most part, this knowledge has been deeply entrenched in the everyday life of Inuit. Of course, among Inuit who produce and share this knowledge, there are recognized experts, by which I mean particular individuals who have devoted countless hours of their lives to studying the properties and complex behavior of sea ice. Earlier scholars of the Inuit world, like Franz Boas (1888), Knud Rasmussen (1927), Richard Nelson (1969), and Milton Freeman (1976), witnessed the knowledge of Inuit experts at work in particular

M.T. Bravo (✉)
Department of Geography, University of Cambridge, Downing Place, Cambridge CB2 3EN, UK
e-mail: mb124@cam.ac.uk

I. Krupnik et al. (eds.), *SIKU: Knowing Our Ice*,
DOI 10.1007/978-90-481-8587-0_20,

Arctic places. But only now has it been extensively codified across the "Inuit universe" in dictionaries and glossaries, explained in terms of strategies and skills, and mapped in relation to trails and routes.

A characteristic of highly skilled practice, in general, is that when it is performed at the highest levels, it often looks simple, as though anyone could do it with just a bit of practice (Collins 1985; Ingold 2000). But that is an illusion. In this volume, some of the intricate knowledge and skills required to inhabit and navigate across the sea ice are opened up for readers with a level of detail and understanding that is unprecedented. What is still more impressive is that the case studies are drawn from places across the Inuit Arctic from Chukotka to the east coast of Greenland. These are accomplishments for which the hunters, travelers, elders, scientists, and other local observers who collaborated in producing this volume can be justly proud of their labors.

The Inuit *Ontology* of Sea Ice

What kind of an object is sea ice? All the contributors agree that it is at the core of traditional coastal Inuit cultures because it is the primary material substrate of their routes *on* or *through* which Inuit travel. It is also a place in which to live or dwell, because it provides shelter and sustenance for the marine animals on which human survival in the Arctic has traditionally depended. Thus, sea ice is a place that is very rich in the social relations of the animals and people that constitute the Inuit world. Although not itself a food for living organisms, it nevertheless is a central productive component in Arctic marine ecosystems that sustain life, for both marine species and peoples. In that sense, sea ice has a profound *social ontology*, an existence as a social object by virtue of the deep-seated meanings and relations that connect it to Inuit life. Perhaps that is one way we should read Theo Ikummaq's statement, quoted in the volume's introduction, that "sea ice has a life of its own."

For many people who know little about the Inuit world, it would seem odd, if not wrong, to say that sea ice is a "social object." How can something that is not a living organism have a social life? To this, we might reply that it is the significance of sea ice for Inuit that binds it to their lives and gives it profound social meaning. Like most types of objects in the natural world, our knowledge of sea ice is derived from both the characteristic qualities through which it is perceived and classified (e.g., hardness, flexibility, color, texture, surface contours) and what we can call its intrinsic qualities (e.g., its composition, age, structure, texture).

Philosophers tell us that the senses (e.g., seeing, hearing, touching) with which human subjects observe and organize their perceptions are an integral part of knowledge. Taken together, these intrinsic qualities, the characteristic features, and the faculties of perception of sea ice are known through its nomenclatures, set of place-names, descriptions, rules, actions, and narratives. In the Inuit world these are incorporated into daily life as tacit knowledge that is practiced with great skill, acquired through a long apprenticeship in situ, but is normally difficult to articulate. In that sense, it is the constant use of skilful knowledge, unremarked upon, that makes sea ice a social object to the Inuit (Lave and Wenger 1991).

This is a familiar argument to those with a background in the humanities and social sciences, and it is also widely appreciated by ecologists and glaciologists who have spent their professional lives observing sea ice. But it is worth remembering that in the minds of many urban consumers who have never been to the Arctic, ice is generally perceived to be frozen water, dispensed to keep drinks cool, and not very much else. It is true that there is a growing public awareness of the role of melting sea ice as an indicator of climate change, but that component of public awareness is easy to overestimate. This may serve to remind us why this International Polar Year needs to be as concerned with public education as with producing specialized research.

Ice, Commerce, and Natural History

The International Polar Year 2007–2008 coincides with a growing recognition that climate change presents a long-term crisis for all humanity. It may therefore be worth asking why sea ice has been popularly perceived as an inert or "dumb" substance. What is it about western civilization that has for centuries caused us to undervalue sea ice? It is tempting to attribute this either to a basic lack of knowledge or alternatively to a western scientific tradition that has been preoccupied with the mechanical properties of nature, ignoring its spiritual qualities. If that is true, then the inclusion of the humanities and northern citizens in this International Polar Year that has yielded this comprehensive study of Inuit sea ice knowledge can be interpreted as an attempt to compensate for the spiritual inadequacy of science's cold rationality, as if to make it more human. There is clearly evidence that indigenous worldviews are sources of insight for educated readers across many cultures, and in some sense, revealing that the rich Inuit *social ontology* of sea ice does compensate for the perception that the western tradition of geophysics is based on an excessively mechanistic rational tradition. However, this would also be to misrepresent the rich intellectual and cultural history of the earth sciences. Since the cultural history of sea ice remains to a large extent overlooked, this epilogue provides an opportunity to present an all-too-brief introduction to the historical context of popular perceptions of sea ice.

To provide some context for early modern studies of the natural history of ice, we need to remember that Europeans brought industrial-scale exploitation to the Arctic in the seventeenth century through the fur trade and whaling. The establishment of the fur trade in North America and Russia and the hunting of whales in the Greenland Seas took place on an industrial scale primarily to satisfy the demand of European markets for furs, oil, and baleen (Rich 1968; Jackson 1978). Sea ice not only presented a danger to navigation for whalers but also offered some security to the whales. Though sea ice also had practical value for whalers, for the most part it represented a source of obstruction and potential damage to their ships (Scoresby

1820). Thus, the overall or net "value" of sea ice to European mariners was clearly negative; it represented a risk and a cost to navigation rather than an opportunity or an asset. The economic value of this cost was calculated using a certain amount of guesswork, to determine the risk ship owners faced to their capital from damage by sea ice. Thus the extent of sea ice was worked into the insurance premiums for expeditions and, in at least one instance, factored into calculations of the economic feasibility of Arctic exploration (Bravo 2006). In other words, sea ice in the European perspective has, for centuries, been counted as a risk and a financial liability for Arctic shipping.

Early modern natural history (1600–1800) reflected similar contemporary philosophical, religious, and economic concerns to those that motivated exploration. Gentlemen of science who collected and classified botanical, zoological, and mineralogical collections were largely sympathetic to the practical uses of knowledge. Transplanting and acclimatizing crops from foreign lands to European soil and experimenting with agricultural techniques to achieve greater yields became seen by many of Europe's leading botanists in the eighteenth century as vehicles for self-sufficiency in food production and increasing national prosperity (Jardine et al. 1996). This philosophy of national economic improvement applied no less to the oceans. Even if breeding fish seemed largely out of the question, the fact that the contents of the oceans were free for the taking, unencumbered by the interests of foreign landowners, was seen as a means of "improving" a nation's balance of trade as well as its food supply (Bravo 2006).

Seen in the light of economic improvement, sea ice was regarded as a hindrance to both navigation and fishing and hence a natural restraint on the growth of commerce that was essential for the inhabitants of Europe. William Paley's *Natural Theology* (1802), which could be found on the bookshelves of most early nineteenth century educated British readers of natural history, set out to show how the natural world provided evidence that the world was designed by a wise and beneficent God. Paley described the circulation of the oceans as a balanced and harmonious system, praising water as the purest of fluids. To justify how a beneficent God would create the polar regions full of sea ice, Paley explained how the ice in the Arctic balances, or compensates for, the heat that makes the temperate regions habitable. Without the "dissolving" power of heat, Paley remarks, "all fluids would be frozen. The ocean itself would be a quarry of ice; universal nature stiff and dead" (Paley 1802:405). Part of this vision of the harmony of nature was to allow that plants and animals are "not capable of interchanging their situations, but are respectively adapted to their own (ibid:417)." He argued, hypothetically, that were a change in the earth's axis of rotation to cause Europe to become more polar, "instead of rejoicing in our temperate zone, and annually preparing for... the rather agreeable succession of seasons... we might come to be locked up in the ice, and darkness of the arctic circle, with bodies neither inured to its rigors, nor provided with shelter or defense against them (ibid)." In Paley's vision, the most that could be said for sea ice was that its proper place was in the polar regions and that it formed part of the global system that sustained the comfortable temperatures suited to the inhabitants of the temperate latitudes.

Twenty years later, when the British navy renewed its search for a Northwest Passage, the idea of "freedom of navigation" was construed in nationalist terms and associated with unrivalled British imperial naval power following the defeat of Napoleon. Controlling the high seas with powerful merchant and military navies was seen as the condition in which British global economic power could flourish. John Barrow, the Admiralty Second Secretary, when building up patriotic support for the idea of a new "modern" program of Arctic exploration, explained to his readers that reports by whalers of relatively open Arctic seas the previous season (1817) was the sign of a "providential improvement" and a permanent state change in the Arctic climate. This improvement, he conjectured, would compensate for the previous "Year without a Summer" and ensuing winter of 1816–1817 that had produced agricultural shortages across northern Europe and parts of North America and perhaps enabled the Northwest Passage or a trans-polar passage to be discovered and navigated successfully for the first time (Barrow 1817; Bravo 1992).

Barrow – for whom the northernmost point of Alaska, Point Barrow, at 71°17′ N has been named – was much more than a proponent of Arctic exploration. He had risen from the ranks of a naturalist and cartographer as a colonial civil servant in South Africa to one of the most powerful advocates in Britain of the national benefits to be accrued from a partnership between science and empire. His close monitoring of whalers' reports for signs of the breakup of a barrier of Arctic sea ice was central to his vision of a Northwest Passage as a "social ontology" rooted in commerce, progress, romance, and nationalism (Bravo; ibid). The extent to which writings like those of Barrow provide a convenient historical tradition to be invoked by today's advocates of globalized Arctic shipping and resource extraction is clear for all to see.

Less well known is the history of rival ontologies of sea ice. When Barrow claimed that anomalies in the seasonal ice reports of 1817 would usher in a new golden age of Arctic navigation, the public's enthusiastic response helped turn the project into a national sensation. Not everyone, however, was convinced. Natural philosophers (what today we might call experimental physicists) decried Barrow's portrayal of ice as hyperbolic romance and an abuse of science. John Leslie, professor of mathematics at Edinburgh University, argued that on the basis of experiments carried out to determine the specific heat and convection properties of ice, the reported sea ice anomalies in the Arctic could not possibly account for the changes to weather witnessed in Europe. Leslie and his Edinburgh colleagues were, in general, advocates of deriving practical benefits from the application of science to commerce, but they were highly critical of the wholesale hijacking of the science of ice for a social ontology so clearly rooted in nationalism and romance (Leslie 1818).

The observed rapid decline of sea ice concentrations in the Arctic today, meticulously documented by scientists, is open to similar kinds of exaggeration that go well beyond the usual accusations of inaccuracy leveled at the media. Traditionally Inuit approaches to managing risk have been based on a capacity to respond flexibly and quickly to changing local conditions (Bates 2006). This is an important factor in the Inuit relationship to *sila*. This is in stark contrast to the complex flows of global capital in which changing environmental conditions are constantly valued in the price of futures in commodities' markets. Even though the price of commodities like oil and gas is currently unremarkable, the prospect of shortages and higher prices in the coming years presents opportunities for powerful nations and commercial interests to undertake strategic planning to prepare themselves for the day when access to Arctic resources, including shipping, becomes more feasible. Today environmental regulation and governance in the Arctic take on the role of instruments for simultaneously creating international stability to enable resource extraction to proceed smoothly, while also enabling nations to seek competitive advantages over each other, by positioning themselves to capitalize on the unintended and unfortunate consequences of climate change (Bravo 2009).

This all-too-brief exploration of the role of sea ice in international commerce is simply meant to illustrate the complex asymmetries in the political, economic, and social systems in which sea ice is valued. In the humanities and social sciences, we are growing accustomed to seeking epistemological equivalence for indigenous knowledge, as if to say this particular traditional system of knowledge is as valid as the dominant paradigm developed by western science. While there is a grain of truth in this form of argument, it is also misleading (Sillitoe 1998, 2007). The real danger is the way in which global markets driven by efficiency and profit employ the knowledge of experts to speculate on the future value of the environment. Neoclassical economic theory argues that more efficient delivery of goods to markets presents the owners of infrastructure and commodities with new opportunities, but this thinking reflects the logic of a social ontology that is radically at odds with the way in which indigenous peoples traditionally value their resources. Shipping may, for example, help to maximize the profit streams for Arctic mining operations, but it may also disrupt the stability of multi-year ice that has for centuries been central to the Arctic marine environment and the coastal cultures that have existed in harmony with it (AMSA 2009).

The Politics of Sea ice Knowledge

If Inuit and other indigenous groups are to succeed in gaining recognition of their social ontologies in the governance structures of the Arctic, collaborating in partnerships in national and international deliberations must be an essential part of their strategy (see ICC quoted in Chapter 19 by Henshaw, this volume). As Henshaw argues, Inuit must continue to develop sufficient capacity to participate fully in the Working Groups of the Arctic Council, to ensure that the status of the indigenous permanent participants remains a central feature. This means finding effective and powerful ways to communicate why and how the Arctic marine environment is a home for Inuit culture and not merely a passage for ships passing through. The petition by Sheila Watt-Cloutier on behalf of 63 Inuit to the Inter-American Commission on Human Rights (2005) concerning the infringement of their human rights is one example of how public awareness can be raised. It also stands to reason that those parties with global shipping interests currently seeking greater access to the Arctic

Council, will be concerned with the importance of sea ice and the coastal environment for Inuit only to the extent that there is compelling evidence that these issues continue to be important for Arctic member states.

Thanks to the research in this book, there now exists a far more detailed understanding of the Inuit and, generally, indigenous social ontologies of sea ice that goes well beyond the general knowledge that previously existed in printed form. The research partners in this book have gone to considerable lengths not only to produce essays that are scholarly and accurate but also to create practical resources to foster further learning and transmission of knowledge within the communities where the knowledge resides. Presenting the knowledge across new media platforms (like the Internet, online atlases, or CD-ROMs) and in a variety of institutions, like land-based schools, is also a vital achievement (e.g., Aporta 2006, Fox 2003). It deserves far more credit than is currently given in our research culture, and the many research partners that produced this volume on the SIKU project are to be congratulated.

In the light of this discussion, the designation of SIKU as an International Polar Year project takes on added importance. It is significant that a project ostensibly dedicated to illuminating an indigenous ontology is carried out under the auspices of a designated program of the International Council for Science (ICSU) and the World Meteorological Organization (WMO). As the editors have remarked, this is the first time in 125 years that the Science Plan for an International Polar Year event has officially recognized the participation of northern citizens, the humanities, and the social sciences. SIKU is at its core a *humanistic* project that is in fact different from the more popular term, the "human dimension," that was written into the official published version of the IPY Science Framework (2004:4, 7). Humanism, of course, is a long and rich family of philosophical traditions with different conceptions of the roots of critical enquiry and empiricism that are at the heart of the quest to challenge received dogma. Particularly relevant here is a humanities tradition of knowledge that emerged in the early nineteenth century (closely linked to Wilhelm von Humboldt) that incorporated a philosophical commitment to education and virtue so as to take account of pluralism and linguistic diversity. A guiding principle behind Humboldt's ideal conception of the humanities was that a unity of all knowledge exists in its multiple forms or ontologies and not as a single framework based on reductionism or standardization.

The SIKU project and this book as its main outcome are not merely a compilation of indigenous knowledge about sea ice. It also provides an opportunity for us to reflect on the philosophical assumptions behind International Polar Years. Previous IPYs in their conception were largely dictated by the needs of geophysical and meteorological observing networks for coordinated observations and standardization. Although that remains a prominent feature of this IPY, the pluralism inherent in its added humanities theme creates an opportunity for the IPY to be far more inclusive. It has the impact of demanding that polar research be made relevant to differing international constructions of citizenship and, ultimately, to recognize the social ontologies that have for millennia allowed the Arctic's changing environments to be a home for their inhabitants.

References

Aporta, C. 2006. *Anijaarniq: Introducing Inuit Landskill and Wayfinding*. Multimedia CD-ROM. Department of Education, Nunavut Government and the Nunavut Research Institute, Iqaluit.

Arctic Council 2009. *Arctic Marine Shipping Assessment Report*.

Barrow, J. 1817. Review of 'Narrative of A voyage to Hudson's Bay in his MS Rosamond...' by Lieut. Chappell. 1817. *Quarterly Review* 18(35): 199–223.

Bates, P. 2006. Knowing Caribou: Inuit, Ecological Science and Traditional Ecological Knowledge in the Canadian North. Ph.D. dissertation, University of Aberdeen.

Boas, F. 1888. *The Central Eskimo*. Smithsonian Institution, Bureau of Ethnology. Sixth Annual Report, Washington.

Bravo, M., 1992. Science and Discovery in the Admiralty's Search for a Northwest Passage. Ph.D. dissertation, University of Cambridge.

Bravo, M. 2006. Geographies of exploration and improvement: William Scoresby and Arctic Whaling (1722–1822). *Journal of Historical Geography* 32(3): 512–538.

Bravo, M. 2009. Community-based monitoring and self-interest. *The Circle* 1: 11–13.

Collins, H. 1985. *Changing Order: Replication and Induction in Scientific Practice*. Chicago and London: University of Chicago Press

Fox, S. 2003. When the Weather Is Uggianaqtuq: Inuit Observations of Environmental Change. Multi-media, interactive CD-ROM. Produced at the Cartography Lab, Department of Geography, University of Colorado at Boulder. Distributed by the National Snow and Ice Data Center (NSIDC) and Arctic System Sciences (ARCSS), National Science Foundation.

Freeman, M. (ed.), 1976. *Inuit Land Use and Occupancy Project*. Ottawa, ON: Department of Indian and Northern Affairs.

ICSU. 2004. *A Framework for the International Polar 2007–2008*. Produced by the ICSU IPY 2007–2008 Planning Group. Paris: International Council for Science (ICSU).

Ingold, T. 2000. *The Perception of the Environment: Essays in Livelihood, Dwelling, and Skill*. New York: Routledge.

Jackson, G. 1978. The northern fishery boom, 1783-c.1808. *The British Whaling Trade*. Hamden, CT: Archon Books, 70–90.

Jardine, N., Secord, J., and Spary, E. 1996. *Cultures of Natural History*. Cambridge: Cambridge University Press.

Lave, J. and Wenger, E. 1991. *Situated Learning: Legitimate Peripheral Participation*. Cambridge: Cambridge University Press.

Leslie, J. 1818. Review of 'The possibility of approaching the north pole asserted... by the Hon. D. Barrington, A new edition. With an appendix, containing papers on the same subject and on a north west passage'. *Edinburgh Review* 30(59): 1–59.

Nelson, R. 1969. *Hunters of the Northern Ice*. Chicago and London: University of Chicago Press.

Paley, W. 1802. *Natural Theology; or evidences of the existence and attributes of the Deity collected from the appearances of nature*. A. Faulder.

Rasmussen, K. 1927. *Across Arctic America: Narrative of the Fifth Thule Expedition*. G. P. Putnam's Sons.

Rich, E. 1968. *The Fur Trade and the Northwest to 1857*. Toronto, ON: McClelland and Stewart.

Scoresby, W., Jr. 1820. *An Account of the Arctic Regions*, 2 vols. Edinburgh: A. Constable.

Sillitoe, P. 1998. The development of indigenous knowledge: A new applied anthropology. *Current Anthropology* 39(2): 223–235.

Sillitoe, P. 2007. *Local Science Vs. Global Science: Approaches to Indigenous Knowledge in International Development*. New York and Oxford: Berghahn.

Appendix A
Nunavimmiut Sea Ice Terminology

Chris Furgal, Martin Tremblay, and Eli Angiyou

Contributors: Annie Baron, Tuumasi Annanack, Sarah Tukkiapik, Peter Tookalook, Annie Kasudluak, Michael Barrett, Laina Grey and Agata Durkalec.

Inuit knowledge constitutes an important tool for adaptation to climate change among Nunavimmiut (Inuit of Nunavik) in Northern Quebec (Nunavik). Since 2003 the Kativik Regional Government (KRG) has been conducting research with the communities of the region on the topic of climate change, impacts, and adaptation with a specific focus on ice and ice safety for community travel and access to land- and sea-based resources (hunting, fishing, and gathering activities). Within the framework of this project, researchers from KRG and Trent University have been working with Nunavik communities to establish local ice monitoring programs as well as document Inuit knowledge. Since 2006, this project has been conducted as part of a larger Inuit sea ice research project called the "Inuit Sea Ice Use and Occupancy Project (ISIUOP)" funded under the International Polar Year Program. Between 2006 and 2008 the Nunavik project team carried out a series of semi-directed interviews with experienced hunters and elders from four Nunavik communities – Umiujaq, Akulivik, Kangiqsualujjuaq, and Kuujjuaq. In addition, supplementary interviews were also conducted as part of cooperative projects in Ivujivik, Nunavik and Sanikiluaq, Nunavut. The interviews were conducted to document and understand changes in sea and lake ice dynamics taking place in the region and to document local knowledge on strategies and approaches to adaptation to these changes. Through a review of the interview transcripts a lexicon of Inuttitut terms used in the communities to describe the various ice formations and related processes of ice formation and breakup was also developed. The initial list of ice terminology was then verified and further developed during return trips to the communities and follow-up interviews with the local ice experts. The list was then enhanced by the addition of terms gathered and included in research interviews done in Sanikiluaq, Nunavut, as part of the *Voices From The Bay* project (McDonald et al., 1997). The impressive terminological richness presented in the list below results from the great knowledge of elders and experienced hunters of the ice-covered land and ocean and of the physical processes structuring their environment. This list of lexicon used in the four Nunavik communities, including additions from Sanikiluaq as presented in McDonald et al. (1997), is one of the many products of the Nunavik research project underway on this topic.

The vocabulary presented in the attached list used to describe the ice varies according to the dialects across the region of Nunavik. Within region differences

I. Krupnik et al. (eds.), *SIKU: Knowing Our Ice*,
DOI 10.1007/978-90-481-8587-0, © Springer Science+Business Media B.V. 2010

can be observed between the terms used by Nunavimmiut from southern and northern areas of Hudson Bay, Hudson Strait, and Ungava Bay. For example, the term *allanuk* that means "mobile ice" among Umiujamiut is replaced by the term *aulaniq* among Ivujivimiut. There are also synonyms used within the same community. The term *pirtutak* used by Akulivimiut to mean ice formed by a fine layer of snow deposited on water can also be called *tuktuyaq* by members of the same community.

The Inuttitut terminology of the ice is rich and precise. Certain words, in addition to describing the actual forms of ice, relate to the processes of their formation. An interesting example comes from the Hudson Bay area where local ice experts use, among others, three Inuttitut terms to describe the process of sea ice melting in spring, *upingasak*, *upingaak*, and *akunaagiq*. *Upingasak* indicates the stage of melting at the beginning of spring, the first phase of the melt. During this phase, the ice, which is white at the end of the winter, adopts a blue color from the snowmelt and eventually turns white once again after the snow has melted on top of it, and the water it forms has drained from its surface. *Akunaagiq* refers to the second phase of melting that follows *upingasak*. From the accumulation of water related to the melting of the ice, the ice adopts a white, blue, and then black color. It is considered not safe to travel on when it is black in appearance. *Upingaak* refers to the third and last phase of melting in spring. During this phase, the ice is not safe to travel or walk-on.

Inuit knowledge of the ice, in particular the terminology of sea ice, formations, and processes, provides valuable insights into the processes of ice formation and breakup in these communities. The value of this knowledge in protecting individuals in the community from unsafe travel or hunting conditions related to ice stability cannot be underestimated. Local ice terminology constitutes a set of structured terms passed down from generation to generation describing a dynamic environment that has always been in a state of change. However, the transmission of this knowledge to younger generations appears to be challenged and perhaps hindered by a number of other changes going on within many Inuit communities. We recommend that this type of traditional knowledge be included in the local school curriculum via the involvement of elders and experienced hunters and the use of student trips or other processes for experiential knowledge transmission. To support the use of traditional Nunavik ice terminology and facilitate its use in schools and other public programs, the research team leading this project produced a color poster with some (50) Inuttitut ice terms and their associated English explanations (*see attached Table for complete list of terms*). We hope that this form of documentation and dissemination will be of some assistance in raising awareness about the importance of this living dictionary of the environment present in the collective knowledge of many Inuit elders and experienced travelers and the value of this knowledge for facing future changes in these regions including those related to climate change and environmental variability.

Acknowledgments The information presented in this document is the knowledge shared by Elders and other experts in the communities of Akulivik, Ivujivik, Kangiqsualujjuaq, Kangiqsujuaq, Kuujjuaq, Sanikiluaq, and Umiujuaq. We are grateful to them for their willingness to participate in the project and share their knowledge of the land and sea with the project team. We also thank the Kativik Regional Government, ArcticNet, Consortium Ouranos, Environment Canada, Natural Resources Canada, the International Polar Year Program (through the ISIUOP Project), the Nasivvik Centre, Transport Québec, Centre d'études nordiques, Laval University, Trent University, and the Makivik Corporation for financial and in-kind support for this project.

Term (community of specific use)	Synonym (community of specific use)	Description	Season	Umiujaq	Akulivik	Kuujjuaq	Kangiqsualujjuaq	Sanikiluaq
Akgutitak	*Qinuarq* (Akulivik)	• Slushy mixture of ice and snow that freezes into flexible ice and moves with the waves	Winter	X	Qinuarq			√
Akimmitavinirk		• Was against the head wind			X			
Aggiqakkuit		• Were lifted onto the ice by sea currents or waves			X			
Aggiraqtavining		• Ice broken by strong currents or high winds	Winter		X			√
Akgitkuit		• Forms when ice is broken up by strong currents or waves colliding against the floe edge and the broken ice is submerged, allowing new ice to form on top	Winter		X			√
Apputainaq		• Thin layer of "false ice" covering open water; a crack covered with snow, without any ice beneath the snow						√
Aqiqakuit	*Akitkuit*	• Piled ice formed on top of the submerged ice; see *akgitkuit*	Winter					√
Aulaning	*Allanuk*	• Moving ice; formed ice continuously moving in currents beyond the floe edge; pack ice often moving near the shore	Winter		X			√
Ikiarik		• Piece of solid ice pushed on top of another during a wind storm or by spring tides. Evidently from *ikiaq* – between two surfaces that adhere to each other	Winter	X				X
Ikiqtiniq	*Immatiniq, Tungirliniq*	• Water in lakes between ice and land during the spring thaw in the west	Spring					√
Iktaniq	*Milutsinik*	• Snow-soaked water that freezes at the floe edge. Unsafe due to sea currents it is avoided by hunters and animals						√

Term (community of specific use)	Synonym (community of specific use)	Description	Season	Umiujaq	Akulivik	Kuujjuaq	Kangiqsualujjuaq	Sanikiluaq
Immatinning		• A pool of melted ice in the tidal area *(Ungava) Iniruvik* – ice crack (joint) that opens and closes continuously like a hinge during high and low tides, but does not shift sideways. *Ivunik* – rough, scrambled ice of varying thickness formed when moving ice collides with the floe edge and piles up	Spring		X			X
Iniruvik		• Ice fissured by changes in the tide and frozen again by cold temperatures. Safe to walk-on except when it is newly formed	Winter		X			
Ittiniq (Umiujaq), Ittinirq (Akulivik, Kuujjuaq)		• Calf ice that piles on the edge of the landfast ice (*tuvaq*) at the tidal line. This ice grows continuously above the rocks, lifted by the sea beneath. This ice can be observed in rivers where a tide exists	Winter	X	X	X		
Ittiniviniit		• Remains of pack ice at the tidal line between the land and the solid shore ice. Can be observed in spring when shore ice is carried off by winds or tides	Spring	X	X			X
Ivujialik	*Sikuttigutjaq*	• Someone who is a victim of ice pressure ridges	Spring		X			√
Ivujiarivait		• The moving ice floes crush him in rough weather	Winter	X				X
Ivujiarivait		• Ice flows that claim a life as they breakup	Winter					√
Ivujut		• The piling up of ice flows under pressure out at sea	Winter	X				X
Ivusijuq		• Strong winds or currents that are forcing ice floes to pile up	Winter		X			√
Killinigursituq		• (*Killinigusiqtuq*) there are ice floes forming						√

Term (community of specific use)	Synonym (community of specific use)	Description	Season	Umiujaq	Akulivik	Kuujjuaq	Kangiqsualujjuaq	Sanikiluaq
Killiniq		• Side of ice closest to open water and furthest from mainland		X	X			X
Kiviniq		• A depression usually formed near shorelines and created by the weight of high-tide water that has risen through the cracks. Water exists on the ice after the ice has cracked		X				X
Maniilaq	*Maniiligaak*	• A surface of pack ice with icicles (smaller than *manittuit*)			X			√
Milutsinik	*Iktaniq*	• Snow-soaked water that freezes at the floe edge. It is unsafe in current areas and is avoided by hunters and animals		X	X			X
Miqiaq		• "Molting ice"; thick pack ice from North Hudson Bay that often crowds areas so there is no open water in sight. It will stay near the shoreline until it is clean			X		X	√
Napakkait		• Newly formed ice only 3–6 mm thick and easily broken by winds, currents, and waves. As the pieces break they move upward or downward and freeze into ice sheets separated by open water						√
Napakkuit		• See *napakkait*						√
Napakutak		• See *napakkait*						√
Pikiatuuk		• Surfaced			X		X	
Puikangajuk		• Land or ice that appears to be suspended in air, like a mirage		X	X			X
Puikkatuq		• When sea water begins to permeate ice softened by warm weather conditions						√

Term (community of specific use)	Synonym (community of specific use)	Description	Season	Umiujaq	Akulivik	Kuujjuaq	Kangiqsualujjuaq	Sanikiluaq
Pullait		• Air pockets with a very thin layer of "false ice" covering open water (see *apputainaq*). They are created by current-formed air bubbles or by air bubbles created by ducks diving for food through the openings beneath the ice during an early freeze-up, a bubble of air rising from the water		X	X			X
Pullaq		• A bubble of air rising from far below the surface of the water		X	X			X
Putaaq		• A lone piece of floating ice		X	X			X
Putatait		• See *putataviniq*			X			X
Putatak	*Sikulirutit*	• New ice						√
Putataviniq		• Older piece of ice separated from other ice by currents, on which new ice, *sikuliak*, has formed						√
Puttaq, Puutak (Kangiqsualujjuaq)		• Autumn floating ice, the first ice floating in ice floes; in summer, the broken pieces of shore ice					X	√
Qainguniq		• Created when slush forms under colder air temperatures offshore and is washed ashore by the wind. It then freezes as beach ice at the high-water mark. *Qainnguq* – border of solid ice stuck to the edge (particularly to the rocky shore) and whose top is never passed except by the highest tides						√
Qainuk		• See *qainguniq*						√
Qalirittinik		• Ice created by thin pieces of ice piling atop one another due to strong currents and moving ice						√
Qamait		• Ice pushed upward; see *puqurniq* – when the pressure that causes ice to break pushes the broken ice upward			X		X	√

Term (community of specific use)	Synonym (community of specific use)	Description	Season	Umiujaq	Akulivik	Kuujjuaq	Kangiqsualujjuaq	Sanikiluaq
Qautsaulittu		• Ice which breaks after having tested it with a harpoon; antonym of *kaqusaruk*	Fall	X				
Qinualuk		• Block of ice formed from sea water that, while compact, is still soft						√
Qiqngurusirtuk		• New ice formed from slush in narrow water bodies, like inlets; it is harder than *qinuk* but still soft and unreliable for travel						√
Quasaq		• Very slippery (black) ice that can be found on ground, sea, lakes, or pack ice. Caused by freezing rain, too slippery to walk-on *(Ungava Bay, Western Hudson Bay)*						√
Quliqiaq		• *Napakkait*. Newly formed ice only 3–6 mm thick and easily broken by winds, currents, and waves. As the pieces break they move upward or downward and freeze into ice sheets separated by open water						√
Qullunirsiutuq		• Hanging shore ice that has broken during low tide leaving a gap in the ice when the tide is out						√
Qullupiaq		• New crack that forms in different directions when ice collides at the floe edge. The force of impact creates cracks in both the solid ice and the incoming ice; see *piquniq* (has many cracks)			X			√
Quluniq		• “Has cracked” (*tukkilik*), deformed by the pressure of high and low tides. Fissure, crack in the ice (wider and deeper than *aajuraq*) in pack ice; a melting of the ice floes along the iceblink			X			√

Term (community of specific use)	Synonym (community of specific use)	Description	Season	Umiujaq	Akulivik	Kuujjuaq	Kangiqsualujjuaq	Sanikiluaq
Qunniq		• *Qungniq* a crack (opening) in the ice on a lake or on the shore (Ungava Bay); only in shore ice (Labrador)						√
Qupugaq		• Striate only on sea ice in formation – (Labrador)						√
Qutitaq		• An ice hole hidden or diminished by frost						√
Sarliarusiq		• Ice on the shore						√
Sarliarutaq		• Ice floes forced onto the shore at the top of a bay						√
Siatuninik		• A grouping of ice pieces moving together in the current						√
Siku		• Ice, icicle (ice at sea, on shore, lakes, rivers)						√
Sikuak, Sikkuaq		• Newly formed thin layer of ice in its earliest stage of formation		X	X		X	√
Sikuarpuq		• The process of ice beginning to form			X			√
Sikuirtuq		• When the ice fully melts after breaking up			X			√
Sikujuaniq		• Dirty, old-looking pack ice that does not melt away in the early summer						√
Sikuliak		• Newly formed ice with no snow on top; thinner than old ice but safe to walk or travel on. Usually formed when *uiguak* and *akgutinik* meet and freeze solid						√
Sikulirutiit	*Tuvaaluk*	• Smooth ice	Winter					
Sikulirutit		• New ice, freshly frozen from part saltwater and part freshwater along shorelines and within inlets. High tide will float it and winds can blow it offshore in broken pieces. Seals like to be on that ice						√
Sikullaq		• Describes one piece of floating ice among many (big or small)						√

Term (community of specific use)	Synonym (community of specific use)	Description	Season	Umiujaq	Akulivik	Kuujjuaq	Kangiqsualujjuaq	Sanikiluaq
Sikuqraaq		• (*Sikuraaq*) new ice, thick enough to carry weight						√
Sikurluk		• Small icicles (not good ice)			X			√
Sikusuilaq		• A place which cannot freeze over (at sea or in a river, because of current = *aukkaniq*)						√
Sikutait		• Ice in small inlets or bays formed before the land-fast ice, *tuvak*, starts developing – solid						√
Sikutak		• New ice that forms from *sikuak,* once inlets are frozen						√
Sikutsiarippuq		• (*Sikutsiarikpuq*) it is very transparent ice			X			√
Sikuvippuq		• The ice is almost frozen						√
Sinaaq		• An ice flow at the edge of the shore: *tuvaup sinaanga*						√
Siqkuitiniq		• A crack (*tukkilik*) deformed by the pressure of high and low tides						√
Siqummaq		• A crack in glass ice, crevasse in pack ice						√
Tuvarlu		• Uneven fast moving ice or bad *tuvak;* (see *umartuq*) (West Hudson Bay)			X			√
Tuvapaq		• At the edge of the sea (West Hudson Bay)		X	X			X
Tuvariirpuq		• The sea or lake ice has finished solidifying (West Hudson Bay)		X	X			X
Tuvarlipuq		• At breakup the sea or lake ice is dangerous	Spring					√
Tuvarpuq		• Fast ice covering the sea or lake		X	X			X
Tuvarqusaq		• (*Tuvaqqusaq*) sea ice, now good for sled trips (Ungava)			X			X
Tuvarsimajuq		• (*Tuvaqsimajuq*) the sea or lake ice had become solid in the past		X	X			X

Term (community of specific use)	Synonym (community of specific use)	Description	Season	Umiujaq	Akulivik	Kuujjuaq	Kangiqsua lujjuaq	Sanikiluaq
Tuvarurpuq		• *(Tuvaruqpuq)* sea or lake ice becomes strong enough not to be broken by wind (Ungava); ice which is thick		X	X			X
Tuvarursimajuq		• (*Tuvaruqsimajuq*) the sea or lake ice has become solid		X	X			X
Tuvasak		• New ice that forms from *sikuak*; see *sikutak; (tuvatsak, tuvaksak*) the first ice that forms on the sea, thin but strong and will be fast ice eventually		X	X			X
Tuvatsaniartuq		• (*Tuvaksaniartuq*) goes hunting on the sea ice that is forming (tries as soon as it can carry him), *Tuvvipuq* – reaches the sea ice			X			X
Tuviatu		• Water space between ice fields		X	X			
Tuviatuk		• Water coming from melting ice	Spring		X			X
Uiguak		• Smooth solid ice formed when wind blows from the land-fast ice		X	X			X
Uiguaq		• Addition of new ice in an open crack; old sea ice	Winter	X	X			X
Uinning		• A wide stretch of open water caused by the breakup of shore ice (parallel to the shore, hence between the shore ice (*tuvaq*) and the floating pack ice)			X			√
Ukiurjait		• New ice that forms from *sikuak*						√

Note: The terms marked with √ come from Miriam McDonald, Lucassie Arragutainaq, Zack Novalinga (1997). Voices from the Bay: Traditional ecological knowledge of Inuit and Cree in the Hudson Bay bioregion. Ottawa, ON: Canadian Arctic Resources Committee; Sanikiluaq, N.W.T.: Environmental Committee of Municipality of Sanikiluaq. All other terms marked with an X were gathered through interviews conducted for this project.

Appendix B
Publications (a) and Posters/Presentations (b) Given at Conferences and Workshops on the Outcomes of the SIKU and Affiliated Projects, 2006–2010

Gita J. Laidler, Claudio Aporta, and A. Chase Morrison

(a) Publications

Aporta, C. 2009. The trail as home: Inuit and their pan-Arctic network of routes. *Human Ecology* 37(2): 131–146.

Arnestad Foote, B. 2009. Point hope. Life on frozen water/Tikigaq: en fotografisk reise blant eskimoene i Point Hope 1959–1962. Chicago and Fairbanks: University of Chicago Press and University of Alaska Press/ Oslo.

Bogoslovskaya, L., Vdovin, B., and Golbtseva, V. 2008. Izmeneniia klimata v regione Beringova proliva: Integratsiia nauchnykh i traditisionnykh znanii (Climate change in the Bering Strait Region: Integration of scientific and indigenous knowledge (SIKU, IPY #166)). Ekonomicheskoe planirovanie i upravlenie 3–4 (8–9):58–68. Moscow (in Russian).

Bogoslovskaya, L.S. and Golbtseva, V.V. 2007. Traditional navigation techniques and safety rules. In *Osnovy morskogo zveroboinogo promysla.* L. Bogoslovskaya, I. Slugin, I. Zagrebin, and I. Krupnik (eds.), Moscow and Anadyr: Russian Institute of Cultural and Natural Heritage and Chukotka Institute of Teachers' Training, pp. 334–341 (in Russian).

Brauen G. and Taylor, D.R.F. 2007. A cybercartographic framework for audible mapping. *Geomatica* (Special Issue on CNC National Report on Cartography) 61(2): 127–136.

Bravo, M.T. 2008. Sea ice mapping: Ontology, mechanics, and human rights at the ice floe edge. In *High Places: Cultural Geographies of Mountains and Ice.* D. Cosgrove and V. della Dora (eds.), London: IB Tauris, pp. 161–176.

Bravo, M.T. 2009. Voices from the sea ice and the reception of climate impact narratives. *Journal of Historical Geography* 35(2): 256–278. Feature issue on Climate Change Narratives.

Druckenmiller, M.L., Eicken, H., Johnson, M.A., Pringle, D.J., and Willliams, C.C. 2009. Towards an integrated coastal sea ice observatory: System components and a case study at Barrow, Alaska. *Cold Regions Science and Technology* 56: 61–72.

Eicken, H., Gradinger, R., Salganek, M., Shirasawa, K., Perovich, D.K., Leppäranta, M. (eds.) 2009. Sea ice field research techniques.Fairbanks, AK: University of Alaska Press, 566pp.

Eicken, H., and Krupnik, I. in press. Learning about sea ice and its use from the Kingikmiut, in Kingikmi Sigum Qanuq Ilitaavut - Wales Inupiaq sea ice dictionary. In: I. Krupnik, H. Anungazuk, and M. Druckenmiller, (eds.) Washington, DC: Arctic Studies Center, Smithsonian Institution.

Eicken, H., Krupnik, I., Weyapuk, W. Jr., and Druckenmiller, M.L. in press. Ice seasons at Wales, 2006–2007, in Kingikmi Sigum Qanuq Ilitaavut - Wales Inupiaq sea ice dictionary. In: I. Krupnik, H. Anungazuk, and M. Druckenmiller (eds.) Washington, DC: Arctic Studies Center, Smithsonian Institution.

Eicken, H., Lovecraft, A.L., and Druckenmiller, M.L. 2009. Sea ice system services: A framework to help identify and meet information needs relevant for Arctic observing networks. *Arctic* 62(2):119–136.

Elders of Kangiqsualujjuaq, Furgal, C., Wilkes, J., Annanak, T., Tremblay, M., and Alain, J. 2007. *Inuit Observations of Climate and Environmental Change: Perspectives from Elders in Kangiqsualujjuaq*. Peterborough: Kativik Regional Government, Kuujjuaq, Nunavik and Trent University.

Fienup-Riordan, Ann. 2010 in press. "Yup'ik perspectives on climate change: 'The world is following its people.' " *Etude/Inuit/Studies* 34(1).

Fienup-Riordan, Ann and Rearden, A. 2010 in press. Ellavut/our world and weather. Seattle: University of Washington Press.

Ford, J.D., Gough, W.A., Laidler, G.J., MacDonald, J., Irngaut, C., and Qrunnut, K. 2009. Sea ice, climate change, and community vulnerability in northern Foxe Basin, Canada. *Climate Research* 38: 137–154.

Gearheard, S. 2008a. A change in the weather. *Natural History*, February, 32–38.

Gearheard, S. (curator). 2008b. *Silavut: Inuit Voices in a Changing World*. Exhibit: University of Colorado Museum of Natural History, April 15, 2008–May 1, 2009.

Gearheard, S., Matumeak, W., Angutikjuaq, I., Maslanik, J., Huntington, H.P., Leavitt, J., Matumeak-Kagak, D., Tigullaraq, G., and Barry, R.G. 2006. "It's not that simple": A comparison of sea ice environments, uses of sea ice, and vulnerability to change in barrow, Alaska, USA, and Clyde River, Nunavut, Canada. *AMBIO* 35(4): 203–211.

Gearheard, S., Pocernich, M., Stewart, R., Sanguya, J., and Huntington. H.P. Linking Inuit knowledge and meteorological station observations to understand changing wind patterns at Clyde River, Nunavut. *Climatic Change*, (in press, published online 23 June 2009, DOI 10.1007/s10584-009-9587-1).

Golbtseva, V.V. 2008. Wind regime and indigenous knowledge abound local winds in uelen. In *Chukotka: Rational Resource Management and Ecological Security. Proceedings of the Chukotka Branch of the SVKNII* 12. Magadan, Russia, pp. 145–161 (in Russian).

Golbtseva, V.V. 2009. Types of sea ice in marine hunters' lexicon and safety measures during sea-mammal hunting. *Kerek Pedagogical Readings* 2. St. Petersburg: Polytechnical University (in Russian).

Henshaw, A. 2009. Sea ice: The sociocultural dimensions of a melting environment in the Arctic. In *Anthropology and Climate Change: From Encounters to Actions*. M. Nuttall and S. Crates (eds.), Walnut Creek, CA: Left Coast Press, pp. 153–165.

Krupnik, I. in press. SIKU project steers to its book and to completion. *ASC Newslett* 17 (2010).

Krupnik, I. 2006. We have seen these warm weathers before. Indigenous observations, archaeology, and the modeling of Arctic climate change. In *Dynamics of Northern Societies. Publications from the National Museum, Studies in Archaeology and History.* J. Arneborg and B. Grønnow (eds.), vol 10. Copenhagen: National Museum of Denmark, pp. 11–21.

Krupnik, I. 2008. Project SIKU: IPY Study of indigenous knowledge of ice. *Arctic Studies Center Newsletter* 15: 16–18.

Krupnik, I. 2009a. "The way we see it coming": Building the legacy of indigenous observations in IPY 2007–2008. In *Smithsonian at the Poles: Contributions to International Polar Year Science*. I. Krupnik, M. Lang, and S. Miller (eds.), Washington: Smithsonian Institution Scholarly Press, pp. 129–142.

Krupnik, I. 2009b. 'The ice we want our children to know': SIKU project in Alaska and Siberia, 2007–2008. *Alaska Park Science* 8(2): 122–127.

Krupnik, I. 2009c. When the ice, photos, and memories come together. In: Berit A. Foote (ed) Point Hope, Alaska: Life on the frozen water. Photographs 1959–1962. Fairbanks: University of Alaska Press, pp. 10–11.

Krupnik, I. 2009d. SIKU (sea ice knowledge and use) study advances in its third year. *Arctic Studies Center Newslett* 16: 26–28.

Krupnik, I. and Bogoslovskaya, L. 2007. Izmeneniie klimata i narody Arktiki. Proekt SIKU v Beringii (Climate change and Arctic People: SIKU Project in Beringia). *Ekologicheskoe planirovanie i upravlenie* 4(5): 77–84.

Krupnik, I. and Bogoslovskaya, L. 2008. International Polar Year 2007–2008. Project SIKU in Alaska and Chukotka. In *Beringia: A Bridge of Friendship*. Tomsk: TSPU Press, pp. 196–204.

Laidler, G.J. and Elee, P. 2008a. Human geographies of sea ice: Freeze/thaw processes around Cape Dorset, Nunavut, Canada. *Polar Record* 44: 51–76.

Laidler, G.J. and Ikummaq, T. 2008b. Human geographies of sea ice: Freeze/thaw processes around Igloolik, Nunavut, Canada. *Polar Record* 44: 127–153.

Laidler, G.J., Dialla, A., and Joamie, E. 2008. Human geographies of sea ice: Freeze/thaw processes around Pangnirtung, Nunavut, Canada. *Polar Record* 44: 335–361.

Laidler, G.J., Ford, J.D., Gough, W.A., Ikummaq, T., Gagnon, A.S., Kowal, S., Qrunnut, K., and Irngaut, C. 2009. Travelling and hunting in a changing Arctic: Assessing Inuit vulnerability to sea ice change in Igloolik, Nunavut. *Climatic Change* 94: 363–397.

Laidler, G.J., and Kapfer, M. 2009. Connecting community observations and expertise with the floe edge service (English/Inuktitut). Ottawa (for distribution in Nunavut): Department of Geography and Environmental Studies, Carleton University, Ottawa, ON. (available online at: http://www.straightupnorth.ca/Sikuliriji/Publications.html)

Lauriault, T.P., Craig, B., Pulsifer, P.L., and Taylor, D.R.F. (2008). Today's data are part of tomorrow's research: Archival issues in the sciences. *Archivaria* 64: 165–186, December.

Lauriault, T.P., Pulsifer, P.L., Taylor, D.R.F. in press. The preservation and archiving of geospatial digital data: Challenges and opportunities for cartographers. In: Markus Jobst (ed.), Prospective cartographic heritage. Berlin: Springer.

Mahoney, A. and Gearheard, S. 2008 Handbook for community-based sea ice monitoring, NSIDC Special Report 14, Boulder, CO, USA: National Snow and Ice Data Center. http://nsidc.org/pubs/special/nsidc_special_report_14.pdf

Mahoney, A., Gearheard, S., Oshima, T., and Qillaq, T. 2009. Sea ice thickness measurements from a community-based observing network. *Bulletin of the American Meteorological Society* 90(3): 370–377.

Pulsifer, P., Hayes, A., Fiset, J.P., and Taylor, D.R.F. 2007. An education outreach atlas based on geospatial infrastructures: Lessons learned from the development of an on-line polar atlas. *Proceedings from the IPY GeoNorth 2007 Conference in Yellowknife*, Northwest Territories.

Pulsifer, P.L., Hayes, A., Fiset, J.P., and Taylor, D.R.F. 2008. An education and outreach Atlas based on geographic infrastructure: Lessons learned from the development of an on-line Polar Atlas. *Geomatica* 62(2): 169–188.

Pulsifer, P.L. and Taylor, D.R.F. 2007. Spatial data infrastructure: Implications for sovereignty in the Canadian Arctic. *Meridian (Canadian Polar Commission).* Spring/Summer: 1–5 (English and French).

Tremblay, M. and Furgal, C. 2008. Les changements climatiques au Nunavik et au Nord du Québec: L'accès au territoire et aux ressources. Final Report to Nothern Ecosystem Initiatives, Environment Canada, 167 p.

Tremblay, M., Furgal, C., Larrivée, C., Annanack, T., Tookalook, P., Qiisik, M., Angiyou, E., Swappie, N., and Barrett, M. 2008. Climate change in Nunavik: Adaptation strategies from community-based research. *Arctic* 61(Suppl. 1): 27–34.

Vdovin, B.I. and Yevstifeev, A.Yu. 2008. Climate change in Eastern Chukotka over the past century, according to the instrumental records. In *Beringia: A Bridge of Friendship*. Tomsk: TSPU Press, pp. 17–24.

Wisniewski, J. 2009. Come on Ugzruk, Have Mercy: Experience, Relationality and Knowing in Kigiqtaamiut Hunting and Ethnography. Unpublished PhD Dissertation, University of Alaska Fairbanks, Department of Anthropology.

(b) Posters/Presentations

Aporta, C. 2007a. Carleton Spring Conference, Guest Speaker.

Aporta, C. 2007b. Guest speaker in a presentation to the Clerk of the Privy Council, Iqaluit.

Aporta, C. 2008a. The Ice is What we Want our Children to Know: Documenting Inuit Sea Ice Use in Canada. Paper Presented at the IPY Open Science Conference in St. Petersburg, Russia.

Aporta, C. 2008b. Overview of the Inuit Sea Ice Use and Occupancy Project. Presentation to the Canadian Circumpolar Ambassadors, Iqaluit.

Aporta, C. 2008c. Inuit and Their Use and Knowledge of the Sea Ice. Presentation to a Delegation of Nordic Journalists, Ottawa.

Aporta, C. 2008d. Overview of ISIUOP. Presentation to the International Polar Day, Canadian Museum of Civilization, Ottawa.

Aporta, C., Gearheard, S., Furgal, C., Taylor, D.R.F., and Laidler, G.J. 2009. Documenting and Representing Inuit Use of the Sea Ice. Annual Meeting of the Canadian Association of Geographers, Ottawa, May 27.

Aporta, C., and Laidler, G.J. 2009. Overview of the inuit sea ice use and occupancy project, presentation to the Cold Regions Technology Group, Canadian Hydraulics Centre, National Research Council of Canada, Ottawa, ON, November 18.

Aporta, C. and Taylor, D.R.F. 2007. Research Works Luncheon, Carleton University, Guest Speakers, Ottawa.

Druckenmiller, M.L. 2008. Whaling trails on landfast sea ice at Barrow, Alaska. Sixth International Congress of Arctic Social Sciences, Nuuk, Greenland, August 26.

Druckenmiller, M.L. 2009. Working with the community of Barrow to monitor and understand landfast sea ice. Barrow young researchers' seminar series, University of Alaska Fairbanks, Fairbanks, AK, March 19.

Druckenmiller, M.L., and Eicken, H. 2009a. Geophysical and Iñupiaq perspectives and observations of shorefast sea ice. Inland Northwest Research Alliance (INRA) Symposium: Lessons from continuity and change in the 4th international Polar year, Fairbanks, AK, March 4–6.

Druckenmiller, M.L., and Eicken, H. 2009b Monitoring Alaska's coastal sea-ice: Some work of the seasonal ice zone observing network. Alaska Forum on the Environment, Anchorage, AK, February 2–6.

Gearheard, S. 2008a. Silavut: Inuit Voices in a Changing World. Presentation for Teachers and Educators, Grand Opening of the *Silavut: Inuit Voices in a Changing World* Exhibit, University of Colorado Museum of Natural History. Boulder, April 19.

Gearheard, S. 2008b. Igliniit ("trails") Project: Combining Inuit Knowledge and GPS Technology to Track Environmental Change. Paper Presented by S. Gearheard at the 6th International Congress of the International Arctic Social Sciences Association. Nuuk, Greenland, August 22–26.

Gearheard, S. 2008c. *Syv-kabale nassarlugu piniariarneq*. News article in Greenland's National newspaper Sermitsiaq. In Greenlanic and Danish, Interview and article by Inge Rasmussen.

Gearheard, S. 2009a. Establishing a Community-Based Sea Ice Observing Network in the Arctic. Invited speaker, Inuit Tapiriit Kanatami (ITK) Inuit Climate Change Working Group Workshop, Ottawa, Canada, January 15.

Gearheard, S. 2009b. Establishing a Community-Based Sea Ice Observing Network in the Arctic. Invited speaker, Climate and Cryosphere (CliC) Sea Ice Workshop. Tromsø, Norway, January 27.

Gearheard, S. 2009c. Siku-Inuit-Hila Project: Sea Ice Change in Greenland. Invited speaker, Greenland Science Summit/Webinar. Kangerlussuaq, Greenland, July 8.

Gearheard, S. 2009d. Siku-Inuit-Hila Project. Invited speaker and participant, Global Long Term Human Ecodynamics Conference. Eagle Hill, Maine, October 15–18.

Gearheard, S. 2009e. Inuit and environmental change research in Nunavut, Canada. Invited speaker, Gettysburg College, Gettysburg, October 20.

Gearheard, S., Angutikjuak, I., and Tigullaraq, G. 2008. *Silavut: Inuit Voices in a Changing World.* Public Lecture for the Grand Opening of the *Silavut: Inuit Voices in a Changing World* Exhibit, University of Colorado Museum of Natural History. Boulder, April 18.

Gearheard, S. and Holm, L. 2008. Issittumi ilisimatusarnermi (Arctic Science) Inuit sulegatigiinnerannik (Human Collaboration) nalliuttorsiutiginninneq (Celebration). Invited Keynote

Address at the 6th International Congress of the International Arctic Social Sciences Association. Nuuk, Greenland, August 22–26.

Golbtseva, V.V. 2009. Sea Ice as a Life Source to the People of the Bering Strait Region. Paper Presented at the 'Shared Beringia Heritage' Conference, Anadyr, Russia, September 15, 2009.

Gough, W.A., Laidler, G.J., and Ford, J.D. 2009. The Changing Weather of Igloolik. Annual Meeting of the Canadian Association of Geographers, Ottawa, May 27.

Karpala, K. 2008a. Preliminary Results of Constructions of Inuit Perspectives on Climate Change. Poster Presented at the Arctic Change conference in Quebec, Canada.

Karpala, K. 2008b. Constructions of Inuit Perspectives on Climate Change. Paper Presented at the Annual University of Ottawa-Carleton University Student Northern Research Symposium in Ottawa, Canada.

Karpala, K., Aporta, C., and Laidler, G.J. 2009. Constructions of Inuit Perspectives of Climate Change. Annual Meeting of the Canadian Association of Geographers, Ottawa, May 27.

Kelley, K. 2008. Preliminary Results on the Politics of Sea Ice in the Canadian Arctic. Poster Presented at the 2008 Annual Ottawa-Carleton Student Northern Research Symposium in Ottawa.

Kelley, K., Laidler, G.J., and Aporta, C. 2009. Politics of Sea Ice in the Canadian Arctic. Annual Meeting of the Canadian Association of Geographers, Ottawa, May 27.

KNR-TV. 2007. Inuit isaannit silaannaq. (Documentary Film on Siku-Inuit-Hila Project (Chapter 11)) in Qaanaaq. http://www.knr.gl/index.php?id=2022

Krupnik, I. 2006. On a Thinning Ice: Indigenous People and Icebreaker Community Face Arctic Climate Change. Paper Presented to the Committee on the Assessment of U.S. Coast Guard Polar Icebreaker Roles and Future Needs, U.S. National Academies, March 2.

Krupnik, I. 2007a. On a Thinning Ice: Arctic Residents Face Climate Change. Paper Presented at the symposium "Impact of an Ice-Diminishing Arctic on Naval and Maritime Operations," Washington, July 12.

Krupnik, I. 2007b. The Way We See It Coming. Building the Legacy of Indigenous Observations in IPY 2007–2008. Paper Presented at the Symposium, "Smithsonian at the Poles: Contributions to IPY Science", Washington, May 4.

Krupnik, I. 2007c. Arctic Residents and Climate Change: Contributions to IPY Science. Presentation at the 'Science Day,' Arctic Science Summit Week, Hanover, March 19.

Krupnik, I. 2008a. "The Ice We Want Our Children to Know": SIKU Project (IPY #166) Overview, with an Emphasis on Alaska and Siberia. Paper Presented at the ICASS-6 Session, Nuuk, August 26.

Krupnik, I. (moderator). 2008b. How We Learned What We Know: Indigenous Experts Document Arctic Ice and Climate Change. Special Panel, with Hajo Eicken, Herbert Anungazuk, Winton Weyapuk, Ronald Brower, Sr., and Joe Leavitt. University of Alaska Fairbanks, Fairbanks, October 15.

Krupnik, I. 2009a. Sea Ice and a Cultural-Ethnographic Landscape. Paper Presented at the Russian Institute of Cultural and Natural Heritage, Moscow, Russia, February 15.

Krupnik, I. 2009b. As the Ice Keeps Thinning: Update on Arctic People-Sea Ice Connections, 2007–2009. Paper presented at the symposium Impact of an Ice-Diminishing Arctic on Naval and Maritime Operations. Annapolis, June 10.

Krupnik, I. and Bogoslovskaya, L. 2007. International Polar Year 2007–2008. Project SIKU in Alaska and Chukotka. Paper Presented at the 'Shared Beringia Heritage' Conference. Anadyr, Russia, September 16.

Krupniki, I. and Weyapuk, W., Jr. 2008. Qanuq Ilitaavut: How We Learned What We Know. Wales Inupiaq Sea Ice Dictionary. Paper Presented at the ICASS-6, Nuuk, August 2008.

Laidler, G., Aporta, C., Gearheard, S., Furgal, C., and Taylor, D.R.F. 2008a. Inuit Sea Ice Use and Occupancy Project (ISIUOP). International Congress of Arctic Social Sciences (ICASS) VI, Nuuk, Greenland, August 26.

Laidler, G., Aporta, C., Gearheard, S., Furgal, C., and Taylor, D.R.F. 2008b. Inuit Sea Ice Use and Occupancy Project (ISIUOP). Annual Meeting of the Canadian Association of Geographers, Québec, May 23.

Laidler, G., DeAbreu, R., Elee, P., Furgal, C., Hirose, T., Ikummaq, T., Joamie, E., Kapfer, M., and Piekarz, D. (Laidler & Ikummaq presenting) 2008c. Connecting Community Observations and Expertise with the Floe Edge Service. Arctic Change 2008, Québec, December 10.

Laidler, G., Elee, P., Ikummaq, T., and Joamie, E. 2008d. Mapping Inuit Sea Ice Knowledge and Use. International Congress of Arctic Social Sciences (ICASS) VI, Nuuk, Greenland, August 26.

Laidler, G., Taylor, D.R.F., Pulsifer, P., Hayes, A., Fiset, J.-P., and Aporta, C. 2008e. Creating an Online Cybercartographic Atlas of Sea Ice. Annual Meeting of the Canadian Association of Geographers, Québec, May 23.

Laidler, G.J. 2006a. SIKU: A Circumpolar Study of Sea Ice Knowledge and Use in Northern Communities. The CRYSYS Decade (1995–2005), Final Annual Science Meeting, Toronto, February 24.

Laidler, G.J. 2006b. The Importance of Sea Ice Processes, Use, and Change to Residents of Cape Dorset, Nunavut. The CRYSYS Decade (1995–2005), Final Annual Science Meeting, Toronto, February 22.

Laidler, G.J. 2007a. Experiencing Change: Observations and Implications of Sea Ice Change Around Three Nunavut Communities. Annual Meeting of the Canadian Association of Geographers, Saskatoon, Saskatchewan (Blackwell Publishing Award Presentation), June 1.

Laidler, G.J. 2007b. Bridging Scales and Cultures: Facilitating Knowledge-Exchanges on Sea Ice and Climate Change. Annual Meeting of the Canadian Association of Geographers, Saskatoon, Saskatchewan (Robin P. Armstrong Memorial Prize Presentation), May 31.

Laidler, G.J. 2007c. Inuit Sea Ice Use and Occupancy Project (ISIUOP). For International Polar Year (IPY) Launch. Iqaluit, March 7.

Laidler, G.J. 2008a. Inuit Sea Ice Knowledge and Use in a Changing Environment. For the Department of Geography and Environmental Studies (Carleton University) Founders Seminar Series. Ottawa, February 1.

Laidler, G.J. 2008b. Inuit Sea Ice Knowledge and Use in a Changing Environment. For the Northumberland Learning Connection speaker series on "Ice", Coburg, April 24.

Laidler, G.J. 2008c. Intersecting Geographies: Hunters' and Scientists' Indicators for SIKU (Sea Ice) Use and Sila (Weather) Forecasting Around Baffin Island. For the Department of Geography (York University) Colloquium Series, October 21.

Laidler, G.J. 2008d. Creating Intersections in Northern Geographies of Sea Ice: Collaboration, Observation, and Education. For the explore! Northern Research Day, Carleton University, Ottawa, December 3.

Laidler, G.J. 2008e. Knowledge-Sharing – Considerations in the Development of a Communication Strategy. For the International Polar Year Researcher's Workshop at Arctic Change 2008, Québec, December 9.

Laidler, G.J. 2009. Knowledge-Sharing – Considerations in the Development of a Communication Strategy. For the Aboriginal Policy Research Conference Session on "Engaging Communities in Research: Dealing with Data and Traditional Knowledge (IPY)", Ottawa, March 10.

Laidler, G.J., Dialla, A., Elee, P., Joamie, E., and Ikummaq, T. 2007. Sea Ice Is Not Just "Ice": A Regional Comparison of Inuktitut Sea Ice Terminology Around Baffin Island, Nunavut. Annual Meeting of the Canadian Association of Geographers, Saskatoon, Saskatchewan, May 31.

O'Keefe, K. 2009. The Igliniit ("trails") Project: Inuit Hunters and Geomatics Engineering Students Collaborating to Develop an Interactive GPS Tracking System in Nunavut, Canada. Paper presented at the Institute of Navigation 2009 International Technical Meeting, Anaheim, CA, January 26–28.

Pulsifer, P.L. 2008. Interactive Atlases for Polar Regions Education and Outreach: Representation Meets Knowledge Management. Paper Presented at the International Polar Year: Researchers Workshop 2008, December 9, Québec.

Pulsifer, P.L. 2009. Documenting inuit knowledge: Data management meets education and outreach. Presentation at the IPY international data meeting, Ottawa, ON, 29 September–1 October.

Pulsifer, P.L., Laidler, G., Taylor, D.R.F., and Hayes, A. 2008a. Representing Inuit Sea Ice Knowledge and Use for Education and Outreach: Creating an IPY Legacy Using Emerging Data Management Strategies. Paper Presented at the Arctic Change 2008 Conference, December 9–12, Québec.

Pulsifer, P.L., Taylor, D.R.F., Hayes, A., and Fiset, J.P. 2008b. Creating an Online Cybercartographic Atlas of Sea Ice. Paper Presented at the 6th International Congress of Arctic Social Sciences, August 22–26, Nuuk, Greenland.

Pulsifer, P.L., Nickels, S., Tomlinson, S., Laidler, G., Aporta, C., Taylor, D.R.F., Hayes, A. 2009a. Documenting inuit knowledge using distributed information and multimedia interfaces: Knowledge preservation and sharing through partnership. Paper Presented at the GeoNorth 2009 Conference, Fairbanks, AK, August 4–6.

Pulsifer, P.L., Pyne S., Lauriault, T.P., Taylor, D.R.F., Hayes, A., Caquard, S. 2009b. The role of cybercartography in exploring, visualizing and preserving The past. Paper Presented at the visualizing the past: Tools and techniques for understanding historical processes workshop, University of Richmond, VA, February 20–21.

Tremblay, M., Furgal, C., Annanack, T., Angiyou, E., Naviaxie, J., and Barrett, M. 2008a. Climate Change in Nunavik (Canada): Adaptation Strategies Developed for a Safe Ice Access. ICASS VI. Nuuk, Groenland, Poster.

Tremblay, M., Furgal, C., Tookalook, P., Annanack, T., Qiisiq, M., Angiyou, E., and Barrett, M. 2008b. Nunavimiut Ice Terminology: Ensuring Safe Access to Territory and Resources. Arctic Change 2008. December 2008, Québec, Poster.

Tremblay, M., Furgal, C., Tookalook, P., Annanack, T., Qiisiq, M., Angiyou, E., and Barrett, M. 2008c. La terminologie inuite de la glace: Une valeur intrinsèque pour un accès sécuritaire aux ressources et au territoire au Nunavik. 3e Symposium scientifique Ouranos. Montréal, Poster.

Vukvukai, N.I., and Golbtseva V.V. 2007. The use of traditional knowledge in monitoring climate change in the Arctic zone of Chukotka. Paper presented at the 'Shared Beringia Heritage Conference,' Anadyr, Russia, September 18.

Wisniewski, J. 2008a. Empathising with Animals: A View Toward Inupiaq Hunting as Knowing. Paper Presented in the Session, "The Ethnography of Relationality: Creating, Experiencing and Being in the World." 35th Annual Meeting of the Alaska Anthropology Association, Anchorage, February 28–March 1.

Wisniewski, J. 2008b. *Kigiqtaamiut* Rules About Traveling on the Sea Ice. Paper Presented at the Alaska Native Language Center Climate, Language, and Indigenous Perspectives Conference. Fairbanks. August 13–15.

Wisniewski, J. 2008c. Learning as Doing, Being as Knowing About *Sigu*: An Inupiaq Hunting Pedagogy of Experience in Shishmaref, Alaska. Paper Presented in Session "SIKU: Polar Residents Document Arctic Ice and Climate Change," ICASS VI, Nuuk, Greenland, August 22–26.

Wisniewski, J. 2008d. "Well, Let's Go Look Around": Kigiqtaamiut Hunting, as a Relational Ecological Knowing. Paper Presented in Session "Animal Subjects Exploring the Shifting Ground of Human/Animal Relationships." 107th American Anthropology Association Annual Meeting, San Francisco, November 19–23.

Color Plates

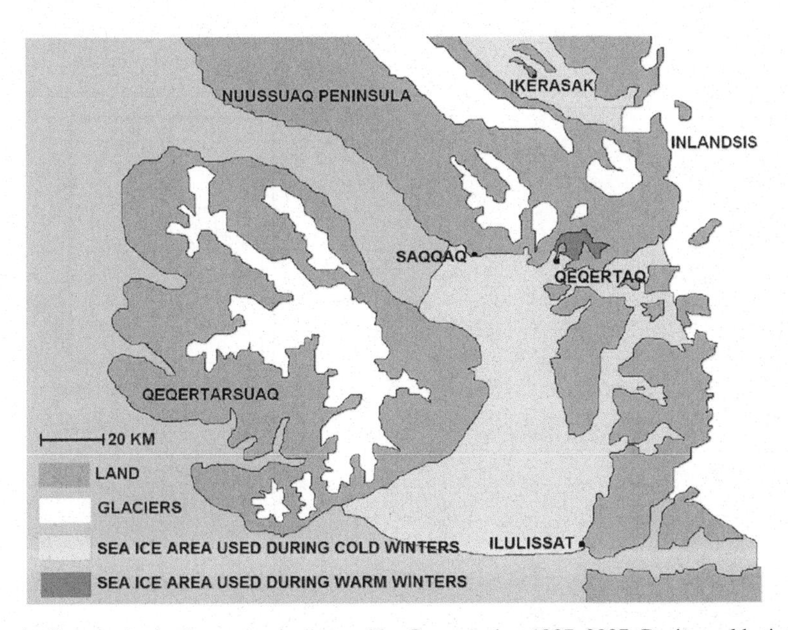

Plate 1 Changes in the ice-covered area used by Qeqertamiut, 1987–2007. During cold winters, sea ice typically filled the bay, stretching north from Ilulissat to Saqqaq. Formerly, the sea ice occupied an area of 600–3,500 km^2 (*yellow*) but in recent years this number has been reduced to an area less than 50–70 km^2 (*red*). (See also Fig. 2.9 on page 40)

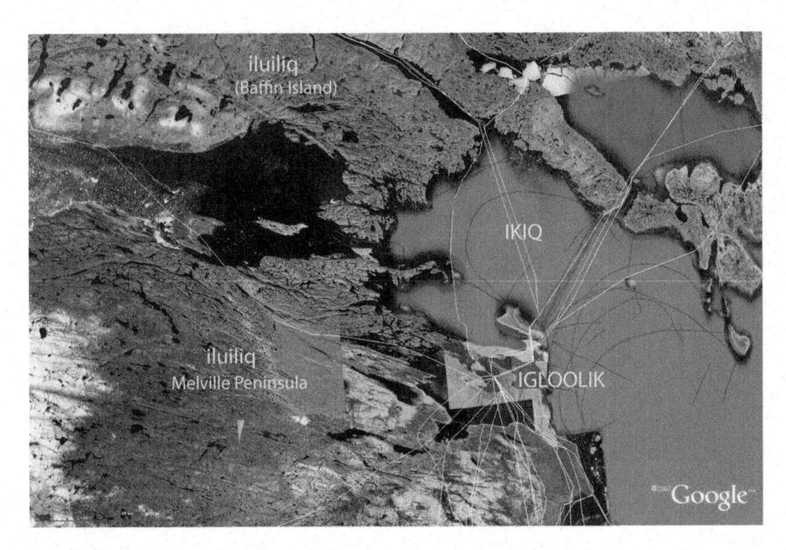

Plate 2 Map of the two land masses (selected sea ice features – in *red* – adapted from Laidler and Ikummaq 2008; selected trails – in *white* – selected from Aporta 2009) (See also Fig. 7.1 on page 168)

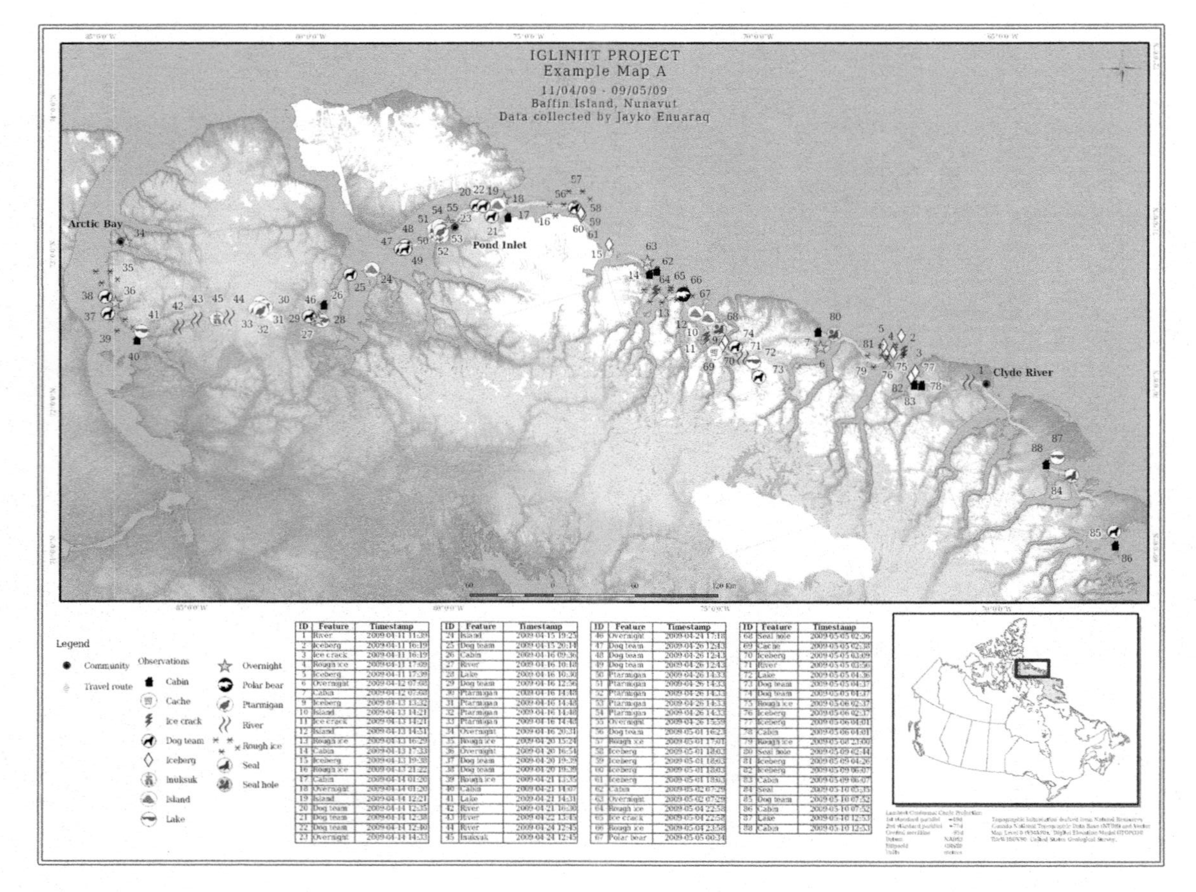

Plate 3 Map from Igliniit test data collected by Jayko Enuaraq while travelling between Clyde River and Arctic Bay in 2009. (See also Fig. 8.10 on page 198)

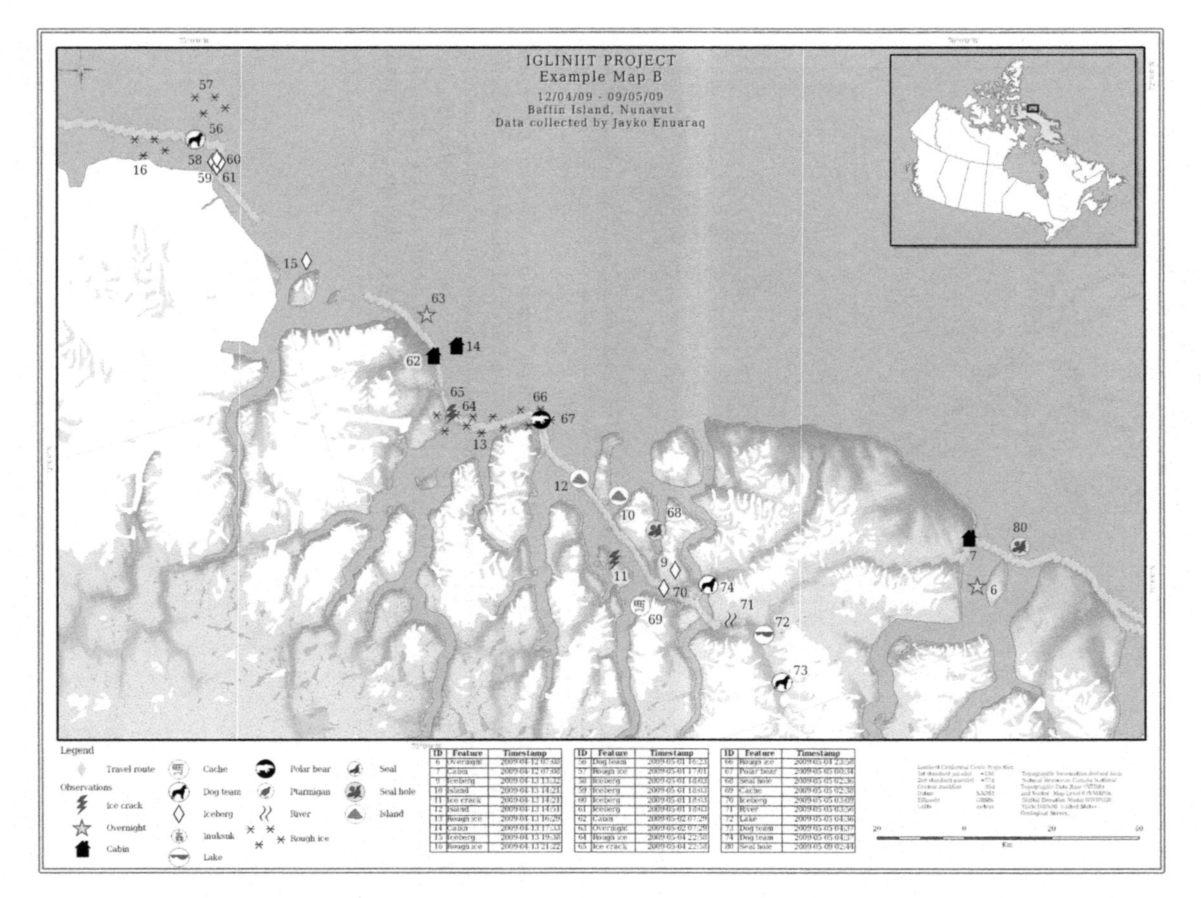

Plate 4 Enlarged image section of the previous map. (See also Fig. 8.11 on page 199)

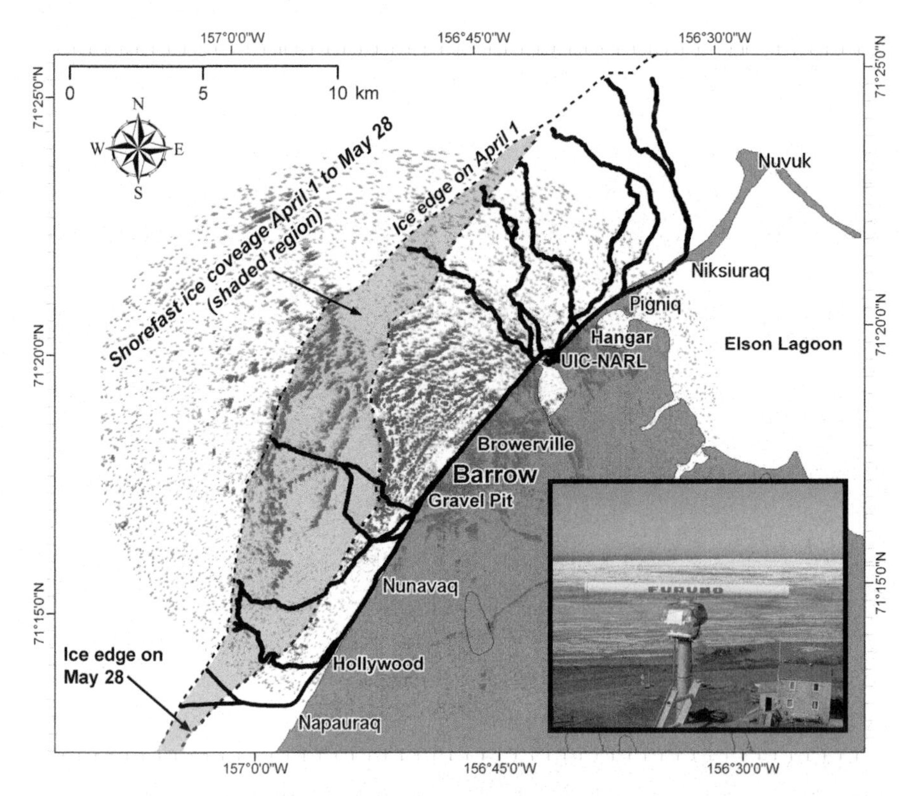

Plate 5 Map of the 2007 whaling trails. Many of the trails shown here traversed the region that existed in the shorefast ice between break-out events on March 31 and May 28. The background in this image shows a sample radar backscatter image (*dark speckles* represent ice features) as recorded during the break-out on May 28. The location of the main trail off Napauraq was hand-drawn after the whaling season ended based on the input from members of the community. The 10 kW X-band Furuno marine radar in downtown Barrow is shown in the *lower right* photo. (See also Fig. 9.9 on page 218)

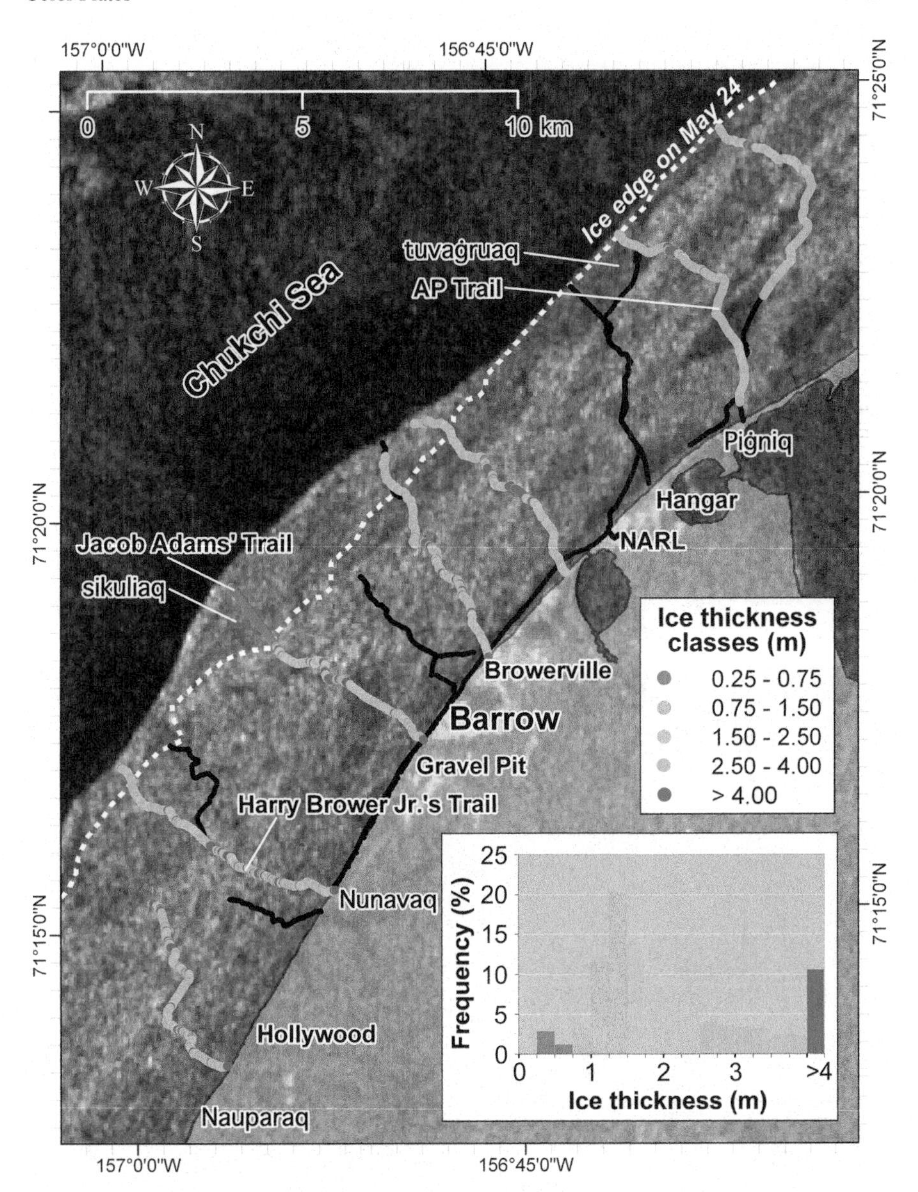

Plate 6 Map of the 2008 whaling trails. Trails are shown here with ice thickness data overlaid on select trails where measurements were made. The two trails south of Nunavaq were not fully mapped since they were incomplete at the time of mapping in early to mid-April. The trail off Barrow was abandoned before making it to the ice edge. The SAR image, acquired by the RADARSAT-1 satellite and provided by the Canadian Space Agency and C.E. Tweedie and A.G. Gaylord, is from April 5, 2008. (See also Fig. 9.11 on page 221)

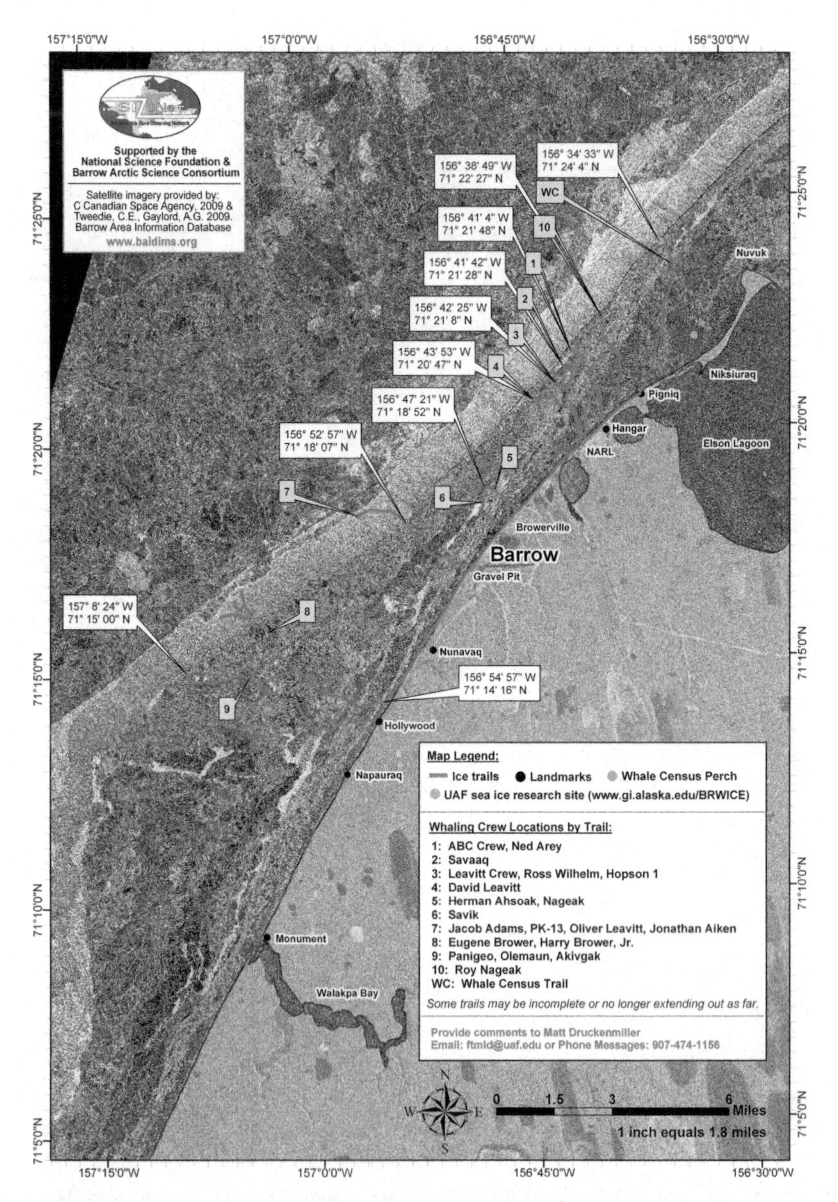

Plate 7 Map of the 2009 whaling trails. This exact map was provided to the community during the whaling season. The SAR image, acquired by the European Remote Sensing satellite ERS-2 and provided by the Canadian Space Agency and C.E. Tweedie and A.G. Gaylord, is from May 16, 2009, just prior to the opening of the lead shown in Fig. 9.1. Various GPS locations are labeled to assist with navigation. Locations are also shown for the camp of the 2009 bowhead whale census orchestrated by the North Slope Borough's Department of Wildlife Management and of our sea ice mass balance site that measured level ice growth and other variables of interest. (See also Fig. 9.13 on page 224)

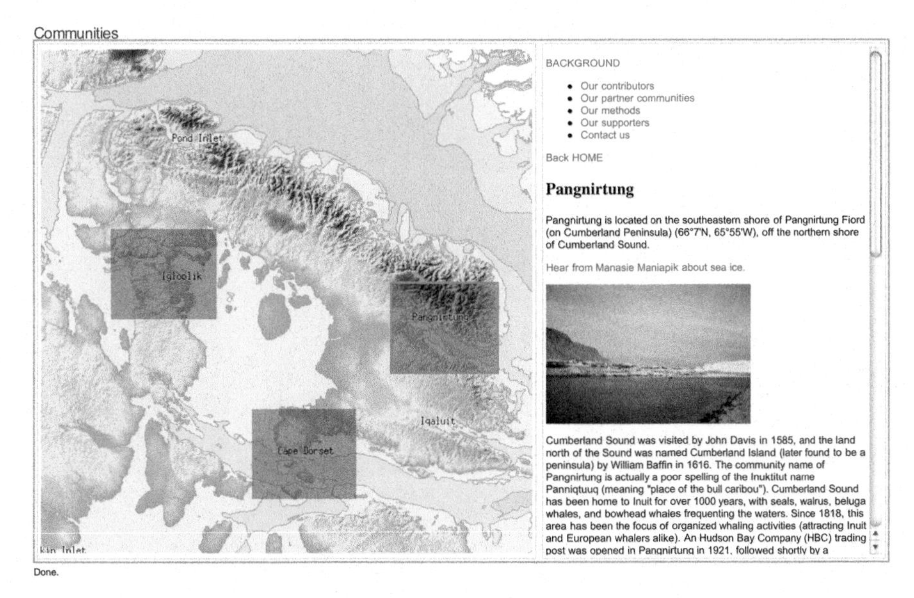

Plate 8 An interactive map provides information about partner communities. The historical, geographical, and sociological aspects of the communities are described using text, photographs, sound, and potentially video (See also Fig. 10.2 on page 237)

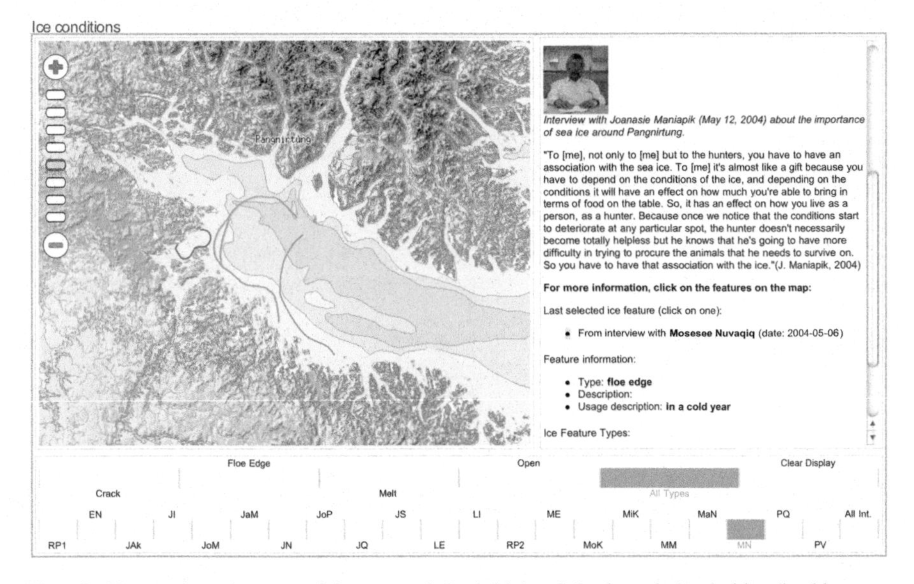

Plate 9 Users select the type of feature and the initials of the knowledge holder (in this case MN [Mosesee Nuvaqiq]) using the selection bars. Clicking on a feature displays more information about that feature in the *right* frame of the window. This information may include multimedia content such as digital photographs, audio clips, or video clips. In this figure an audio clip (*top* of information box) is provided along with a photograph of the knowledge contributor (See also Fig. 10.4 on page 239)

Index

T

W

CPSIA information can be obtained
at www.ICGtesting.com
Printed in the USA
LVOW01s2312151216
517434LV00003B/6/P